T0301041

Fundamentals of Digital Communication Systems

This dynamic textbook provides students with a concise and accessible introduction to the fundamentals of modern digital communications systems. Building from first principles, its comprehensive approach equips students with all of the mathematical tools, theoretical knowledge, and practical understanding they need to excel.

Key features

- Equips students with a strong mathematical foundation spanning signals and systems, probability, random variables, and random processes.
- Introduces students to key concepts in digital information sources, analog-to-digital conversion, digital modulation, power spectra, multicarrier modulation, and channel coding.
- Includes over 85 illustrative examples, and more than 270 theoretical and computational end-of-chapter problems, allowing students to connect theory to practice.
- Accompanied by downloadable MATLAB code, and a digital solutions manual for instructors.

Suitable for a single-semester course, this succinct textbook is an ideal introduction to the field of digital communications for senior undergraduate students in electrical engineering.

Tolga M. Duman is a Professor of Electrical Engineering at Bilkent University, Ankara and a former Professor of Electrical Engineering at Arizona State University, with a focus on digital communications, wireless and mobile communications, underwater acoustic communications, channel coding/modulation, signal processing, and information theory. He is a recipient of the IEEE Third Millennium Medal, a former Editor-in-Chief of *IEEE Transactions on Communications*, and a Fellow of the IEEE.

"This is a concise textbook on digital communications which successfully meets the need in terms of scope and depth of treatment. The approach the author provides will appeal to a broad range of instructors and students."

Ender Ayanoglu, University of California, Irvine

"This book is an excellent textbook for undergraduate students, based on the years' teaching experience of the author. It very well explains and illustrates all the fundamental concepts in modern digital communications and has a good balance between mathematical fundamentals and practical applications. It also has excellent exercise problems at the end of each chapter. This book is well written and enjoyable to read."

Xiang-Gen Xia, University of Delaware

Fundamentals of Digital Communication Systems

Tolga M. Duman
Bilkent University, Ankara

Shaftesbury Road, Cambridge CB2 8EA, United Kingdom

One Liberty Plaza, 20th Floor, New York, NY 10006, USA

477 Williamstown Road, Port Melbourne, VIC 3207, Australia

314–321, 3rd Floor, Plot 3, Splendor Forum, Jasola District Centre, New Delhi – 110025, India

103 Penang Road, #05–06/07, Visioncrest Commercial, Singapore 238467

Cambridge University Press is part of Cambridge University Press & Assessment, a department of the University of Cambridge.

We share the University's mission to contribute to society through the pursuit of education, learning and research at the highest international levels of excellence.

www.cambridge.org
Information on this title: www.cambridge.org/highereducation/isbn/9781009318099

DOI: 10.1017/9781009318129

When citing this work, please include a reference to the DOI 10.1017/9781009318129

First published 2025

A catalogue record for this publication is available from the British Library

A Cataloging-in-Publication data record for this book is available from the Library of Congress

ISBN 978-1-009-31809-9 Hardback

Additional resources for this publication at www.cambridge.org/duman

To

Dilek, Mert Bora, and Berk Arda

Contents

Preface

This book is intended as a first course on digital communications at the undergraduate level. While providing a very basic and short description of analog communications as part of a chapter and emphasizing the difference between analog and digital transmission, it deliberately omits any significant coverage; instead, it focuses on detailed coverage of digital communication systems starting from first principles.

Analog and digital communications material coverage has evolved significantly over the last few decades. In the early 1990s, when I was an undergraduate student, the introductory course on communications was on analog transmission techniques. Digital communications was only covered if there was a second senior-level course. The communications textbooks were written with this in mind: roughly half devoted to analog and the other half to digital communications. Also, in many universities, digital communications were covered only at the graduate level. Many things have changed since then. In particular, typically, there is a single undergraduate-level course on communications as part of the undergraduate curriculum. Given the advances over the past three decades, we advocate that it is now necessary to cover the more widely used, and arguably the more important, digital communications material. This book is written with this philosophy in mind.

This book is intended solely as a textbook (rather than a reference). As such, there is a strong emphasis on the basics and illustrative examples in each chapter, along with many theoretical and computer problems. At the same time, the coverage is concise so that it can be covered in a one-semester course in almost its entirety. We provide occasional references to applications of different digital communications techniques; for instance, we emphasize that orthogonal frequency-division multiplexing is used in 4G and 5G wireless communication systems as well as asymmetric digital subscriber lines; however, there is no attempt to cover communications standards or specific applications in any detail. After studying this book, the student would be equipped to delve into the details of any communication standard or application.

The book is roughly divided into four major parts: (1) mathematical preliminaries (on signals and systems, probability, random variables, and random processes) in one chapter; (2) information sources and analog-to-digital conversion in two chapters; (3) digital modulation including bandpass transmission techniques, power spectra of transmitted signals, and multicarrier modulation in four chapters; and (4) channel-coding basics in one chapter. Each chapter contains solved problems as well as a large number of chapter-end problems. Also included at the end of each chapter is a rich set of computer problems to illustrate different concepts through Monte Carlo simulations and other computational techniques.

The material in this book is organized as follows.

Chapter 1 provides a brief introduction to digital communications and gives an overview of elements of a communication system, and in particular, a digital communication system. The objective is to introduce the basic communications setup and highlight the different types of problems that need to be solved. Different components introduced in this introductory chapter are detailed in the rest of the book.

The necessary mathematical preliminaries are given in **Chapter 2**. Specifically, we cover the relevant concepts from signals and systems, probability, random variables, and random processes. Gaussian processes are also covered in sufficient depth to allow for a thorough study of noise at the receiver of a communication system. It is assumed that the student has taken a course on signals and systems as well as a course on probability and random variables. Therefore, particularly in some parts, the coverage is brief and is aimed as a refresher. Other parts, including the coverage of random processes, are given in more depth as they may not be covered in the introductory course on probability and random variables.

Chapter 3 presents a brief coverage of analog communications, including descriptions of amplitude and angle modulation methods, their time-domain and frequency-domain characterization, and their pros and cons. Also reviewed is the signal quality at the receiver output. This is the only coverage of analog communications in this book, as the main objective is to clarify the distinctions and similarities between analog and digital communication techniques. The rest of the chapter focuses on the conversion of analog sources into a digital form through sampling and quantization and a brief coverage of waveform coding techniques. Scalar and vector quantization is described in some depth, and the quality of the quantizer output is examined using techniques from probability and random variables. With the representation of analog sources in digital form as a sequence of bits, the rest of the book considers only digital information sources and digital transmission techniques.

Chapter 4 is about digital information sources. The concept of the information content of a source is detailed, and this concept is formalized through the source-coding theorem for discrete memoryless sources. The proof of the theorem is omitted as it is beyond our scope; however, intuitive explanations are provided to explain how it follows. The second part of the chapter deals with practical source-coding techniques, including Huffman coding and variations of Lempel–Ziv coding. They are explained through different examples, which explicitly illustrate how these techniques can provide compression of digital data.

Digital modulation fundamentals are studied in **Chapter 5**. Building upon the mathematical preliminaries given in Chapter 2, white Gaussian noise is described, and the widely used additive white Gaussian noise (AWGN) channel model is detailed. Also covered is the concept of signal space, which allows for representation of the transmitted signals corresponding to different symbols as vectors and accordingly, the derivation of the optimal receiver structure. The case of binary antipodal signaling is given in full detail as a simpler introduction to the concept of modulation, and then the more general case of M-ary signaling is studied. Both maximum a posteriori (MAP) and maximum likelihood (ML) receivers are given. Receiver implementations with a bank of correlators and with matched filtering are described. Also studied are some basic properties of matched filtering, which appear in different contexts such as signal processing and detection/estimation. In addition, the idea of union bound is introduced so as to provide a performance assessment of different modulation schemes over an AWGN channel. As exemplary modulation techniques, M-ary pulse amplitude modulation (PAM) along with orthogonal signaling and its variants are given. At the end of the chapter, symbol timing recovery is briefly discussed.

We cover digital transmission via single-carrier bandpass signals in **Chapter 6**. The channel effects on bandpass signals are described along with basic bandpass modulation techniques (that include phase-shift keying (PSK) and quadrature amplitude modulation (QAM)). Lowpass-equivalent forms of bandpass signals and noise processes are given, and it is clarified that digital communication systems can also be studied through these equivalent lowpass models. The differences among the coherent, non-coherent, and differentially coherent receivers are clarified, and the differential phase-shift keying scheme is briefly described. The chapter concludes with a short study of carrier phase-synchronization techniques.

Chapter 7 focuses on the spectral content of digitally modulated signals and discusses the related concept of transmission over bandlimited channels. The effects of non-ideal channels on digitally transmitted signals are explained, and the transmit and receive pulse design to avoid intersymbol interference is described. Using these basic results, the spectral efficiencies of different baseband and single-carrier bandpass transmission schemes are determined. In addition, power spectral density calculation for digitally modulated signals for both baseband and bandpass transmissions is illustrated.

As examples of other important and widely used modulation schemes, frequency-shift keying (and minimum-shift keying) along with orthogonal frequency-division multiplexing (OFDM) are covered in **Chapter 8**. This is the fourth and final chapter on digital modulation.

Chapter 9 is on channel coding. First, the need for channel coding is explained through specific examples, and it is shown that performance improvements can be obtained by adding controlled redundancy into the transmitted sequence. Then, a brief coverage of the ultimate limits for reliable transmission over noisy channels is given, namely, Shannon's celebrated channel-coding theorem is presented for binary symmetric channels and AWGN channels. The rest of the chapter is on practical channel-coding techniques, specifically linear block codes and convolutional codes. The coverage is at a high level for the case of linear block codes, where basic code description and code properties are given, and a performance analysis over binary symmetric channels is carried out. For the case of convolutional codes, the coverage is more detailed: different representations of convolutional codes, decoding algorithms over binary symmetric and AWGN channels, and performance analysis results are described.

Finally, **Chapter 10** is a short coverage of some communication system design issues. Specifically, the topics of link budget analysis, the effects of transmission losses, and the use of both analog and regenerative repeaters to combat them are discussed. Also studied are methods of sharing a common communication medium by different transmitters, namely, frequency-division multiple access, time-division multiple access, and code-division multiple access.

The material in this book is designed as a succinct introduction to digital communications, and as such it is intended for use as a textbook for a senior undergraduate-level course on the subject. It should be possible to cover this book in its entirety in a one-semester course. In this case, presenting the material in the same order as given in the book would be a good choice. However, it is also possible to omit or shorten the coverage of some parts and place more emphasis on others. For instance, the focus in Chapter 7 can be on spectral efficiency with less emphasis on pulse design for no intersymbol interference or power spectral density of digitally modulated signals. Also, in Chapter 8, the emphasis could be on FSK.

Finally, we highlight that solutions to end-of-chapter problems, including related MATLAB codes, are made available as online resources for both students and instructors.

Acknowledgments

Many friends and colleagues provided feedback on different versions of this manuscript. In particular, I am grateful to John G. Proakis (Professor Emeritus, Northeastern University) for providing an in-depth review of an early version of the manuscript and for his encouragement in its completion, and to Masoud Salehi (Northeastern University) for his input at different stages of its preparation. I am also indebted to Nghi Tran (University of Akron), Mojtaba Vaezi (Villanova University), Serdar Yuksel (Queen's University), Cihan Tepedelenlioglu (Arizona State University), and Junshan Zhang (University of California Davis) for providing detailed comments, corrections, and recommendations on the manuscript. My Bilkent University colleagues Sinan Gezici and Nail Akar have also provided feedback and donated some end-of-chapter problems, for which I am grateful. I am also thankful to Busra Tegin, who helped generate some of the figures in the book. I would also like to thank Ender Ayanoglu (University of California Irvine), Sudharman Jayaweera (University of New Mexico), Xiang-Gen Xia (University of Delaware), and other anonymous reviewers for their constructive comments and criticisms, which have significantly improved the content and presentation of the manuscript.

Notation

E_x	energy of signal $x(t)$
P_x	power content of signal $x(t)$
$*$	convolution
$\Pi(t)$	rectangular function (defined as 1 on $[-1/2, 1/2]$, and 0 otherwise)
$\Lambda(t)$	triangular function
$A \cup B$	union of events/sets A and B
$A \cap B$	intersection of events/sets A and B
$\mathcal{F}$	Fourier transform operator
$\mathbb{P}(A)$	probability of event A
$\mathbb{P}(A\|B)$	conditional probability of A given B
$\mathbb{E}[X]$	statistical expectation of the random variable X
$\mathbb{E}[X\|Y]$	conditional expectation of X given Y
$F_X(x)$	cumulative distribution function of the random variable X
$F_{X,Y}(x,y)$	joint CDF of X and Y
$f_X(x)$	probability density function of the random variable X
$f_{X,Y}(x,y)$	joint PDF of X and Y
$f_{Y\|X}(y\|x)$	conditional PDF of Y given X
m_X	mean of the random variable X
$\boldsymbol{\mu}$	mean vector of a random vector
$\boldsymbol{C}$	covariance matrix of a random vector
Var(X)	variance of the random variable X
σ_X	standard deviation
ρ	correlation coefficient of two random variables
$\Phi(x)$	CDF of the standard Gaussian random variable
$Q(x)$	Q-function (the right tail probability of the standard Gaussian random variable)
$\text{Cov}(X, Y)$	covariance of the random variables X and Y
$\mathcal{N}(m, \sigma^2)$	Gaussian random variable with mean m and variance σ^2
$\mu_X(t)$	mean function of the random process $X(t)$
$R_X(t_1, t_2),\ R_X(\tau)$	autocorrelation function of the random process $X(t)$
$C_X(t_1, t_2),\ C_X(\tau)$	autocovariance function of the random process $X(t)$
$\bar{R}_X(\tau)$	average autocorrelation function of a cyclostationary random process
$R_{XY}(t_1, t_2),\ R_{XY}(\tau)$	cross-correlation function of the random processes $X(t)$ and $Y(t)$
$R_W(\tau)$	autocorrelation function of white noise

$S(f),\ S_X(f)$	power spectral density of a random process
$S_n(f)$	power spectral density of thermal noise
$S_W(f)$	power spectral density of white noise
k_a	amplitude sensitivity constant (in conventional AM)
k_f	frequency sensitivity constant (in FM)
k_p	phase sensitivity constant (in PM)
μ	modulation index (in conventional AM)
η	modulation efficiency (in conventional AM)
$\beta_{FM},\ \beta_{PM}$	modulation index in FM and in PM
$\hat{m}(t)$	Hilbert transform of the signal $m(t)$
$\mathrm{SNR_{bb}}$	basesband signal-to-noise ratio
$c(x),\ e(x)$	compressor/expander in non-uniform PCM
$\binom{n}{k}$	n choose k
log	logarithm in base 2
$n!$	n factorial
ln	natural logarithm
$H(X)$	entropy of the discrete random variable X
$H_b(p)$	binary entropy function
P_e	average symbol error probability
$P_{e,m}$	conditional error probability for symbol m
P_b	bit error probability
$\mathcal{D}_m$	decision region for symbol m
$\langle x(t), y(t)\rangle$	inner product of $x(t)$ and $y(t)$
$\boldsymbol{x}\cdot\boldsymbol{y}$	dot product of the vectors $\boldsymbol{x}$ and $\boldsymbol{y}$
$\angle a$	angle of the complex number a
E_s	energy per symbol
E_b	energy per bit
$\bar{E}_s$	average energy per symbol of a constellation
$d_{\min}$	minimum Euclidean distance; minimum distance of a linear block code
γ_s	signal-to-noise ratio per symbol (E_s/N_0)
γ_b	signal-to-noise ratio per bit (E_b/N_0)
$x_+(t)$	pre-envelope of the bandpass signal $x(t)$
$\tilde{x}(t),\ x_l(t)$	lowpass equivalent of the bandpass signal $x(t)$
$x_I(t)$	in-phase component of the bandpass signal $x(t)$
$x_Q(t)$	quadrature component of the bandpass signal $x(t)$
$\mathcal{R}$	real part
$g_{rc}(t)$	raised cosine pulse in time domain
$G_{rc}(t)$	raised cosine pulse in frequency domain
$r_{M\text{-PSK}}$	spectral efficiency of M-PSK (also used for other modulation schemes with different subscripts)
C	channel capacity
$\boldsymbol{G}$	generator matrix of a linear block code
$\boldsymbol{H}$	parity-check matrix of a linear block code

$\boldsymbol{s}$	syndrome of a received sequence
$A(z)$	weight-enumerating polynomial of a linear block code
$T(X, Y, Z),\ T(X, Y)$	transfer function of a convolutional code
d_{free}	free distance of a convolutional code
$d_H(\boldsymbol{x}, \boldsymbol{y})$	Hamming distance of the binary vectors $\boldsymbol{x}$ and $\boldsymbol{y}$
$w_H(\boldsymbol{x})$	Hamming weight of the binary vector $\boldsymbol{x}$
B_{neq}	noise-equivalent bandwidth
F	noise figure of an amplifier

Abbreviations

AM	amplitude modulation
AMPS	Advanced Mobile Phone System
AWGN	additive white Gaussian noise
BFSK	binary frequency-shift keying
BPSK	binary phase-shift keying
BSC	binary symmetric channel
CDF	cumulative distribution function
CDMA	code division multiple access
CPFSK	continuous-phase frequency-shift keying
CPM	continuous-phase modulation
dB	decibel
DFT	discrete Fourier transform
DMS	discrete memoryless source
DPCM	differential pulse-code modulation
DPSK	differential phase-shift keying
DSB	double sideband
DSB-SC	double sideband suppressed carrier
DSSS	direct-sequence spread spectrum
FDMA	frequency division multiple access
FFT	fast Fourier transform
FHSS	frequency-hopping spread spectrum
FM	frequency modulation
FSK	frequency-shift keying
GSM	Global System for Mobile
HDD	hard-decision decoding
Hz	Hertz
ICI	intercarrier interference
IDFT	inverse discrete Fourier transform
IFFT	inverse fast Fourier transform
ISI	intersymbol interference
i.i.d.	independent and identically distributed
LDPC	low-density parity check
LHS	left-hand side
LSB	lower sideband
LTI	linear time invariant
LZ	Lempel–Ziv
LZW	Lempel–Ziv–Welch

MAP	maximum a posteriori
ML	maximum likelihood
MSK	minimum-shift keying
OFDM	orthogonal frequency-division multiplexing
OQPSK	offset quadrature phase-shift keying
PAM	pulse amplitude modulation
PAPR	peak-to-average power ratio
PCM	pulse-code modulation
PDF	probability density function
PLL	phase-locked loop
PM	phase modulation
PMF	probability mass function
PPM	pulse position modulation
PSD	power spectral density
PSK	phase-shift keying
QAM	quadrature amplitude modulation
QPSK	quadrature phase-shift keying
RHS	right-hand side
SDD	soft-decision decoding
SNR	signal-to-noise ratio
SQNR	signal-to-quantization noise ratio
SSB	single sideband
SSS	strict-sense stationary
TDMA	time-division multiple access
USB	upper sideband
VSB	vestigial sideband
WSS	wide-sense stationary

1 Introduction

Communications refer to the transfer of information from a sender to a receiver. Examples of practical communication systems abound. For instance, we use our wireless phones to make calls, send text messages, or browse the internet; we watch our favorite shows delivered to our TVs via satellite signals; we listen to our radios, send emails from our computers or mobile devices, upload files to the cloud, store our photos or videos on hard drives, burn CDs with our favorite music, and so on. These examples and many others are facilitated through the communication technologies developed over many decades. The objective of this book is to cover communication system basics, with particular attention to the mathematical models and analysis in a rigorous fashion, to enable a comprehensive understanding of the fundamental principles of communication systems.

Communications may take place over vastly different media. For instance, we can use radio signals (electromagnetic waves) to transmit over wireless links, electrical signals over copper wirelines, optical signals through fiber, or acoustic signals underwater. Examples of communication systems also include storage devices. For instance, we write on a CD or DVD through a *burning* process, which is nothing but a transmitter operation, storing the digital data in the physical medium which might change it (channel), reading the recorded signal, and recovering the stored bits at a later time (receiver operation). Similarly, we use magnetic storage to write on a hard drive (transmitter), store the data (transmission through the channel), and recover the stored bits later on through a read head (receiver). As a further example, we store digital information on solid-state drives using electronic storage (employing different charge levels).

While the technologies enabling transmission across various media differ significantly, the basic principles of communications are the same; hence, we can study the general principles of communications without reference to the underlying technology. This chapter aims to introduce the basic communication system block diagram, describe the components involved, and distinguish between analog and digital communication systems.

The chapter is organized as follows. We provide an overview of a general communication system in Section 1.1. We then focus on digital communication systems and list the components involved and their functionalities in Section 1.2. The rest of the book will cover the elements of a digital communication system introduced here in great detail. We briefly compare analog versus digital communications and highlight the motivation for employing digital communication techniques in Section 1.3. We conclude the chapter with an outline of the rest of the book in Section 1.4.

1.1 Elements of a Communication System

Elements of a basic communication system include an information source, a transmitter, a channel, a receiver, and an output transducer, as depicted in Fig. 1.1.

The information source produces the message signal to be transmitted. This signal can be continuous-time, continuous-amplitude (analog), such as a speech signal, or digital (i.e., a sequence of symbols (or bits)). For the latter case, the signal can be natively digital (as in an email or text) or obtained by converting an analog signal into a digital form. Information sources include voice signals, video signals, images, sensor measurements, and computer data, among others. The input transducer converts the output of the information source into an electrical signal; for instance, a microphone converts sound waves into electrical signals representing the output of a voice source.

The message signal is input to a transmitter whose function is to perform modulation. Modulation is the process of generating a signal suitable for transmission over the communication channel derived from the message signal. In other words, it matches the message signal to the channel in such a way that it is still possible to recover it from the modulated waveform.

There are two main categories of modulation: analog modulation and digital modulation. For analog modulation, the message is represented by a continuous-time signal, and this message signal modulates the amplitude, phase, or frequency of a carrier signal (a high-frequency sinusoidal signal) to produce the modulated waveform. Analog modulation is classified into amplitude and angle modulation, where the latter refers to phase and frequency modulation. For digital modulation, however, a sequence of symbols represents the message, and the modulated waveform is constructed by embedding them into continuous-time signals. Examples of digital modulation schemes include pulse amplitude modulation (PAM), for which the message symbols are used to select the amplitude levels of a basic pulse, phase-shift keying (PSK), for which the message symbols determine the phase of the carrier signal (sinusoidal signal), and frequency-shift keying (FSK), for

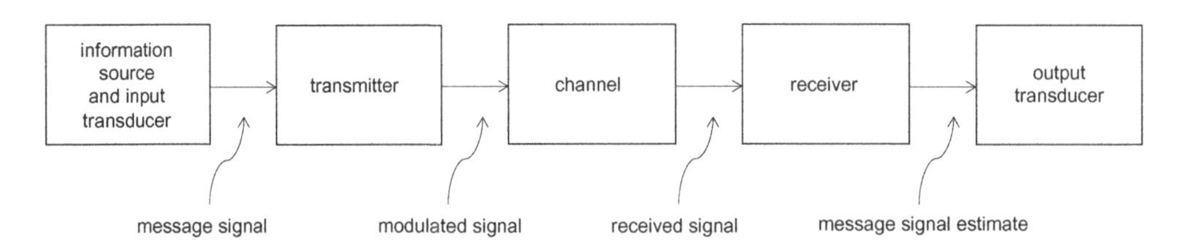

Figure 1.1 A basic communication system.

which the message signals determine the frequency of the sinusoidal signal for the corresponding transmission interval.

There are many reasons why modulation is needed. First, the channel may not be suitable for directly transmitting the message signal. For instance, the frequency of signals in the audible range is from 20 Hz to 20 kHz, which would require huge antenna dimensions for wireless transmission (simply impossible). Modulating message signals using much higher carrier frequencies makes it possible to transmit them through radio waves via practical means. Through modulation, it becomes possible to multiplex different messages over the same communication medium or to share the communication medium by multiple users, referred to as multiple access. Furthermore, modulation can also provide better noise immunity at the receiver (depending on the specific modulation scheme being used).

The communication channel is the physical medium over which the modulated signal gets transmitted. For instance, wireless radio signals are transmitted through the atmosphere; optical signals are transmitted through optical fiber, air, or even underwater; electrical signals can be transmitted through copper, and so on. The channel corrupts the transmitted signal. For instance, the signal gets attenuated depending on the distance between the transmitter and the receiver, and a noise signal is added. There could also be interference from other transmissions in the medium, multipath propagation effects, and unpredictable time variations, resulting in *fading* of transmitted signals.

The receiver observes the (corrupted) signal at the channel output and performs demodulation, referring to the recovery of an estimate of the message signal from the noisy received signal.

Transmit signal power and the available channel bandwidth (in the frequency domain) are the fundamental resources for communications. With increasing transmit power or channel bandwidth or both, when the communication system is adequately designed, the message signal reproduction at the receiver will be better quality. Furthermore, there is a trade-off between power and bandwidth. Stated differently, for the same quality of signal reproduction at the receiver, if the transmit power is reduced, the bandwidth will need to be increased, and vice versa. These ideas will be detailed and made precise in subsequent chapters.

1.2 Digital Communication Systems

While analog communication systems found widespread use in the past, more recently, digital transmission techniques have taken over due to their many advantages. Indeed, most analog communication systems are now being abandoned (including analog TV transmission in most parts of the world). AM and FM radio broadcasting are the only major commercial analog communications applications remaining. Various communication systems widely used worldwide, such as satellite TV, cellular telephony, Wi-Fi, and transmissions over the internet, are all digital.

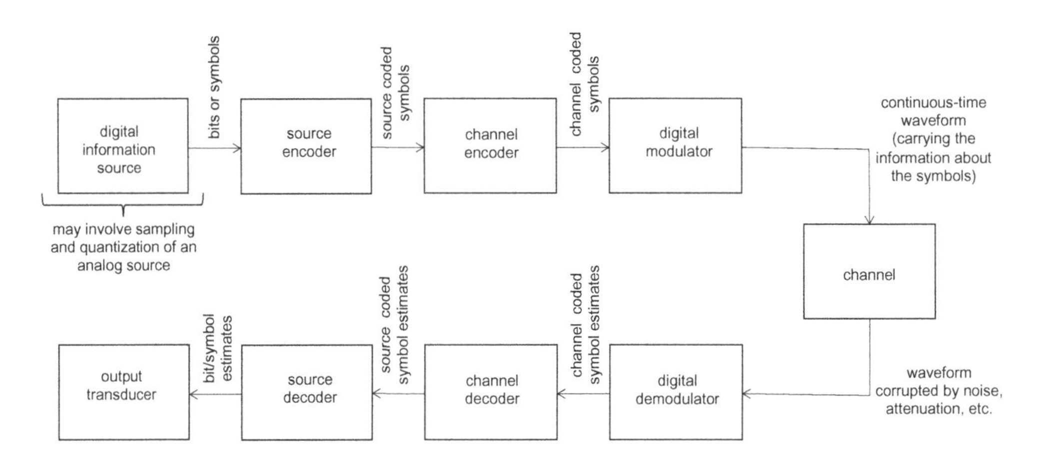

Figure 1.2 Components of a digital communication system.

The basic block diagram of a digital communication system is depicted in Fig. 1.2. The components include a digital information source, source encoder, channel encoder, digital modulator on the transmitter side, and the corresponding digital demodulator, channel decoder, and source decoder blocks on the receiver side.

Let us briefly go through the different components of a digital communication system.

Source Encoder/Decoder. A digital information source produces nothing but a sequence of symbols (or, for the binary case, 0s and 1s, referred to as bits) representing the underlying message. For instance, a digital image is represented as a sequence of pixel values (each represented by a sequence of bits). As another example, a text message is a sequence of letters and special characters. While some information sources are inherently digital (e.g., a text message), others may be obtained by converting an originally analog signal into a digital form. For example, an analog speech signal can be sampled, and the samples can be quantized to obtain a digital representation of the source.

The output of a digital information source typically contains a significant amount of redundancy. For example, neighboring pixels in an image are usually similar to each other, or consecutive samples of a speech source take on similar values. Also, when writing an English sentence, one can omit many letters or symbols while keeping the message intelligible. Therefore, the sequences produced at the source output may be (unnecessarily) long. The source encoder removes the redundancy in the digital information sequence and compresses it, reducing the number of symbols/bits that must be transmitted to the destination.

As a simple illustration, the output of an information source may be a long sequence of bits (e.g., corresponding to the contents of this book on a computer):

$$1, 0, 1, 1, 1, 0, 0, 0, 1, 1, 1, 0, 1, 1, \ldots, 0.$$

The output of the source encoder block will then be a (shorter) sequence of bits, for example,

$$0, 1, 0, 0, 1, 1, 0, 1, \ldots, 1,$$

which is obtained by an algorithm that removes the redundancy in the original (long) sequence. The source-encoded version is then input to a channel encoder. That is, this compressed sequence of bits is channel coded, as explained below. The input to the source decoder is the encoded version (obtained at the output of the channel decoder), and the output of the source decoder is the original information sequence.

The above description refers to *lossless* source coding, that is, it assumes no information loss in the encoding process, and the source decoder output is the same as the original information sequence. We need to utilize lossless compression techniques in applications where data integrity is essential, and even a one-bit error cannot be tolerated. For instance, the usual compression algorithms used on a computer (i.e., various versions of *zip* programs) implement lossless compression techniques. On the other hand, not all applications require lossless compression; in other words, source coding can also be lossy since, in certain applications, only an approximate reproduction of the message may be sufficient. For example, a high-definition video played on a standard-definition screen will lose some details while still conveying the message with an acceptable quality.

Channel Encoder/Decoder. The output of the source encoder is fed to a channel encoder whose role is to add *controlled* redundancy to the sequence to protect the transmitted bits against errors introduced by the channel. This can be done by an algorithm that maps the input sequence to a (longer) channel-coded sequence (e.g., by appending parity bits to the input bit sequence).

As an illustration, when the source encoder output given above is input to the channel encoder, the result will be a longer sequence of bits (compared to the output of the source encoder), for example,

$$1, 1, 0, 0, 1, 0, 1, 1, 0, 1, 0, 1, 1, 0, 1, \ldots, 1,$$

as redundancy is added to the sequence. These channel-encoded bits are then modulated and transmitted over the noisy communication channel.

The demodulator output bits (or soft information about the transmitted bits) are produced at the receiver side. The demodulator output may contain errors due to the noisy communication medium. The decoder uses the code constraints to correct these errors and obtains an estimate of the transmitted bits/symbols (that were input to the channel encoder).

The channel-coding operation appears to perform the opposite function of a source encoder: it adds redundancy while the source encoder removes the redundancy present in the original data. The channel encoder increases the size of the sequence to be transmitted, which may be counterintuitive. This is explained as follows: while there is redundancy in the source, it is not easy to exploit it since we do not have control over it. However, with channel coding, we have a controlled redundancy. We know precisely the relationship among the transmitted bits, and we can use this information for error detection or correction purposes at the receiver side in an effective manner.

Digital Modulator/Demodulator and Channel Effects. The next block at the transmitter side is the digital modulator, which takes the channel-coded bits/symbols as input. It maps the sequence into a signal suitable for transmission over the channel. For example, the signal generated for transmission over a wireless medium is a radio wave, while for transmission over an underwater acoustic channel, the signal generated is a sound wave. This modulated waveform is then transmitted over the channel.

Let us give a simple illustration. Assume that the input bit sequence to a digital modulator (i.e., the output of the channel encoder) is

$$0, 0, 1, 1, 1, 0, 1, 0, 0.$$

Also, assume that the modulator employs a rectangular pulse of duration T to transmit each of these bits; it selects the pulse amplitude as positive to represent the bit "1" and as negative for the bit "0." The resulting digital modulator output, which is the transmitted signal over the noisy communication channel, is depicted in Fig. 1.3.

Due to various impairments caused by the communication channel, the received signal will not be the same as the transmitted one. Such impairments may include signal attenuation, natural or artificial interference, noise, and so on. Figure 1.4 depicts a simple illustration of the received signal corresponding to the transmitted signal in Fig. 1.3.

The noise-corrupted (received) signal is fed to a digital demodulator, and an estimate of the channel-coded bits/symbols is obtained. This process involves different steps, including timing recovery. The receiver must determine the symbol boundaries and identify the part of the received signal corresponding to the transmission of each separate bit. Assuming this is established precisely, the received signal corresponding to each transmission period is input to the demodulator to generate an estimate for the transmitted bit/symbol. These estimates are then fed to the channel

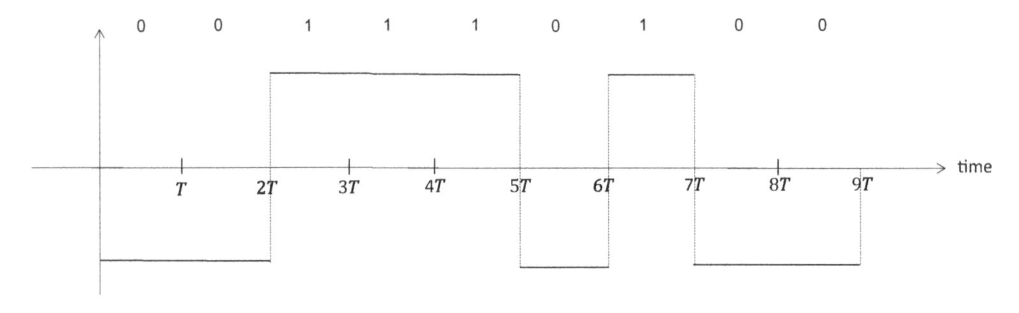

Figure 1.3 A sample digital modulator output.

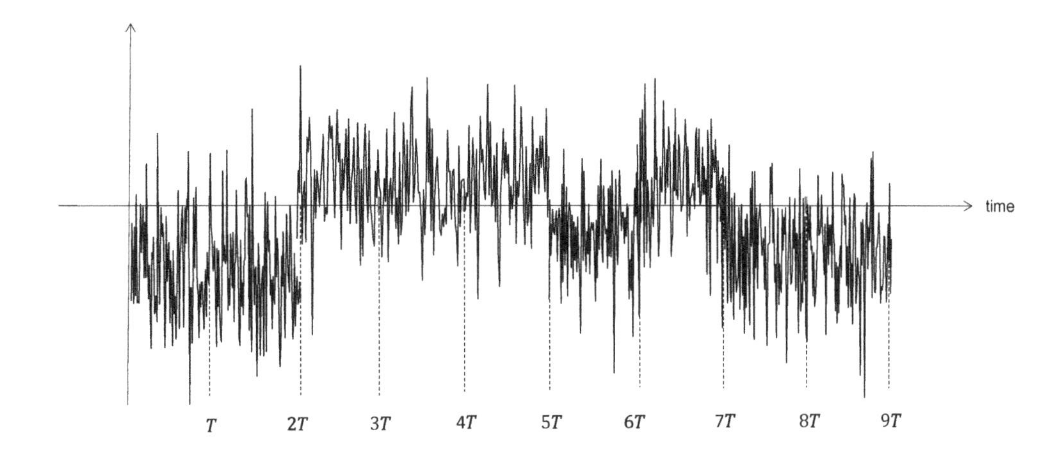

Figure 1.4 Depiction of a received signal through a noisy communication channel.

decoder, which produces an estimate of the channel encoder input (i.e., the source encoder output). These estimates are input to the source decoder, which produces an estimate of the message sequence.

Two critical metrics for a digital communication system are the symbol or bit error probability at the receiver output and the transmission rate in symbols or bits per unit of time. There is, in general, a trade-off between these two measures: for a given transmit power and channel bandwidth, if the error probability is to be improved, then the transmission rate needs to be reduced, or if the transmission rate is to be increased, the error probability will deteriorate. The exact nature of this trade-off will become apparent in Chapter 9. Nevertheless, some simple comments can be made without any in-depth study. For instance, everything else being the same, if the transmit power is increased, the error probability will reduce. Also, the data rate can be increased if a higher channel bandwidth is available.

The above description of a general digital communication system is somewhat simplified. For instance, the digital demodulator and the channel decoder work jointly in many communication systems. Similarly, it is possible to perform source coding and channel coding together, resulting in joint source–channel-coding schemes. On the other hand, this level of abstraction will help us study different components of a digital communication system in depth and develop an understanding of the basic principles. In the following chapters, some extensions beyond this basic block diagram will also become apparent.

1.3 Why Digital Communications?

Let us close this section by highlighting the necessity for digital communications and summarizing some of its advantages compared to analog communications. First, many signals, such as email messages, paging texts, or computer data, are natively digital. Hence, one would need to employ digital transmission techniques to transmit them. In other words, we have no option in certain applications but to use digital communications.

A significant advantage of digital transmission over analog is that it offers better signal fidelity and robustness against noise. Since the information being transmitted is a sequence of symbols or bits, the function of the receiver is to observe the noisy signal and decide which symbol is transmitted in each transmission period. In other words, it is feasible to effectively obliterate the channel noise effects (using powerful channel-coding techniques). This is not the case with analog communications, as some noise will still be present in the estimated message signal.

Another advantage of digital transmission is that it becomes feasible to do compression before transmission, facilitated using source-coding algorithms. This way, the redundancy in the transmitted signal can be eliminated, reducing the number of symbols being transmitted considerably. This enables more efficient use of the channel bandwidth, a precious resource. To give a concrete example, while the first-generation cellular systems (AMPS) required bandwidth of 30 kHz for voice calls using analog communication techniques, it became possible to employ a similar bandwidth to accommodate around 20 voice calls in the second-generation digital cellular systems – a 20-fold increase in the number of users served.

Digital transmission techniques also offer more options than analog communications when sharing a common medium. We can use time division, code division, or frequency division for multiplexing different messages or for multiple access; on the other hand, the only option with analog transmission is frequency division.

Digital transmission also enables the use of secure communications via encryption algorithms. For example, in a system where the transmitter and receiver share a common key sequence, the transmitter can encrypt the information transmitted via this key through an encryption algorithm. This makes the data unintelligible to unintended receivers who do not have the key, while the legitimate receiver can decrypt it. It is not possible to accomplish this via analog communication techniques.

There are further advantages of digital communications over analog communications in terms of implementation complexity and reliability of the communication circuitry. Digital circuits are more reliable than analog ones and are easier and cheaper to implement. The implementation is more flexible, and the same/similar components can be reused for different purposes. Their power consumption is also less.

1.4 Overview of the Book

In the rest of this book, we will go over the different components of a digital communication system in detail to develop a thorough understanding of the basic concepts. In Chapter 2, we will present the necessary mathematical preliminaries.

We will summarize the essential concepts from signals and systems, and we will review probability, random variables, and random processes. All these concepts will be used heavily in the subsequent chapters.

In Chapter 3, we will briefly review analog transmission techniques, cover the conversion of analog signals into digital form, and highlight the differences between analog and digital communications. We will discuss the processes of sampling and quantization (analog-to-digital conversion), and we will assess the quality of the resulting digital representation. We will also cover pulse-code modulation and its variants.

Chapter 4 is devoted to an understanding of the information content of a digital information source. We will define the concept of entropy and discuss the ultimate limits of lossless compression (source coding). We will also present several practical source-coding algorithms to demonstrate how compression takes place.

Chapters 5 and 6 are devoted to the fundamentals of digital modulation. In Chapter 5, we will describe the basics of linear modulation. Specifically, we will consider additive white Gaussian noise channels, a standard and realistic model for many communication channels. We will characterize the noise process and discuss its properties. We will then present the signal space concepts and develop the optimal receiver structures for binary and non-binary signaling with linear modulation. We will also analyze the error probability at the receiver output to assess the system performance. Chapter 6 is devoted to single-carrier bandpass transmission techniques. We will cover representations of deterministic and random bandpass signals, and we will explain how these representations can be used for the study and development of single-carrier bandpass communication systems. We will also highlight the relevant channel effects and describe different receiver structures based on how the channel phase is handled. Synchronization issues will also be discussed in some detail, timing recovery in Chapter 5 and carrier phase synchronization in Chapter 6.

Spectral occupancy of communication signals will be studied in Chapter 7. Specifically, we will discuss ways of designing communication pulses suitable for transmission over bandlimited channels and analyze the spectral efficiencies of different modulation techniques. We will also compute the power spectral density of modulated signals, that is, we will determine the density of the transmitted power as a function of frequency. This will enable us to assess whether the transmitted signals are interfering with other communication systems or if they are impacted by interference from other systems.

Multicarrier digital communication systems are the subject of Chapter 8. Specifically, we will cover frequency-shift keying and the important special case of minimum-shift keying. Also, we will study orthogonal frequency-division multiplexing, which is a highly practical modulation scheme adopted in 4G and 5G wireless standards, as well as for certain wireline transmissions.

Channel-coding techniques are studied in Chapter 9. We start with some basic ideas from information theory and determine the ultimate transmission limits.

We then go over two basic classes of codes in detail: linear block codes and convolutional codes.

Finally, we provide brief coverage of link budget analysis, methods of combatting transmission losses in a communication system, and multiple-access techniques in Chapter 10.

2 Mathematical Preliminaries

This chapter deals with the basics of signals and systems, probability theory, random variables, and random processes. These subjects will be heavily needed for our coverage of different topics in digital communications. For example, using basic signals and systems knowledge, we will address conversion of analog sources into a digital form. With basic probability and random variables, we will be able to determine the information content of a source, figure out the fundamental limits of compression, and quantify the amount of loss due to the quantization process needed for analog-to-digital conversion. With the extensions to random processes, we can characterize noise in a communication system and determine the error probabilities for different transmission schemes. Using these mathematical fundamentals, we will also be able to find the spectral content of random signals, including those of the transmitted signals; hence, we will be able to compute the amount of interference caused by a specific transmission system on others operating in adjacent frequency bands.

We start with a crash course on signals and systems in Section 2.1. The coverage includes a review of linear time-invariant systems, Fourier series expansion of periodic signals, and Fourier transform, including several basic examples and essential properties. Our coverage of signals and systems is concise, with the assumption that the student has already taken a course on the subject. We aim to briefly review the material pertinent to communication systems and establish the general notation.

We then provide a probability review in Section 2.2, taking an axiomatic approach. Along with the basic definitions and some important properties, we also review conditional probability, the total probability theorem, and Bayes' rule, which will be repeatedly used in the subsequent chapters. We go over the concept of random variables in Section 2.3, where we introduce the cumulative distribution function, probability density function, and probability mass function, along with functions of random variables as well as averages, including mean and variance of random variables. Also presented are several important examples, including Gaussian random variables. Extensions to multiple random variables (random vectors) and the important particular case of jointly Gaussian random variables are also studied. Our coverage of probability and random variables is also relatively short and to the point, with the understanding that the student has also been exposed to this material in a different course. Nevertheless, the level of detail provided is sufficient for understanding the rest of the material in the book without difficulty.

Section 2.4 is devoted to random processes. The autocorrelation function of a random process, extensions to multiple random processes, and different notions of stationarity are studied along with the basic definitions and examples. Particular attention is paid to wide-sense stationary processes, and the concept of power spectral density is introduced. Also explored is the filtering of wide-sense stationary random processes, including the essential properties of their autocorrelation function and power spectral density. Due to their significance in modeling noise in a communication system, we devote Section 2.5 to Gaussian random processes. The chapter is concluded in Section 2.6.

2.1 Signals and Systems

In this section, we will review the fundamentals of (deterministic) signals and linear systems as they relate to different concepts in communications that will be covered in this book.

A signal is nothing but a function of time. For example, a voltage level across a resistor could represent a speech signal. Systems operate on input signals to produce output signals. For instance, an amplifier whose input is a speech signal in electrical form produces an amplified (electrical) signal at its output. In the following, we cover several important concepts in signals and systems that will be needed throughout the book. Specifically, we focus on the case of continuous-time signals and systems (omitting altogether their discrete-time counterparts) as this will be sufficient for our purposes.

We will use the following definitions pertaining to signals in our study of communication systems.

The energy of a (real) signal $x(t)$ is defined as

$$E_x = \lim_{T \to \infty} \int_{-T/2}^{T/2} x^2(t) dt. \tag{2.1}$$

Similarly, we define the power content of $x(t)$ as

$$P_x = \lim_{T \to \infty} \frac{1}{T} \int_{-T/2}^{T/2} x^2(t) dt. \tag{2.2}$$

Similar definitions apply if the signal is complex-valued by replacing $x^2(t)$ with $|x^2(t)|$. If the signal's energy is finite, then its power content is 0. Or, equivalently, if the signal's power content is non-zero, it has infinite energy.

2.1.1 Linear Time-Invariant Systems

We call a system linear if its output to a linear combination of two input signals is the linear combination of the individual outputs, that is, the output of the system to the input signal $\alpha x_1(t) + \beta x_2(t)$ is given by

$$\mathcal{L}\{\alpha x_1(t) + \beta x_2(t)\} = \alpha y_1(t) + \beta y_2(t), \tag{2.3}$$

where $y_i(t)$ is the output of the system to $x_i(t)$, $i = 1, 2$, and $t \in \mathbb{R}$ denotes the time variable. We call a system time-invariant if the response to a time-shifted version of an input signal is the time-shifted version of the original output by the same amount, that is, for any $t_0 \in \mathbb{R}$

$$\mathcal{L}\{x(t - t_0)\} = y(t - t_0), \quad t \in \mathbb{R}, \tag{2.4}$$

where $y(t)$ is the output of the system to $x(t)$ at its input. A system that is both linear and time-invariant is called a linear time-invariant (LTI) system.

LTI systems are most easily characterized through their response to an *impulse* function applied at their input. The impulse function, also known as Dirac delta, denoted by $\delta(t)$, takes on the value 0 for all $t \neq 0$; however, it integrates to 1. Therefore, its value at the origin is not finite, and it is not a function in the usual sense.[1] For a continuous function $x(t)$, we also have

$$\int_{-\infty}^{\infty} x(t)\delta(t - t_0)dt = x(t_0). \tag{2.5}$$

We define the response of an LTI system to $\delta(t)$ at its input as its *impulse response*, denoted by $h(t)$. That is,

$$h(t) = \mathcal{L}\{\delta(t)\}. \tag{2.6}$$

See Fig. 2.1 for a depiction.

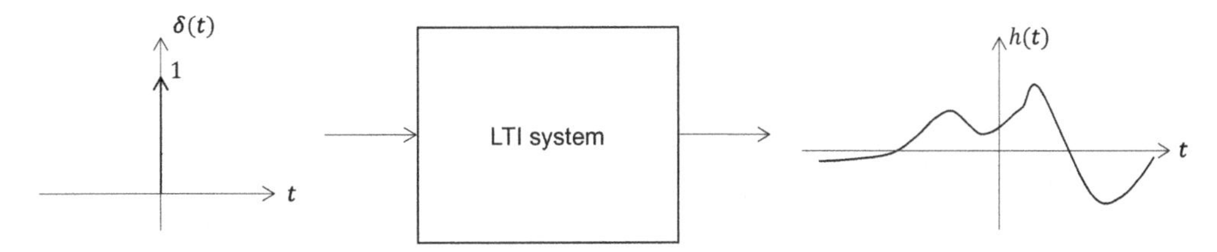

Figure 2.1 Illustration of an impulse function $\delta(t)$ input to an LTI system, resulting in the impulse response $h(t)$ at the output.

[1] We note here that the Dirac delta function is not a regular function, and a rigorous study requires one to employ distribution theory. Nonetheless, for all practical purposes, the only cautionary remark for studying such a generalized signal is that it should be studied under an integral sign (e.g., as an input to a linear system whose output is characterized by a convolution integral as done in the following).

Using the impulse response of an LTI system, we can determine its output $y(t)$ to an arbitrary input $x(t)$ through the convolution integral

$$y(t) = x(t) * h(t) = \int_{-\infty}^{\infty} x(\tau)h(t-\tau)d\tau. \tag{2.7}$$

This result can be established by first noticing that the input signal can be written as

$$x(t) = \int_{-\infty}^{\infty} x(\tau)\delta(t-\tau)d\tau, \tag{2.8}$$

and then, using the linearity of the system, we obtain

$$y(t) = \mathcal{L}\{x(t)\} = \int_{-\infty}^{\infty} x(\tau)\mathcal{L}\{\delta(t-\tau)\}d\tau. \tag{2.9}$$

Since the system is time-invariant, we can write $\mathcal{L}\{\delta(t-\tau)\} = h(t-\tau)$, resulting in the form of the convolution integral given above.

Note that a simple change of variables shows that the convolution operation is commutative, that is,

$$x(t) * h(t) = h(t) * x(t). \tag{2.10}$$

2.1.2 Periodic Signals and Fourier Series Expansion

We call a signal $x(t)$ periodic if

$$x(t) = x(t + T_0) \tag{2.11}$$

holds for all time instances t and for some T_0. T_0 is referred to as the period of the signal. The smallest possible period is called the fundamental period. Important examples of periodic signals include *sine* and *cosine* functions, and the closely related, complex exponential functions.

An important result is the following. Let $x(t)$ be a periodic signal with fundamental period T_0. Then, under some relatively mild conditions, it can be represented as a linear combination of complex exponentials with frequencies k/T_0 for $k = \ldots, -2, -1, 0, 1, 2, \ldots$. That is, we can write

$$x(t) = \sum_{k=-\infty}^{\infty} a_k \exp\left(j2\pi\frac{k}{T_0}t\right) \tag{2.12}$$

with

$$a_k = \frac{1}{T_0}\int_{T_0} x(t)\exp\left(-j2\pi\frac{k}{T_0}t\right)dt, \tag{2.13}$$

where the integral is taken over one full period.

The above expansion of a periodic signal is called the Fourier series expansion, and the sequence of coefficients $\{a_k\}$ are referred to as the Fourier series coefficients.

Example 2.1

Determine the Fourier series expansion coefficients of the sinusoidal signal

$$x(t) = A\cos(2\pi f_0 t + \phi_0). \tag{2.14}$$

Solution

We can write

$$x(t) = \frac{1}{2}Ae^{j\phi_0}e^{j2\pi f_0 t} + \frac{1}{2}Ae^{-j\phi_0}e^{-j2\pi f_0 t}, \tag{2.15}$$

which is nothing but the Fourier series expansion of the periodic signal with period $T_0 = \frac{1}{f_0}$. The Fourier series coefficients are given by $a_1 = \frac{1}{2}Ae^{j\phi_0}$, $a_{-1} = \frac{1}{2}Ae^{-j\phi_0}$, and $a_k = 0$ for $k \neq -1, 1$.

Example 2.2

Determine the Fourier series expansion of the periodic impulse train

$$x(t) = \sum_{n=-\infty}^{\infty} \delta(t - nT_0), \tag{2.16}$$

as depicted in Fig. 2.2.

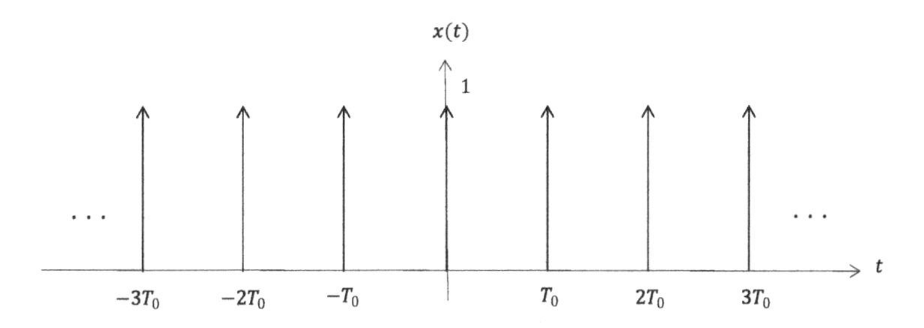

Figure 2.2 A periodic impulse train with period T_0.

Solution

The Fourier series expansion coefficients can be computed as

$$a_k = \frac{1}{T_0}\int_{-T_0/2}^{T_0/2} \delta(t)\exp\left(-j2\pi\frac{k}{T_0}t\right)dt = \frac{1}{T_0}, \tag{2.17}$$

for all $k \in \mathbb{Z}$. Hence, we obtain

$$x(t) = \frac{1}{T_0}\sum_{k=-\infty}^{\infty} \exp\left(j2\pi\frac{k}{T_0}t\right) \tag{2.18}$$

as the Fourier series expansion of the periodic impulse train.

Let us now consider the response of an LTI system to a complex exponential as input, and use this result to determine the system's response to arbitrary periodic signals. The response of an LTI system with impulse response $h(t)$ to the input $\exp(j2\pi f_0 t)$ is given by the convolution integral

$$\mathcal{L}\{\exp(j2\pi f_0 t)\} = \int_{-\infty}^{\infty} h(\tau)\exp(j2\pi f_0(t-\tau))d\tau \tag{2.19}$$

$$= \exp(j2\pi f_0 t)\int_{-\infty}^{\infty} h(\tau)\exp(-j2\pi f_0 \tau)d\tau \tag{2.20}$$

$$= H(f_0)\exp(j2\pi f_0 t) \tag{2.21}$$

where $H(f)$, defined as

$$H(f) = \int_{-\infty}^{\infty} h(\tau)\exp(-j2\pi f\tau)d\tau, \tag{2.22}$$

is the frequency response of the LTI system. In other words, the response to a complex exponential with frequency f_0 (i.e., period $T_0 = 1/f_0$) is again a complex exponential with the same frequency; however, there is an amplitude scaling and phase shift that can be readily determined from the impulse response of the LTI system. This point is illustrated in Fig. 2.3.

Using the above observation, we can determine the response of an LTI system with impulse response $h(t)$ (and frequency response $H(f)$) to an arbitrary periodic signal $x(t)$ with period T_0 whose Fourier series expansion is given by

$$x(t) = \sum_{k=-\infty}^{\infty} a_k \exp\left(j2\pi\frac{k}{T_0}t\right). \tag{2.23}$$

Denoting the response signal as $y(t)$ and using the linearity of the system, we obtain

$$y(t) = \sum_{k=-\infty}^{\infty} a_k H\left(\frac{k}{T_0}\right)\exp\left(j2\pi\frac{k}{T_0}t\right). \tag{2.24}$$

We see that the output signal is also periodic with the same period, and the expression in (2.24) can be thought of as its Fourier series expansion with Fourier series coefficients $a_k H(k/T_0)$.

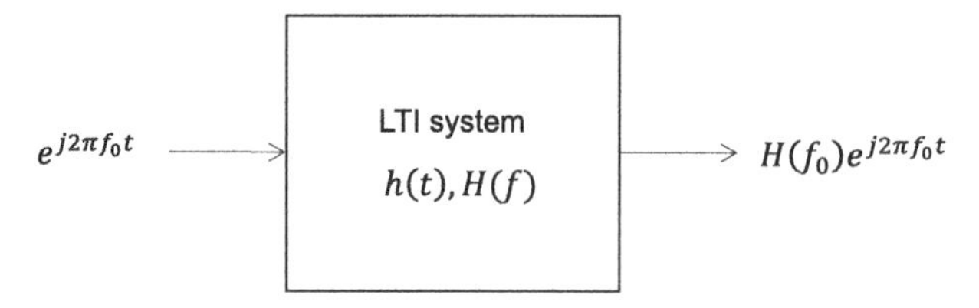

Figure 2.3 Compex exponential with frequency f_0 input to an LTI system with frequency response $H(f)$.

2.1.3 Fourier Transform

The concept of Fourier series expansion can be extended by taking the period of the signal approaching infinity, resulting in the definition of Fourier transform. That is, given a continuous-time signal, denoted by $x(t)$, its Fourier transform is given by the integral

$$\mathcal{F}\{x(t)\} = X(f) = \int_{-\infty}^{\infty} x(t)e^{-j2\pi ft}dt, \tag{2.25}$$

provided that the signal is square integrable, that is, it has finite energy. With the time t given in seconds, the frequency f is measured in Hertz (Hz). This expression is also referred to as the Fourier transform analysis equation.

Note that the Fourier transform definition also applies to periodic signals (whose energies are not finite) through the use of impulses.

The support of the Fourier transform of a signal is a measure of its spectral occupancy. If the signal has rapid time variations, the magnitude of $X(f)$ is significant for higher frequencies. In contrast, if the signal is slowly time-varying, the content of $X(f)$ is concentrated around $f \approx 0$.

Given the Fourier transform of a square-integrable signal, we can compute its time-domain representation through the inverse Fourier transform relationship, which is given by

$$x(t) = \int_{-\infty}^{\infty} X(f)e^{j2\pi ft}df. \tag{2.26}$$

The inverse transform is also referred to as the Fourier transform synthesis equation.

Let us give several examples.

Example 2.3

Compute the Fourier transform of $\delta(t)$.

Solution

Using the definition of the Fourier transform integral, we obtain

$$\mathcal{F}\{\delta(t)\} = \int_{-\infty}^{\infty} \delta(t)e^{-j2\pi ft}dt = 1. \tag{2.27}$$

In addition, using the inverse transform relationship, we can argue that the Fourier transform of a constant is an impulse function. That is, $\mathcal{F}\{1\} = \delta(f)$.

The above symmetric relationship is no coincidence, as clarified by examining the form of the Fourier transform analysis and synthesis equations and observing that they are almost the same except for a sign change in the argument of the exponential function in the integrals.

We can also readily obtain the following Fourier transform pairs:

$$\mathcal{F}\{\delta(t - t_0)\} = e^{-j2\pi f t_0} \tag{2.28}$$

and

$$\mathcal{F}\left\{e^{j2\pi f_0 t}\right\} = \delta(f - f_0). \tag{2.29}$$

Example 2.4

Determine the Fourier transforms of the sinusoidal functions $\cos(2\pi f_0 t)$ and $\sin(2\pi f_0 t)$.

Solution

Writing the sine and cosine functions as linear combinations of complex exponentials, that is,

$$\cos(2\pi f_0 t) = \frac{1}{2}\left(e^{j2\pi f_0 t} + e^{-j2\pi f_0 t}\right) \tag{2.30}$$

and

$$\sin(2\pi f_0 t) = \frac{1}{2j}\left(e^{j2\pi f_0 t} - e^{-j2\pi f_0 t}\right), \tag{2.31}$$

and applying the Fourier transform integral, we obtain

$$\mathcal{F}\{\cos(2\pi f_0 t)\} = \frac{1}{2}(\delta(f - f_0) + \delta(f + f_0)) \tag{2.32}$$

and

$$\mathcal{F}\{\sin(2\pi f_0 t)\} = \frac{1}{2j}(\delta(f - f_0) - \delta(f + f_0)). \tag{2.33}$$

As further examples, we define the rectangular function (denoted by $\Pi(t)$) and triangular function (denoted by $\Lambda(t)$) as follows:

$$\Pi(t) = \begin{cases} \frac{1}{2}, & \text{if } -\frac{1}{2} < t < \frac{1}{2}, \\ 0, & \text{else} \end{cases} \tag{2.34}$$

and

$$\Lambda(t) = \begin{cases} t + 1, & \text{if } -1 < t \leq 0, \\ -t + 1, & \text{if } 0 < t < 1, \\ 0, & \text{else.} \end{cases} \tag{2.35}$$

By direct integration, we can show that

$$\mathcal{F}\{\Pi(t)\} = \text{sinc}(f), \text{ and } \mathcal{F}\{\Lambda(t)\} = \text{sinc}^2(f), \tag{2.36}$$

where

$$\text{sinc}(f) = \begin{cases} \frac{\sin \pi f}{\pi f}, & \text{if } f \neq 0, \\ 1, & \text{if } f = 0. \end{cases} \tag{2.37}$$

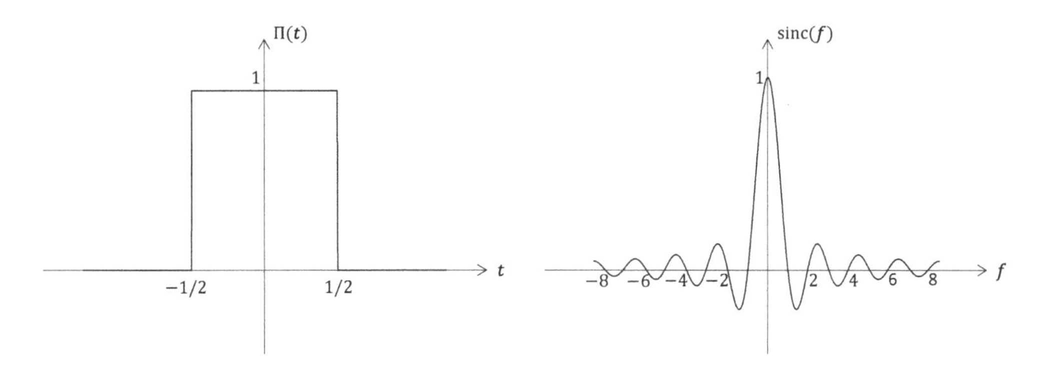

Figure 2.4 Rectangular function and its Fourier transform (sinc) function.

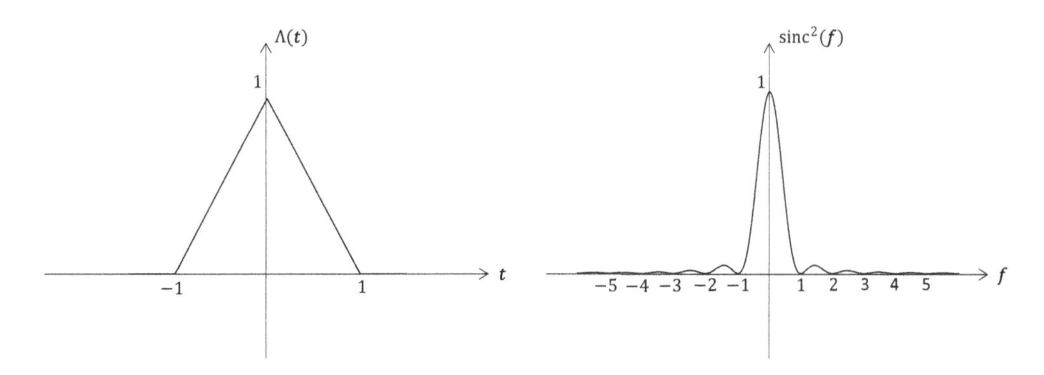

Figure 2.5 Triangular function and its Fourier transform.

Figures 2.4 and 2.5 depict the rectangular and triangular functions along with their Fourier transforms sinc(f) and sinc^2(f), respectively.

Recall from the previous subsection that the frequency response of an LTI system $H(f)$ is nothing but the Fourier transform of its impulse response $h(t)$.

Tables 2.1 summarizes some common Fourier transform pairs.

We close this subsection with a simple illustration of the relationship between the time-domain and frequency-domain representations of signals. In Fig. 2.6, we give three different sinusoidal signals along with their Fourier transforms. The sinusoid with a lower frequency (i.e., the more slowly varying one) has a smaller range of frequencies for which the Fourier transform has a significant magnitude. As expected, the more rapidly varying one occupies a broader spectrum. This is true in general: the Fourier transform of a more rapidly changing signal has components in higher frequencies.

2.1.4 Properties of Fourier Transform

There are several important Fourier transform properties that we will need in our coverage of communication systems.

Table 2.1 Some common Fourier transform pairs.

Time-domain function	Corresponding Fourier transform
1	$\delta(f)$
$\delta(t)$	1
$\exp(j2\pi f_0 t)$	$\delta(f - f_0)$
$\cos(2\pi f_0 t)$	$\frac{1}{2}\left(e^{j2\pi f_0 t} + e^{-j2\pi f_0 t}\right)$
$\sin(2\pi f_0 t)$	$\frac{1}{2j}\left(e^{j2\pi f_0 t} - e^{-j2\pi f_0 t}\right)$
$\Pi(t)$	$\text{sinc}(f)$
$\Lambda(t)$	$\text{sinc}^2(f)$
$\text{sinc}(t)$	$\Pi(f)$
$\text{sinc}^2(t)$	$\Lambda(f)$

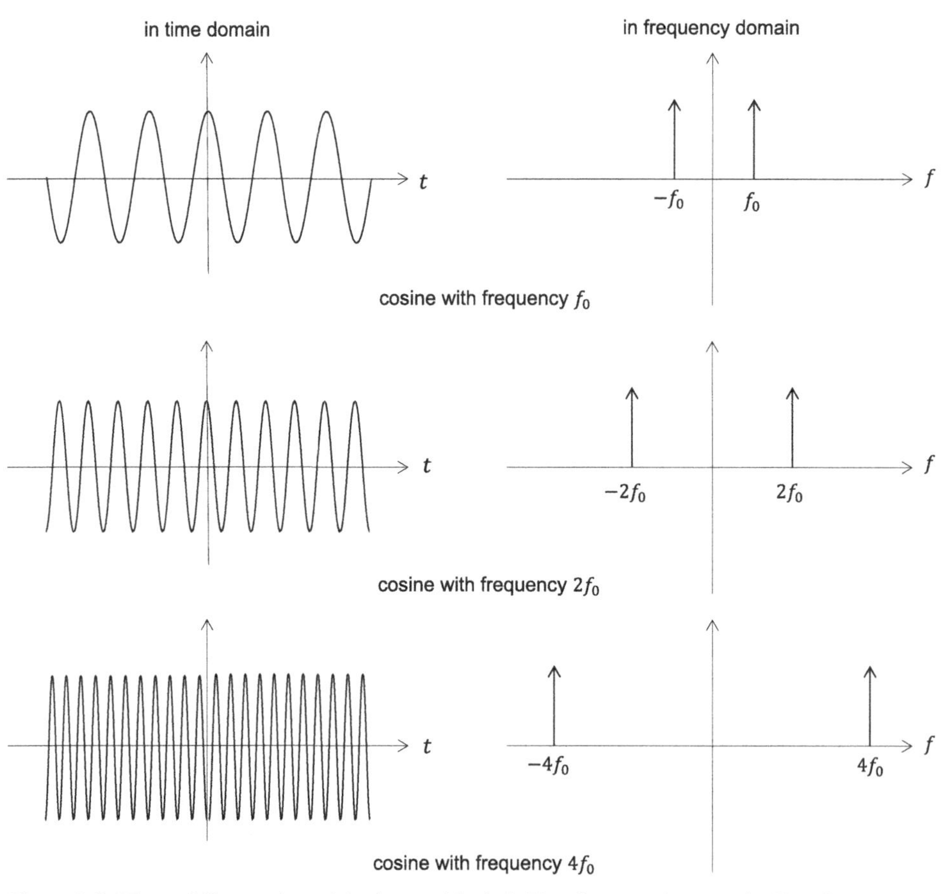

Figure 2.6 Three different sinusoids along with their Fourier transforms. The Fourier transform of a more rapidly varying signal has a higher frequency content.

In the following, to ease the notational presentation, we write

$$\mathcal{F}\{x(t)\} = X(f) \tag{2.38}$$

to signify that the function $x(t)$ on the left-hand side is a time-domain function (with index "t") while the term on the right-hand side is a frequency-domain function (with index "f").

Linearity. Given two signals $x_1(t)$ and $x_2(t)$ with Fourier transforms $X_1(f)$ and $X_2(f)$, respectively, the Fourier transform of their linear combination is the same linear combination of the Fourier transforms. That is,

$$\mathcal{F}\{\alpha x_1(t) + \beta x_2(t)\} = \alpha X_1(f) + \beta X_2(f) \tag{2.39}$$

for arbitrary complex numbers α and β. The proof of the statement follows directly from the Fourier transform integral.

Duality. The form of the Fourier transform analysis and synthesis equations being identical except for a sign difference in the argument of the exponential function in the integrals gives rise to the following property: if a given signal $x(t)$ has a Fourier transform $X(f)$, then the signal of the form $X(t)$ has a Fourier transform of $x(-f)$. As an immediate application, using the fact that

$$\mathcal{F}\{\Pi(t)\} = \text{sinc}(f), \tag{2.40}$$

we obtain

$$\mathcal{F}\{\text{sinc}(t)\} = \Pi(-f) = \Pi(f). \tag{2.41}$$

Shift in Time Domain. Given a signal $x(t)$ with Fourier transform $X(f)$, the Fourier transform of a time-shifted version of the signal can be obtained easily as

$$\mathcal{F}\{x(t - t_0)\} = e^{-j2\pi f t_0} X(f). \tag{2.42}$$

In other words, a shift in the time domain is a multiplication by a complex exponential function in the frequency domain.

Modulation. Given $\mathcal{F}\{x(t)\} = X(f)$, we have

$$\mathcal{F}\left\{x(t)e^{j2\pi f_0 t}\right\} = X(f - f_0). \tag{2.43}$$

In other words, the Fourier transform of a signal multiplied by a complex exponential with frequency f_0 is equal to a shifted version of the original Fourier transform by the same frequency.

Time Scaling. Given $\mathcal{F}\{x(t)\} = X(f)$, we have

$$\mathcal{F}\{x(at)\} = \frac{1}{|a|} X(f/a). \tag{2.44}$$

That is, *compression* in time results in *expansion in frequency* and vice versa.

Differentiation. With $\mathcal{F}\{x(t)\} = X(f)$, we have

$$\mathcal{F}\left\{\frac{d}{dt}x(t)\right\} = j2\pi f X(f). \tag{2.45}$$

Integration. With $\mathcal{F}\{x(t)\} = X(f)$, we have

$$\mathcal{F}\left\{\int_{-\infty}^{t} x(\tau)d\tau\right\} = \frac{X(f)}{j2\pi f} + \frac{1}{2}X(0)\delta(f). \tag{2.46}$$

Convolution in Time Domain. The Fourier transform of convolution of two functions $x(t)$ and $h(t)$ is the multiplication of the individual Fourier transforms, that is,

$$\mathcal{F}\{x(t) * h(t)\} = X(f)H(f). \tag{2.47}$$

This can easily be proved by using the definition of convolution and the Fourier transform integrals. Due to this property, we observe for an LTI system that the Fourier transform of the output signal is nothing but the multiplication of the Fourier transform of the input signal and the system's frequency response.

Plancherel–Parseval Theorem. Given two functions $x(t)$ and $y(t)$ with Fourier transforms $X(f)$ and $Y(f)$, respectively, we have

$$\int_{-\infty}^{\infty} x(t)y^*(t)dt = \int_{-\infty}^{\infty} X(f)Y^*(f)df \tag{2.48}$$

and, by taking $y(t) = x(t)$, we can also write

$$\int_{-\infty}^{\infty} |x(t)|^2 dt = \int_{-\infty}^{\infty} |X(f)|^2 df. \tag{2.49}$$

This theorem shows that the signal energy can also be computed using its Fourier transform, which may simplify certain computations considerably. For instance, one can readily obtain the energy of $\mathrm{sinc}(t)$ as 1 using

$$\int_{-\infty}^{\infty} \mathrm{sinc}^2(t)dt = \int_{-\infty}^{\infty} \Pi^2(f)df = \int_{-1/2}^{1/2} 1df = 1. \tag{2.50}$$

2.2 Probability

We will need concepts and tools from probability, random variables, and random processes for a thorough study of digital communication systems. For instance, we will be computing error probabilities of various modulation schemes and characterizing the information content of a source, which require basic knowledge of probability and random variables. We will also be modeling the noise effects in transmission over a channel and determining the power content of a transmitted signal as a function of frequency, which requires concepts from random processes.

In this section, we will cover the basics of probability. In the subsequent sections, we will extend our coverage to random variables and random processes, and then, we will go over the particularly important case of Gaussian processes.

2.2.1 Axioms of Probability

Let Ω be the sample space. An element $\omega \in \Omega$ is called an outcome, and a subset $E \subset \Omega$ is an event. Probability $\mathbb{P}$ is a set function that satisfies three axioms:

(1) $\mathbb{P}(E) \geq 0$;
(2) $\mathbb{P}(\Omega) = 1$;
(3) for a finite or countable collection of disjoint events $E_1, E_2, E_3, \ldots, E_n, \ldots$ (namely, $E_i \cap E_j = \phi$ where ϕ denotes the empty set or the null event, for all i, j with $i \neq j$), we have

$$\mathbb{P}\left(\bigcup_{i=1}^{\infty} E_i\right) = \sum_{i=1}^{\infty} \mathbb{P}(E_i). \tag{2.51}$$

It is important to note that at this point, we are brushing over some technicalities, and indeed, the definition above is not mathematically precise. In particular, not all subsets of a sample space are necessarily measurable (i.e., not all subsets can be assigned a probability). On the other hand, for all intents and purposes throughout this book, or even when covering more advanced material in signal processing or communications, this is sufficient; hence, we will overlook this issue and proceed with the above definition.

Several properties follow easily from the definition of probability. For instance:

- for any event E, $\mathbb{P}(E) \leq 1$;
- the probability of the complement of the event E, denoted by E^c, is given by

$$\mathbb{P}(E^c) = 1 - \mathbb{P}(E); \tag{2.52}$$

- the probability of the null event ϕ is zero;
- for two events such that $E_1 \subset E_2$, we have

$$\mathbb{P}(E_1) \leq \mathbb{P}(E_2); \tag{2.53}$$

- for two arbitrary events A and B

$$\mathbb{P}(A \cup B) = \mathbb{P}(A) + \mathbb{P}(B) - \mathbb{P}(A \cap B). \tag{2.54}$$

As a simple exercise, let us prove the last property. From set theory, we know that

$$A \cup B = A \cup (A^c \cap B). \tag{2.55}$$

Since the events A and $A^c \cap B$ are mutually exclusive, using the third axiom of probability, we can write

$$\mathbb{P}(A \cup B) = \mathbb{P}(A) + \mathbb{P}(A^c \cap B). \tag{2.56}$$

On the other hand, we also have $B = (A \cap B) \cup (A^c \cap B)$, and since $A \cap B$ and $A^c \cap B$ are mutually exclusive

$$\mathbb{P}(B) = \mathbb{P}(A \cap B) + \mathbb{P}(A^c \cap B). \tag{2.57}$$

Combining this with (2.56), we arrive at the result in (2.54), as desired.

2.2.2 Conditional Probability and Bayes' Rule

In many applications, we encounter problems involving computation of the probability of a particular event given an observation. For instance, consider a basic binary communication system where we are transmitting 0 or 1, each with a probability of 1/2. Given the observation through a noisy channel, we will often be interested in the probability of the transmitted bit being 0 (or 1), requiring the computation of a conditional probability. While the unconditional probability of the transmitted bit being 0 (or 1) is 1/2, the conditional probability of the transmitted bit given the channel output will be very different.

We formally define the conditional probability of an event A given the event B as

$$\mathbb{P}(A|B) = \frac{\mathbb{P}(A \cap B)}{\mathbb{P}(B)}, \tag{2.58}$$

whenever $\mathbb{P}(B) \neq 0$. We simply define $\mathbb{P}(A|B) = 0$ when $\mathbb{P}(B) = 0$.

It is easy to check that the conditional probability is a valid probability assignment, that is, it satisfies the axioms of probability. Therefore, for an event B with a non-zero probability:

(1) for any event A, $\mathbb{P}(A|B) \geq 0$;

(2) for the conditional probability of the sure event, we have

$$\mathbb{P}(\Omega|B) = \frac{\mathbb{P}(\Omega \cap B)}{\mathbb{P}(B)} = \frac{\mathbb{P}(B)}{\mathbb{P}(B)} = 1; \tag{2.59}$$

(3) for any (finite or countably infinite) set of mutually exclusive events $E_1, E_2, \ldots, E_n, \ldots$, we have

$$\mathbb{P}\left(\bigcup_{i=1}^{\infty} E_i \,\middle|\, B\right) = \sum_{i=1}^{\infty} \mathbb{P}(E_i|B), \tag{2.60}$$

which follows from the definition of the conditional probability and the fact that $E_1 \cap B, E_2 \cap B, \ldots, E_n \cap B, \ldots$ are also mutually exclusive.

Example 2.5

Consider the roll of a fair die. The sample space of the experiment is $\Omega = \{1, 2, \ldots, 6\}$. The probability of the event $A = \{1\}$, that is, the probability of the outcome being 1, is simply $\mathbb{P}(A) = 1/6$.

Define the event $B = \{1, 3, 5\}$, that is, B is the event that the experiment's outcome is an odd number. The conditional probability of A given B is then

$$\mathbb{P}(A|B) = \frac{\mathbb{P}(A \cap B)}{\mathbb{P}(B)} = \frac{\mathbb{P}(\{1\})}{\mathbb{P}(\{1, 3, 5\})} = \frac{1/6}{3/6} = 1/3, \tag{2.61}$$

which is different from the unconditional probability of the event A.

Independence. Two events A and B are called independent if

$$\mathbb{P}(A \cap B) = \mathbb{P}(A)\mathbb{P}(B). \tag{2.62}$$

We can observe that, for independent events, conditioning on one does not give any new information about the other, that is, $\mathbb{P}(A|B) = \mathbb{P}(A)$.

We say that m events $A_1, A_2, \ldots, A_m$ are mutually independent if for any k of these events, denoted by $B_1, B_2, \ldots, B_k$, we have

$$\mathbb{P}\left(\bigcap_{i=1}^{k} B_i\right) = \prod_{i=1}^{k} \mathbb{P}(B_i). \tag{2.63}$$

Note that it is possible to have pairwise independent but mutually dependent events. On the other hand, mutual independence implies pairwise independence.

The computation of the probability of an event is often simplified by conditioning on specific events that make up the sample space. For example, assume that a digital communication system operates over a channel which is in either a *good* state or a *bad* state, each with a certain probability. The problem of computing the bit error probability may be simplified by solving the subproblems of the probability of error given that the channel is in the good state and the probability of error given that the channel is in the bad state, and then combining the results.

Assume that the events $E_1, E_2, \ldots, E_n$ form a partition of the sample space, namely, they are mutually exclusive ($E_i \cap E_j = \phi$ for $i, j \in \{1, 2, \ldots, n\}$, $i \neq j$), and they are collectively exhaustive (i.e., their union is the sample space Ω). Then, *the total probability theorem* states that the probability of an arbitrary event is given by

$$\mathbb{P}(A) = \sum_{i=1}^{n} \mathbb{P}(A \cap E_i) \tag{2.64}$$

$$= \sum_{i=1}^{n} \mathbb{P}(E_i)\mathbb{P}(A|E_i). \tag{2.65}$$

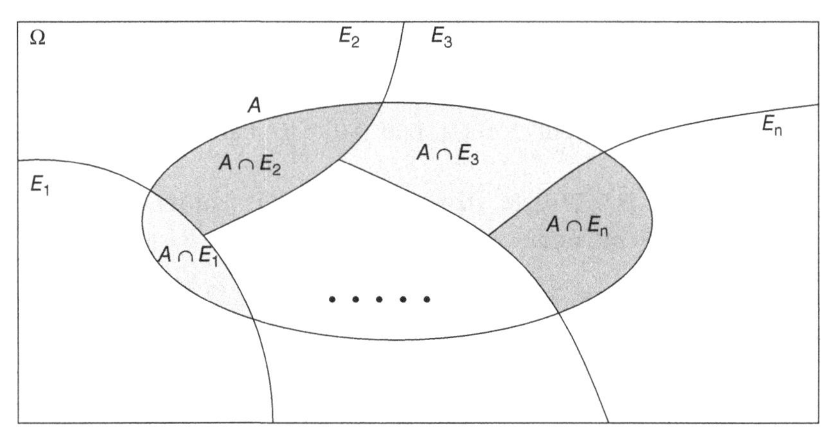

Figure 2.7 The event A is nothing but the union of mutually exclusive events $A \cap E_i$, $i = 1, 2, \ldots, n$, hence its probability $\mathbb{P}(A)$ is the sum of the probabilities $\mathbb{P}(A \cap E_i)$ for $i = 1, 2, \ldots, n$.

The proof of the total probability theorem is quite simple and intuitive because the events $A \cap E_i$ are mutually exclusive, and their union is simply the event A. A graphical illustration is given in Fig. 2.7.

Example 2.6

Consider transmission of a bit (0 or 1) over a noisy channel. Assume the channel has two possible states: *good* and *bad*. The conditional probability of error given that the channel is in the good state is 0.001, while the conditional probability of error in transmission when the channel is in the bad state is 0.2. At any given time instance, the probability that the channel is in the good state is 0.9. Compute the probability that a transmitted bit will be received in error.

Solution

Let E denote the error event, S_g denote the event that the channel is in the good state, and S_b denote the event that the channel is in the bad state. From the total probability theorem, we can write

$$\begin{aligned} \mathbb{P}(E) &= \mathbb{P}(S_g)\mathbb{P}(E|S_g) + \mathbb{P}(S_b)\mathbb{P}(E|S_b) && (2.66) \\ &= 0.9 \times 0.001 + 0.1 \times 0.2 && (2.67) \\ &= 0.0209. && (2.68) \end{aligned}$$

We will utilize the total probability theorem for similar calculations throughout the book.

We close this section with another important result, namely, Bayes' rule, which states that

$$\mathbb{P}(A|B) = \frac{\mathbb{P}(B|A)\mathbb{P}(A)}{\mathbb{P}(B)}, \tag{2.69}$$

where A and B are arbitrary events whose probabilities are not zero. The proof is immediate from the definition of conditional probability.

We will rely on Bayes' rule throughout our coverage of digital communications to solve important problems, as some seemingly complicated conditional probability calculations will be greatly simplified by reversing the roles of the two events and looking at the problem differently. A simple example is given below.

Example 2.7

Consider the setting in Example 2.6. Determine the conditional probability that the channel was in the good state (event S_g) given that the transmission of a particular bit was in error (i.e., calculate $\mathbb{P}(S_g|E)$).

Solution

Using Bayes' rule, we can write

$$\mathbb{P}(S_g|E) = \frac{\mathbb{P}(E|S_g)\mathbb{P}(S_g)}{\mathbb{P}(E)} \tag{2.70}$$

$$= \frac{0.001 \times 0.9}{0.0209} \tag{2.71}$$

$$= 0.043. \tag{2.72}$$

Notice that while the unconditional probability of S_g is 90%, conditioned on the event that the transmission was unsuccessful, the conditional probability reduces to only about 4.3%.

2.3 Random Variables

2.3.1 Basic Definitions and Properties

A random variable is defined as a mapping from the sample space Ω to the set of real numbers $\mathbb{R}$. For instance, the next bit or symbol that will be transmitted is modeled as a random variable, and so is the amount of error introduced at the output of an analog-to-digital converter for each input sample.

When we deal with random variables, we will always suppress the underlying sample space, experiment, probability assignment, and explicit mapping from the experimental outcomes (elements of the sample space) to the real numbers. Instead, we will characterize them by using functions that summarize their statistical properties, namely, cumulative distribution function (CDF), probability density function (PDF), and probability mass function (PMF).

Cumulative Distribution Function. The CDF for an arbitrary random variable is defined as

$$F_X(x) = \mathbb{P}(X \leq x), \tag{2.73}$$

where the subscript X denotes the random variable to which the CDF belongs, and $x \in \mathbb{R}$. Note that we write this expression with some abuse of notation. More precisely, what we mean is the following:

$$F_X(x) = \mathbb{P}\Big(\{\omega : \omega \in \Omega \text{ and } X(\omega) \leq x\} \Big). \tag{2.74}$$

Nevertheless, we will use this simplified notation throughout the book.

Some properties of the CDF are given as follows:

- $0 \leq F_X(x) \leq 1$ for any $x \in \mathbb{R}$;
- $F_X(x)$ is a non-decreasing function of x;
- $\lim_{x \to -\infty} F_X(x) = 0$;
- $\lim_{x \to \infty} F_X(x) = 1$;
- $F_X(x)$ is continuous from the right (its limit from the right is equal to the value of the function), namely,

$$F_X(x) = \lim_{t \to x^+} F_X(t). \tag{2.75}$$

The probability that the random variable will take on a value in an interval (left-open, right-closed) is given by

$$\mathbb{P}(a < X \leq b) = F_X(b) - F_X(a). \tag{2.76}$$

One must be careful when the interval is left-closed, right-open, or both. For instance,

$$\mathbb{P}(a \leq X \leq b) = F_X(b) - F_X(a^-), \tag{2.77}$$

where $F_X(a^-)$ is the limit of the CDF at a approaching from the left.

Similarly, the probability that the random variable X takes on a specific value a is given by

$$\mathbb{P}(X = a) = F_X(a) - F_X(a^-). \tag{2.78}$$

Namely, when there is a mass point at a, the CDF is discontinuous, and the amount of the "jump" in the CDF is equal to the probability of the specific mass point.

We classify a random variable X into three types:

- *discrete random variables*, for which X takes on finite or countably many values;
- *continuous random variables*, for which X takes on only a continuum of values with no mass points;
- *mixed random variables*, for which X both takes on only a continuum of values and has mass points.

The CDF is a continuous function for continuous random variables since there are no mass points, while it is a staircase function for discrete random variables. The CDFs of different types of random variables are illustrated in Fig. 2.8.

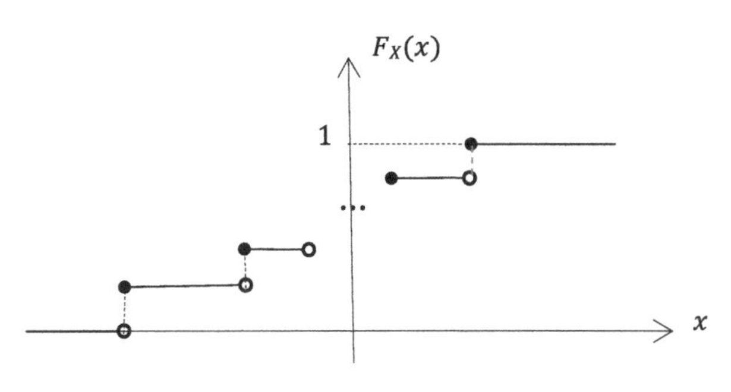

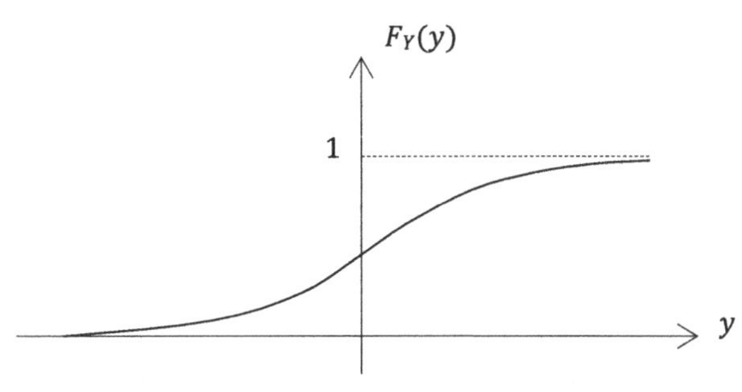

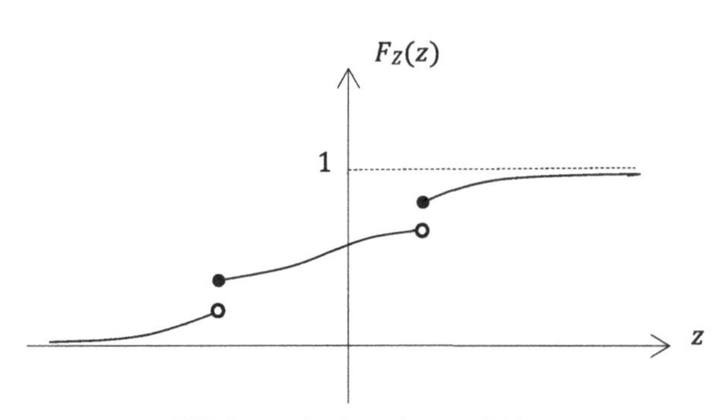

Figure 2.8 Illustration of CDF for discrete, continuous, and mixed random variables.

Probability Density Function. We define the PDF of a random variable X as the derivative of its CDF, namely,

$$f_X(x) = \frac{dF_X(x)}{dx}. \tag{2.79}$$

The derivative may not exist; therefore, the PDF of a random variable may not exist. However, we will allow *impulses* in the PDF to keep it more general. Namely, whenever there is a mass point with probability p, the CDF will have a jump of amount p, and the PDF will include an impulse of weight p at the mass point.

Some properties of the PDF are given as follows:

- $f_X(x) \geq 0$;
- $\int_{-\infty}^{\infty} f_X(x)dx = 1$;
- $F_X(x) = \int_{-\infty}^{x} f_X(u)du$;
- $\mathbb{P}(a < X \leq b) = \int_{a}^{b} f_X(x)dx$.

Note that when computing the probability of a random variable being in a certain interval, as in computations using the CDF, one has to be careful about the interval's endpoints when there are mass points at the boundaries.

Probability Mass Function. For discrete random variables, we also define their probability mass function as

$$P_X(x_i) = \mathbb{P}(X = x_i) = p_i, \tag{2.80}$$

where x_i are the values taken by the random variable X and p_i are their probabilities.

2.3.2 Functions of a Random Variable and Averages

Consider a random variable X and an arbitrary function $y = g(x)$. The function of the random variable X, defined as $Y = g(X)$, is also a random variable whose characterization can be obtained from the PDF or CDF of the original random variable.

Let us give several examples.

Example 2.8

Consider a discrete random variable X taking on the values $\{1, 2, \ldots, 10\}$ with equal probabilities. Define a new random variable Y as

$$Y = X \mod 2. \tag{2.81}$$

Then Y is a discrete random variable taking on the values 0 and 1. We have

$$\mathbb{P}(Y = 0) = \mathbb{P}(X \in \{2, 4, 6, 8, 10\}) = 5 \times \frac{1}{10} = \frac{1}{2}, \tag{2.82}$$

hence, the PMF of Y is $P_Y(0) = P_Y(1) = \frac{1}{2}$.

Example 2.9

Let X be a continuous random variable with PDF

$$f_X(x) = \begin{cases} 1, & \text{if } x \in (0, 1), \\ 0, & \text{else.} \end{cases} \tag{2.83}$$

Determine the PDF of a new random variable defined as $Y = -\ln X$.

Solution

The CDF of the new random variable Y evaluated at any $y \in \mathbb{R}$ can be written as

$$F_Y(y) = \mathbb{P}(Y \leq y) \tag{2.84}$$
$$= \mathbb{P}(-\ln X \leq y) \tag{2.85}$$
$$= \mathbb{P}(X \geq e^{-y}). \tag{2.86}$$

For $y \leq 0$, this probability is simply 0 as the random variable X always takes on values less than 1; and, for $y > 0$, we have

$$F_Y(y) = \int_{e^{-y}}^{1} 1dx \tag{2.87}$$
$$= 1 - e^{-y}. \tag{2.88}$$

Differentiating the CDF with respect to y, we obtain the PDF of Y as $f_Y(y) = e^{-y}$ for $y > 0$, and 0 otherwise.

In the above example, we have obtained a continuous random variable from another continuous random variable through a transformation. It is also possible to obtain a discrete or mixed random variable from a continuous random variable. The following example illustrates this point.

Example 2.10

Consider the continuous random variable Y with PDF $f_Y(y) = e^{-y}$ for $y \geq 0$, and 0 otherwise. Define a new random variable $Z = \min(Y, 1)$. Determine the PDF of Z.

Solution

We can write

$$F_Z(z) = \mathbb{P}(Z \leq z) \tag{2.89}$$
$$= \mathbb{P}(\min(Y, 1) \leq z) \tag{2.90}$$
$$= \begin{cases} F_Y(z), & \text{if } z < 1, \\ 1, & \text{else} \end{cases} \tag{2.91}$$
$$= \begin{cases} 0, & \text{if } z < 0, \\ 1 - e^{-z}, & \text{if } 0 \leq z < 1, \\ 1, & \text{if } z \geq 1. \end{cases} \tag{2.92}$$

It is clear that Z is a mixed random variable as its CDF is discontinuous at $z = 1$.

We also use statistical averages to characterize random variables. We define the *mean* (also called the *expected value*) of a random variable X as

$$\mathbb{E}[X] = m_X = \int_{-\infty}^{\infty} x f_X(x) dx \tag{2.93}$$

assuming that the integral exists. We note that there is a law of large numbers, which states that the average of a large number of realizations of the random variable will approach its expected value.

Note that for discrete random variables, we can also compute the expectation using the probability mass functions. That is, we can write

$$\mathbb{E}[X] = m_X = \sum_x x P_X(x). \tag{2.94}$$

We are often interested in the average of a function of a random variable. The mean of $Y = g(X)$ can be determined by finding the PDF of the new random variable Y and then applying the definition of the expected value. On the other hand, this is not necessary. Instead, the expectation can also be determined using the PDF of the original random variable. That is, we can write

$$\mathbb{E}[g(X)] = \int_{-\infty}^{\infty} g(x) f_X(x) dx. \tag{2.95}$$

We define the nth moment of a random variable X as

$$\mathbb{E}[X^n] = \int_{-\infty}^{\infty} x^n f_X(x) dx \tag{2.96}$$

whenever the integral exists, where n is a positive integer. The nth central moment of X is defined as $\mathbb{E}[(X - m_X)^n]$, that is, the mean is subtracted first, and then the nth moment is computed.

The second-order central moment of a random variable is defined as its *variance*, denoted by $\text{Var}(X)$ or σ_X^2. That is,

$$\text{Var}(X) = \mathbb{E}[(X - m_X)^2] \tag{2.97}$$

$$= \mathbb{E}[X^2] - m_X^2, \tag{2.98}$$

where the second equation follows easily by expanding the square term and rewriting the expectation integral in an equivalent form.

The square root of the variance, denoted by σ_X, is called the standard deviation of the random variable.

The variance (or standard deviation) of a random variable indicates how much the realizations will deviate from the mean. If the variance is small, we can expect that different realizations of X will be near its mean. Conversely, if it is large, the realizations may differ significantly from the mean value. Intuitively, the first-order and second-order averages of a random variable (mean and variance) provide some information about it; however, this characterization is weaker compared to the one obtained from the PDF or CDF. The characterization becomes more and more accurate by considering additional moments of the random variable. Indeed, under certain conditions, if the moments of all orders are known, then the PDF and CDF can be obtained, that is, we can achieve a complete statistical characterization of the random variable.

It is easy to see that the expected value of the sum of two different functions of X is equal to the sum of the individual expected values, that is,

$$\mathbb{E}[g_1(X) + g_2(X)] = \mathbb{E}[g_1(X)] + \mathbb{E}[g_2(X)]. \tag{2.99}$$

For the case of a linear transformation, we have

$$\mathbb{E}[aX + b] = a\mathbb{E}[X] + b \tag{2.100}$$

where a and b are real numbers. It can also readily be shown that

$$\mathrm{Var}(aX + b) = a^2\mathrm{Var}(X). \tag{2.101}$$

2.3.3 Examples of Random Variables

We encounter certain random variables repeatedly throughout our study of digital communication systems. For example, transmitted bits can be modeled using a Bernoulli random variable, which is a discrete random variable taking on only two values, 1 or 0, with probabilities $\mathbb{P}(X = 1) = p$ and $\mathbb{P}(X = 0) = 1 - p$, where $0 \leq p \leq 1$. The mean of a Bernoulli random variable is simply $\mathbb{E}[X] = p$, and its variance can be computed as

$$\mathrm{Var}(X) = \mathbb{E}[X^2] - m_X^2 \tag{2.102}$$

$$= 1^2 \times p + 0^2 \times (1 - p) - p^2 \tag{2.103}$$

$$= p(1 - p). \tag{2.104}$$

As another example, a binomial random variable is characterized by two parameters, n and p, where n is a positive integer, while p is a real number between 0 and 1. The random variable is integer-valued with a PMF

$$\mathbb{P}(X = k) = \binom{n}{k} p^k (1 - p)^{n-k}, \tag{2.105}$$

for $k = 0, 1, 2, \ldots, n$, and 0 otherwise. Consider n independent symbol transmissions over a noisy channel. If the error probability in each transmission is p, the binomial random variable with the above PMF represents the total number of errors.

The mean of a binomial random variable is $\mathbb{E}[X] = np$, and its variance is $\mathrm{Var}(X) = np(1 - p)$.

A simple example of a continuous random variable is the uniform random variable on the interval (a, b), which has a PDF given by

$$f_X(x) = \begin{cases} \frac{1}{b-a}, & \text{if } a < x < b, \\ 0, & \text{else.} \end{cases} \tag{2.106}$$

The mean of the uniform random variable is

$$\mathbb{E}[X] = \int_a^b x \frac{1}{b - a} dx = \frac{a + b}{2}, \tag{2.107}$$

while its variance can be computed as

$$\mathrm{Var}(X) = \frac{(b-a)^2}{12}. \tag{2.108}$$

Another example of a continuous random variable is the exponential random variable with parameter λ defined by the probability density function

$$f_X(x) = \begin{cases} \lambda e^{-\lambda x}, & \text{if } x > 0, \\ 0, & \text{else.} \end{cases} \tag{2.109}$$

We can easily compute its mean and variance as $1/\lambda$ and $1/\lambda^2$, respectively.

In Example 2.9, we have seen that realizations of an exponential random variable can be obtained from those of a uniform random variable through a certain function. In fact, this is a more general property: one can generate realizations of an arbitrary continuous random variable from those of a uniform random variable. The following example illustrates this point.

Example 2.11

Let U be a uniform random variable on $(0, 1)$. Show that the random variable obtained through the transformation $X = F^{-1}(U)$, where F is an arbitrary continuous distribution function, has CDF F.

Solution

The result follows by directly computing the CDF of the random variable X, that is,

$$\begin{aligned} F_X(x) &= \mathbb{P}(X \leq x) && (2.110) \\ &= \mathbb{P}(F^{-1}(U) \leq x) && (2.111) \\ &= \mathbb{P}(U \leq F(x)) && (2.112) \\ &= F(x), && (2.113) \end{aligned}$$

where the third line follows since the CDF of a random variable is non-decreasing, and the last line follows since the random variable U is uniform on the interval $(0, 1)$.

The above result can be used for simulations of a communication system to produce realizations of random variables that are not directly available.

Note also that for any continuous random variable X with CDF $F_X(x)$, the random variable obtained by the transformation $Y = F_X(X)$ is uniform on the interval $(0, 1)$. To see this, for any $y \in [0, 1]$, we can write

$$\begin{aligned} F_Y(y) &= \mathbb{P}(Y \leq y) && (2.114) \\ &= \mathbb{P}(F_X(X) \leq y) && (2.115) \\ &= \mathbb{P}(X \leq F_X^{-1}(y)) && (2.116) \\ &= F_X(F_X^{-1}(y)) && (2.117) \\ &= y. && (2.118) \end{aligned}$$

Noting also that $F_Y(y) = \mathbb{P}(Y \leq y) = 0$ for any $y < 0$, and $F_Y(y) = \mathbb{P}(Y \leq y) = 1$ for any $y > 1$, we arrive at the conclusion that Y is a uniform random variable on the interval $(0, 1)$. From this result and Example 2.11, we can argue that any continuous random variable can be obtained from any other continuous random variable through a suitable transformation.

Another random variable commonly encountered in communications is the Gaussian random variable, which will be studied in the following subsection.

2.3.4 Gaussian Random Variables

A *Gaussian* random variable (also called a *normal* random variable) has two parameters, $\mu \in \mathbb{R}$ and $\sigma^2 \in \mathbb{R}^+$. It is represented by the notation $X \sim \mathcal{N}(\mu, \sigma^2)$, and its PDF is given by

$$f_X(x) = \frac{1}{\sqrt{2\pi}\sigma} \exp\left(-\frac{(x-\mu)^2}{2\sigma^2}\right), \tag{2.119}$$

for all $x \in \mathbb{R}$.

One can verify that this function integrates to 1; hence, it is a valid probability density function. To see this, we write the integral of $f_X(x)$ from $-\infty$ to ∞ as

$$I = \frac{1}{\sqrt{2\pi}\sigma} \int_{-\infty}^{\infty} \exp\left(-\frac{(x-\mu)^2}{2\sigma^2}\right) dx \tag{2.120}$$

$$= \frac{1}{\sqrt{2\pi}} \int_{-\infty}^{\infty} \exp(-u^2/2) du, \tag{2.121}$$

which follows by a change of variables $u = \frac{x-\mu}{\sigma}$. By writing the integral I in two different forms with different integration variables, we obtain

$$I^2 = \frac{1}{2\pi} \int_{-\infty}^{\infty} \exp(-u^2/2) du \int_{-\infty}^{\infty} \exp(-v^2/2) dv \tag{2.122}$$

$$= \frac{1}{2\pi} \int_{-\infty}^{\infty} \int_{-\infty}^{\infty} \exp(-(u^2 + v^2)/2) du dv \tag{2.123}$$

$$= \frac{1}{2\pi} \int_{0}^{2\pi} \int_{0}^{\infty} \exp(-r^2/2) r dr d\theta \tag{2.124}$$

$$= \frac{1}{2\pi} \int_{0}^{2\pi} d\theta \int_{0}^{\infty} \exp(-r^2/2) r dr \tag{2.125}$$

$$= 1. \tag{2.126}$$

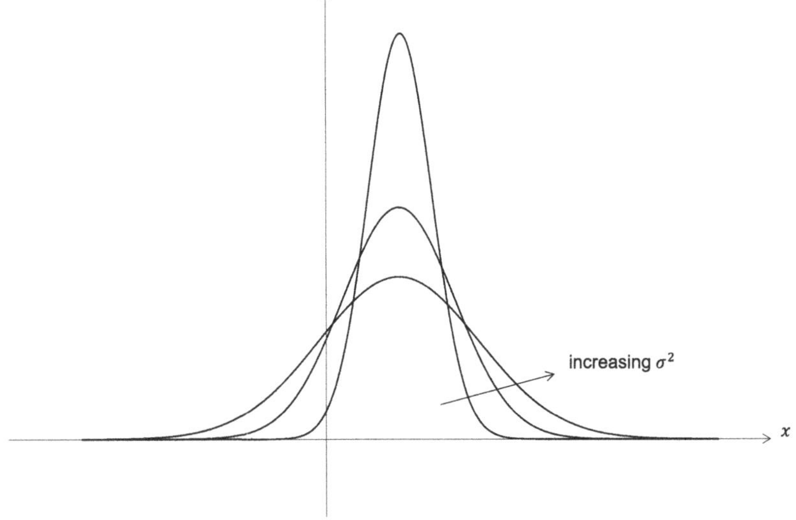

Figure 2.9 Gaussian PDF examples (same mean, different variances).

Note that the third line follows by switching to the polar coordinates and using $u^2 + v^2 = r^2$, and $dudv = rdrd\theta$. Therefore, the value of the integral becomes $I = 1$, demonstrating that $f_X(x)$ above integrates to 1, as claimed.

The parameters μ and σ^2 in the definition of the PDF of a Gaussian random variable are, in fact, the mean and the variance of a Gaussian random variable, respectively. This fact can be checked by direct computation (and using the above integration trick).

The PDFs of three different Gaussian random variables with the same mean but different variances are depicted in Fig. 2.9. As the variance of the Gaussian random variable is increased, the PDF becomes shallower and distributes over a wider region, which means that the random variable takes values over a broader range with a higher probability.

Gaussian random variables are encountered in practice quite often. This is due to the central limit theorem, which roughly states that under certain general conditions, the probability density function of the sum of N random variables (normalized with $\sqrt{N}$) as N increases approaches the PDF of a Gaussian random variable. We will discuss this point further and its application in the context of digital communications when we cover the characterization and effects of noise in Chapter 5.

Standard Gaussian Random Variable. A Gaussian random variable X with zero mean and unit variance is called the standard Gaussian random variable. We write it as $X \sim \mathcal{N}(0, 1)$. Its PDF can be expressed as

$$f_X(x) = \frac{1}{\sqrt{2\pi}} e^{-x^2/2}, \quad x \in \mathbb{R}. \tag{2.127}$$

The CDF of a standard Gaussian random variable is denoted by $\Phi(x)$. That is,

$$\Phi(x) = \int_{-\infty}^{x} \frac{1}{\sqrt{2\pi}} e^{-u^2/2} du. \tag{2.128}$$

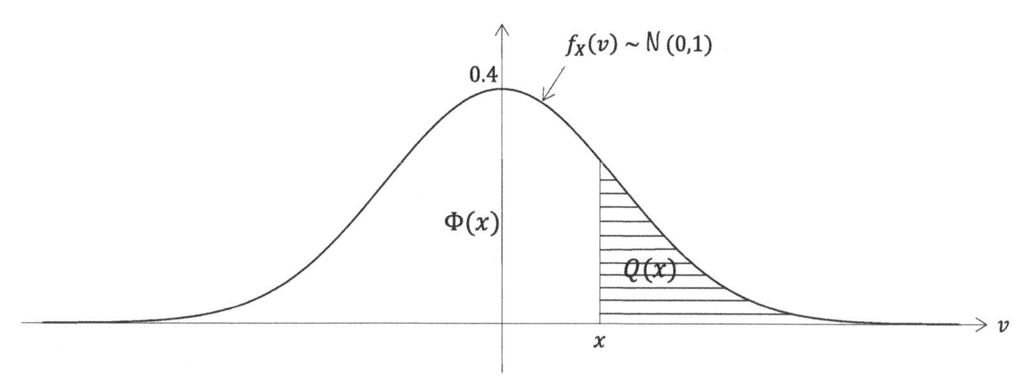

Figure 2.10 Interpretation of the Q- and Φ-functions as areas under the standard Gaussian PDF.

There is no closed-form expression for $\Phi(x)$; however, efficient numerical routines can be used to evaluate its value for any $x \in \mathbb{R}$.

We also define a closely related function, referred to as the Q-function:

$$Q(x) = 1 - \Phi(x) = \int_{x}^{\infty} \frac{1}{\sqrt{2\pi}} e^{-u^2/2} du, \tag{2.129}$$

which is the right-tail probability of a standard Gaussian random variable, namely, $Q(x) = \mathbb{P}(X > x)$ for $X \sim \mathcal{N}(0, 1)$. This function does not admit a closed form either.

Note also that the Q-function is related to the complementary error function $\operatorname{erfc}(x)$, defined as

$$\operatorname{erfc}(x) = \frac{2}{\sqrt{\pi}} \int_{x}^{\infty} e^{-t^2} dt, \tag{2.130}$$

via

$$Q(x) = \frac{1}{2} \operatorname{erfc}\left(\frac{x}{\sqrt{2}}\right). \tag{2.131}$$

Figure 2.10 illustrates $\Phi(x)$ and $Q(x)$ as areas under the PDF of a standard Gaussian random variable.

It is straightforward to show the following:

- $\Phi(x) + Q(x) = 1$;
- $Q(-x) = 1 - Q(x)$;
- $Q(0) = \frac{1}{2}$, $Q(\infty) = 0$, $Q(-\infty) = 1$.

Table 2.2 provides some values of the Q-function that can be used for solving probability problems involving Gaussian random variables.

There are also easily computable tight upper and lower bounds on the Q-function. In particular, we have the following simple upper bound:

$$Q(x) \leq \frac{1}{2} e^{-x^2/2}, \tag{2.132}$$

Table 2.2 ***Q*-function table.**

x	$Q(x)$	x	$Q(x)$	x	$Q(x)$
0	0.5	2.0	2.28×10^{-2}	4.0	3.17×10^{-5}
0.1	0.4602	2.1	1.79×10^{-2}	4.1	2.07×10^{-5}
0.2	0.4207	2.2	1.4×10^{-2}	4.2	1.33×10^{-5}
0.3	0.3821	2.3	1.07×10^{-2}	4.3	8.54×10^{-6}
0.4	0.3446	2.4	8.2×10^{-3}	4.4	5.41×10^{-6}
0.5	0.3085	2.5	6.21×10^{-3}	4.5	3.4×10^{-6}
0.6	0.2743	2.6	4.66×10^{-3}	4.6	2.11×10^{-6}
0.7	0.2420	2.7	3.47×10^{-3}	4.7	1.3×10^{-6}
0.8	0.2119	2.8	2.56×10^{-3}	4.8	7.93×10^{-7}
0.9	0.1841	2.9	1.87×10^{-3}	4.9	4.79×10^{-7}
1.0	0.1587	3.0	1.35×10^{-3}	5.0	2.87×10^{-7}
1.1	0.1357	3.1	9.68×10^{-4}	5.1	1.7×10^{-7}
1.2	0.1151	3.2	6.87×10^{-4}	5.2	9.96×10^{-8}
1.3	0.0968	3.3	4.83×10^{-4}	5.3	5.79×10^{-8}
1.4	0.0808	3.4	3.37×10^{-4}	5.4	3.33×10^{-8}
1.5	0.0668	3.5	2.33×10^{-4}	5.5	1.9×10^{-8}
1.6	0.0548	3.6	1.59×10^{-4}	5.6	1.07×10^{-8}
1.7	0.0446	3.7	1.08×10^{-4}	5.7	5.99×10^{-9}
1.8	0.0359	3.8	7.23×10^{-5}	5.8	3.32×10^{-9}
1.9	0.0287	3.9	4.81×10^{-5}	5.9	1.82×10^{-9}

which is particularly tight for large values of x.

We close this subsection by noting that the CDF of an arbitrary Gaussian random variable can be written in terms of the Φ-function. That is, for a Gaussian random variable $Y \sim \mathcal{N}(\mu, \sigma^2)$, a simple change of variables shows that

$$F_Y(y) = \Phi\left(\frac{y-\mu}{\sigma}\right). \tag{2.133}$$

In other words, any problem involving Gaussian random variables can be solved using the Φ- or Q-functions in a straightforward manner.

2.3.5 Multiple Random Variables

In many settings, we will encounter scenarios involving multiple random variables that must be jointly characterized. For instance, in a wireless communications setting, the received signal and the random channel gain are two different random variables that must be considered together.

Consider, first, the case of two random variables X and Y defined on the same sample space. We define their joint CDF as

$$F_{X,Y}(x,y) = \mathbb{P}(X \leq x, Y \leq y) \tag{2.134}$$

and their joint PDF as

$$f_{X,Y}(x,y) = \frac{\partial^2}{\partial x \partial y} F_{X,Y}(x,y). \tag{2.135}$$

Since the joint CDF is non-decreasing in both x and y, $f_{X,Y}$ is non-negative. We can determine the joint CDF from the joint PDF using

$$F_{X,Y}(x,y) = \int_{-\infty}^{x} \int_{-\infty}^{y} f_{X,Y}(u,v)\,du\,dv. \tag{2.136}$$

It is also clear that the joint PDF must integrate to 1, that is,

$$\int_{-\infty}^{\infty} \int_{-\infty}^{\infty} f_{X,Y}(u,v)\,du\,dv = 1. \tag{2.137}$$

We can also obtain the marginal CDF and marginal PDF of X and Y from their joint CDF or joint PDF. We can write

$$F_X(x) = F_{X,Y}(x,\infty) \text{ and } F_Y(y) = F_{X,Y}(\infty,y), \tag{2.138}$$

and by integrating over the other variable, we obtain

$$f_X(x) = \int_{-\infty}^{\infty} f_{X,Y}(x,y)\,dy \text{ and } f_Y(y) = \int_{-\infty}^{\infty} f_{X,Y}(x,y)\,dx. \tag{2.139}$$

We can deduce the marginal PDFs and CDFs of the individual random variables from their joint PDF and joint CDF; however, the converse is not true. When only the marginal PDFs or marginal CDFs are available, we have less information about the random variables, and we cannot obtain a complete characterization without any additional knowledge.

We can determine the probability that the pair of random variables will take on values in a specific region denoted by A by integrating the joint PDF over the given region. That is,

$$\mathbb{P}((X, Y) \in A) = \iint_{(x,y)\in A} f_{X,Y}(x,y)\,dx\,dy. \tag{2.140}$$

Conditional PDF. We define the conditional PDF of the random variable Y given the random variable X as

$$f_{Y|X}(y|x) = \frac{f_{X,Y}(x,y)}{f_X(x)} \tag{2.141}$$

if $f_X(x) \neq 0$, and 0 otherwise.

Independence. Two random variables X and Y are called statistically independent if all possible events involving X only and those involving Y only are independent. Based on this definition, we obtain a result that allows us to check for the independence of random variables using their joint PDF. That is, two random variables X

and Y are statistically independent if and only if their joint PDF is the product of the marginal PDFs, that is,

$$f_{X,Y}(x,y) = f_X(x) f_Y(y). \tag{2.142}$$

It is clear that for statistically independent random variables, observing the realizations of one random variable does not provide any information about the other. Therefore, the related conditional probabilities are the same as the unconditional ones. In other words, for independent random variables X and Y, we have

$$f_{Y|X}(y|x) = f_Y(y) \text{ and } f_{X|Y}(x|y) = f_X(x). \tag{2.143}$$

Multiple Random Variables. All the above concepts generalize to the case of more than two random variables. For instance, for N random variables $X_1, X_2, \ldots, X_N$, the joint CDF is defined as

$$F_{X_1,X_2,\ldots,X_N}(x_1, x_2, \ldots, x_N) = \mathbb{P}(X_1 \leq x_1, X_2 \leq x_2, \ldots, X_N \leq x_N), \tag{2.144}$$

and their joint PDF is given by

$$f_{X_1,X_2,\ldots,X_N}(x_1, x_2, \ldots, x_N) = \frac{\partial^N}{\partial x_1 \partial x_2 \ldots \partial x_N} F_{X_1,X_2,\ldots,X_N}(x_1, x_2, \ldots, x_N). \tag{2.145}$$

The random variables $X_1, X_2, \ldots, X_N$ are statistically independent if and only if

$$f_{X_1,X_2,\ldots,X_N}(x_1, x_2, \ldots, x_N) = f_{X_1}(x_1) f_{X_2}(x_2) \ldots f_{X_N}(x_N). \tag{2.146}$$

We define conditional PDFs, conditional CDFs, marginals, and so on, similarly to the case of two random variables.

The expected value of a function of multiple random variables is computed similarly to the case of one random variable. Namely, we can write .

$$E[g(X_1, \ldots, X_N)] = \int_{-\infty}^{\infty} \ldots \int_{-\infty}^{\infty} g(x_1, \ldots, x_N) f_{X_1,\ldots,X_N}(x_1, \ldots, x_N) dx_1 \ldots dx_N. \tag{2.147}$$

Covariance. Covariance of two random variables X and Y is defined as

$$\mathrm{Cov}(X, Y) = \mathbb{E}[(X - m_X)(Y - m_Y)], \tag{2.148}$$

which can also be written as $\mathrm{Cov}(X, Y) = \mathbb{E}[XY] - m_X m_Y$, where m_X and m_Y are the means of the random variables X and Y, respectively.

Denoting the standard deviation of X and Y by σ_X and σ_Y, respectively, we define the correlation coefficient $\rho_{X,Y}$ as

$$\rho_{X,Y} = \frac{\mathrm{Cov}(X, Y)}{\sigma_X \sigma_Y}. \tag{2.149}$$

The correlation coefficient satisfies $-1 \leq \rho_{X,Y} \leq 1$. To see this:

$$
\begin{aligned}
\left|\mathrm{Cov}(X, Y)\right| &= \left|\mathbb{E}[(X - m_X)(Y - m_Y)]\right| && (2.150)\\
&= \left|\int_{-\infty}^{\infty}\int_{-\infty}^{\infty} (x - m_X)(y - m_Y) f_{X,Y}(x,y)dxdy\right| && (2.151)\\
&\leq \left(\int_{-\infty}^{\infty}\int_{-\infty}^{\infty} (x - m_X)^2 f_{X,Y}(x,y)dxdy\right)^{1/2} \\
&\quad \left(\int_{-\infty}^{\infty}\int_{-\infty}^{\infty} (y - m_Y)^2 f_{X,Y}(x,y)dxdy\right)^{1/2} && (2.152)\\
&= \left(\int_{-\infty}^{\infty} (x - m_X)^2 f_X(x)dx\right)^{1/2} \left(\int_{-\infty}^{\infty} (y - m_Y)^2 f_Y(y)dy\right)^{1/2} && (2.153)\\
&= \sigma_X \sigma_Y, && (2.154)
\end{aligned}
$$

where the third line follows from the Cauchy–Schwarz inequality.[2] Dividing both sides by $\sigma_X\sigma_Y$, we establish that $|\rho_{X,Y}| \leq 1$.

Two random variables X and Y are called uncorrelated if their covariance is zero, that is, if $\mathrm{Cov}(X, Y) = 0$, or equivalently, $\mathbb{E}[XY] = \mathbb{E}[X]\mathbb{E}[Y]$.

Let X and Y be two random variables with a covariance of $\mathrm{Cov}(X, Y)$. Let us compute the variance of their sum $Z = X + Y$. Since $\mathbb{E}[Z] = \mathbb{E}[X] + \mathbb{E}[Y]$, we obtain

$$
\begin{aligned}
\mathrm{Var}(Z) &= \mathbb{E}\left[(X + Y - \mathbb{E}[X] - \mathbb{E}[Y])^2\right] && (2.155)\\
&= \mathbb{E}\left[(X - \mathbb{E}[X])^2 + (Y - \mathbb{E}[Y])^2 + 2(X - \mathbb{E}[X])(Y - \mathbb{E}[Y])\right] && (2.156)\\
&= \mathrm{Var}(X) + \mathrm{Var}(Y) + 2\,\mathrm{Cov}(X, Y), && (2.157)
\end{aligned}
$$

depicting that the covariance of two random variables plays an important role in determining the variance of their sum.

The concepts of covariance and correlation coefficient appear naturally in estimation problems. Consider the following problem: there are two random variables, and we are interested in producing an estimate of the realization of one of them given the realization of the other. For instance, the two random variables could be a vehicle's speed and the Doppler shift observed by radar bouncing off a radio signal on the vehicle. We would directly observe the realization of the Doppler shift, and we might be interested in determining the vehicle's speed. If the estimator is a linear function of the observation random variable, we say that it is a linear estimator.

The optimal linear estimator of the realization of a random variable based on the observation of another (which minimizes the average squared error between the actual and estimated values) involves the covariance of the two random variables.

[2] The Cauchy–Schwarz inequality states that for two real integrable functions $f(x)$ and $g(x)$, the following holds:

$$
\int_a^b f(x)g(x)dx \leq \left(\int_a^b f^2(x)dx\right)^{1/2} \left(\int_a^b g^2(x)dx\right)^{1/2}.
$$

Indeed, when the correlation coefficient is 1 or -1, that is, the random variables are fully correlated, it is possible to perfectly estimate one from an observation of the other (in the mean squared error sense). On the other hand, when the covariance (or the correlation coefficient) is 0, the observation of the second random variable is not helpful: the best linear estimator will simply be the expected value of the random variable itself, disregarding the observed random variable's realization.

Independence Implies Being Uncorrelated. If two random variables X and Y are statistically independent, then they are uncorrelated. To see this, consider two statistically independent random variables X and Y, and let us compute their covariance:

$$\begin{aligned}
\mathrm{Cov}(X, Y) &= \mathbb{E}[XY] - m_X m_Y && (2.158)\\
&= \int_{-\infty}^{\infty}\int_{-\infty}^{\infty} xy f_{X,Y}(x,y)dxdy - m_X m_Y && (2.159)\\
&= \int_{-\infty}^{\infty} x f_X(x)dx \int_{-\infty}^{\infty} y f_Y(y)dy - m_X m_Y && (2.160)\\
&= m_X m_Y - m_X m_Y && (2.161)\\
&= 0, && (2.162)
\end{aligned}$$

where the third line follows from the independence of the random variables (i.e., using $f_{X,Y}(x, y) = f_X(x)f_Y(y)$) and by splitting the double integral into two integrals involving only x and y separately. Since the resulting covariance is zero, the random variables are uncorrelated.

We close this subsection by noting that the converse of the above statement is generally not valid. Namely, if two random variables are uncorrelated, they are not necessarily independent. There is an important special case for which the converse also becomes true, which will be covered shortly.

Let us give an example of a pair of random variables which are uncorrelated but not independent. Let $X \sim \mathcal{N}(0, 1)$ and define $Y = X^2$. It is easy to check that these two random variables are uncorrelated; however, they are obviously not independent. Indeed, for this particular example, while a linear estimator will not provide a good result, one can estimate Y from X with no error by simply squaring the observed realization of X.

2.3.6 Jointly Gaussian Random Variables

In our study of communication systems, particularly when dealing with noise, we frequently encounter a set of random variables called *jointly Gaussian* or *multivariate Gaussian*. Consider an $N \times 1$ random vector

$$\boldsymbol{X} = \begin{bmatrix} X_1 \\ X_2 \\ \vdots \\ X_N \end{bmatrix}. \qquad (2.163)$$

Denote its expected value by $\boldsymbol{\mu}$, that is,

$$\boldsymbol{\mu} = \begin{bmatrix} \mathbb{E}[X_1] \\ \mathbb{E}[X_2] \\ \vdots \\ \mathbb{E}[X_N] \end{bmatrix}, \tag{2.164}$$

and define its $N \times N$ covariance matrix as

$$\boldsymbol{C} = \mathbb{E}\left[(\boldsymbol{X} - \boldsymbol{\mu})(\boldsymbol{X} - \boldsymbol{\mu})^T\right]. \tag{2.165}$$

Clearly, the ijth element of $\boldsymbol{C}$ is the covariance of X_i and X_j when $i \neq j$, and its diagonal elements are the variances of the random variables X_i.

It is readily seen that the covariance matrix is symmetric since $\mathrm{Cov}(X_i, X_j) = \mathrm{Cov}(X_j, X_i)$. Also, the covariance matrix is positive semidefinite.[3] To see this, consider an arbitrary (deterministic) $N \times 1$ vector $\boldsymbol{a}$. We can write

$$\boldsymbol{a}^T\boldsymbol{C}\boldsymbol{a} = \boldsymbol{a}^T\mathbb{E}[(\boldsymbol{X} - \boldsymbol{\mu})(\boldsymbol{X} - \boldsymbol{\mu})^T]\boldsymbol{a} \tag{2.166}$$

$$= \mathbb{E}[\boldsymbol{a}^T(\boldsymbol{X} - \boldsymbol{\mu})(\boldsymbol{X} - \boldsymbol{\mu})^T\boldsymbol{a}] \tag{2.167}$$

$$= \mathbb{E}[Y^2] \tag{2.168}$$

where $Y = \boldsymbol{a}^T(\boldsymbol{X} - \boldsymbol{\mu})$ is a real-valued (scalar) random variable. Hence, it follows that $\boldsymbol{a}^T\boldsymbol{C}\boldsymbol{a} \geq 0$, that is, the matrix $\boldsymbol{C}$ is positive semidefinite.

We say that the random vector $\boldsymbol{X}$ is jointly Gaussian or multivariate Gaussian if any linear combination of its components, that is, $a_1X_1 + a_2X_2 + \cdots + a_NX_N$ for any set of arbitrary real numbers $a_1, a_2, \ldots, a_N$, is a Gaussian random variable. We denote jointly Gaussian random variables as $\boldsymbol{X} \sim \mathcal{N}(\boldsymbol{\mu}, \boldsymbol{C})$. When the covariance matrix is positive definite, hence invertible, the joint PDF of $X_1, X_2, \ldots, X_N$ is given by

$$f_{\boldsymbol{X}}(\boldsymbol{x}) = \frac{1}{(2\pi)^{N/2}\sqrt{\det \boldsymbol{C}}} \exp\left\{-\frac{1}{2}(\boldsymbol{x} - \boldsymbol{\mu})^T\boldsymbol{C}^{-1}(\boldsymbol{x} - \boldsymbol{\mu})\right\}, \tag{2.169}$$

where $\boldsymbol{x} = [x_1 \;\; x_2 \;\; \ldots \;\; x_N]^T$, $x_i \in \mathbb{R}$ for $i = 1, 2, \ldots, N$, and det denotes the determinant of a square matrix. As expected, this expression boils down to the usual Gaussian PDF for $N = 1$.

The joint PDF is referred to as the bivariate PDF for the special case of two jointly Gaussian random variables. Let X and Y be jointly Gaussian, and assume that their means are μ_X and μ_Y, and variances are σ_X^2 and σ_Y^2. Also, assume that their correlation coefficient is ρ (with $|\rho| < 1$). Namely, the mean vector and the covariance matrix are given as

$$\boldsymbol{\mu} = \begin{bmatrix} \mu_X \\ \mu_Y \end{bmatrix}, \quad \boldsymbol{C} = \begin{bmatrix} \sigma_X^2 & \rho\sigma_X\sigma_Y \\ \rho\sigma_X\sigma_Y & \sigma_Y^2 \end{bmatrix}, \tag{2.170}$$

[3] We say that a symmetric matrix $\boldsymbol{A}$ with real elements is positive semidefinite if, for any real column vector $\boldsymbol{b}$, $\boldsymbol{b}^T\boldsymbol{A}\boldsymbol{b} \geq 0$. The matrix is called positive definite if the inequality is strict, that is, if $\boldsymbol{b}^T\boldsymbol{A}\boldsymbol{b} > 0$ for all non-zero $\boldsymbol{b}$.

respectively. Plugging these into the multivariate Gaussian PDF expression and simplifying, we can write the joint PDF in a more explicit form as

$$f_{X,Y}(x,y) = \frac{1}{2\pi\sigma_X\sigma_Y\sqrt{1-\rho^2}} \exp\left\{-\frac{1}{2(1-\rho^2)}\left(\frac{(x-m_X)^2}{\sigma_X^2} - \frac{2\rho(x-m_X)(y-m_Y)}{\sigma_X\sigma_Y} + \frac{(y-m_Y)^2}{\sigma_Y^2}\right)\right\}. \tag{2.171}$$

Jointly Gaussian random vectors possess some very nice properties. For instance, if $X_1, X_2, \ldots, X_N$ are jointly Gaussian, then any subset of them is also jointly Gaussian. Also, any subset of the random variables conditioned on any other subset is jointly Gaussian.

Example 2.12

A jointly Gaussian random vector $\begin{bmatrix} X \\ Y \\ Z \end{bmatrix}$ has a mean vector $\begin{bmatrix} 1 \\ -2 \\ -1 \end{bmatrix}$ and a covariance matrix

$$\mathbf{C} = \begin{bmatrix} 2 & -1 & 1 \\ -1 & 3 & -1 \\ 1 & -1 & 2 \end{bmatrix}. \tag{2.172}$$

Determine

(1) the mean vector and covariance matrix of $\begin{bmatrix} X \\ Z \end{bmatrix}$,

(2) the conditional PDF of Z given X, $f_{Z|X}(z|x)$.

Solution

(1) We noted that any subset of jointly Gaussian random variables is also jointly Gaussian. Therefore, it is immediate that the random vector $\begin{bmatrix} X \\ Z \end{bmatrix}$ is jointly Gaussian. To completely specify it, we only need to determine its mean vector and covariance matrix. We can readily obtain these quantities from those of the original random vector as

$$\begin{bmatrix} X \\ Z \end{bmatrix} \sim \mathcal{N}\left(\begin{bmatrix} 1 \\ -1 \end{bmatrix}, \begin{bmatrix} 2 & 1 \\ 1 & 2 \end{bmatrix}\right). \tag{2.173}$$

Noting that the determinant of the covariance matrix is 3, and its inverse is

$$\begin{bmatrix} 2/3 & -1/3 \\ -1/3 & 2/3 \end{bmatrix}, \tag{2.174}$$

by direct calculation, we can write the joint PDF of X and Z in an explicit form as

$$f_{X,Z}(x,z) = \frac{1}{2\pi\sqrt{3}} \exp\left(-\frac{1}{3}(x^2 + z^2 - xz - 3x + 3z + 3)\right). \tag{2.175}$$

(2) Using the result of the previous part, and the fact that $X \sim \mathcal{N}(1, 2)$, we obtain

$$f_{Z|X}(z|x) = \frac{f_{X,Z}(x,z)}{f_X(x)} \tag{2.176}$$

$$= \frac{\frac{1}{2\pi\sqrt{3}} \exp\left(-\frac{1}{3}(x^2 + z^2 - xz - 3x + 3z + 3)\right)}{\frac{1}{\sqrt{2\pi}\sqrt{2}} \exp\left(-\frac{1}{4}(x-1)^2\right)} \tag{2.177}$$

$$= \frac{1}{\sqrt{2\pi}\sqrt{3/2}} \exp\left(-\frac{1}{3}\left(z - \frac{x-3}{2}\right)^2\right). \tag{2.178}$$

In other words, conditioned on $X = x$, Z is a Gaussian random variable with mean $\frac{x-3}{2}$ and variance $3/2$, that is, $Z|X = x \sim \mathcal{N}(\frac{x-3}{2}, \frac{3}{2})$.

Another important property is that any linear combination of jointly Gaussian random variables $X_1, X_2, \ldots, X_N$ is also jointly Gaussian. Let $\boldsymbol{X} \sim \mathcal{N}(\boldsymbol{\mu}_X, \boldsymbol{C}_X)$, $\boldsymbol{A}$ be an $M \times N$ matrix, and $\boldsymbol{b}$ be an $M \times 1$ vector. Define the new vector of random variables obtained through the linear combination $\boldsymbol{Y} = \boldsymbol{A}\boldsymbol{X} + \boldsymbol{b}$. Then $\boldsymbol{Y} \sim \mathcal{N}(\boldsymbol{\mu}_Y, \boldsymbol{C}_Y)$ with $\boldsymbol{\mu}_Y = \boldsymbol{A}\boldsymbol{\mu}_X + \boldsymbol{b}$ and $\boldsymbol{C}_Y = \boldsymbol{A}\boldsymbol{C}_X\boldsymbol{A}^T$.

Example 2.13

Consider the previous example. Define the random variables $W = 2X + Z$ and $V = -Y + 2Z$. Determine the joint distribution of the random vector $\begin{bmatrix} W \\ V \end{bmatrix}$ and the marginal distributions of W and V.

Solution

This is a linear transformation of jointly Gaussian random variables; hence, the random vector $\begin{bmatrix} W \\ V \end{bmatrix}$ is also jointly Gaussian. We write

$$\begin{bmatrix} W \\ V \end{bmatrix} = \begin{bmatrix} 2 & 0 & 1 \\ 0 & -1 & 2 \end{bmatrix} \begin{bmatrix} X \\ Y \\ Z \end{bmatrix}. \tag{2.179}$$

Hence, we have

$$E\left\{\begin{bmatrix} W \\ V \end{bmatrix}\right\} = \begin{bmatrix} 2 & 0 & 1 \\ 0 & -1 & 2 \end{bmatrix} \begin{bmatrix} 1 \\ -2 \\ -1 \end{bmatrix} = \begin{bmatrix} 1 \\ 0 \end{bmatrix}. \tag{2.180}$$

The covariance matrix of the random vector $\begin{bmatrix} W \\ V \end{bmatrix}$ is given by

$$\begin{bmatrix} 2 & 0 & 1 \\ 0 & -1 & 2 \end{bmatrix}\begin{bmatrix} 2 & -1 & 1 \\ -1 & 3 & -1 \\ 1 & -1 & 2 \end{bmatrix}\begin{bmatrix} 2 & 0 & 1 \\ 0 & -1 & 2 \end{bmatrix}^T = \begin{bmatrix} 14 & 11 \\ 11 & 15 \end{bmatrix}. \quad (2.181)$$

Namely,

$$\begin{bmatrix} W \\ V \end{bmatrix} \sim \mathcal{N}\left(\begin{bmatrix} 1 \\ 0 \end{bmatrix}, \begin{bmatrix} 14 & 11 \\ 11 & 15 \end{bmatrix}\right). \quad (2.182)$$

Furthermore, we can readily note that $W \sim \mathcal{N}(1, 14)$ and $V \sim \mathcal{N}(0, 15)$.

Finally, we note that uncorrelated and jointly Gaussian random variables are also independent. Recall that the independence of a pair of random variables implies that they are uncorrelated; this property states that the converse of this statement is also true when the random variables are jointly Gaussian.

Let us prove the above result. Assume that two jointly Gaussian random variables X and Y are uncorrelated, hence the covariance matrix of the random vector $\begin{bmatrix} X \\ Y \end{bmatrix}$ is diagonal, that is,

$$\boldsymbol{C} = \begin{bmatrix} \sigma_X^2 & 0 \\ 0 & \sigma_Y^2 \end{bmatrix}. \quad (2.183)$$

Its determinant is simply $\sigma_X^2\sigma_Y^2$, and its inverse is given by

$$\boldsymbol{C}^{-1} = \begin{bmatrix} 1/\sigma_X^2 & 0 \\ 0 & 1/\sigma_Y^2 \end{bmatrix}. \quad (2.184)$$

Therefore, the joint PDF of X and Y can be written as

$$f_{X,Y}(x,y) = \frac{1}{2\pi\sigma_X\sigma_Y}\exp\left(-\frac{1}{2}[x-\mu_X \;\; y-\mu_Y]\begin{bmatrix} 1/\sigma_X^2 & 0 \\ 0 & 1/\sigma_Y^2 \end{bmatrix}\begin{bmatrix} x-\mu_X \\ y-\mu_Y \end{bmatrix}\right) \quad (2.185)$$

$$= \frac{1}{\sqrt{2\pi}\sigma_X}\exp\left(-\frac{(x-\mu_X)^2}{2\sigma_X^2}\right)\frac{1}{\sqrt{2\pi}\sigma_Y}\exp\left(-\frac{(y-\mu_Y)^2}{2\sigma_Y^2}\right). \quad (2.186)$$

Namely, we can write $f_{X,Y}(x,y) = f_X(x)f_Y(y)$, that is, the joint PDF is equal to the product of the marginal PDFs of X and Y. Hence, the random variables X and Y are independent.

2.4 Random Processes

We now extend the concept of random variables further. We define a *random process* (also called a *stochastic process*) as a mapping from the sample space to real functions, denoted by $X(\omega, t)$, where $\omega \in \Omega$ is an outcome of the underlying experiment, and t denotes time. For a fixed outcome $\omega_0 \in \Omega$, $X(\omega_0, t)$ is a deterministic function; while for a specified time instance t_0, $X(\omega, t_0)$ becomes a random variable. As in our coverage of random variables, we suppress the dependence on the experiment's outcome, and denote a random process as $X(t)$. The time index can be discrete or continuous. When it is discrete, we have a discrete-time random process or a sequence of random variables; when it is continuous, we have a continuous-time random process.

For a complete statistical characterization of a random process $X(t)$, we need the joint CDFs (or joint PDFs, when they exist) of random variables $X(t_1), X(t_2), \ldots, X(t_n)$ for all $(t_1, t_2, \ldots, t_n) \in \mathbb{R}^n$, and all positive integers n, namely,

$$F_{X(t_1),X(t_2),\ldots,X(t_n)}(x_1, x_2, \ldots, x_n). \tag{2.187}$$

We define the mean function of a random process as

$$\mu_X(t) = \mathbb{E}[X(t)] = \int_{-\infty}^{\infty} x f_{X(t)}(x)dx, \tag{2.188}$$

and its autocorrelation function as

$$R_X(t_1, t_2) = \mathbb{E}[X(t_1)X(t_2)] \tag{2.189}$$

$$= \int_{-\infty}^{\infty}\int_{-\infty}^{\infty} x_1 x_2 f_{X(t_1),X(t_2)}(x_1, x_2)dx_1 dx_2. \tag{2.190}$$

A similar second-order average is the autocovariance function, defined as

$$C_X(t_1, t_2) = E[(X(t_1) - \mu_X(t_1))(X(t_2) - \mu_X(t_2))] \tag{2.191}$$

$$= R_X(t_1, t_2) - \mu_X(t_1)\mu_x(t_2). \tag{2.192}$$

Let us give a simple example.

Example 2.14

Define a random process $X(t)$ as

$$X(t) = A\cos(2\pi f_c t + \Theta), \tag{2.193}$$

where A is a positive constant, and Θ is a uniform random variable on $[0, 2\pi)$. Several realizations of this random process are depicted in Fig. 2.11. Determine the mean and autocorrelation functions of $X(t)$.

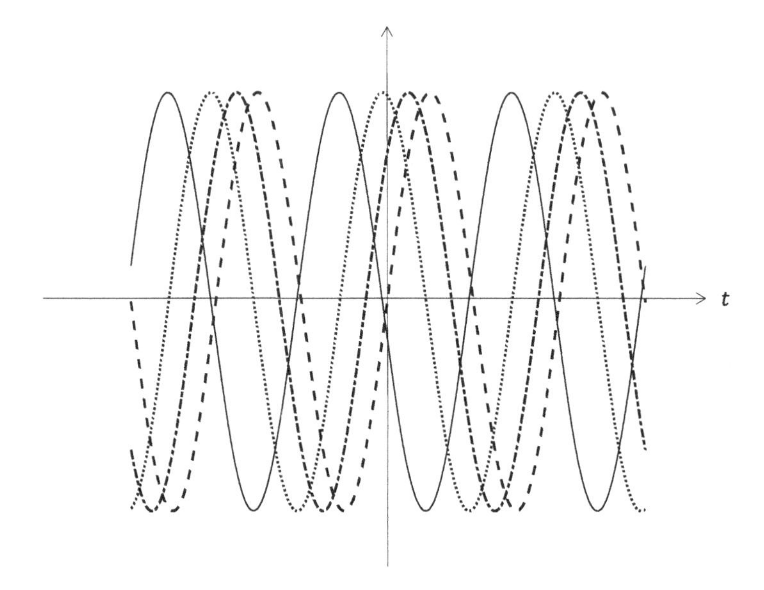

Figure 2.11 Several realizations of the random-phased sinusoid.

Solution

The mean function of $X(t)$ is given by

$$\mu_X(t) = \mathbb{E}[A\cos(2\pi f_c t + \Theta)] \tag{2.194}$$

$$= \int_0^{2\pi} \frac{1}{2\pi} A\cos(2\pi f_c t + \theta)d\theta \tag{2.195}$$

$$= \frac{A}{2\pi}\sin(2\pi f_c t + \theta)\Big|_0^{2\pi} \tag{2.196}$$

$$= 0, \tag{2.197}$$

and its autocorrelation function is calculated as

$$R_X(t_1, t_2) = \mathbb{E}\Big[\Big(A\cos(2\pi f_c t_1 + \Theta)\Big)\Big(A\cos(2\pi f_c t_2 + \Theta)\Big)\Big] \tag{2.198}$$

$$= \frac{A^2}{2}\mathbb{E}\Big[\cos(2\pi f_c(t_1 - t_2)) + \cos(2\pi f_c(t_1 + t_2) + 2\Theta)\Big] \tag{2.199}$$

$$= \frac{A^2}{2}\cos(2\pi f_c(t_1 - t_2)) + \frac{A^2}{2}\int_0^{2\pi} \frac{1}{2\pi}\cos(2\pi f_c(t_1 + t_2) + 2\theta)d\theta \tag{2.200}$$

$$= \frac{A^2}{2}\cos(2\pi f_c(t_1 - t_2)) + \frac{A^2}{8\pi}\sin(2\pi f_c(t_1 + t_2) + 2\theta)\Big|_0^{2\pi} \tag{2.201}$$

$$= \frac{A^2}{2}\cos(2\pi f_c(t_1 - t_2)). \tag{2.202}$$

We will return to this random-phased sinusoid example to illustrate several other concepts later.

2.4.1 Stationary Random Processes

Roughly speaking, random processes with statistical properties that do not change with time are referred to as stationary processes. Otherwise, the process is called non-stationary. There are different notions of stationarity, among which strict-sense stationarity, wide-sense stationarity, and cyclostationarity are important for our purposes.

Strict-Sense Stationary (SSS) Processes. A random process is called stationary in the strict sense if the joint CDFs of the random variables obtained by sampling the process at different time instances are identical with a shift of the time axis, that is, if

$$F_{X(t_1+\tau),X(t_2+\tau),\ldots,X(t_k+\tau)}(x_1,x_2,\ldots,x_k) = F_{X(t_1),X(t_2),\ldots,X(t_k)}(x_1,x_2,\ldots,x_k), \tag{2.203}$$

for all $\tau \in \mathbb{R}$, $k \in \mathbb{Z}^+$, $(t_1,t_2,\ldots,t_k) \in \mathbb{R}^k$.

Wide-Sense Stationary (WSS) Processes. A random process is called wide-sense stationary if its mean function is a constant and its autocorrelation function is only a function of the time difference, that is,

(1) $\mu_X(t) = \mu$,
(2) $R_X(t_1,t_2) = R_X(\tau)$, with $\tau = t_1 - t_2$ for some function $R_X(\cdot)$.

Example 2.15

Consider the random-phased sinusoid studied in Example 2.14. We have determined its mean function as $\mu_X(t) = 0$, which is a constant, and its autocorrelation function as $R_X(\tau) = \frac{A^2}{2}\cos(2\pi f_c\tau)$ (with $\tau = t_1 - t_2$), which is a function of the time difference only. Therefore, this process is WSS.

Wide-sense stationarity is a weaker notion than strict-sense stationarity. Indeed, if a process is SSS, then it is also WSS. To see this, consider an SSS process $X(t)$. Due to its strict-sense stationarity, $F_{X(t)}(x) = F_{X(t+\tau)}(x)$ for any t and τ, $\forall x \in \mathbb{R}$, hence the first-order CDF is not a function of time, that is, $F_{X(t)}(x) = F_X(x)$. Clearly, the first-order PDF is also independent of t, $f_{X(t)}(x) = f_X(x)$, and therefore

$$\mu_X(t) = \int_{-\infty}^{\infty} x f_{X(t)}(x)dx = \mu, \tag{2.204}$$

that is, the mean function does not depend on the time instant t (i.e., it is a constant). From the strict-sense stationarity of $X(t)$, we also have $F_{X(t_1),X(t_2)}(x_1,x_2) = F_{X(t_1-t_2),X(0)}(x_1,x_2)$, that is, the joint CDF of $X(t_1)$ and $X(t_2)$ depends on the time instances only through their difference $\tau = t_1 - t_2$. Therefore, the autocorrelation function

$$R_X(t_1, t_2) = \int_{-\infty}^{\infty} \int_{-\infty}^{\infty} x_1 x_2 f_{X(t_1),X(t_2)}(x_1, x_2) dx_1 dx_2 \tag{2.205}$$

is only a function of τ. In summary, for an SSS random process, the mean function is constant, and the autocorrelation function depends only on the time difference, that is, the process is also WSS.

Cyclostationary Random Processes. A random process is called cyclostationary (in the wide sense) if

(1) $\mu_X(t + T_0) = \mu_X(t)$,
(2) $R_X(t_1 + T_0, t_2 + T_0) = R_X(t_1, t_2)$

for some period T_0. It should be clear that cyclostationarity is a weaker notion than wide-sense stationarity.

We also define the average autocorrelation function for a cyclostationary process:

$$\bar{R}_X(\tau) = \frac{1}{T_0} \int_0^{T_0} R_X(t + \tau, t) dt. \tag{2.206}$$

It is interesting to note that given a cyclostationary process $X(t)$ with period T_0, a new random process defined as

$$Y(t) = X(t - \kappa), \tag{2.207}$$

where κ is a uniform random variable on the interval $[0, T_0)$ independent of the random process $X(t)$, is WSS with mean function

$$E[Y(t)] = \mu_Y = \frac{1}{T_0} \int_0^{T_0} \mu_X(t) dt, \tag{2.208}$$

and autocorrelation function $R_Y(\tau) = \bar{R}_X(\tau)$, i.e., the average autocorrelation function of the original cyclostationary process $X(t)$.

Example 2.16

Consider a random process defined as

$$X(t) = A \cos(2\pi f_0 t), \tag{2.209}$$

where f_0 is a constant and A is a uniform random variable on the interval $[0, 1]$. Realizations of this random process are sinusoids with the same frequency and phase but with different (random) amplitudes. Determine the mean and autocorrelation functions of $X(t)$ and show that it is cyclostationary.

Solution

The mean function of $X(t)$ is given by

$$\mu_X(t) = \mathbb{E}[X(t)] = \int_0^1 a\cos(2\pi f_0 t)da = \cos(2\pi f_0 t)\frac{a^2}{2}\Big|_0^1 = \frac{1}{2}\cos(2\pi f_0 t), \quad (2.210)$$

which is periodic with period $T_0 = 1/f_0$. The autocorrelation function of $X(t)$ is calculated as

$$R_X(t_1, t_2) = \mathbb{E}[X(t_1)X(t_2)] \quad (2.211)$$

$$= \mathbb{E}[A^2\cos(2\pi f_0 t_1)\cos(2\pi f_0 t_2)] \quad (2.212)$$

$$= \mathbb{E}[A^2]\cos(2\pi f_0 t_1)\cos(2\pi f_0 t_2) \quad (2.213)$$

$$= \left(\int_0^1 a^2 da\right)\frac{1}{2}\Big(\cos(2\pi f_0(t_1+t_2)) + \cos(2\pi f_0(t_1-t_2))\Big) \quad (2.214)$$

$$= \frac{1}{6}\cos\Big(2\pi f_0(t_1+t_2)\Big) + \frac{1}{6}\cos\Big(2\pi f_0(t_1-t_2)\Big). \quad (2.215)$$

One can easily verify that autocorrelation is periodic with period $\frac{T_0}{2}$, that is,

$$R_X\left(t_1 + \frac{T_0}{2}, t_2 + \frac{T_0}{2}\right) = R_X(t_1, t_2). \quad (2.216)$$

Using the common period of the mean and autocorrelation functions, we deduce that the process is cyclostationary with period T_0.

We can also determine the average autocorrelation function of this cyclostationary process as follows:

$$\bar{R}_X(\tau) = \frac{1}{T_0}\int_0^{T_0} R_X(t+\tau, t)dt \quad (2.217)$$

$$= \frac{1}{T_0}\int_0^{T_0} \frac{1}{6}\cos\Big(2\pi f_0(2t+\tau)\Big) + \frac{1}{6}\cos(2\pi f_0\tau)dt \quad (2.218)$$

$$= \frac{1}{6}\cos(2\pi f_0\tau). \quad (2.219)$$

2.4.2 Complex Random Processes

So far, we have limited our discussion to the case of real random processes, that is, mappings from the sample space of an experiment to real functions of time. It is also useful to define complex random processes in a similar way. A complex random process $X(t)$ is described by two real random processes: one representing its real part and the other representing its imaginary part. In later chapters, we will make significant use of such processes when dealing with bandpass communication systems. In the rest of this section, we will consider complex random processes for more generality, unless otherwise stated.

All the concepts we discussed for the case of real random processes straightforwardly generalize to the complex case. However, we need to remember that when dealing with a complex random variable (by sampling the process at a specific time), we indeed have two random variables – one for the real part and one for the imaginary part. One slight generalization we need is for the definition of autocorrelation function:

$$R_X(t_1, t_2) = \mathbb{E}[X(t_1)X^*(t_2)], \tag{2.220}$$

which is consistent with the earlier autocorrelation function definition (without the conjugation) when the process is real.

2.4.3 Properties of Autocorrelation Function for WSS Processes

We now review several important properties of the autocorrelation function for WSS processes. We consider the more general case of complex random processes, that is, the autocorrelation function is defined as

$$R_X(\tau) = \mathbb{E}[X(t+\tau)X^*(t)]. \tag{2.221}$$

- The autocorrelation function is conjugate symmetric, that is, $R_X(\tau) = R_X^*(-\tau)$, which readily follows since

$$R_X^*(-\tau) = \Big(\mathbb{E}[X(t-\tau)X^*(t)]\Big)^* \tag{2.222}$$
$$= \mathbb{E}[X(t)X^*(t-\tau)] \tag{2.223}$$
$$= R_X(\tau). \tag{2.224}$$

- $R_X(0)$ is real-valued and it is non-negative, that is, $R_X(0) \geq 0$. This follows simply by setting $\tau = 0$, and noting that $R_X(0) = \mathbb{E}[|X(t)|^2]$. This value is nothing but the average power of the WSS random process.
- $|R_X(\tau)| \leq R_X(0)$, that is, the autocorrelation function attains its maximum for $\tau = 0$. This is true whether the process is complex or real, however, the proof for the real case is particularly simple:

$$\mathbb{E}[(X(t+\tau) - X(t))^2] \geq 0, \tag{2.225}$$

which implies that

$$\mathbb{E}[X^2(t+\tau)] + \mathbb{E}[X^2(t)] - 2\mathbb{E}[X(t+\tau)X(t)] \geq 0. \tag{2.226}$$

From this inequality, we deduce that $2R_X(0) \geq 2R_X(\tau)$, that is, $R_X(0) \geq R_X(\tau)$. Similarly, starting with $\mathbb{E}[(X(t+\tau)+X(t))^2] \geq 0$, we find that $R_X(0) \geq -R_X(\tau)$. Combining these two results, we obtain $|R_X(\tau)| \leq R_X(0)$.

2.4.4 Multiple Random Processes

We can also consider two or more random processes at a time. Take, for instance, two random processes $X(t)$ and $Y(t)$. For a complete statistical description for this pair of random processes, we refer to the joint CDF of the random variables $X(t_1), X(t_2), \ldots, X(t_n), Y(t'_1), Y(t'_2), \ldots, Y(t'_m)$ for all non-negative integers n, m, and for all time instances $(t_1, t_2, \ldots, t_n) \in \mathbb{R}^n$, $(t'_1, t'_2, \ldots, t'_m) \in \mathbb{R}^m$.

The random processes $X(t)$ and $Y(t)$ are called independent if the random vectors $(X(t_1), X(t_2), \ldots, X(t_n))$ and $(Y(t'_1), Y(t'_2), \ldots, Y(t'_m))$ are independent for all positive integers n, m, and for all time instances $(t_1, t_2, \ldots, t_n) \in \mathbb{R}^n$, $(t'_1, t'_2, \ldots, t'_m) \in \mathbb{R}^m$.

The cross-correlation function of two processes $X(t)$ and $Y(t)$ is defined as

$$R_{XY}(t, u) = \mathbb{E}[X(t)\, Y^*(u)], \tag{2.227}$$

where $t, u \in \mathbb{R}$ are two arbitrary time instances. It should be clear that we have $R_{XY}(t, u) = R^*_{YX}(u, t)$.

Two processes $X(t)$ and $Y(t)$ are called jointly WSS if

(1) they are individually WSS,
(2) $R_{XY}(t_1, t_2)$ is a function of $\tau = t_1 - t_2$ only.

When the cross-correlation function depends only on the time difference $\tau = t_1 - t_2$, denoted by $R_{XY}(\tau)$, we have the conjugate symmetry $R_{XY}(\tau) = R^*_{YX}(-\tau)$.

Two random processes $X(t)$ and $Y(t)$ are called uncorrelated if the random variables $X(t_1)$ and $Y(t_2)$ are uncorrelated $\forall t_1, t_2 \in \mathbb{R}$. It is also easy to show that if the two processes are independent, they are also uncorrelated. However, the converse of this statement is not true in general.

Let us go through an example.

Example 2.17

Consider a WSS random process $X(t)$ with autocorrelation function $R_X(\tau)$. Define two new processes as

$$X_1(t) = X(t) \sin(2\pi f_c t + \Theta), \tag{2.228}$$

$$X_2(t) = X(t) \cos(2\pi f_c t + \Theta), \tag{2.229}$$

where Θ is a uniform random variable on $[0, 2\pi)$. Assume that the random variable Θ and the random process $X(t)$ are independent.

Determine the autocorrelation functions of $X_1(t)$ and $X_2(t)$. Also, calculate their cross-correlation functions and argue that the two processes are jointly WSS.

Solution

We can write

$$R_{X_1}(t_1, t_2) = \mathbb{E}[X(t_1) \sin(2\pi f_c t_1 + \Theta) X^*(t_2) \sin(2\pi f_c t_2 + \Theta)] \tag{2.230}$$

$$= \mathbb{E}[X_1(t)X^*(t_2)]\mathbb{E}[\sin(2\pi f_c t_1 + \Theta)\sin(2\pi f_c t_2 + \Theta)] \tag{2.231}$$

$$= R_X(t_1, t_2)E\left[\frac{1}{2}\left(\cos(2\pi f_c(t_1 - t_2)) - \cos(2\pi f_c(t_1 + t_2) + 2\Theta)\right)\right] \tag{2.232}$$

$$= \frac{1}{2}R_X(t_1 - t_2)\cos(2\pi f_c(t_1 - t_2)), \tag{2.233}$$

where the second line follows from the independence of Θ and $X(t)$; the third line is a simple trigonometric identity; and the last line follows by computing the expectation of the cosine terms similar to the computation in Example 2.14. In other words,

$$R_{X_1}(\tau) = \frac{1}{2}R_X(\tau)\cos(2\pi f_c \tau). \tag{2.234}$$

By following the same steps, we can also show that

$$R_{X_2}(\tau) = \frac{1}{2}R_X(\tau)\cos(2\pi f_c \tau). \tag{2.235}$$

It is also straightforward to determine the mean functions of the two random processes: $\mathbb{E}[X_1(t)] = \mathbb{E}[X_2(t)] = 0$. Hence, the mean functions are constant for both processes, and their autocorrelation functions depend only on the time difference τ. Therefore, we can argue that $X_1(t)$ and $X_2(t)$ are individually WSS.

Let us also compute their cross-correlation function

$$R_{X_1,X_2}(t_1, t_2) = \mathbb{E}[X(t_1)\sin(2\pi f_c t_1 + \Theta)X^*(t_2)\cos(2\pi f_c t_2 + \Theta)] \tag{2.236}$$

$$= \mathbb{E}[X_1(t)X^*(t_2)]\mathbb{E}[\sin(2\pi f_c t_1 + \Theta)\cos(2\pi f_c t_2 + \Theta)] \tag{2.237}$$

$$= R_X(t_1, t_2)\mathbb{E}\left[\frac{1}{2}\left(\sin(2\pi f_c(t_1 - t_2)) + \sin(2\pi f_c(t_1 + t_2) + 2\Theta)\right)\right] \tag{2.238}$$

$$= \frac{1}{2}R_X(t_1 - t_2)\sin(2\pi f_c(t_1 - t_2)), \tag{2.239}$$

or $R_{X_1,X_2}(\tau) = \frac{1}{2}R_X(\tau)\sin(2\pi f_c \tau)$, that is, their cross-correlation function depends only on the time difference τ. Combining this result with the earlier observation that the two processes are individually WSS, we can argue that $X_1(t)$ and $X_2(t)$ are jointly WSS.

2.4.5 LTI Filtering of a WSS Random Process

In our study of transmission techniques over noisy channels, we will encounter the problem of characterizing the noise (which will be modeled as a random process) at the output of a filter. Of particular interest is the filtering of WSS processes by an LTI system. Assume that a WSS process $X(t)$ with mean function μ_X and autocorrelation function $R_X(\tau)$ is input to an LTI system with impulse response $h(t)$ as shown in Fig. 2.12.

As $X(t)$ is a mapping from the sample space of an experiment to the set of (complex) functions of time, the output of the system, denoted by $Y(t)$, is also a mapping from the sample space to complex-valued functions; hence it is also a random process. We are interested in determining its mean and autocorrelation functions.

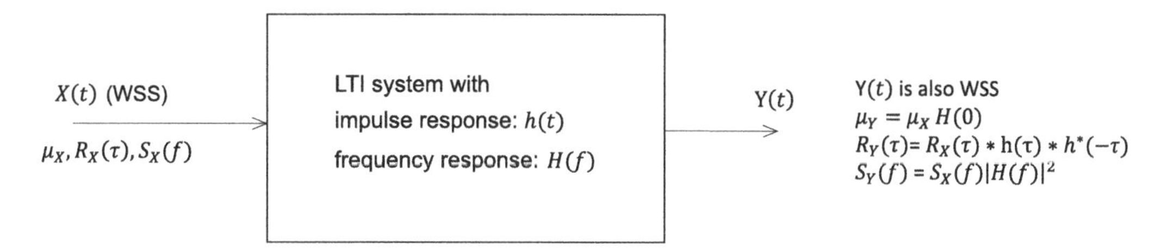

Figure 2.12 Illustration of a WSS random process $X(t)$ input to an LTI system.

The input–output relationship of an LTI system is given by the convolution integral; hence, the output process can be written as

$$Y(t) = \int_{-\infty}^{\infty} h(u)X(t-u)du. \tag{2.240}$$

This should be interpreted as integrating different realizations of the random process $X(t)$ (corresponding to different outcomes of the underlying experiment) to determine the corresponding realizations of the output random process. The mean function of $Y(t)$ is given by

$$\mu_Y(t) = \mathbb{E}[Y(t)] \tag{2.241}$$

$$= \mathbb{E}\left[\int_{-\infty}^{\infty} h(u)X(t-u)du\right] \tag{2.242}$$

$$= \int_{-\infty}^{\infty} h(u)\mathbb{E}[X(t-u)]du \tag{2.243}$$

$$= \mu_X \int_{-\infty}^{\infty} h(u)du, \tag{2.244}$$

where the third line follows by changing the order of the integral and the expectation operation, and noting that the impulse response of the system is deterministic and the expectation applies only to the random process $X(t)$. Changing the orders of the integration and expectation is justified assuming that $\mathbb{E}[X(t)]$ exists and the impulse response of the LTI system is absolutely integrable. In the following, we will encounter similar scenarios, and we will change orders of integrals and expectations without explicitly justifying them.

The mean function of $Y(t)$ is a constant. Also, since

$$\int_{-\infty}^{\infty} h(u)du = H(0), \tag{2.245}$$

where $H(f)$ is the frequency response of the LTI system, we can write

$$\mu_Y(t) = \mu_Y = \mu_X H(0). \tag{2.246}$$

We can determine the autocorrelation function of the output random process as follows:

$$R_Y(t_1,t_2) = \mathbb{E}[Y(t_1)Y^*(t_2)] \tag{2.247}$$

$$= \mathbb{E}\left[\int_{-\infty}^{\infty} h(u)X(t_1-u)du \int_{-\infty}^{\infty} h^*(v)X^*(t_2-v)dv\right] \tag{2.248}$$

$$= \int_{-\infty}^{\infty} h^*(v)\left[\int_{-\infty}^{\infty} h(u)\mathbb{E}[X(t_1-u)X^*(t_2-v)]du\right]dv \tag{2.249}$$

$$= \int_{-\infty}^{\infty} h^*(v)\left[\int_{-\infty}^{\infty} h(u)R_X(t_1-t_2+v-u)du\right]dv \tag{2.250}$$

$$= \int_{-\infty}^{\infty} h^*(v)g(t_1-t_2+v)dv, \tag{2.251}$$

where $g(t)$ is defined as the convolution of the impulse response of the LTI system with the autocorrelation function of the input process, that is, $g(t) = h(t) * R_X(t)$.

At this point, it is clear that the autocorrelation function of the output process $Y(t)$ is a function of the time difference $t_1 - t_2$ only. Recalling that its mean function is a constant, we conclude that $Y(t)$ is WSS. To reiterate: if the input to an LTI system is WSS, so is the output.

Letting $\tau = t_1 - t_2$, we can write

$$R_Y(\tau) = \int_{-\infty}^{\infty} h^*(v)g(\tau+v)dv \tag{2.252}$$

$$= \int_{-\infty}^{\infty} h^*(-z)g(\tau-z)dz \tag{2.253}$$

$$= h^*(-\tau) * g(\tau), \tag{2.254}$$

where the second line follows through a change of variables, and the third line is simply the definition of the convolution integral. In other words, we obtain the output autocorrelation function of $Y(t)$ in a convenient form as

$$R_Y(\tau) = R_X(\tau) * h(\tau) * h^*(-\tau). \tag{2.255}$$

It is also straightforward to show that the input and output processes $X(t)$ and $Y(t)$ are jointly WSS. To prove this result, as the processes are individually WSS, one only needs to compute their cross-correlation function and show that its time dependence is only on $\tau = t_1 - t_2$ using a line of argument similar to the one above.

2.4.6 Power Spectral Density of a WSS Process

Consider a WSS random process $X(t)$. We define its power spectral density (PSD), denoted by $S_X(f)$, as the Fourier transform of its autocorrelation function $R_X(\tau)$. That is, the PSD is defined as

$$S_X(f) = \int_{-\infty}^{\infty} R_X(\tau) e^{-j2\pi f\tau} d\tau, \tag{2.256}$$

which is given in watts/Hz (W/Hz). We will establish shortly that, as the name implies, the PSD shows the power content of the process as a function of frequency. Clearly, $R_X(\tau)$ can be determined from $S_X(f)$ using the inverse Fourier transform formula.

Let us go through several important properties of power spectral density.

- It is immediate from the definition of the autocorrelation function and the PSD, and from the Fourier transform properties, that the average power of the WSS process $X(t)$ is given by

$$\mathbb{E}[|X(t)|^2] = R_X(0) = \int_{-\infty}^{\infty} S_X(f) df. \tag{2.257}$$

In other words, integrating the PSD over the entire frequency range gives the average power of the process.

- It is also immediate that

$$S_X(0) = \int_{-\infty}^{\infty} R_X(\tau) d\tau. \tag{2.258}$$

- If the input to an LTI system with frequency response $H(f)$ is a WSS process with PSD $S_X(f)$, then the output process has a PSD given by

$$S_Y(f) = S_X(f)|H(f)|^2. \tag{2.259}$$

This is easily verified by taking the Fourier transform of the output autocorrelation function given in (2.255). Since convolution in the time domain is multiplication in the frequency domain, and the Fourier transform of $h^*(-\tau)$ is $H^*(f)$, the result in (2.259) readily follows.

- The PSD is real and non-negative. The fact that $S_X(f)$ is real is easily seen from the Fourier transform properties by recalling that the autocorrelation function $R_X(\tau)$ is conjugate symmetric, that is, $R_X(\tau) = R_X^*(-\tau)$. Let us now show that it cannot be negative.

  Consider a WSS process $X(t)$ with PSD $S_X(f)$ as input to a hypothetical LTI system with frequency response $|H(f)| = 1$ for $f \in [f_0, f_0 + \Delta f]$ for a very small Δf, and $H(f) = 0$ otherwise, that is, an ideal bandpass filter with a very narrow passband of $[f_0, f_0 + \Delta f]$. Then, the output process is WSS and has a PSD of

$$S_Y(f) = S_X(f)|H(f)|^2 \tag{2.260}$$

$$= \begin{cases} S_X(f), & \text{if } f \in [f_0, f_0 + \Delta f], \\ 0, & \text{else.} \end{cases} \quad (2.261)$$

Assuming that the bandwidth of the filter Δf is very small, the average power of the output random process is approximately given by

$$\mathbb{E}[|Y(t)|^2] \approx S_X(f_0)\Delta f. \quad (2.262)$$

As the output power cannot be negative, we establish that $S_X(f_0) \geq 0$. This is true for any value of f_0; hence we conclude that $S_X(f) \geq 0$ for all f.

Only the frequency components of the input process that are in the frequency band $[f_0, f_0 + \Delta f]$ survive at the output of this hypothetical filter. In other words, the entire output power is due to the frequency components in this narrow frequency band. With $\Delta f \approx 0$, we can write

$$S_X(f_0) \approx \frac{\text{power content of } X(t) \text{ in } [f_0, f_0 + \Delta f]}{\Delta f}, \quad (2.263)$$

which indicates that the PSD is nothing but the power density of a WSS process as a function of frequency, as stated earlier.

- If $X(t)$ is real, then the PSD is an even function, that is, $S_X(f) = S_X(-f)$. This follows easily from

$$S_X(-f) = \int_{-\infty}^{\infty} R_X(\tau)e^{j2\pi f\tau} d\tau \quad (2.264)$$

$$= \int_{-\infty}^{\infty} R_X(-u)e^{-j2\pi fu} du \quad (2.265)$$

$$= \int_{-\infty}^{\infty} R_X(u)e^{-j2\pi fu} du \quad (2.266)$$

$$= S_X(f), \quad (2.267)$$

where the second line results from a change of variables, and the third line follows from $R_X(u) = R_X(-u)$ using the assumption that $X(t)$ is real.

- The average power content of a WSS random process in a particular frequency band can be computed by integrating the power spectral density over the given frequency band. Typically, we will use this concept for real processes (whose power spectral densities are even functions of f). Hence, when we refer to the power content in the frequency band $[f_1, f_2]$ (where $0 \leq f_1 \leq f_2$), we will mean the contributions in the frequency bands $[f_1, f_2]$ and $[-f_2, -f_1]$. In other words, the average power content of a real process $X(t)$ in the frequency band $[f_1, f_2]$ is given by

$$\int_{-f_2}^{-f_1} S_X(f)df + \int_{f_1}^{f_2} S_X(f)df = 2\int_{f_1}^{f_2} S_X(f)df. \quad (2.268)$$

Let us give several examples of power spectral density computation.

Example 2.18

Consider the random process given in Example 2.14. We have previously shown that the process is WSS, and it has an autocorrelation function of $R_X(\tau) = \frac{A^2}{2}\cos(2\pi f_c \tau)$. Taking the Fourier transform, we obtain its PSD as

$$S_X(f) = \frac{A^2}{4}\Big(\delta(f - f_c) + \delta(f + f_c)\Big). \tag{2.269}$$

Since $X(t)$ is nothing but a randomly phased sinusoid with a constant frequency f_c, it is no surprise that the entire power content of the process is at the frequencies $\pm f_c$.

Example 2.19

Consider the random process given in Example 2.17. We have the autocorrelation function of $X_1(t)$ as $R_{X_1}(\tau) = \frac{1}{2}R_X(\tau)\cos(2\pi f_c \tau)$. Hence, it follows that

$$S_{X_1}(f) = \frac{1}{2}S_X(f) * \left\{\frac{1}{2}\Big(\delta(f - f_c) + \frac{1}{2}\delta(f + f_c)\right\} \tag{2.270}$$

$$= \frac{1}{4}\left(S_X(f - f_c) + S_X(f + f_c)\right). \tag{2.271}$$

We see that multiplication by a sinusoidal process with a frequency of f_c (and a uniform phase) translates the PSD of the original random process to frequencies around f_c and $-f_c$.

Similarly to the definition of the PSD, we define the cross-spectral density of two jointly WSS random processes as the Fourier transform of their cross-correlation function $R_{XY}(\tau)$.

Example 2.20

Consider two zero mean jointly WSS random processes $X(t)$ and $Y(t)$. Determine the PSD of the random process $Z(t) = X(t) + Y(t)$.

Solution

The autocorrelation function of $Z(t)$ can be written as

$$R_Z(t_1, t_2) = \mathbb{E}[(X(t_1) + Y(t_1))(X(t_2) + Y(t_2))^*] \tag{2.272}$$

$$= \mathbb{E}[X(t_1)X^*(t_2)] + \mathbb{E}[Y(t_1)Y^*(t_2)] + \mathbb{E}[X(t_1)Y^*(t_2)] + \mathbb{E}[Y(t_1)X^*(t_2)]. \tag{2.273}$$

As $X(t)$ and $Y(t)$ are jointly WSS, we have

$$R_Z(\tau) = R_X(\tau) + R_Y(\tau) + R_{XY}(\tau) + R_{YX}(\tau). \tag{2.274}$$

Combining this result with $\mathbb{E}[Z(t)] = \mathbb{E}[X(t)] + \mathbb{E}[Y(t)] = 0$, that is, the mean of $Z(t)$ being a constant, we establish that $Z(t)$ is WSS. Taking the Fourier transform of

$R_Z(\tau)$, we obtain

$$S_Z(f) = S_X(f) + S_Y(f) + S_{XY}(f) + S_{YX}(f). \tag{2.275}$$

Therefore, the power spectral density of the sum process depends on the individual power spectral densities and the cross-spectral densities of the two processes.

Example 2.21

Consider the previous example, but assume that the two processes $X(t)$ and $Y(t)$ are also uncorrelated. In this case

$$R_{XY}(\tau) = \mathbb{E}[X(t+\tau)Y^*(t)] = \mathbb{E}[X(t+\tau)]\mathbb{E}[Y^*(t)] = 0. \tag{2.276}$$

Similarly, $R_{YX}(\tau) = 0$. Hence, $S_{XY}(f) = S_{YX}(f) = 0$ and the PSD of the sum process $Z(t)$ becomes

$$S_Z(f) = S_X(f) + S_Y(f). \tag{2.277}$$

We will encounter scenarios in which two or more uncorrelated, zero mean, and WSS random processes are summed to form another random process (e.g., the aggregate noise process due to disturbances from different noise sources). The result of this example shows that, in such cases, we need to add up the individual PSDs to determine the PSD of the new process.

2.5 Gaussian Processes

This section introduces a class of random processes of paramount importance in our study of communication systems: Gaussian random processes. Such processes will be used to model the electronic noise at a communication receiver; hence, their characterization and properties are critical for determining the optimal receiver structures and assessing system performance.

We refer to a random process $X(t)$ as a Gaussian random process if the random variables obtained by sampling it at different time instances, $X(t_1), X(t_2), \ldots, X(t_n)$, are jointly Gaussian $\forall n \in \mathbb{Z}^+$ and $\forall (t_1, t_2, \ldots, t_n) \in \mathbb{R}^n$.

Recall the properties of jointly Gaussian random vectors from Section 2.3.6. In particular, any subset of jointly Gaussian random variables is also jointly Gaussian, marginal PDFs are Gaussian, linear combinations are jointly Gaussian, and any subset of jointly Gaussian random variables conditioned on any other subset is also jointly Gaussian.

An alternate (and equivalent) definition of a Gaussian random process is as follows: $X(t)$ is referred to as a Gaussian random process if the random variable obtained by

$$Y = \int_0^T g(t)X(t)dt \tag{2.278}$$

(for $T \in \mathbb{R}$) is a Gaussian random variable for any function $g(t)$. We will use both definitions throughout our coverage as appropriate.

We say that two random processes $X(t)$ and $Y(t)$ are jointly Gaussian processes if the random variables $X(t_1), X(t_2), \ldots, X(t_n)$, $Y(\tau_1)$, $Y(\tau_2), \ldots, Y(\tau_m)$ are jointly Gaussian for any $n, m \in \mathbb{Z}^+$, $\forall(t_1, t_2, \ldots, t_n) \in \mathbb{R}^n$ and $\forall(\tau_1, \tau_2, \ldots, \tau_m) \in \mathbb{R}^m$.

Several properties of Gaussian random processes are stated next.

- The first-order and second-order averages $\mu_X(t)$ and $R_X(t_1, t_2)$ (or $C_X(t_1, t_2)$) fully characterize a Gaussian random process. This follows because the joint PDF of a jointly Gaussian random vector is determined by its mean vector and covariance matrix (which are readily obtained from the mean function and the autocorrelation (or autocovariance) function) for any set of random variables obtained by sampling the process at different time instances.
- If a Gaussian random process is WSS, it is also SSS. To see this, consider a WSS Gaussian process $X(t)$ with mean $\mu_X(t) = \mu_X$ and autocovariance function $C_X(t_1, t_2) = C_X(\tau)$ with $\tau = t_1 - t_2$. For any random vector $\mathbf{X} = [X(t_1), X(t_2), \ldots, X(t_n)]^T$, the mean vector is simply a constant $n \times 1$ vector, and the covariance matrix is an $n \times n$ matrix with the ijth element being $C_X(t_i - t_j)$. Therefore, a vector obtained by a time-shifted samples of the process $\mathbf{X}' = [X(t_1 + \tau), X(t_2 + \tau), \ldots, X(t_n + \tau)]^T$ has the same mean vector and the same covariance matrix. Since the mean vector and the covariance matrix fully determine the statistics of a jointly Gaussian random vector, we argue that the joint PDF of $X(t_1), X(t_2), \ldots, X(t_n)$ remains invariant against a time shift; hence the random process $X(t)$ is SSS.

  Recall that strict-sense stationarity implies wide-sense stationarity for any random process. The above result establishes that the converse is also valid for the case of Gaussian random processes, that is, strict-sense stationarity and wide-sense stationarity are equivalent.
- If the input to a linear filter is a Gaussian random process, then the output is also a Gaussian random process.[4] This is readily derived using the convolution integral and the second definition of Gaussian processes introduced earlier.
- For a Gaussian process $X(t)$, if $X(t_1), X(t_2), \ldots, X(t_n)$ are uncorrelated, then they are also independent (see Section 2.3.6).

Finally, we also note that a particular case of Gaussian random processes, namely, the *white Gaussian random process*, is of particular importance for our coverage of digital communication systems. The white Gaussian random process is a zero-mean, WSS (hence also SSS) Gaussian random process with a constant power spectral density; therefore, its autocorrelation function is an impulse function at the origin. Specifically, we will model the additive noise process at a communication receiver as a white Gaussian random process, as detailed in Chapter 5.

Let us give a couple of examples of Gaussian random processes.

[4] This is true assuming that the linear filter is stable. Since we did not discuss the stability of linear systems, we brush over this technical detail.

Example 2.22

A zero-mean stationary Gaussian random process $X(t)$ with autocorrelation function $R_X(\tau) = 5\delta(\tau)$ is input to an LTI system with impulse response $h(t) = e^{-t}u(t)$. Assume that the filter output $Y(t)$ is sampled at $t = 1$, and the random variable $Z = Y(1)$ is obtained. Determine the mean and variance of the resulting random variable, its distribution, and the probability that its value will exceed a threshold of 3, that is, $\mathbb{P}(Z > 3)$.

Solution

Since the input to the LTI system is a zero-mean random process, so is the output, as determined earlier in the chapter. That is, $\mathbb{E}[Y(t)] = 0$. Hence, $\mathbb{E}[Z] = 0$. The autocorrelation function of the output process is given by

$$R_Y(\tau) = R_X(\tau) * h(\tau) * h^*(-\tau), \tag{2.279}$$

as also obtained earlier. Using $R_X(\tau) = 5\delta(\tau)$, we obtain

$$\begin{aligned} R_Y(\tau) &= 5\delta(\tau) * h(\tau) * h^*(-\tau) && (2.280)\\ &= 5h(\tau) * h^*(-\tau) && (2.281)\\ &= 5(e^{-\tau}u(\tau)) * (e^{\tau}u(-\tau)) && (2.282)\\ &= \frac{5}{2}e^{-|\tau|}. && (2.283) \end{aligned}$$

Hence the variance of the random variable Z is given by $R_Y(0) = \frac{5}{2}$.

Since the random process $X(t)$ is a Gaussian process, the filter output is also a Gaussian process; hence, its sample at time $t = 1$ is a Gaussian random variable. In other words, $Z \sim \mathcal{N}(0, 5/2)$. Hence, we obtain

$$\mathbb{P}(Z > 3) = Q\left(\frac{3}{\sqrt{5/2}}\right) = Q\left(\sqrt{\frac{18}{5}}\right) \approx 0.029 \tag{2.284}$$

as the desired probability.

Example 2.23

Consider a zero-mean Gaussian random process $X(t)$ with autocorrelation function $R_X(\tau) = \frac{5}{2}e^{-|\tau|}$. Define two random variables obtained by sampling the process at two different time instances as $Y = X(1)$ and $Z = X(2)$. Determine the joint PDF of Y and Z, and the probability $\mathbb{P}(2Y + Z > 1)$.

Solution

Since $X(t)$ is a Gaussian process, the samples Y and Z are jointly Gaussian random variables. Therefore, to find the joint PDF of the random vector $\begin{bmatrix} Y \\ Z \end{bmatrix}$, we need to determine its mean vector and covariance matrix. The process is zero mean; therefore, the mean vector is $\begin{bmatrix} 0 \\ 0 \end{bmatrix}$. Noting that the variances are

$$\mathrm{Var}(Y) = \mathrm{Var}(Z) = R_X(0) = \frac{5}{2}, \tag{2.285}$$

and the covariance of Y and Z is

$$\mathbb{E}[YZ] = \mathbb{E}[X(2)X(1)] = R_X(2-1) = R_X(1) = \frac{5}{2e}, \tag{2.286}$$

the covariance matrix of $\begin{bmatrix} Y \\ Z \end{bmatrix}$ is obtained as

$$\begin{bmatrix} \frac{5}{2} & \frac{5}{2e} \\ \frac{5}{2e} & \frac{5}{2} \end{bmatrix}. \tag{2.287}$$

In other words,

$$\begin{bmatrix} Y \\ Z \end{bmatrix} \sim \mathcal{N}\left(\begin{bmatrix} 0 \\ 0 \end{bmatrix}, \begin{bmatrix} \frac{5}{2} & \frac{5}{2e} \\ \frac{5}{2e} & \frac{5}{2} \end{bmatrix} \right). \tag{2.288}$$

To determine the probability $\mathbb{P}(2Y + Z > 1)$, we note that a linear combination of jointly Gaussian random variables is Gaussian. Hence, $2Y + Z$ is a Gaussian random variable. Its mean is simply

$$\mathbb{E}[2Y + Z] = 2\mathbb{E}[Y] + \mathbb{E}[Z] = 0, \tag{2.289}$$

and its variance is given by

$$\begin{aligned} \mathrm{Var}(2Y + Z) &= \mathbb{E}[(2Y + Z)^2] && (2.290) \\ &= 4\mathbb{E}[Y^2] + 4\mathbb{E}[YZ] + \mathbb{E}[Z^2] && (2.291) \\ &= 4 \times \frac{5}{2} + 4 \times \frac{5}{2e} + \frac{5}{2} && (2.292) \\ &\approx 16.18. && (2.293) \end{aligned}$$

Therefore,

$$\mathbb{P}(2Y + Z > 1) \approx Q\left(\frac{1}{\sqrt{16.18}}\right) \approx 0.4 \tag{2.294}$$

is obtained.

2.6 Chapter Summary

This chapter has reviewed basic signals and systems, probability, random variables, and random processes. These concepts will be used repeatedly throughout our study of digital communication systems. For instance, we will heavily rely on probability and random variables when determining the optimal receiver structures and the average error performance of a digital communication system. Special attention was paid to Gaussian random variables and Gaussian random processes. This coverage was motivated by the fundamental channel model encountered in communication systems, as will become apparent in the subsequent chapters.

PROBLEMS

2.1 Determine whether or not the systems described by the input–output relationships below are linear and/or time-invariant.
(a) $y(t) = 2x(t-3) + x(t-5)$.
(b) $y(t) = -x(2-t) + x(t)$.
(c) $y(t) = \int_0^t x(\tau - 2)d\tau$.
(d) $y(t) = x(2t+3) - x(3t)$.

2.2 Determine whether or not the systems described by the input–output relationships below are linear and/or time-invariant.
(a) $y(t) = \frac{dx(t)}{dt}$.
(b) $y(t) = \int_0^{2t} x(\tau - 1)d\tau$.
(c) $y(t) = t^2x(t) - tx(t-1)$.
(d) $y(t) = x(t) * x(t)$.

2.3 Determine the impulse response and the frequency response of the following LTI systems.
(a) $y(t) = \int_t^{t+2} 2x(\tau - 3) - x(\tau - 4)d\tau$.
(b) $y(t) = x(t) * (u(t-1) + \delta(t+1))$, where $u(t)$ is the unit step function.
(c) $y(t) = 3x(t-3) + 2x(t-4)$.

2.4 Determine the output of an LTI system with the impulse response given by

$$h(t) = e^{-t}u(t)$$

(where $u(t)$ is the unit step function) to the following input signals.
(a) $x(t) = e^{-2t}u(t)$.
(b) $x(t) = u(t-1) - u(t+1)$.

2.5 Determine the frequency responses of the LTI systems with the impulse responses given below.
(a) $h(t) = \cos(2\pi f_0 t) + 2\sin(4\pi f_0 t)$.
(b) $h(t) = e^{-2t}u(t)$.
(c) $h(t) = 2u(t-1) - u(t+1)$.
(d) $h(t) = 3\delta(t-2) + 2\delta(t-1) + \delta(t)$.

2.6 Assume that the response of an LTI system to a unit step function is given by $g(t)$. Determine its impulse response.

2.7 Let $y(t)$ be the output of an LTI system to the input signal $x(t)$. Show that the response of the system to the input signal $\frac{dx(t)}{dt}$ is given by $\frac{dy(t)}{dt}$.

2.8 Determine the Fourier transform of the periodic impulse train

$$x(t) = \sum_{n=-\infty}^{\infty} \delta(t - nT_0),$$

where T_0 is its period.

2.9 Consider an LTI system with impulse response given by

$$h(t) = 5e^{-2t}u(t - 1)$$

(where $u(t)$ is the unit step function). Determine the output of the system to the following inputs.

(a) $x(t) = \cos(20\pi t) + 2\sin(10\pi t)$.

(b) $x(t) = e^{j2\pi f_0 t}$.

(c) $x(t) = \sum_{n=-\infty}^{\infty} \Pi(t - 3n)$.

2.10 Prove the modulation property of the Fourier transform, which states that the Fourier transform of $x(t)e^{j2\pi f_0 t}$ is given by $X(f - f_0)$.

2.11 Prove that the Fourier transform of convolution of two functions is the product of their Fourier transforms.

2.12 Prove the Plancherel–Parseval theorem, which states that

$$\int_{-\infty}^{\infty} x(t)y^*(t)dt = \int_{-\infty}^{\infty} X(f)Y^*(f)df.$$

2.13 Prove that the Fourier transform of the signal $x(at)$, where a is a non-zero real number, is given by $\frac{1}{|a|}X(f/a)$.

2.14 Assume that the prevalence of a disease in a certain population is 0.01%. We have a test to determine if someone has the disease or not. The test result is accurate with a probability of 99%.

What is the probability that a randomly selected person who tests positive for the disease is indeed sick?

2.15 Consider the transmission of packets over a channel, which can be in one of three different states: state A, state B, and state C. The probability that a transmitted packet will be received correctly is 99% for state A, 90% for state B, and 80% for state C. The channel is in state A with probability 0.2, state B with probability 0.3, and state C with probability 0.5.

(a) Determine the average probability of error in the transmission of a packet.

(b) Given that a packet is received correctly, what is the (conditional) probability that the channel was in state C?

2.16 Let U be a uniform random variable on the interval $(1, 5)$. Determine the probability density function of the following random variables.

(a) $Y = \sqrt{\ln U}$.

(b) $Z = \max(U, 3)$.

2.17 Assume that X is an exponential random variable with a mean of 2. Determine

(a) $\mathbb{P}(X > 3)$.
(b) $\mathbb{P}(X > 5|X > 3)$.
(c) $\mathbb{E}[X|X > 3]$.
(d) $\mathbb{E}[X^2|3 < X < 5]$.

2.18 Let X and Y be two independent exponential random variables with means $1/2$ and $1/3$, respectively. Determine
(a) The probability density function of $Z = \min(X, Y)$.
(b) The probability density function of $W = \max(X, Y)$.
(c) The joint probability density function of $Z = \min(X, Y)$ and $W = \max(X, Y)$.

2.19 Let X and Y be independent uniform random variables on $(0, 1)$. Define $Z = 2X + Y$ and $W = X - Y$.
(a) Determine the joint probability density function of Z and W.
(b) Find the covariance of Z and W.
(c) Compute the conditional expectation $\mathbb{E}[Z|W]$.

2.20 Let X and Y be independent exponential random variables with parameters 1 and 2, respectively.
(a) Determine the probability $\mathbb{P}(X + 3Y > 5)$.
(b) Determine the probability density function of $Z = 2X + Y$.
(c) Determine the conditional probability density function of X given Z, where $Z = 2X + Y$.

2.21 Consider the probability density function of a Gaussian random variable

$$f_X(x) = \frac{1}{\sqrt{2\pi}\sigma} \exp\left(-\frac{(x-\mu)^2}{2\sigma^2}\right), \quad \forall x \in \mathbb{R},$$

where $\mu \in \mathbb{R}$ and $\sigma \in \mathbb{R}^+$. Prove that the parameter μ is the random variable's mean, and the parameter σ^2 is its variance.

2.22 A jointly Gaussian random vector $[X \;\; Y \;\; Z]^T$ has a mean vector $[2 \;\; 0 \;\; 1]^T$ and a covariance matrix

$$\mathbf{C} = \begin{bmatrix} 4 & -2 & 1 \\ -2 & 3 & -1 \\ 1 & -1 & 4 \end{bmatrix}.$$

(a) Determine the mean vector and covariance matrix of the random vector $\begin{bmatrix} X \\ Z \end{bmatrix}$.
(b) Two random variables W and V are defined as $W = 3Y - 2Z$ and $V = -X + 2Y$. Determine their joint distribution.

2.23 Let $\begin{bmatrix} X \\ Y \end{bmatrix}$ be a jointly Gaussian random vector with mean vector $\begin{bmatrix} 0 \\ 0 \end{bmatrix}$ and covariance matrix

$$C = \begin{bmatrix} 2 & -1 \\ -1 & 2 \end{bmatrix}.$$

Define a new random variable $Z = X + Y$.
Determine the conditional probability density function of X given Z.

2.24 Consider a wide-sense stationary Gaussian random process $X(t)$ with zero mean and autocorrelation function $R_X(\tau) = 2 - |\tau|$, for $|\tau| < 2$, and 0 otherwise. This process is input to a linear system with the input–output relationship given as

$$\mathcal{L}\{x(t)\} = tx(t) + x(t-2).$$

(a) Determine the power spectral density of the input process $X(t)$.
(b) Is the output process $Y(t)$ Gaussian? Is it WSS? Why or why not?
(c) Determine the probability $\mathbb{P}(Y(1) > 3Y(2) + 5)$.

2.25 A WSS random process $X(t)$ is input to an LTI system. Prove that $X(t)$ and the output of the system $Y(t)$ are jointly WSS.

2.26 Consider jointly wide-sense stationary (WSS) random processes $X(t)$ and $Y(t)$ with cross-spectral density $S_{XY}(f)$. Assume that $Z(t)$ is the output of a linear time-invariant (LTI) system with frequency response $H_1(f)$ when $X(t)$ is applied at the input. Similarly, the process $W(t)$ is the output of an LTI system with frequency response $H_2(f)$ when $Y(t)$ is applied at its input. Since the inputs to the LTI systems are WSS, both $Z(t)$ and $W(t)$ are individually WSS.

Show that $Z(t)$ and $W(t)$ are also jointly WSS and determine their cross-spectral density $S_{ZW}(f)$ in terms of $S_{XY}(f)$, $H_1(f)$, and $H_2(f)$.

2.27 Consider a random process defined as

$$X(t) = Y\cos(2\pi f_0 t) + Z\sin(2\pi f_0 t),$$

where f_0 is a constant, and Y and Z are zero-mean random variables. Prove that the random process $X(t)$ is wide-sense stationary if and only if Y and Z are uncorrelated random variables with equal variance.

2.28 Define a random process as

$$X(t) = \cos(2\pi t + \Theta) + \sin(2\pi t + \Theta),$$

where Θ is a uniform random variable on the interval $[0, 2\pi)$.

(a) Find the mean and autocorrelation functions of the process. Is the process wide-sense stationary?
(b) Assume that the process $X(t)$ is input to a system with frequency response $H(f) = \frac{1}{1-j4f}$. What is the power spectral density of the output process? What is the average power content of the output process?

2.29 A random process $X(t)$ is given by

$$X(t) = A\sin(2\pi f_0 t + \Theta),$$

where A is a uniform random variable on $[0, 1]$ and Θ is a mixed random variable with probability density function $f_\Theta(\theta) = \frac{1}{2}\delta(\theta - \frac{\pi}{2}) + \frac{1}{4\pi}(u(\theta) - u(\theta - 2\pi))$. *Note*: $u(\cdot)$ is the unit step function. A and Θ are independent random variables.

(a) Determine the mean and autocorrelation functions of $X(t)$, and argue that the random process is cyclostationary.
(b) Determine the average autocorrelation function of $X(t)$.

(c) For a cyclostationary process, the power spectral density is defined as the Fourier transform of the average autocorrelation function. Determine the PSD of $X(t)$.

2.30 Consider two jointly stationary (real) random processes $X(t)$ and $Y(t)$ with autocorrelation functions R_X and R_Y, respectively, and the cross-correlation function R_{XY}. Prove that

$$|R_{XY}(\tau)| \leq \sqrt{R_X(0)R_Y(0)} \leq \frac{1}{2}\Big(R_X(0) + R_Y(0)\Big).$$

2.31 Let $\begin{bmatrix} X \\ Y \end{bmatrix}$ be a jointly Gaussian random vector with mean $\begin{bmatrix} 0 \\ 0 \end{bmatrix}$ and covariance matrix

$$\boldsymbol{C} = \begin{bmatrix} 3 & 1 \\ 1 & 2 \end{bmatrix}.$$

Determine the conditional probability density function of X given Y, and the conditional expectation $\mathbb{E}[X|\,Y]$.

2.32 A random process is given by

$$X(t) = A_1 \cos(2\pi 20k\; t + \Theta) + A_2 \cos(2\pi 30k\; t + \Theta),$$

where $\begin{bmatrix} A_1 \\ A_2 \end{bmatrix}$ is a random vector with mean $\begin{bmatrix} 1 \\ 2 \end{bmatrix}$ and covariance matrix $\mathbf{C} = \begin{bmatrix} 4 & 1 \\ 1 & 4 \end{bmatrix}$, and Θ is a uniform random variable on the interval $[0, 2\pi]$. The random vector $\begin{bmatrix} A_1 \\ A_2 \end{bmatrix}$ and the random variable Θ are independent.

(a) Determine the mean and autocorrelation functions of the random process $X(t)$. Is the process wide-sense stationary? Is it cyclostationary?

(b) The power spectral density of a cyclostationary process is defined as the Fourier transform of its average autocorrelation function, and it has the same interpretation as the one defined for wide-sense stationary processes.

Determine the power spectral density and the average power content of the random process $X(t)$.

2.33 Consider a wide-sense stationary Gaussian random process $X(t)$ with mean $\mu_X = 1$ and autocorrelation function $R_X(\tau) = 2 + \Lambda(\tau/2)$. Assume that $X(t)$ is input to a system, and the corresponding output is $Y(t)$, where

$$Y(t) = tX(t) + X(t-1).$$

(a) Is the system linear? Is it time-invariant? Is the process $Y(t)$ Gaussian?

(b) Determine the mean and autocorrelation functions of the output process $Y(t)$. Is the process WSS?

(c) Compute the probabilities $\mathbb{P}(Y(1) > 2)$ and $\mathbb{P}(Y(1) > Y(2) - 1)$.

2.34 Let $X(t)$ and $Y(t)$ be two uncorrelated random processes. The first process is defined as

$$X(t) = A\sin(2\pi f_0 t + \Theta),$$

where A is a uniform random variable on $(-1, 1)$ and Θ is a uniform random variable on $[0, 2\pi)$. Assume that A and Θ are independent. The second process is wide-sense stationary (WSS), and its power spectral density is a rectangular function given by

$$S_Y(f) = \frac{1}{f_0}\Pi\left(\frac{f}{f_0}\right).$$

(a) Determine the mean and autocorrelation functions of $X(t)$. Is this process WSS? Are the two processes $X(t)$ and $Y(t)$ jointly WSS?

(b) Define a new process $Z(t) = X(t) + Y(t)$.

(i) Determine the average power content of $Z(t)$.

(ii) The process $Z(t)$ is input to an ideal lowpass filter with unit gain and cut-off frequency $f_0/2$. Is the output process (denoted by $W(t)$) WSS? What is its power spectral density? What is its autocorrelation function? What is its average power content?

2.35 Define a random process $X(t)$ as $X(t) = A + \cos(4\pi t + \Theta)$, where A and Θ are independent random variables. Θ is uniform over $(0, 2\pi)$, and A is a discrete random variable taking on values in the set $\{-2, 0, 4\}$ with probabilities $\mathbb{P}(A = -2) = \mathbb{P}(A = 4) = 1/4$ and $\mathbb{P}(A = 0) = 1/2$.

(a) Determine the probability density function of the random variable $Y = X(t_0)$ for a given time instance $t = t_0$.

(b) Find the mean and autocorrelation functions of $X(t)$. Is $X(t)$ wide-sense stationary?

(c) What is the power spectral density of $X(t)$?

(d) Assume that the process $X(t)$ is input to a linear time-invariant system with frequency response $H(f) = \frac{1}{1-f^2+jf}$. What is the power spectral density of the output process?

(e) Find the autocorrelation function of the output process in part (d).

2.36 A random process $X(t)$ is given by

$$X(t) = A\cos(2\pi f_0 t + \Theta),$$

where A and Θ are independent random variables. A is a uniform random variable on the interval $[1, 2]$, and Θ is a discrete random variable taking on three different values with $\mathbb{P}(\Theta = 0) = \mathbb{P}(\Theta = \frac{\pi}{2}) = \mathbb{P}(\Theta = \pi) = 1/3$.

Determine the mean and autocorrelation functions of the process $X(t)$ and show that it is cyclostationary. Also, determine the period and the average autocorrelation function of the process.

2.37 $X(t)$ denotes a zero-mean WSS Gaussian random process with an autocorrelation function

$$R_X(\tau) = 4\,\text{sinc}^2(10^4\tau)\ \text{W}.$$

(a) What is the average power P_X of the random process $X(t)$?
(b) Obtain and plot the power spectral density $S_X(f)$ of the random process $X(t)$ in units of W/Hz.
(c) What is the bandwidth of the random process $X(t)$?
(d) Compute the probability $\mathbb{P}(X(0.00005) - 2X(-0.00001) > 4)$.
(e) Assume that the process $X(t)$ is passed through an ideal lowpass filter with bandwidth 5 kHz with output denoted by $Y(t)$. Plot the power spectral density $S_Y(f)$ as a function of f, and determine the average power P_Y.
(f) The process $Z(t)$ is defined as

$$Z(t) = \frac{X(t) + X(t-\delta) + X(t+\delta)}{3},$$

where δ is a positive real number. Determine the expression for the average power of this process P_Z in terms of the parameter δ. Also, find the minimum value of δ such that P_Z is the lowest.

2.38 A random process $X(t)$ is given by

$$X(t) = A_1 \cos(1000\pi t + \Theta_1) - A_2 \sin(1000\pi t + \Theta_2),$$

where A_1, A_2 are constants, and Θ_1 and Θ_2 are independent uniform random variables on the interval $[0, 2\pi)$.
(a) Determine the mean and autocorrelation functions of the process $X(t)$. Is it wide-sense stationary?
(b) This process is input to an LTI system with the impulse response $h(t) = 4000\,\text{sinc}^2(2000t)$. Determine the power spectral density of the output process $Y(t)$.
(c) What are the average powers of $X(t)$ and $Y(t)$? What are their average powers in the frequency band 900–1100 Hz?

2.39 Consider a zero-mean process $X(t)$, which is stationary in the strict sense with the following first-order probability density function:

$$f_{X(t)}(x) = \begin{cases} 1+x, & \text{if } -1 \le x < 0, \\ 1-x, & \text{if } 0 < x \le 1, \\ 0, & \text{else.} \end{cases}$$

Also, $X(t_1)$ and $X(t_2)$ are independent whenever $t_1 \neq t_2$. Define a new process $Y(t)$ as

$$Y(t) = 3X(t) - X^2(t).$$

(a) Calculate the mean function of $Y(t)$.
(b) Calculate the autocorrelation function of $Y(t)$.
(c) Is $Y(t)$ wide-sense stationary? Why or why not?

2.40 $X(t)$ is a Gaussian random process with mean function $\mu(t) = 2t - 3$ and autocovariance function

$$C_X(t_1, t_2) = 2\exp(-|t_1 - t_2|).$$

Determine

(a) $\mathbb{P}(X(2) < -1)$.

(b) $\mathbb{P}(X(3) - 2X(2) > 5)$.

(c) $\mathbb{E}[(2X(2) + X(0))^2]$.

2.41 Let $X(t)$ be a zero-mean WSS Gaussian random process with autocorrelation function $R_X(\tau) = 4e^{-2|\tau|}$. We form the random variables $Z = X(t+1)$ and $W = X(t-1)$.

(a) Find $\mathbb{E}(ZW)$ and $\mathbb{E}[(W+Z)^2]$.

(b) Find the PDF of Z and the probability $\mathbb{P}(Z < 1)$.

(c) How is the vector $[Z \ \ W]^T$ distributed?

2.42 Let $X(t)$ be a zero-mean WSS Gaussian random process with autocorrelation function

$$R_X(\tau) = 10\exp(-|\tau|).$$

This process is input to an LTI system with impulse response $h(t) = 2e^{-t}u(t)$.

(a) Determine the covariance of the random variables $X(5)$ and $Y(7)$.

(b) Compute the conditional expectation $\mathbb{E}[Y(7)|X(5) = x]$.

2.43 Consider a zero-mean WSS Gaussian random process $X(t)$ with power spectral density

$$S_X(f) = 0.002 - 0.001\left|\frac{f}{1000}\right|$$

for $|f| < 2000$ Hz, and 0 otherwise. This process is input to a linear time-invariant system with frequency response

$$H(f) = 2\Pi\left(\frac{f}{2000}\right).$$

The filter output is denoted by $Y(t)$.

(a) Determine the power spectral density of $Y(t)$.

(b) Determine the average power content of the output process $Y(t)$.

(c) Determine the probability $\mathbb{P}(-3 < Y(1) < 2)$.

COMPUTER PROBLEMS

2.44 Generate 1 million realizations of a uniform random variable U on $[0, 1]$. Apply the function

$$g(x) = -\ln(ax)$$

to each sample for three different values of the parameter a. You may take $a = 1, 2$, and 3.

(a) Plot the normalized histogram of the resulting (transformed) samples for the three cases to estimate the PDF of $g(U)$. Is the result consistent with your theoretical expectations?

(b) Estimate the probability $\mathbb{P}(1 < g(U) < 2)$ for the three cases using the realizations by determining the fraction of samples that fall in the interval $(1, 2)$, and compare your estimates with the theoretical expectations.

(c) Estimate the mean and variance of $g(U)$ for the three cases using the realizations. To estimate the mean, compute the average of the realizations. To estimate the variance, compute the average of the square of the samples' differences with the estimated mean. Compare the results with the theoretical expectations.

2.45 We are interested in estimating the probability of an event by Monte Carlo simulation. Consider three independent exponential random variables X, Y, Z with parameters $1, 2, 3$, respectively. Generate 10,000 realizations of these three random variables and use them to estimate the following probabilities.

(a) $\mathbb{P}(XY > W)$.

(b) $\mathbb{P}(X + 2Y < W)$.

(c) $\mathbb{P}(X + Y > W \text{ and } X + W < Y)$.

(d) Repeat the estimates in (a)–(c) with a different set of 10,000 realizations. What do you observe?

(e) Repeat the estimates with two different sets of 1 million realizations (generate two different estimates for each case). Comment on your results.

2.46 We are interested in estimating certain quantities related to the random variable defined as

$$Y = X^3 + e^{-X/10} - \ln |X|,$$

where $X \sim \mathcal{N}(2, 1)$.

Generate 1 million realizations of the random variable X and apply the above transformation to the samples to generate different realizations of the random variable Y. Use these realizations to estimate the following quantities.

(a) $\mathbb{P}(Y > 2)$.

(b) $\mathbb{E}[Y]$.

(c) $\mathrm{Var}(Y)$.

2.47 Let $F_X(x)$ denote the cumulative distribution function of the random variable X. Assume that $F_X(x)$ is invertible and its inverse function is $x = F_X^{-1}(y)$. F_X^{-1} is a non-decreasing function since F_X is a non-decreasing function.

Let U be a uniform random variable on $(0, 1)$. Define a new random variable $Y = F_X^{-1}(U)$, what is the CDF of Y in terms of the CDF of X?

How can you generate a random variable with CDF

$$F_X(x) = \begin{cases} \frac{1}{6}(x + 1), & \text{for } -1 \leq x \leq 0, \\ \frac{1}{6} + \frac{1}{3}x, & \text{for } 0 < x \leq 1, \\ \frac{1}{2}x, & \text{for } 1 < x \leq 2, \\ 0, & \text{else}, \end{cases}$$

using realizations of a uniform random variable in MATLAB? Generate 10,000 realizations of the random variable with the given CDF and plot their normalized histogram. Comment on your results.

2.48 Consider two independent standard Gaussian random variables X and Y. We use the linear transformations $Z = 2X + Y$ and $W = X - 3Y$ to obtain a pair of jointly Gaussian random variables Z and W.

(a) Generate 1 million realizations of X and Y, and use them to obtain as many realizations for the pair Z, W. Show a scatter plot of the generated Z, W pairs. Also, generate an estimate of the joint PDF of Z and W.

(b) Estimate $\mathbb{E}[WZ]$ using the generated realizations and compare the result with the exact value of the expectation.

(c) Estimate the probability $\mathbb{P}(2Z + W > 2)$ using the realizations by computing the fraction of samples that satisfy the inequality, and compare it with the actual value of the desired probability.

(d) Estimate $\mathbb{E}[W^2Z^3]$ using the realizations by computing the average of the function whose expectation is being computed over the generated samples. Repeat the same calculation for another set of realizations. Are the two results consistent?

2.49 This problem aims to study some aspects of Gaussian noise processes through computer simulations using MATLAB. We will work with sampled versions of continuous-time (random) signals.

(a) Generate and plot two sample realizations of zero-mean Gaussian random processes with a constant power spectral density. Since the corresponding autocorrelation function is simply an impulse at the origin, that is, the process at any two time instances are independent, this can done by generating sequences of independent and identically distributed Gaussian random variables. (The assumption is that there is an (implicit) ideal lowpass filter removing the components of noise outside the frequency band $[-f_s/2, f_s/2]$, where f_s is the sampling frequency.)

(b) Input the noise realizations in part (a) to a system described by

$$y[n] = \frac{1}{20}(x[n] + x[n-1] + \cdots + x[n-19]),$$

which is nothing but a moving-average filter. Plot the filter outputs.

For both cases, zoom into a smaller range of samples and show a section of your plot. What do you observe? Are there any (noticeable) differences between the plots in parts (a) and (b)? Comment on your observation.

(c) Select a different filter (other than the moving-average filter) and repeat the previous part.

(d) There are different ways to estimate the power spectral density of a random process in MATLAB (using a long realization of the random process). Use one of these functions on the realizations of the processes generated in the previous three parts to estimate the power spectral densities. Comment on your results.

2.50 Consider a Poisson point process $N(t)$ with density λ, which is described as follows: (1) $N(0) = 0$; (2) $N(t)$ has independent increments; (3) the number of points in any interval of length t is a Poisson random variable, whose probability mass function is

$$\mathbb{P}(N(t) = k) = \frac{(\lambda t)^k}{k!} e^{-\lambda t}. \tag{2.295}$$

The points at which the increments occur are called Poisson points. Denoting the Poisson points by t_i, $i = 1, 2, 3, \ldots,$ the random variables t_1, $t_2 - t_1$, $t_3 - t_2$, ... are independent and identically distributed exponential random variables with parameter λ. Using this fact, generating realizations of a Poisson process becomes straightforward.

A random telegraph signal $X(t)$ takes on the value $+1$ if the number of points on the interval $(0, t)$ is even, and -1 if the number of points is odd.

(a) Generate many realizations of a random telegraph signal and estimate the mean function of the process. Give plots of two sample realizations and estimates of the mean function. Perform this experiment with two different values of the Poisson point density.

(b) Define a new random process as $Y(t) = AX(t)$, where A is a binary random variable taking on $+1$ and -1 with equal probabilities (independent of $X(t)$). Generate a long realization of this random process and estimate its autocorrelation function for two different values of λ. You may use appropriate MATLAB functions to do this (i.e., there is no need to write a complete code that estimates the autocorrelation function of a given sequence). Plot the autocorrelation function estimates and comment on your results.

3 Digital versus Analog Transmission

In this chapter, our objective is to briefly describe analog transmission techniques and then discuss the conversion of analog sources into a digital form suitable for transmission using digital means. Recall from our discussion in the introductory chapter that modulation refers to varying some carrier signal characteristics according to the message signal. When the information-bearing message signal is continuous-time and continuous-amplitude, we have the option of transmitting it directly using analog modulation techniques (as will be discussed shortly) or first converting it into a digital form (again, as will be discussed in this chapter) and then transmitting through digital modulation techniques, which will be studied in detail in the subsequent chapters.

The distinction between analog and digital transmission is critical for understanding communication systems. For analog transmission, the information-bearing signal is analog, and the transmitted signal is also an analog signal, matched to the characteristics of the communication channel. For digital transmission, the information-bearing signal is digital, which may be obtained from an analog signal by sampling and quantization, or it may initially be in digital form. In other words, it is represented by a sequence of symbols. On the other hand, the actual signal being transmitted over the communication medium is still a continuous-time waveform. The primary goal of this chapter is to make this point abundantly clear and to explain in detail the process of converting an analog signal into a digital form.

The chapter is organized as follows. We begin with a review of analog modulation techniques, and introduce amplitude modulation (AM) and angle modulation in Section 3.1. We cover the basic principles of conventional AM, double-sideband suppressed carrier AM, single-sideband AM, and vestigial sideband AM, and provide a description both in the time domain and the frequency domain. We briefly review frequency and phase modulation and argue their equivalence. We devote Section 3.2 to the conversion of analog signals into a digital form through sampling and quantization. We prove the sampling theorem, which states that bandlimited signals can be perfectly reconstructed from their samples as long as the sampling frequency is sufficiently high. We study quantization in some depth, including scalar versus vector quantization and uniform versus non-uniform quantization. We also quantify the amount of loss introduced by a scalar quantizer by computing the signal-to-quantization noise ratio and go over the optimal scalar quantizer design. Section 3.3 covers pulse-code modulation as a waveform coding technique along with its variants, including differential PCM and delta modulation. A brief discussion of digital versus analog transmission is given in Section 3.4, and the chapter is concluded with a summary in Section 3.5.

3.1 Analog Modulation Techniques

In this section, we provide brief coverage of analog transmission techniques. We denote the lowpass (also called baseband) message signal to be transmitted by $m(t)$. We assume that its highest frequency content is W Hz, namely, the Fourier transform of the message signal $|M(f)| = 0$ for $|f| \geq W$. Hence, the message signal has a bandwidth of W. For example, if $m(t)$ represents telephone-quality speech signals, the message bandwidth W is approximately 3.5 kHz.

Assume that we are given a particular frequency band for transmission of the message signal $m(t)$. We need to generate and transmit a signal (which carries the message signal) whose frequency content matches the allocated transmission band. For this purpose, we utilize a sinusoidal waveform denoted by

$$c(t) = A_c \cos(2\pi f_c t) \tag{3.1}$$

as the carrier signal, where A_c is the carrier amplitude and f_c is the carrier frequency. Note that the carrier signal phase is not critical at this stage; hence, we take it as 0. We will comment more on this point later. Our objective is to identify ways of embedding the message signal onto the carrier wave and determine ways of recovering the message signal from the modulated waveform received at the destination. We have two options: (1) we can embed the message signal into the amplitude of the carrier; and (2) we can embed the message into the angle of the carrier. Therefore, there are two primary analog modulation techniques: amplitude modulation and angle modulation, as detailed in the rest of this section.

3.1.1 Amplitude Modulation

There are four different amplitude modulation techniques, namely, conventional AM (also called full AM), double-sideband suppressed carrier (DSB-SC) AM, single-sideband (SSB) AM, and vestigial sideband (VSB) AM. We review these individually, discuss the respective modulation and demodulation schemes, and highlight their merits and demerits.

3.1.1.1 Conventional AM

A conventional AM-modulated signal is given by

$$x(t) = A_c(1 + k_a m(t)) \cos(2\pi f_c t), \tag{3.2}$$

where k_a is a constant called amplitude sensitivity. If the message signal $m(t)$ is a voltage level (in volts, V), then k_a has the units of 1/V, making the product $k_a m(t)$

dimensionless. In conventional AM, the message signal is carried in the amplitude of the carrier signal after proper normalization; however, a pure sinusoid is also transmitted along with the message-carrying portion.

The amplitude sensitivity is selected in such a way as to make the $1+k_a m(t)$ term non-negative for reasons that will become clear shortly. Note that typical message signals (e.g., speech signals) have a dynamic range symmetric around 0. With this in mind, the condition $1 + k_a m(t) \geq 0$ can also be stated as $|k_a m(t)| \leq 1$ for all t.

We define the *modulation index* (also called the modulation factor) as

$$\mu = \max_t |k_a m(t)|, \tag{3.3}$$

and the closely related quantity *percentage modulation* as $100 \max_t |k_a m(t)|$. If $\mu > 1$, the signal is referred to as overmodulated. For overmodulated signals, $1 + k_a m(t) < 0$ for some t.

An illustration of conventional AM signals for two cases, with $\mu \leq 1$ and with $\mu > 1$, is depicted in Fig. 3.1. It is clear from the illustration that if the modulation index is less than or equal to 1, the envelope of the modulated signal $x(t)$ is a scaled version of $(1 + k_a m(t))$. Hence, if we can detect the signal envelope, we can obtain the message signal by simply removing its mean.

Demodulation. Conventional AM signals are demodulated using a simple envelope detector as depicted in Fig. 3.2. With $\mu \leq 1$ and $f_c \gg W$, the output voltage $v_{out}(t)$ can be made close to the envelope of the signal $x(t)$ provided that $(R_s + r_f)C \ll \frac{1}{f_c}$ (to make sure that the capacitor charges sufficiently quickly) and $\frac{1}{W} \gg R_l C \gg \frac{1}{f_c}$ (to make sure that the capacitor does not discharge too fast or too slowly). A typical envelope detector output is depicted in Fig. 3.3.

Spectrum of Conventional AM Signals. The Fourier transform of the modulated signal $x(t)$ is given by

$$X(f) = \frac{A_c}{2}\Big(\delta(f - f_c) + \delta(f + f_c)\Big) + \frac{k_a A_c}{2}\Big(M(f - f_c) + M(f + f_c)\Big), \tag{3.4}$$

where $M(f)$ is the Fourier transform of the message signal $m(t)$.

Figure 3.4 illustrates the Fourier transforms of a message signal and the corresponding modulated signal for a toy example. Note that, in general, one would need to provide magnitude and phase plots of the Fourier transforms; however, we will not worry about these details here as we are only interested in a high-level picture in the frequency domain. Note also that we have utilized a cosine wave with phase 0 as the carrier signal, which is not essential. One can use an arbitrary phased sinusoidal signal with frequency f_c and perform conventional AM modulation. In this case, however, one would need to be more careful when calculating the Fourier transforms of the signals involved.

We can determine the transmission bandwidth required for a given modulated signal by its positive frequency content. Since, for conventional AM signals, the

Figure 3.1 Exemplary conventional AM signals with and without overmodulation.

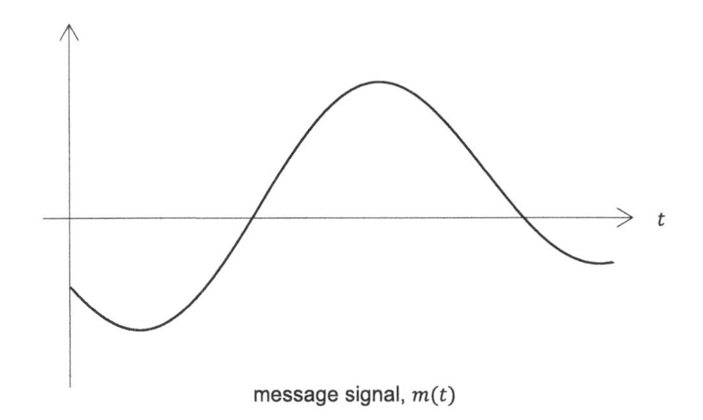

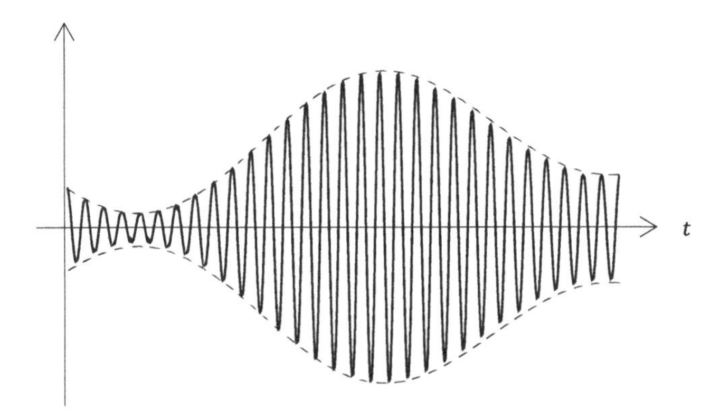

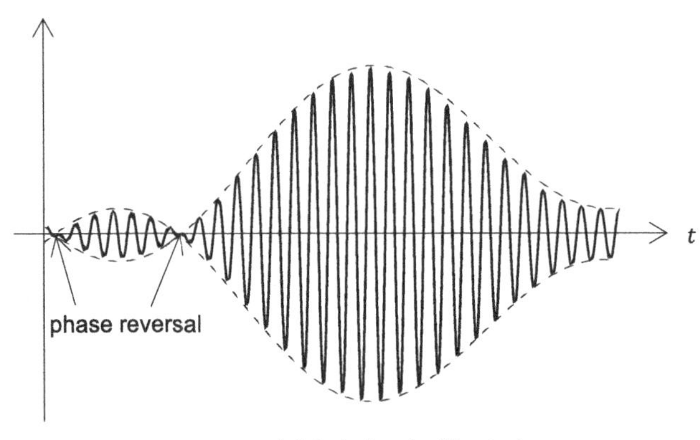

modulated signal occupies the frequency band $f_c - W$ to $f_c + W$, a channel bandwidth of $2W$ is required to transmit the modulated waveform. The portion of the signal with frequency content above f_c Hz is referred to as the upper-sideband (USB) signal, and the part with frequency content below f_c is referred to as the

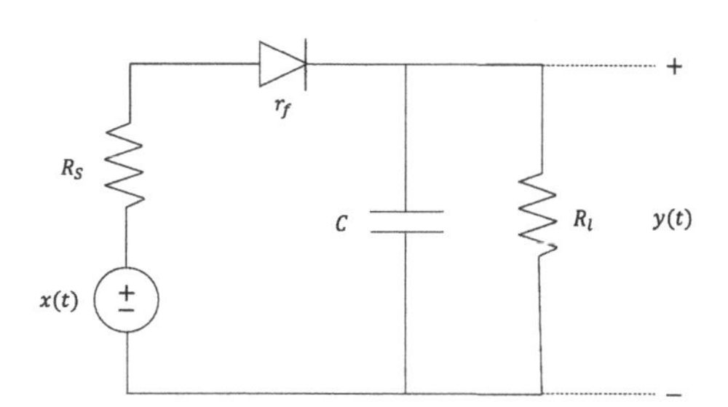

Figure 3.2 A simple envelope detector.

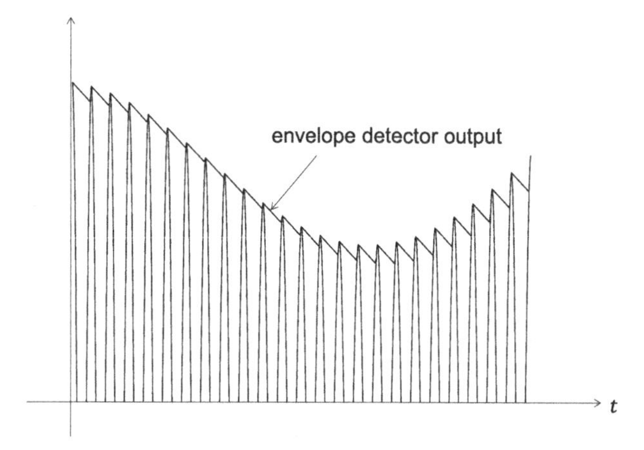

Figure 3.3 Typical output of an envelope detector.

lower-sideband (LSB) signal. The USB and LSB signals are labeled in Fig. 3.4 for the example under consideration.

Advantages and Disadvantages of Conventional AM. The main advantage of conventional AM modulation is the simplicity of its demodulation using a simple envelope detector. However, there is a significant waste of power as a substantial part of the transmitted signal is only a pure sinusoid. The pure sinusoidal portion does not carry any message; however, it is required to ensure that demodulation can be performed using an envelope detector. Commercial AM radio broadcast employs conventional AM because it is essential to have cheap receivers since there are millions of them in any given market while there are only a few transmitters, that is, one can easily tolerate the power inefficiency at the transmitter side.

To get a sense of how much power is wasted with conventional AM, consider the following simple example. Assume that the message signal is given by $m(t) = A_m \cos(2\pi f_m t)$, that is, it is a pure sinusoid with frequency f_m and amplitude A_m. The corresponding conventional AM signal is obtained as

$$x(t) = A_c(1 + k_a A_m \cos(2\pi f_m t)) \cos(2\pi f_c t) \tag{3.5}$$

$$= A_c(1 + \mu \cos(2\pi f_m t)) \cos(2\pi f_c t), \tag{3.6}$$

where $\mu = k_a A_m$ is the modulation index. Using simple trigonometric identities, we can also write

$$x(t) = A_c \cos(2\pi f_c t) + \frac{\mu A_c}{2} \cos(2\pi(f_c + f_m)t) + \frac{\mu A_c}{2} \cos(2\pi(f_c - f_m)t). \tag{3.7}$$

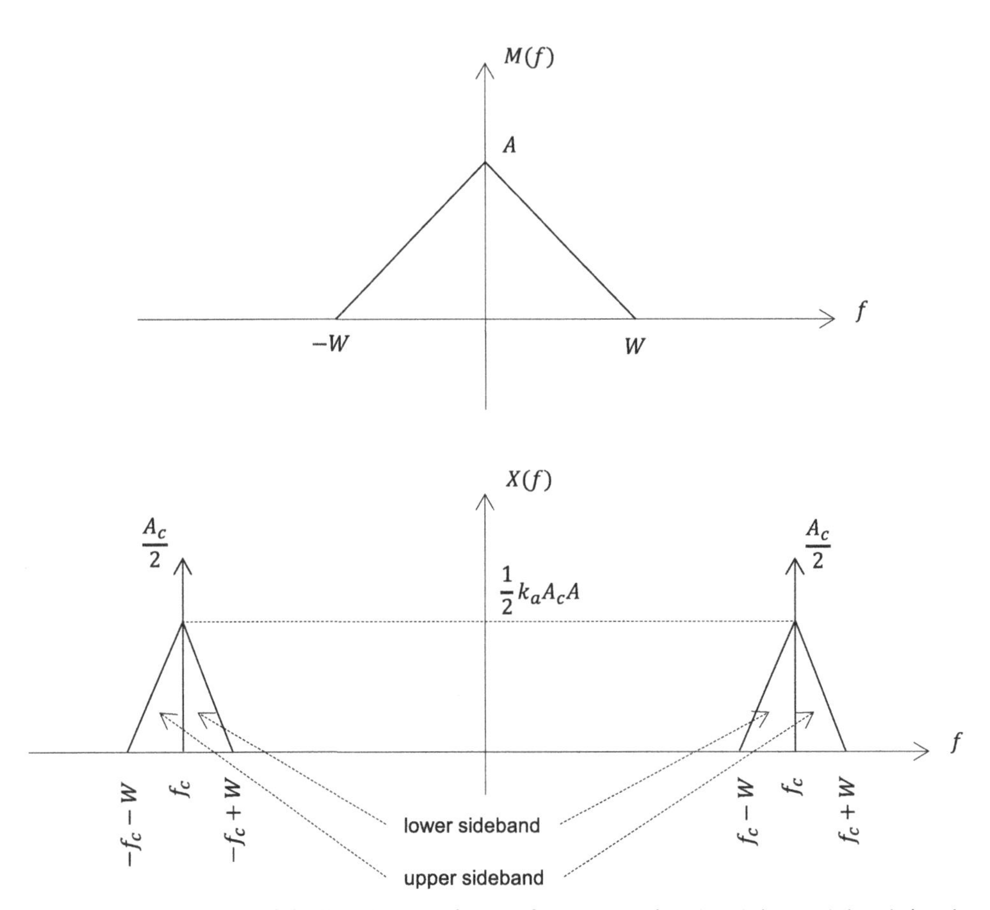

Figure 3.4 Illustration of the Fourier transforms of a message signal and the modulated signal using conventional AM.

Here, the first term is the pure carrier component, the second is the USB signal, and the third is the LSB signal. Recalling the definition of signal power in the previous chapter, we can compute the power of a sinusoidal signal with amplitude A and period T_s as follows:

$$P = \lim_{T\to\infty} \frac{1}{T} \int_{-T/2}^{T/2} A^2 \cos^2\left(\frac{2\pi t}{T_s}\right) dt \tag{3.8}$$

$$= \frac{1}{T_s} \int_{-T_s/2}^{T_s/2} A^2 \cos^2\left(\frac{2\pi t}{T_s}\right) dt \tag{3.9}$$

$$= \frac{1}{T_s} \int_{-T_s/2}^{T_s/2} \frac{A^2}{2}\left(1 + \cos\left(\frac{4\pi t}{T_s}\right)\right) dt \tag{3.10}$$

$$= \frac{A^2}{2}. \tag{3.11}$$

Using this basic result, we can determine the average powers of the USB and LSB signals as $\frac{1}{8}\mu^2 A_c^2$ each, and the average power content of the carrier component as $\frac{1}{2}A_c^2$. Therefore, the ratio of the power in the sidebands (which carry the message signal) to the total power in $x(t)$ is

$$\frac{\frac{1}{4}\mu^2 A_c^2}{\frac{1}{4}\mu^2 A_c^2 + \frac{1}{2}A_c^2} = \frac{\mu^2}{2+\mu^2}. \tag{3.12}$$

Since $\mu \leq 1$, this ratio is certainly less than or equal to 1/3. In other words, for this simple pure tone modulation example, at least two-thirds of the total power is spent on transmitting the carrier component.

The situation is more dramatic for practical conventional AM systems. The lower amplitude levels are more frequent for speech signals, and hence the average power of a typical speech signal is quite small. As a result, for typical speech signal transmission through conventional AM, about 90% of power is wasted in transmitting the carrier component.

3.1.1.2 Double-Sideband Suppressed Carrier (DSB-SC) AM

A major drawback of conventional AM is the transmission of the separate carrier component in the modulated signal. To save power, we remove the pure sinusoidal component and obtain a DSB-SC AM signal as

$$x(t) = A_c m(t) \cos(2\pi f_c t). \tag{3.13}$$

In the frequency domain, we have

$$X(f) = \frac{A_c}{2}\left(M(f - f_c) + M(f + f_c)\right). \tag{3.14}$$

An exemplary message signal and the corresponding DSB-SC AM-modulated version are illustrated in Fig. 3.5. It is clear from the figure that an envelope detector will not work as a demodulator, since the envelope of the signal is not a scaled and offset version of the message signal, unlike the conventional AM signal. This is because, in DSB-SC AM, there is no large pure carrier component in the modulated signal.

Frequency-domain interpretation of a DSB-SC AM signal is provided in Fig. 3.6. The required bandwidth for transmission is $2W$, as in the case of conventional AM.

Demodulation of DSB-SC AM signals requires a *coherent* (or *synchronous*) receiver. In other words, the receiver generates a sinusoidal signal with frequency f_c, multiplies this signal with the modulated signal (ignoring the channel or noise effects), and passes the resulting signal through a lowpass filter to recover the message signal. It is important to note that the phase of the sinusoidal signal employed should match the phase of the sinusoid used for modulation (as observed at the receiver).

The carrier phase recovery at the receiver can be accomplished using a phase-locked loop (PLL), which will be described in Chapter 6. The carrier phase can also be obtained by transmitting a small pilot tone along with the DSB-SC AM

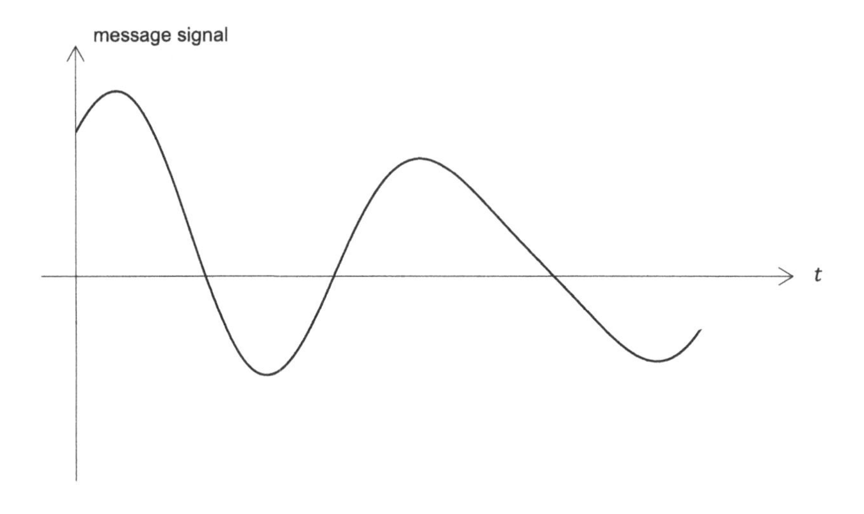

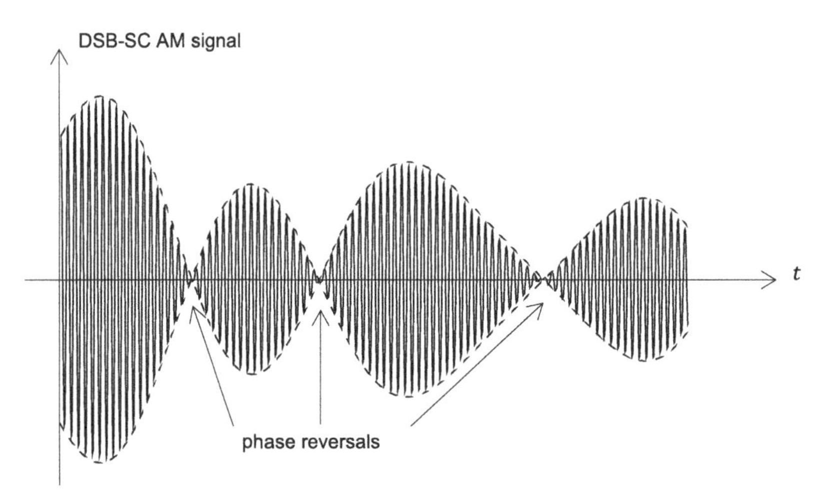

Figure 3.5 Illustration of a message signal and the corresponding DSB-SC AM signal in the time domain.

signal and employing a narrowband bandpass filter at the receiver side. Since the receiver requires the carrier phase information, it is called a coherent or synchronous receiver. The receiver structure is depicted in Fig. 3.7.

To see how this receiver works, denoting the lowpass filter input by $v(t)$, we can write

$$v(t) = x(t)A'_c \cos(2\pi f_c t + \phi) \tag{3.15}$$

$$= A_c A'_c m(t) \cos(2\pi f_c t) \cos(2\pi f_c t + \phi) \tag{3.16}$$

$$= \frac{A_c A'_c}{2} \cos(4\pi f_c t + \phi) + \frac{A_c A'_c}{2} \cos(\phi) m(t), \tag{3.17}$$

where A'_c is the amplitude of the reference carrier signal generated at the receiver, and ϕ is its phase (with respect to the exact phase of the carrier as observed at the receiver side). The signal $v(t)$ in the frequency domain is illustrated in Fig. 3.8.

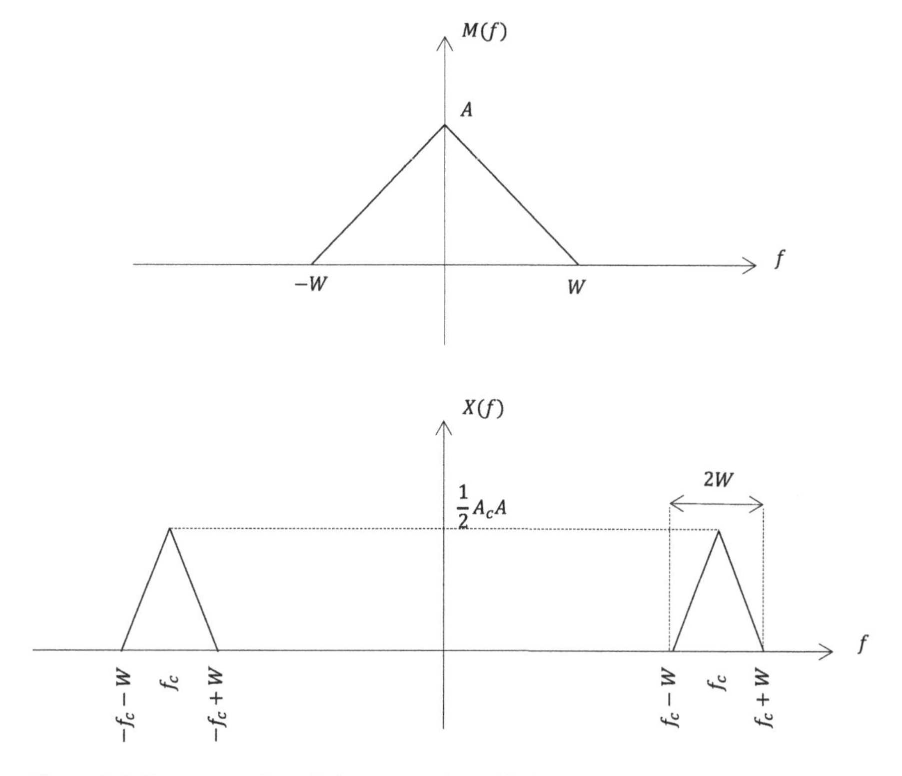

Figure 3.6 Frequency-domain interpretation of DSB-SC AM.

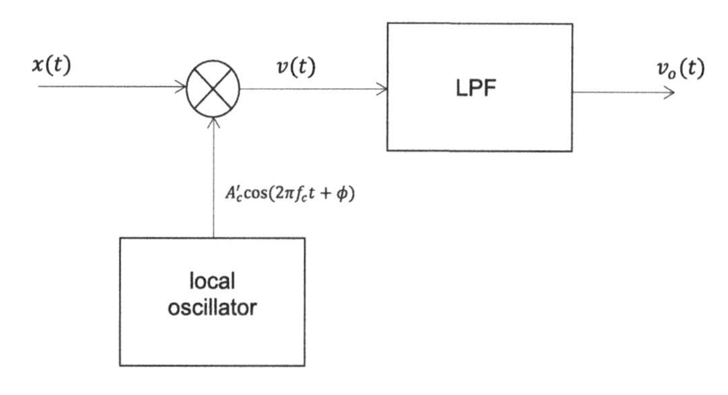

Figure 3.7 Synchronous (or coherent) receiver for DSB-SC AM.

Filtering this signal using a lowpass filter with a cut-off frequency of W, we obtain

$$v_o(t) = \frac{A_c A_c'}{2} \cos(\phi) m(t), \tag{3.18}$$

which is a scaled version of the message signal. If there is no error in the carrier phase estimation, we have $\phi = 0$, and the demodulator output is given by $v_o(t) = \frac{A_c A_c'}{2} m(t)$. If there is some error in the phase estimate, that is, $\phi \neq 0$, the amplitude of the message signal estimate at the receiver output is reduced by a factor of $\cos(\phi)$.

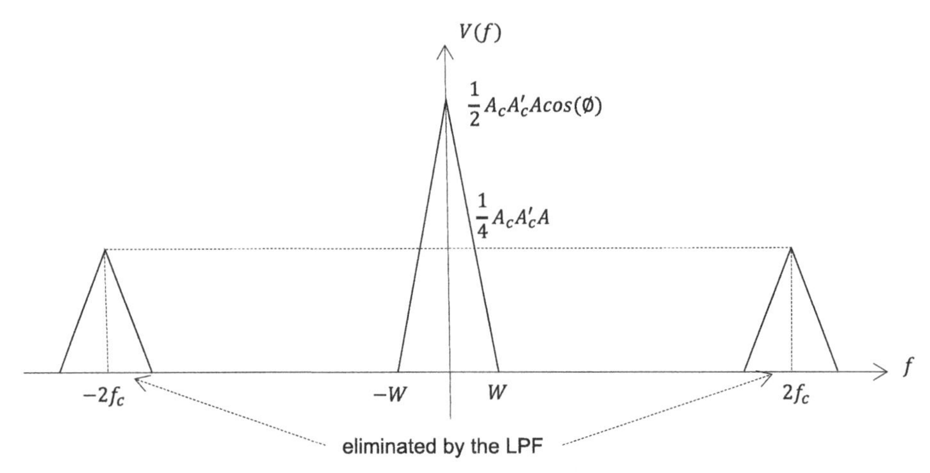

Figure 3.8 Illustration of the Fourier transform of the signal $v(t)$ is DSB-SC AM demodulation in the frequency domain. The lowpass filter eliminates the double-frequency components; therefore, a scaled version of the message signal is obtained.

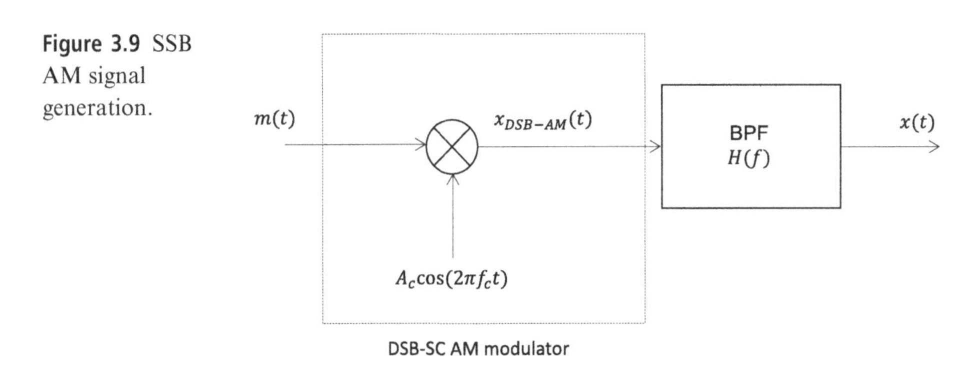

Figure 3.9 SSB AM signal generation.

While some level of phase error may be tolerable in a practical system, there would be a reduction in the output signal quality.

3.1.1.3 Single-Sideband (SSB) AM and Vestigial Sideband (VSB) AM

In conventional AM and DSB-SC AM, we transmit both the upper and lower sidebands of the modulated signal, resulting in a required transmission bandwidth of $2W$. Noting that only one of the sidebands is sufficient to recover the message signal using the same coherent receiver as in the demodulation of DSB-SC AM signals, we can filter out one of the sidebands and transmit an SSB AM-modulated signal, as illustrated in Fig. 3.9. Since only one of the sidebands is transmitted, the required bandwidth for transmission is the same as the message bandwidth W.

We can utilize a BPF with passband $[f_c, f_c + W]$ to keep the USB signal, and one with passband $[f_c - W, f_c]$ in order to retain the LSB (see Fig. 3.10).

We can obtain the SSB AM-modulated signal in the time domain using the message signal and its *Hilbert transform*, $\hat{m}(t)$, which is defined as

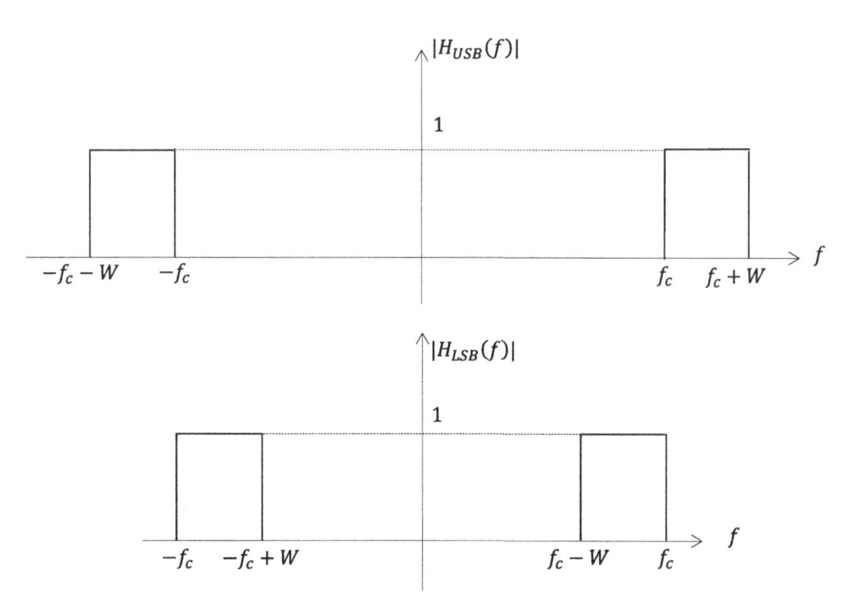

Figure 3.10 Bandpass filters for SSB AM-modulated signal generation (top: to retain the USB signal, bottom: to retain the LSB signal).

$$\hat{m}(t) = m(t) * \frac{1}{\pi t}. \tag{3.19}$$

In other words, the Hilbert transform is the output of an LTI system with impulse response $\frac{1}{\pi t}$ when the input is $m(t)$. Taking the Fourier transform of both sides, we can write

$$\hat{M}(f) = M(f)H(f), \tag{3.20}$$

where

$$H(f) = \begin{cases} -j, & \text{if } f > 0, \\ 0, & \text{if } f = 0, \\ j, & \text{if } f < 0. \end{cases} \tag{3.21}$$

An SSB AM signal (for which the USB is retained) can be written as

$$x(t) = m(t)\cos(2\pi f_c t) - \hat{m}(t)\sin(2\pi f_c t). \tag{3.22}$$

To see this, we consider the frequency-domain representation and write

$$X(f) = M(f) * \mathcal{F}\{\cos(2\pi f_c t)\} - \hat{M}(f) * \mathcal{F}\{\sin(2\pi f_c t)\} \tag{3.23}$$

$$= M(f) * \left(\frac{1}{2}\delta(f - f_c) + \frac{1}{2}\delta(f + f_c)\right)$$
$$- \hat{M}(f) * \left(\frac{1}{2j}\delta(f - f_c) - \frac{1}{2j}\delta(f + f_c)\right) \tag{3.24}$$

$$= \frac{1}{2}\left(M(f - f_c) + M(f + f_c)\right) - \frac{1}{2j}\left(\hat{M}(f - f_c) - \hat{M}(f + f_c)\right). \tag{3.25}$$

Noting that $\hat{M}(f) = -jM(f)$ for $f > 0$ and $\hat{M}(f) = jM(f)$ for $f < 0$, we obtain $\hat{M}(f - f_c) = -jM(f - f_c)$ for $f > f_c$ and $\hat{M}(f - f_c) = jM(f - f_c)$ for $f < f_c$. Similarly, $\hat{M}(f + f_c) = -jM(f + f_c)$ for $f > -f_c$ and $\hat{M}(f + f_c) = jM(f + f_c)$ for $f < -f_c$. Combining these, we arrive at the Fourier transform of the modulated signal as

$$X(f) = \begin{cases} M(f - f_c), & \text{if } f > f_c, \\ M(f + f_c), & \text{if } f < -f_c, \\ 0, & \text{else,} \end{cases} \tag{3.26}$$

which is the SSB AM signal retaining the USB part only.

In the same fashion, we can write the SSB AM signal, which retains the LSB part as

$$x(t) = m(t)\cos(2\pi f_c t) + \hat{m}(t)\sin(2\pi f_c t). \tag{3.27}$$

The demodulation of SSB AM signals is very similar to that of DSB-SC AM signals. Namely, we need a synchronous receiver, that is, we need to multiply the received signal with the carrier signal (with the correct phase) and pass the resulting signal through an LPF with bandwidth W.

We observe that the accuracy of the carrier phase at the receiver is more crucial for SSB AM than in the case of DSB-SC AM demodulation. When the carrier phase $\phi \neq 0$, the receiver output will be (a scaled version of)

$$\cos(\phi)m(t) \pm \sin(\phi)\hat{m}(t). \tag{3.28}$$

Hence, in addition to a decrease in the signal amplitude by the factor of $\cos(\phi)$ (as in the case of demodulation of DSB-SC AM signals), there will also be distortion due to the Hilbert transform of the message signal (scaled with $\sin(\phi)$) added at the receiver output, reducing the demodulated signal quality.

Generating an SSB AM signal is costly because an ideal BPF is needed. As a compromise between the bandwidth requirements and complexity, alternatively, we can utilize what is called *vestigial sideband* (VSB) AM modulation, which transmits a sideband along with a small component (or vestige) of the other. That is, to generate a VSB AM signal, we use a similar approach as in Fig. 3.9; however, instead of using an ideal BPF, we utilize a more practical bandpass filter as depicted in Fig. 3.11.

The filter characteristics are more relaxed in VSB AM, so its implementation becomes more practical. Due to its implementation advantages, VSB-AM is utilized for transmitting video signals in commercial (analog) TV signal transmission.

Demodulation of VSB AM signals is performed similarly to that of DSB-SC AM and SSB AM, namely, using a coherent receiver. Despite the BPF not having a brickwall-type passband, as long as we have

$$H(f - f_c) + H(f + f_c) = \text{constant} \tag{3.29}$$

for $|f| \leq W$, the demodulator output becomes a scaled version of the message signal, and the message signal is recovered at the receiver output.

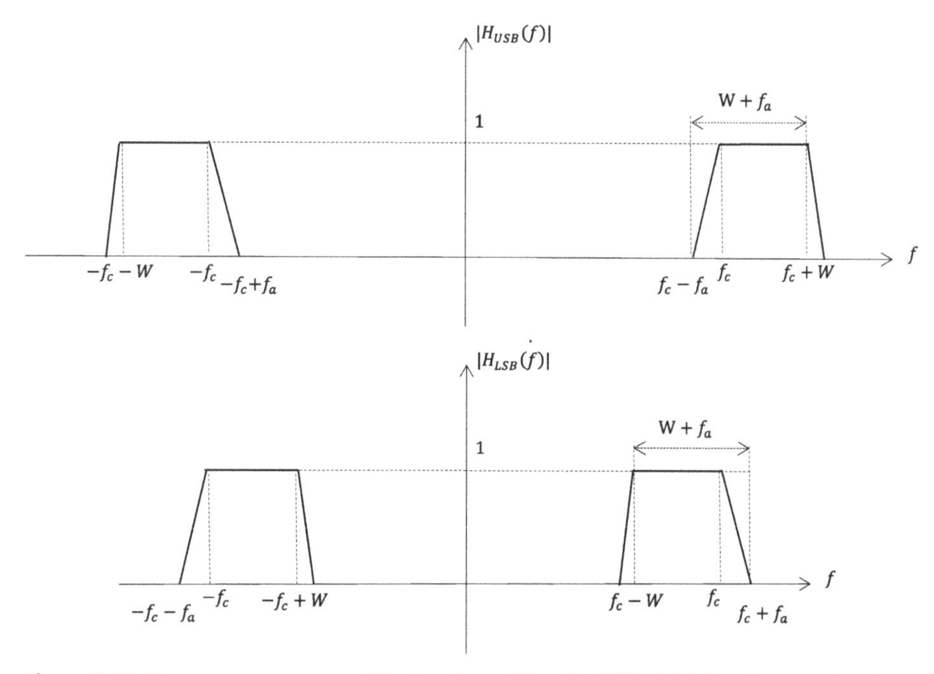

Figure 3.11 Frequency response of the bandpass filter for VSB AM signal generation (top: to retain the USB signal, bottom: to retain the LSB signal).

3.1.2 Angle Modulation

In the previous subsection, we covered methods of embedding the message signal onto the amplitude of a carrier wave. Here, as an alternative to amplitude modulation techniques, we will describe angle modulation schemes, namely, phase modulation (PM) and frequency modulation (FM).

A phase-modulated signal is given by

$$x(t) = A_c \cos(2\pi f_c t + k_p m(t)), \tag{3.30}$$

where k_p is the phase-sensitivity constant in radians/volt (rad/V) (assuming that the message signal is in volts). On the other hand, an FM signal is given by

$$x(t) = A_c \cos\left(2\pi f_c t + 2\pi k_f \int_0^t m(\tau)d\tau\right) \tag{3.31}$$

assuming that $m(t) = 0$ for $t < 0$, where k_f is the frequency-sensitivity constant in hertz/volt (Hz/V). We can determine the instantaneous frequency of the FM signal as

$$f_i(t) = \frac{1}{2\pi}\frac{d}{dt}\left\{2\pi f_c t + 2\pi k_f \int_0^t m(\tau)d\tau\right\} \tag{3.32}$$

$$= f_c + k_f m(t). \tag{3.33}$$

We define the modulation index for FM as

$$\beta_{FM} = \frac{k_f \max |m(t)|}{W}, \tag{3.34}$$

and the modulation index for PM as

$$\beta_{PM} = k_p \max |m(t)|. \tag{3.35}$$

Clearly, both the PM and FM signals have a constant envelope.

Example 3.1

We provide an example of PM and FM modulator outputs for a message signal in the form of a ramp in Fig. 3.12.

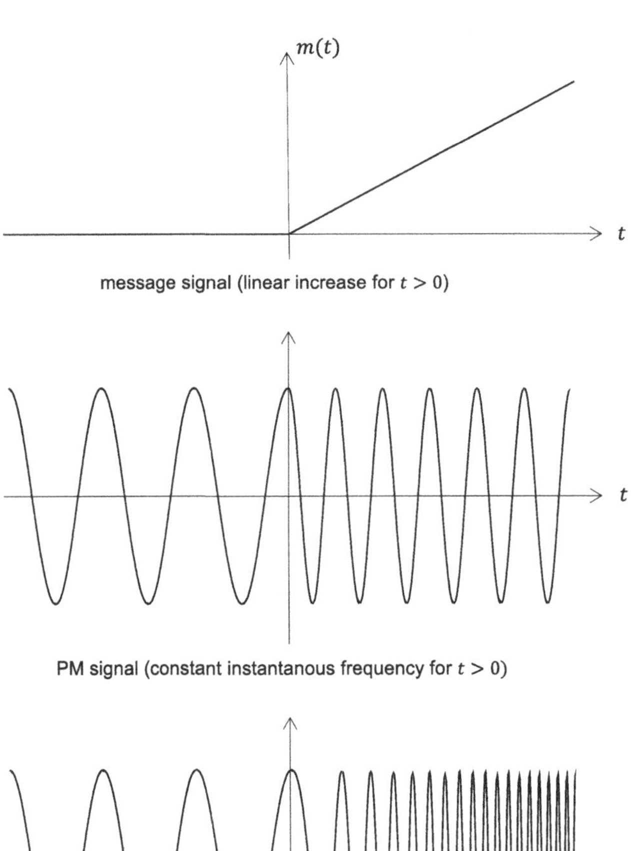

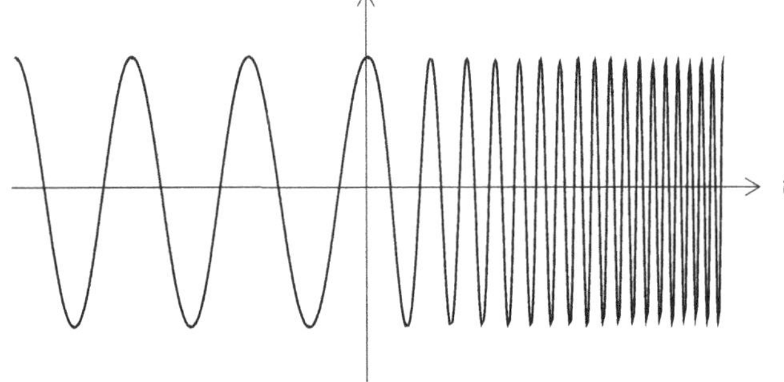

Figure 3.12 Illustration of PM and FM signals corresponding to a ramp-like message signal.

The instantaneous frequency of the modulated signal is f_c for $t < 0$ for both the PM and FM waveforms. However, since the message signal is linearly increasing as a function of time for $t > 0$, the instantaneous frequency becomes $f_c + \frac{k_p}{2\pi}$ for $t > 0$ (still a constant) for the PM signal, while it continues to increase (linearly) for $t > 0$ for FM. The fact that the PM and FM signals have a constant envelope is also observed.

We note that the characterization of PM- or FM-modulated signals in the frequency domain is quite challenging, contrary to the case of AM modulation, for which the Fourier transform of the modulated waveforms can be readily obtained. Therefore, it is not straightforward to determine the bandwidth required to transmit angle-modulated signals. Nonetheless, there is a simple approximate characterization. *Carson's rule* states that the effective bandwidth (which contains at least 98% of the signal power) for angle-modulated signals is given by $2(\beta + 1)W$, where β is the modulation index. With larger values of the modulation index, the required bandwidth for transmission is increased. In practice, β can be much larger than unity; hence, the bandwidth requirements of the PM and FM schemes could be significantly larger than those of AM signals (i.e., there may be significant bandwidth expansion). One last point is that due to this bandwidth expansion, the signal quality at the receiver output becomes superior, as will be discussed in the next subsection.

We also note that an FM modulator can generate a PM-modulated signal and vice versa; hence, these two modulation schemes are equivalent in some sense. See Fig. 3.13 for a depiction of this point.

Commercial FM employs *pre-emphasis/de-emphasis filtering*, which boosts the higher-frequency components of the message signal before FM modulation. Namely, a pre-emphasis filter (which is a highpass filter) is employed at the transmitter, and its inverse (a de-emphasis filter) is used at the receiver. In other words, effectively, the low-frequency components of the message signal are transmitted through FM, while the high-frequency components are transmitted through PM. This is done to improve the signal quality at the receiver output against channel noise.

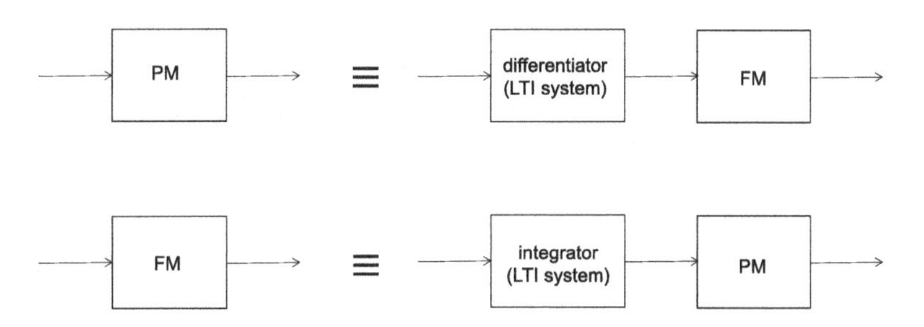

Figure 3.13 A PM modulator can generate an FM signal and vice versa.

3.1.3 Effects of Noise in Analog Modulation

In this subsection, we briefly assess the receiver output quality for amplitude and angle modulation schemes. Let us denote the received signal as

$$r(t) = \alpha x(t) + n(t), \tag{3.36}$$

where $x(t)$ is the analog modulated transmitted waveform, α is the channel attenuation factor, and $n(t)$ is white noise, that is, a random process with a constant power spectral density $S_n(f) = \frac{N_0}{2}$. (Note that this description of white noise is sufficient for our purposes here. We will study it more deeply in Chapter 5 when covering digital modulation.)

Denoting the received signal power level, that is, the average power of $\alpha x(t)$, by P_R, and the message signal bandwidth by W, we define the baseband signal-to-noise ratio as

$$\text{SNR}_{bb} = \frac{P_R}{N_0 W}. \tag{3.37}$$

This is nothing but the signal-to-noise ratio with no modulation. We also define P_{M_n} as the normalized message signal power, that is, the average power of $\frac{m(t)}{\max|m(t)|}$.

Omitting the derivations or details, we provide the ratio of the signal power to the noise power at the receiver output for different analog modulation schemes in Table 3.1.

We define the modulation efficiency η for conventional AM with modulation index μ as

$$\eta = \frac{\mu^2 P_{M_n}}{1 + \mu^2 P_{M_n}}. \tag{3.38}$$

Reading from the table, the signal-to-noise ratio at the receiver output is ηSNR_{bb}. Since the normalized message power $P_{M_n} \leq 1$, with no overmodulation $\mu \leq 1$, the modulation efficiency η is certainly less than 1. Therefore, the output signal-to-noise ratio for the conventional AM scheme is significantly less than the baseband

Table 3.1 Signal-to-noise ratio at the receiver output for different analog modulation schemes. η is the modulation efficiency of conventional AM. The results for FM and PM are valid only if the baseband SNR is above the threshold SNR.

Modulation scheme	Output signal-to-noise ratio
DSB-SC AM	SNR_{bb}
SSB-SC AM	SNR_{bb}
Conventional AM	ηSNR_{bb}
PM	$\beta_{PM}^2 P_{M_n}\text{SNR}_{bb}$
FM	$3\beta_{FM}^2 P_{M_n}\text{SNR}_{bb}$

signal-to-noise ratio. This is because, in conventional AM, a large amount of power is spent transmitting a pure carrier signal (which does not contain any information about the message signal). On the other hand, both the DSB-SC AM and the SSB AM have the same output signal-to-noise ratio as the baseband SNR. Namely, while conventional AM reduces the signal quality, the DSB-SC AM or SSB AM schemes neither improve nor deteriorate the signal quality at the receiver output.

Again referring to the table, the angle modulation schemes may have a significantly improved signal quality at the receiver output compared to the baseband SNR (and, hence, compared to the amplitude modulation schemes), depending on the modulation index and the normalized message power. A caution is the following: the output SNRs for FM and PM schemes given in the table are only valid if the baseband SNR is above a certain threshold. For instance, this threshold is approximately $20(\beta_{FM}+1)$ for FM. Below this value of the baseband SNR, the output becomes indistinguishable from noise. This is called the *threshold effect.* Note that there is no such effect in amplitude modulation.

Provided that the baseband SNR is above the threshold SNR, the angle-modulated schemes offer a dramatic increase in signal quality at the receiver output. For instance, considering FM, with a normalized message power of $1/3$ and a modulation index $\beta_{FM} = 10$, there will be a 100-fold (or 20 dB) increase with respect to the baseband SNR. This increase in the signal quality comes at the cost of bandwidth expansion: with $\beta_{FM} = 10$, Carson's rule states that the effective bandwidth is $22W$, which is 22 times more than that of SSB AM, and 11 times more than those of conventional AM and DSB-SC AM.

3.2 Analog-to-Digital Conversion

Instead of transmitting an analog signal using analog modulation methods, we can first obtain a digital representation of the signal and then transmit the resulting sequence of bits (or symbols) using digital modulation methods. We perform sampling and quantization to convert an analog signal into a digital form. In this section, we detail these two processes.

3.2.1 Sampling

A fundamental result from signals and systems is the *sampling theorem*, which states that if a signal $x(t)$ is bandlimited to W Hz, that is, the Fourier transform of the signal $X(f) = 0$ for $|f| \geq W$, then it can be reconstructed from its samples $\{x(nT_s)\}$ ($n = \ldots, -2, -1, 0, 1, 2, \ldots$) provided that the sampling period $T_s \leq 1/2W$, or equivalently, the sampling frequency $f_s = \frac{1}{T_s} \geq 2W$. This minimum sampling frequency of $2W$, which guarantees a perfect reconstruction of the signal $x(t)$, $t \in \mathbb{R}$, from its samples, is referred to as the Nyquist rate.

The above fundamental result is of paramount importance in practice as typical information sources are bandlimited. Hence, there is no loss of information by considering only their samples, provided that the sampling frequency is sufficiently high. For instance, a typical telephone-quality speech signal is bandlimited to 3.5 kHz; if we sample it using a rate $f_s \geq 7$ kHz, we can have a perfect reconstruction.

In the rest of this subsection, we will prove the sampling theorem.

Let $x(t)$ be bandlimited to W, and assume that we have access to its samples taken with period T_s, denoted by $\{x(nT_s)\}$, $\forall n \in \mathbb{Z}$. Define the impulse-modulated signal obtained from these samples as a new signal

$$x_\delta(t) = \sum_{n=-\infty}^{\infty} x(nT_s)\delta(t - nT_s). \tag{3.39}$$

We can write

$$x_\delta(t) = \sum_{n=-\infty}^{\infty} x(t)\delta(t - nT_s) \tag{3.40}$$

$$= x(t) \sum_{n=-\infty}^{\infty} \delta(t - nT_s). \tag{3.41}$$

Our objective is to obtain the continuous-time signal $x(t)$ from the impulse-modulated signal $x_\delta(t)$. To accomplish this, we take the Fourier transform of both sides. Using the fact that multiplication in the time domain is convolution in the frequency domain, we obtain

$$X_\delta(f) = X(f) * \mathcal{F}\left\{ \sum_{n=-\infty}^{\infty} \delta(t - nT_s) \right\}. \tag{3.42}$$

To determine the Fourier transform of the signal in brackets (also called the periodic impulse train), we notice that it is periodic with the fundamental period of T_s. Therefore, recalling from Chapter 2, this signal (let us denote it by $y(t)$) has a Fourier series expansion in the form of

$$\sum_{k=-\infty}^{\infty} a_k e^{j2\pi t/T_s}, \tag{3.43}$$

where the Fourier series coefficients $\{a_k\}$ are given by

$$a_k = \frac{1}{T_s} \int_{-T_s/2}^{T_s/2} y(t) e^{-j2\pi t/T_s} dt \tag{3.44}$$

$$= \frac{1}{T_s} \int_{-T_s/2}^{T_s/2} \delta(t) e^{-j2\pi t/T_s} dt \tag{3.45}$$

$$= \frac{1}{T_s}. \tag{3.46}$$

In other words,

$$\sum_{n=-\infty}^{\infty} \delta(t - nT_s) = \frac{1}{T_s} \sum_{k=-\infty}^{\infty} e^{j2\pi t/T_s}. \tag{3.47}$$

Hence, the Fourier transform of the signal on the left-hand side is given by

$$\mathcal{F}\left\{\sum_{n=-\infty}^{\infty} \delta(t-nT_s)\right\} = \frac{1}{T_s}\sum_{k=-\infty}^{\infty} \delta\left(f-\frac{k}{T_s}\right). \tag{3.48}$$

Therefore, we obtain

$$X_\delta(f) = X(f) * \left(\frac{1}{T_s}\sum_{k=-\infty}^{\infty} \delta\left(f-\frac{k}{T_s}\right)\right) \tag{3.49}$$

$$= \frac{1}{T_s}\sum_{k=-\infty}^{\infty} X(f) * \delta\left(f-\frac{k}{T_s}\right) \tag{3.50}$$

$$= \frac{1}{T_s}\sum_{k=-\infty}^{\infty} X\left(f-\frac{k}{T_s}\right). \tag{3.51}$$

The relationship between the Fourier transforms of the impulse-modulated signal that can be obtained from the samples only, $x_\delta(t)$, and the original signal $x(t)$ is illustrated in Fig. 3.14 for two different cases: (1) when $f_s = 1/T_s \geq 2W$; and (2) when $f_s = 1/T_s < 2W$.

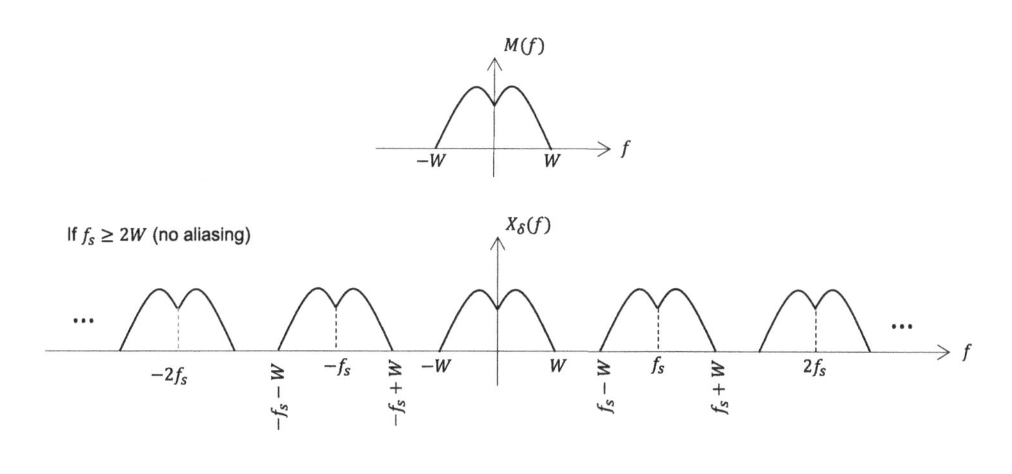

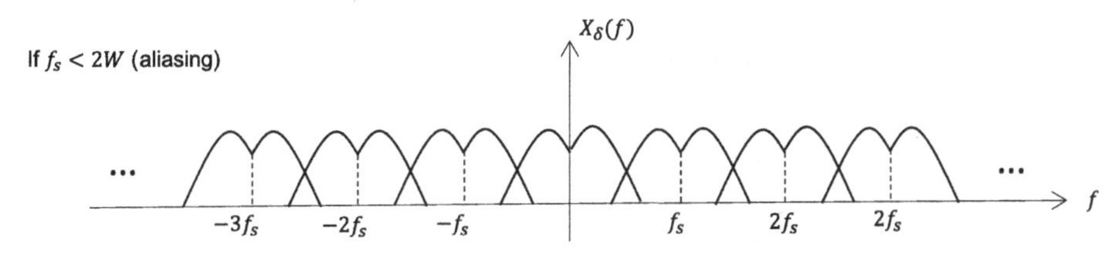

Figure 3.14 Fourier transform of the impulse-modulated signal obtained from the samples of $x(t)$. If the sampling frequency is larger than the Nyquist rate (i.e., $f_s \geq 2W$), there is no aliasing, and $x(t)$ can be recovered from $x_\delta(t)$ through lowpass filtering. If the sampling frequency is not sufficiently high (i.e., $f_s < 2W$), there is aliasing, and the original signal cannot be recovered.

It is clear from the illustration that when the sampling frequency is larger than the Nyquist rate $2W$, $X_\delta(f)$ and $X(f)$ match for frequencies below $f_s - W$ (except for a scaling factor). Hence, by using a lowpass filter to eliminate the higher-frequency components in $X_\delta(f)$, we can obtain the original signal $x(t)$ in an exact form. This completes the proof of the sampling theorem.

There is aliasing if the sampling frequency is smaller than the Nyquist rate. In other words, we do not have a clean replica of $X(f)$ in $X_\delta(f)$, and it is not possible to recover the original continuous-time signal from its samples only with no loss. Nonetheless, via Parseval's theorem, one can often estimate the loss in the recovery even in the presence of aliasing: the corrupted frequency-domain signal (as a square-integrated error between the original signal and the reconstructed signal) leads to an equivalent corruption in the time domain.

3.2.2 Quantization

Analog signals take on a continuum of values; however, to represent them digitally, we only have a finite number of bits (or signal levels) available. Therefore, after sampling an analog signal (i.e., converting it into a discrete-time signal), we need to perform *quantization*.

We classify quantizers as *scalar quantizers* and *vector quantizers*. In scalar quantization, samples are quantized individually, whereas in vector quantization, multiple source samples are quantized jointly.

3.2.2.1 Scalar Quantization

For a scalar quantizer, each source sample is quantized separately. An N-level quantizer is specified by N quantization regions (disjoint intervals) whose union is the real line and a reconstruction value for each. Denoting the boundaries of the quantization regions by $a_1, a_2, \ldots, a_{N-1}$, and the reconstruction levels by $\hat{x}_1, \hat{x}_2, \ldots, \hat{x}_N$, we describe the quantization process for the source sample with the value $x \in \mathbb{R}$ as follows:

$$\begin{aligned} \text{if } x \in (-\infty, a_1], \text{ then } Q(x) &= \hat{x}_1, \\ \text{if } x \in (a_1, a_2], \text{ then } Q(x) &= \hat{x}_2, \\ &\vdots \\ \text{if } x \in (a_{N-2}, a_{N-1}], \text{ then } Q(x) &= \hat{x}_{N-1}, \\ \text{if } x \in (a_{N-1}, \infty), \text{ then } Q(x) &= \hat{x}_N, \end{aligned} \tag{3.52}$$

where $Q(x)$ is the quantizer output corresponding to the value x. Figure 3.15 illustrates the quantization regions and the reconstruction levels for a scalar quantizer.

Typically, the number of quantization levels N is selected as a power of 2, and the quantizer outputs are represented as a sequence of bits. For instance, for $N = 16$, each quantization level is represented by four bits.

Quantization is an irreversible process; once we quantize a source output sample, there is no way we can determine the exact value of the original sample. Since

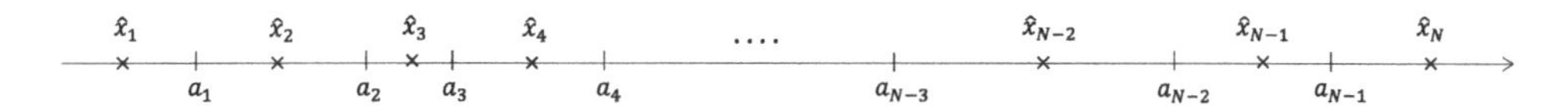

Figure 3.15 A scalar quantizer.

we can only obtain an approximation of the source output with quantization, it is important to assess the quality of a quantizer. To accomplish this, we consider the source samples as realizations of a random variable (denoted by X) and compute the average distortion due to the quantization process. As the input to the quantizer is a random variable, and the quantization process is simply a function, the quantizer output is also a random variable, denoted by $Q(X)$. We define the difference between the actual sample value and its quantized version as the quantization error, denoted by $\tilde{X}$, namely,

$$\tilde{X} = X - Q(X). \tag{3.53}$$

We define the mean squared error distortion (denoted by D) as the expected value of the square of this quantity, namely,

$$D = \mathbb{E}[\tilde{X}^2] = \mathbb{E}[(X - Q(X))^2]. \tag{3.54}$$

As a more meaningful (scale-free) measure of the quality of a quantizer, we define the *signal-to-quantization noise ratio (SQNR)* as

$$\text{SQNR} = \frac{\mathbb{E}[X^2]}{\mathbb{E}[(X - Q(X))^2]}. \tag{3.55}$$

Note that one can adopt other distortion measures as well. For instance, the mean absolute value distortion measure can also be used instead of the mean squared distortion. However, the mean squared distortion measure is often preferred as it results in tractable solutions owing to the fact that squaring is a differentiable operator.

3.2.2.2 Uniform versus Non-uniform Quantization

We classify scalar quantizers as uniform and non-uniform. For a uniform quantizer, all the quantization intervals (except for the first and last) are of equal length, and the corresponding reconstruction levels are midpoints of the intervals, as depicted in Fig. 3.16.

In other words, a uniform quantizer is specified by the number of quantization levels N, the first boundary point a_1, and the quantization interval length Δ. We have

$$a_i = a_1 + (i-1)\Delta, \quad i = 2, 3, \ldots, N-1 \tag{3.56}$$

for the quantization interval boundaries, and

$$\hat{x}_i = a_1 + (i-1)\Delta - \frac{\Delta}{2}, \quad i = 1, 2, \ldots, N \tag{3.57}$$

for the reconstruction levels.

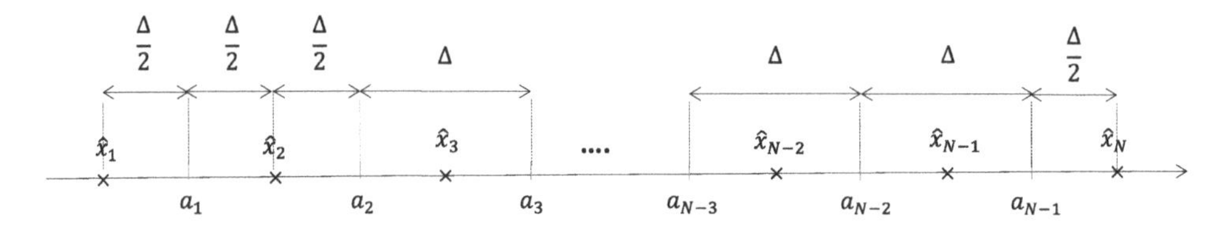

Figure 3.16 A uniform scalar quantizer.

On the other hand, for a non-uniform quantizer, the quantization interval lengths and the reconstruction levels are arbitrary. Hence, an N-level non-uniform quantizer is specified by $N-1$ boundary points, and N reconstruction levels, that is,

$$a_1, a_2, \ldots, a_{N-1}, \hat{x}_1, \hat{x}_2, \ldots, \hat{x}_N, \tag{3.58}$$

as illustrated in Fig. 3.15.

Example 3.2

Assume that a source whose samples X follow the PDF shown in Fig. 3.17 is being quantized using a six-level uniform quantizer on the interval $(-3, 3]$ as shown in the same figure. Determine the resulting MSE and SQNR.

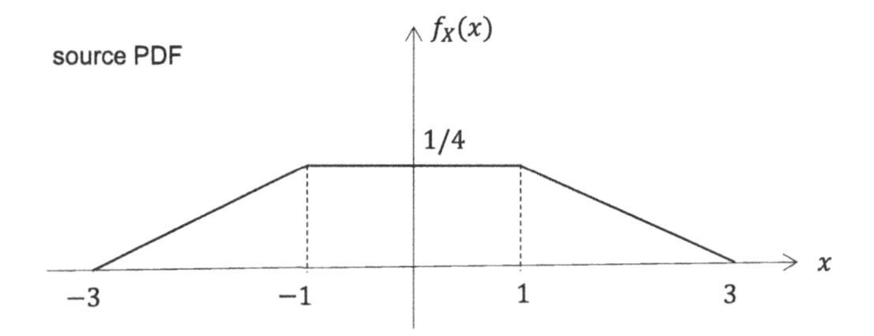

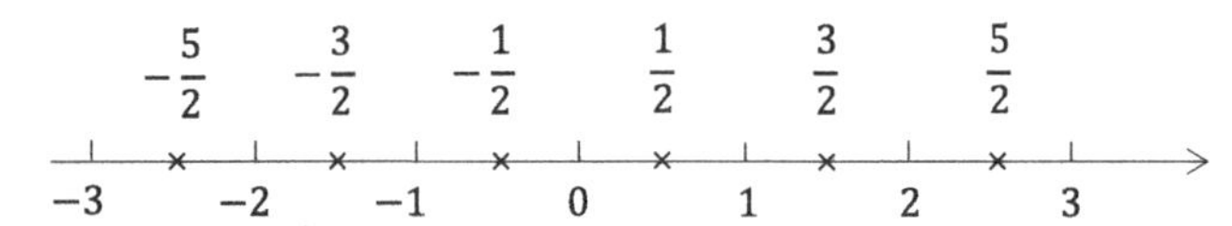

Figure 3.17 PDF of the source samples and the quantizer for the example.

Solution

We are given the source PDF as

$$f_X(x) = \begin{cases} \frac{1}{8}(x+3), & \text{if } -3 < x \leq -1, \\ \frac{1}{4}, & \text{if } -1 < x \leq 1, \\ -\frac{1}{8}(x-3), & \text{if } 1 < x \leq 3, \\ 0, & \text{else.} \end{cases} \tag{3.59}$$

Therefore, using the quantizer structure given, we can write the resulting MSE as

$$D = \sum_{i=1}^{6} \int_{a_{i-1}}^{a_i} (x - \hat{x}_i)^2 f_X(x) dx, \tag{3.60}$$

with $a_0 = -3, a_1 = -2, \ldots, a_6 = 3$, and $\hat{x}_1 = -5/2, \hat{x}_2 = -3/2, \ldots, \hat{x}_6 = 5/2$. Due to the complete symmetry of the source PDF and the quantizer structure, we obtain

$$D = 2\left(\int_0^1 \left(x - \frac{1}{2}\right)^2 \frac{1}{4} dx - \int_1^2 \left(x - \frac{3}{2}\right)^2 \frac{1}{8}(x-3)dx - \int_2^3 \left(x - \frac{5}{2}\right)^2 \frac{1}{8}(x-3)dx\right) \tag{3.61}$$

$$= 2\left(\frac{1}{48} + \frac{1}{64} + \frac{1}{192}\right) \tag{3.62}$$

$$= \frac{1}{12} \tag{3.63}$$

as the MSE.

We compute the source power as

$$\mathbb{E}[X^2] = \int_{-3}^{3} x^2 f_X(x) dx \tag{3.64}$$

$$= \frac{5}{3}. \tag{3.65}$$

The resulting SQNR is then

$$\text{SQNR} = \frac{\mathbb{E}[X^2]}{D} = 20, \tag{3.66}$$

or $10 \log_{10}(20) = 13.01$ dB.

In the above example, the quantization levels are the midpoints of the quantization intervals as a uniform quantizer is used. For the quantization regions $[-1, 0)$ and $[0, 1)$, since the source PDF is flat, the midpoints are the optimal reconstruction levels. However, for the remaining quantization regions, the selection of the midpoints is not optimal.

3.2.2.3 Scalar Quantizer Design

As our main objective is to obtain a good approximation of the source samples, we are interested in designing a quantizer that results in the lowest possible distortion. The problem definition is as follows: Given the probability density function of the source sample X, denoted by $f_X(x)$, and the number of quantization levels N, determine the optimal scalar quantizer that results in the lowest mean squared error distortion.

Uniform Quantization. Recall that an N-level uniform quantizer is specified by only two parameters, a_1 and Δ. Therefore, we can express the distortion using these two parameters as

$$D = \mathbb{E}[(X - Q(X))^2] \tag{3.67}$$

$$= \int_{-\infty}^{a_1} \left(x - \left(a_1 - \frac{\Delta}{2}\right)\right)^2 f_X(x)dx \tag{3.68}$$

$$+ \sum_{i=1}^{N-2} \int_{a_1+(i-1)\Delta}^{a_1+i\Delta} \left(x - \left(a_1 + i\Delta - \frac{\Delta}{2}\right)\right)^2 f_X(x)dx \tag{3.69}$$

$$+ \int_{a_1+(N-2)\Delta}^{\infty} \left(x - \left(a_1 + (N-1)\Delta - \frac{\Delta}{2}\right)\right)^2 f_X(x)dx. \tag{3.70}$$

The objective is to determine a_1 and Δ values that minimize D. A closed-form solution is not feasible, on the other hand, since there are only two parameters in the distortion expression, numerical techniques can be efficiently utilized to find the optimal uniform quantizer.

Note also that uniform quantizers are known to be near optimal at high rates if followed by entropy coding (suitable compression) methods.

Non-uniform Quantization. In this case, we can write the distortion expression as

$$D = \sum_{i=1}^{N} \int_{a_{i-1}}^{a_i} (x - \hat{x}_i)^2 f_X(x)dx, \tag{3.71}$$

with the understanding that $a_0 = -\infty$ and $a_N = \infty$.

Since the boundaries and the reconstruction levels are arbitrary, we have a total of $2N - 1$ unknowns to be optimized. For instance, if we want to design an 8-bit (256-level quantizer), we need to determine 511 parameters that minimize the distortion. This is very different from the uniform quantizer design scenario, and it cannot simply be done using brute-force search; therefore, we need a more systematic approach.

While it may not be obvious, we can make two observations: (1) it becomes tractable to find the optimal reconstruction levels if the quantization interval boundaries are fixed; and (2) it becomes tractable to determine the optimal boundaries if the reconstruction levels are fixed. Based on these observations, a Lloyd-Max quantizer is designed iteratively, as described below.

Lloyd-Max Quantizer Design Algorithm

(1) Start with an initial guess for the boundary points $a_1, a_2, \ldots, a_{N-1}$.
(2) For the current values of $a_1, a_2, \ldots, a_{N-1}$, determine the optimal values of the reconstruction levels $\hat{x}_1, \hat{x}_2, \ldots, \hat{x}_N$ to minimize the mean squared distortion D.
(3) For the current set of reconstruction levels $\hat{x}_1, \hat{x}_2, \ldots, \hat{x}_N$, determine the optimal boundary points $a_1, a_2, \ldots, a_{N-1}$.
(4) Compute the corresponding distortion for the present values of the boundary points and the reconstruction levels. If the change in distortion with respect to that in the previous iteration is below a prespecified (small) value, stop. Otherwise, go back to Step 2.

This is a very efficient algorithm that is guaranteed to converge. This is because the distortion at each step is non-increasing and is bounded from below (by 0). However, the solution is only guaranteed to be locally optimal in general.

We need to specify the details of the second and third steps for a complete description of the Lloyd-Max quantizer design algorithm.

In Step 2, the quantization intervals are fixed, since $a_1, a_2, \ldots, a_{N-1}$ are given, and we are tasked with finding the optimal reconstruction level for each. We can consider each quantization interval separately and determine the corresponding $\hat{x}_i$. More precisely, to determine the value of $\hat{x}_i$, we can differentiate the distortion expression with respect to $\hat{x}_i$ and set the result to 0. Noting that $\hat{x}_i$ appears in only one of the terms in the summation, we can write

$$\frac{\partial D}{\partial \hat{x}_i} = \frac{\partial}{\partial \hat{x}_i} \left\{ \sum_{j=1}^{N} \int_{a_{j-1}}^{a_j} (x - \hat{x}_j)^2 f_X(x) dx \right\} \tag{3.72}$$

$$= \frac{\partial}{\partial \hat{x}_i} \int_{a_{i-1}}^{a_i} (x - \hat{x}_i)^2 f_X(x) dx \tag{3.73}$$

$$= - \int_{a_{i-1}}^{a_i} 2(x - \hat{x}_i) f_X(x) dx. \tag{3.74}$$

Setting this to 0, and solving for $\hat{x}_i$, we determine the ith optimal reconstruction level as

$$\hat{x}_i = \frac{\int_{a_{i-1}}^{a_i} x f_X(x) dx}{\int_{a_{i-1}}^{a_i} f_X(x) dx}. \tag{3.75}$$

Defining the event that the quantizer input falls in the ith quantization interval as A, that is, $A = \{X \in (a_{i-1}, a_i]\}$, we can write the conditional probability density function of X given A as

$$f_{X|A}(x|A) = \frac{f_X(x)}{\int_{a_{i-1}}^{a_i} f_X(u) du}. \tag{3.76}$$

Therefore, we have

$$\hat{x}_i = \int_{a_{i-1}}^{a_i} x f_{X|A}(x|A) dx \tag{3.77}$$

$$= \mathbb{E}[X | X \in (a_{i-1}, a_i]]. \tag{3.78}$$

In other words, the optimal reconstruction level is the conditional mean of the random variable denoting the input to the quantizer given that it falls in the given quantization interval, that is, it is the centroid of the quantization region.

In Step 3, we are given the reconstruction levels and are asked to find the optimal quantization interval boundaries. Recalling that the cost function is simply the squared error, we need to map any given source sample to the nearest quantization level to minimize the cost. Therefore, it is easy to see that the boundaries should be selected as the midpoints of the consecutive reconstruction levels.

We can also arrive at the same conclusion by differentiating the distortion expression with respect to a_is and setting the derivatives to zero. That is, for the ith boundary point a_i, we have

$$\frac{\partial D}{\partial a_i} = \frac{\partial}{\partial a_i}\left\{\sum_{j=1}^{N}\int_{a_{j-1}}^{a_j}(x-\hat{x}_j)^2 f_X(x)dx\right\} \tag{3.79}$$

$$= \frac{\partial}{\partial a_i}\left\{\int_{a_{i-1}}^{a_i}(x-\hat{x}_i)^2 f_X(x)dx + \int_{a_i}^{a_{i+1}}(x-\hat{x}_{i+1})^2 f_X(x)dx\right\} \tag{3.80}$$

$$= f_X(a_i)((a_i-\hat{x}_i)^2-(a_i-\hat{x}_{i+1})^2). \tag{3.81}$$

The second step follows since only two terms of the summation involve the variable a_i, and the third step follows by using the Leibnitz rule for differentiation. Setting this result to 0, and simplifying, we obtain

$$a_i = \frac{1}{2}(\hat{x}_i+\hat{x}_{i+1}), \tag{3.82}$$

which confirms our intuition.

As we mentioned earlier, the Lloyd-Max algorithm is only guaranteed to be locally optimal. However, if the source is *log-concave*,[1] then there exists a unique locally optimal solution (satisfying the optimality conditions noted above). Therefore, under log-concavity, Lloyd-Max iterations converge to the globally optimal quantizer. We note that distributions such as Gaussian, exponential, Laplace, Rayleigh, and uniform, commonly encountered in communications applications, are log-concave.

Let us go through a specific example illustrating the optimization of the reconstruction levels for given quantization interval boundaries (as done in the second step of the Lloyd-Max quantizer).

Example 3.3

Consider Example 3.2. Assume that a six-level quantizer with the same boundary points is utilized. Determine

(a) The conditional PDF of X given that the source sample falls in each quantization interval.
(b) The conditional PDF of the quantization error given that the source sample falls in a specific interval.
(c) The optimal reconstruction levels to minimize the mean squared error.

[1] A function being log-concave means that its logarithm is a concave function.

Solution

(a) The conditional PDF of X given that $X \in (-3, -2]$ is

$$\frac{f_X(x)}{\mathbb{P}(X \in (-3, -2])} = 2(x+3) \tag{3.83}$$

if $x \in (-3, -2]$, and 0 otherwise. Similarly, the conditional PDF of X given that $X \in (-2, -1]$ is

$$\frac{f_X(x)}{\mathbb{P}(X \in (-2, -1])} = \frac{2}{3}(x+3) \tag{3.84}$$

if $x \in (-2, -1]$, and 0 otherwise. The others can be obtained in the same way.

(b) We have $\tilde{X} = X - Q(X) = X + \frac{5}{2}$. Therefore, the conditional PDF of $\tilde{X}$ given that $X \in (-3, -2]$ is

$$f_{\tilde{X}|X\in(-3,-2]}(x) = \begin{cases} \frac{f_X(x-5/2)}{\mathbb{P}(X\in(-3,-2])}, & \text{if } -\frac{1}{2} < x \le \frac{1}{2}, \\ 0, & \text{else,} \end{cases} \tag{3.85}$$

$$= \begin{cases} 2x+1, & \text{if } -\frac{1}{2} < x \le \frac{1}{2}, \\ 0, & \text{else.} \end{cases} \tag{3.86}$$

Similarly, the conditional PDFs of quantization errors, given that X falls in the other quantization intervals, can be computed as follows:

$$f_{\tilde{X}|X\in(-2,-1]}(x) = \begin{cases} \frac{2}{3}x+1, & \text{if } -\frac{1}{2} < x \le \frac{1}{2}, \\ 0, & \text{else;} \end{cases} \tag{3.87}$$

$$f_{\tilde{X}|X\in(-1,0]}(x) = \begin{cases} 1, & \text{if } -\frac{1}{2} < x \le \frac{1}{2}, \\ 0, & \text{else;} \end{cases} \tag{3.88}$$

$$f_{\tilde{X}|X\in(0,1]}(x) = \begin{cases} 1, & \text{if } -\frac{1}{2} < x \le \frac{1}{2}, \\ 0, & \text{else;} \end{cases} \tag{3.89}$$

$$f_{\tilde{X}|X\in(1,2]}(x) = \begin{cases} -\frac{2}{3}x+1, & \text{if } -\frac{1}{2} < x \le \frac{1}{2}, \\ 0, & \text{else;} \end{cases} \tag{3.90}$$

$$f_{\tilde{X}|X\in(2,3]}(x) = \begin{cases} -2x+1, & \text{if } -\frac{1}{2} < x \le \frac{1}{2}, \\ 0, & \text{else.} \end{cases} \tag{3.91}$$

Using the total probability theorem, we can determine the PDF of the overall quantization error. We can write

$$f_{\tilde{X}}(x) = \sum_{n=-3}^{2} f_{\tilde{X}|X\in(n,n+1]}(x)\mathbb{P}(X \in (n, n+1]), \tag{3.92}$$

which evaluates to unity for the interval $-\frac{1}{2} < x \le \frac{1}{2}$, and is zero otherwise. In other words, $\tilde{X}$ is a uniform random variable on the interval $(-1/2, 1/2]$. While it is not asked for, let us also calculate the resulting MSE:

$$D = E[\tilde{X}^2] = \int_{-1/2}^{1/2} x^2 dx = \frac{1}{12}, \tag{3.93}$$

which is, of course, the same value obtained in Example 3.2.

(c) We now determine the optimal reconstruction levels (corresponding to the given quantization boundaries). Using the conditional PDFs of X computed above, for the quantization interval $(-3, -2]$, we obtain

$$\hat{x}_1 = \int_{-3}^{-2} x2(x+3)dx = -\frac{7}{3}. \tag{3.94}$$

Similarly, for the second quantization interval, we find

$$\hat{x}_2 = \int_{-2}^{-1} x\frac{2}{3}(x+3)dx = -\frac{13}{9}. \tag{3.95}$$

Similarly, we can also obtain $\hat{x}_3 = -\frac{1}{2}$, and, by symmetry, we can readily write $\hat{x}_4 = \frac{1}{2}$, $\hat{x}_5 = \frac{13}{9}$, and $\hat{x}_6 = \frac{7}{3}$.

For further illustration, we plot the conditional PDF of the source sample given that it falls in a specific interval for two of the quantization regions, that is, for $(1, 2]$ and $(2, 3]$ in Fig. 3.18.

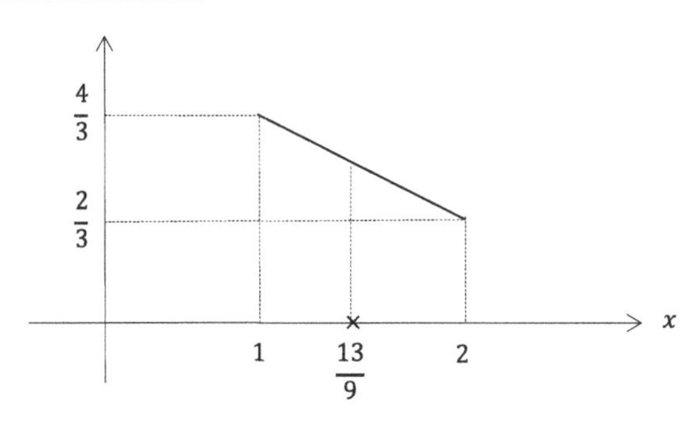

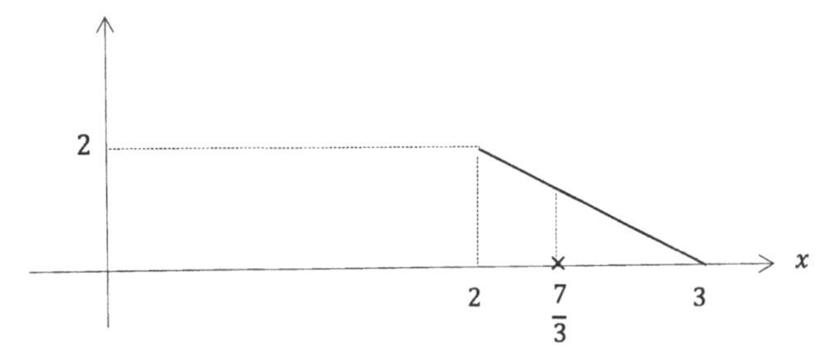

Figure 3.18 Conditional PDF of the source sample given that it falls in a specific interval for two cases and the corresponding optimal reconstruction levels.

One can determine the corresponding MSE with these reconstruction levels in a similar way as done in Example 3.2. We omit this calculation.

3.2.2.4 Quantization of a Gaussian Source

An important case is the quantization of a Gaussian source. Assuming that the source samples are standard Gaussian, that is, $X \sim \mathcal{N}(0,1)$, the optimal uniform and non-uniform quantizers, as well as the resulting mean squared errors, are given in Tables 3.2 and 3.3, respectively. The tables are constructed by following the design rules described earlier. Specifically, the Lloyd-Max quantizer design steps are employed for the non-uniform quantization case. Note that for the specific case of Gaussian sources, since the PDF of the source is log-concave (i.e., its logarithm is a concave function), there is a unique locally optimal quantizer. Therefore, the Lloyd-Max quantizer design algorithm will converge to this unique optimal solution. In other words, the quantizers given in Tables 3.2 and 3.3 are globally optimal.

We can obtain the optimal quantizers for a Gaussian source with any mean and variance using the optimal quantizers for the standard Gaussian source given in Tables 3.2 and 3.3. If $Y \sim \mathcal{N}(\mu, \sigma^2)$ is being quantized using N levels, all we need to

Table 3.2 Optimal uniform quantizer for a Gaussian source (with $\sim \mathcal{N}(0,1)$) for up to 10 levels.

N	Δ	D
2	1.596	0.3634
3	1.224	0.1902
4	0.9957	0.1188
5	0.843	0.0822
6	0.7334	0.06065
7	0.6508	0.04686
8	0.586	0.03744
9	0.5338	0.03069
10	0.4908	0.02568

Table 3.3 Optimal non-uniform quantizer for a Gaussian source (with $\sim \mathcal{N}(0,1)$) for up to 10 levels.

N	$\pm a_i$	$\pm \hat{x}_i$	D
2	0	0.798	0.3634
3	0.612	0, 1.224	0.1902
4	0, 0.9816	0.4528, 1.51	0.1175
5	0.3823, 1.244	0, 0.7646, 1.724	0.07994
6	0, 0.6589, 1.447	0.3177, 1, 1.894	0.05798
7	0.2803, 0.8744, 1.611	0, 0.561, 1.188, 2.033	0.044
8	0, 0.5006, 1.05, 1.748	0.245, 0.756, 1.344, 2.152	0.03454
9	0.2218, 0.6812, 1.198, 1.866	0, 0.4436, 0.9188, 1.476, 2.255	0.02785
10	0, 0.4047, 0.8339, 1.325, 1.968	0.1996, 0.6099, 1.058, 1.591, 2.345	0.02293

do is find the corresponding quantization regions and reconstruction levels for the N-level quantizer for the standard Gaussian source, and then scale them with the standard deviation σ and add the mean μ to obtain the new quantizer parameters. The corresponding mean squared error is easily obtained by scaling the given MSE values in the tables by the variance σ^2.

Example 3.4

Consider a Gaussian source with mean 4 and variance 25. Determine the optimal uniform and non-uniform quantizers with 10 levels.

Solution

Let us first consider uniform quantization. From Table 3.2, we read the optimal output spacing for a standard Gaussian source as 0.4908. Therefore, we obtain the optimal spacing for the source with variance 25 as $\Delta = \sqrt{25} \times 0.4908 = 2.454$. Considering that the mean is 4, we obtain the corresponding quantization region boundaries as

$$-5.816,\ -3.362,\ -0.908,\ 1.546,\ 4,\ 6.454,\ 8.908,\ 11.362,\ 13.816. \tag{3.96}$$

The reconstruction levels are the midpoints of the quantization intervals, that is, they are given by

$$-7.043,\ -4.589,\ -2.135,\ 0.319,\ 2.773,\ 5.227,\ 7.681,\ 10.135,\ 12.589,\ 15.043. \tag{3.97}$$

The corresponding mean squared error is obtained by multiplying the MSE given in the table by the source variance, that is, $25 \times 0.02568 = 0.642$. Hence, the resulting SQNR is $10 \log_{10}\left(\frac{25}{0.642}\right) \approx 15.9$ dB.

We can determine the optimal non-uniform quantizer using Table 3.3. By scaling the quantization interval boundaries by the standard deviation of 5 and adding the mean of 4, we obtain the optimal boundaries as

$$-5.84,\ -2.625,\ -0.1695,\ 1.9765,\ 4,\ 6.0235,\ 8.1695,\ 10.625,\ 13.84. \tag{3.98}$$

In a similar fashion, we obtain the optimal reconstruction levels as

$$-7.725,\ -3.955,\ -1.29,\ 0.9505,\ 3.002,\ 4.998,\ 7.0495,\ 9.29,\ 11.955,\ 15.725. \tag{3.99}$$

The corresponding MSE is $25 \times 0.02293 = 0.57325$, and the SQNR is $10 \log_{10}\left(\frac{25}{0.57325}\right) = 16.399$ dB.

We observe that the non-uniform quantizer performance is superior to the uniform one. The mean squared error is reduced from 0.642 to 0.57325 for the same number of quantization levels, and the SQNR is improved by about 0.5 dB.

3.2.2.5 Vector Quantization

In vector quantization, the idea is to consider multiple source outputs at a time and quantize them jointly. Vector quantization is advantageous compared to scalar

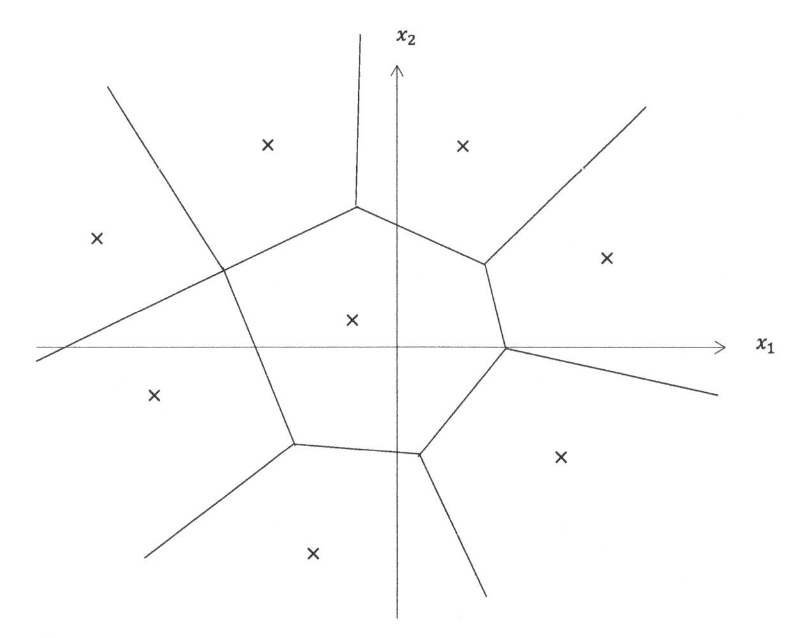

Figure 3.19 Illustration of a vector quantizer considering two source outputs simultaneously. The Voronoi cells and reconstruction levels for each cell are shown.

quantization as it can exploit further structure in the source samples. For instance, for typical source outputs (which are lowpass signals, changing slowly with time), the consecutive samples will take on similar values; hence, certain pairs of outputs become more likely than others. Such structures can be exploited by vector quantization.

As an illustration, Fig. 3.19 depicts a vector quantizer considering two samples simultaneously. The cells shown are the quantization regions, and the points marked with "×" are the corresponding quantization levels. Note that if we employ two separate scalar quantizers for each source output, the quantization regions corresponding to the pair of source outputs will be "rectangular". On the other hand, with vector quantization, the quantization regions are allowed to be more general; hence, the source can be represented with a smaller error using the same number of quantization levels.

Let us give a specific vector quantization example and see how it can provide improvements over scalar quantization.

Example 3.5

Consider a pair of random variables whose joint PDF is given as

$$f_{X,Y}(x,y) = \begin{cases} \frac{1}{2}, & \text{if } 0 < x, y \leq 1 \text{ or } -1 \leq x, y \leq 0, \\ 0, & \text{else.} \end{cases} \tag{3.100}$$

We can deduce that both X and Y take on values in the interval $[-1, 1]$, and they have the same sign. We want to quantize (X, Y) using a four-level quantizer. We define the average distortion as $D = \frac{1}{2}\left[(X - \hat{X})^2 + (Y - \hat{Y})^2\right]$, where $\hat{X}$ is the reconstruction value for X, and $\hat{Y}$ is the reconstruction value for Y.

(a) In this part, suppose that two scalar quantizers are used to quantize these random variables individually. Each quantizer uses two levels; hence, the overall number of quantization levels is four. Observe that the marginal PDFs of both X and Y are uniform on $[-1, 1]$, hence the optimal scalar quantizers for each random variable are described by the quantization intervals $[-1, 0]$ and $(0, 1]$, with the corresponding quantization levels $-\frac{1}{2}$ and $\frac{1}{2}$, respectively.

Determine the average distortion.

(b) As an alternative to the scalar quantizer in part (a), we use a vector quantizer to quantize the two random variables jointly. Specifically, the four quantization regions and the corresponding reconstruction levels for the pair are described as follows:

$$(\hat{x}, \hat{y}) = \begin{cases} \left(-\frac{2}{3}, -\frac{1}{3}\right), & \text{if } -1 \leq x \leq y \leq 0, \\ \left(-\frac{1}{3}, -\frac{2}{3}\right), & \text{if } -1 \leq y < x \leq 0, \\ \left(\frac{1}{3}, \frac{2}{3}\right), & \text{if } 0 < x \leq y \leq 1, \\ \left(\frac{2}{3}, \frac{1}{3}\right), & \text{if } 0 < y < x \leq 0. \end{cases} \tag{3.101}$$

Determine the average distortion and compare it with that of the previous part.

Solution

Figure 3.20 illustrates the support of the joint PDF of X and Y, and the quantization regions and reconstruction levels. For the vector quantization, the quantization regions are triangular. The effective reconstruction regions are the four squares separated by the horizontal and vertical axes for the separate scalar quantization case. Also, two of the four pairs of reconstruction levels are wasted with scalar quantization, since X and Y take on only values with the same sign.

First, one can easily verify that the marginal PDFs of X and Y are uniform on $[-1, 1]$. For instance, $f_X(x)$ can be computed as follows:

$$f_X(x) = \int_{-\infty}^{\infty} f_{X,Y}(x, y)dy \tag{3.102}$$

$$= \begin{cases} \int_{-1}^{0} \frac{1}{2}dy, & \text{if } -1 \leq x \leq 0 \\ \int_{0}^{1} \frac{1}{2}dy, & \text{if } 0 < x \leq 1 \end{cases} \tag{3.103}$$

$$= \frac{1}{2}, \tag{3.104}$$

for $x \in [-1, 1]$, and 0 otherwise.

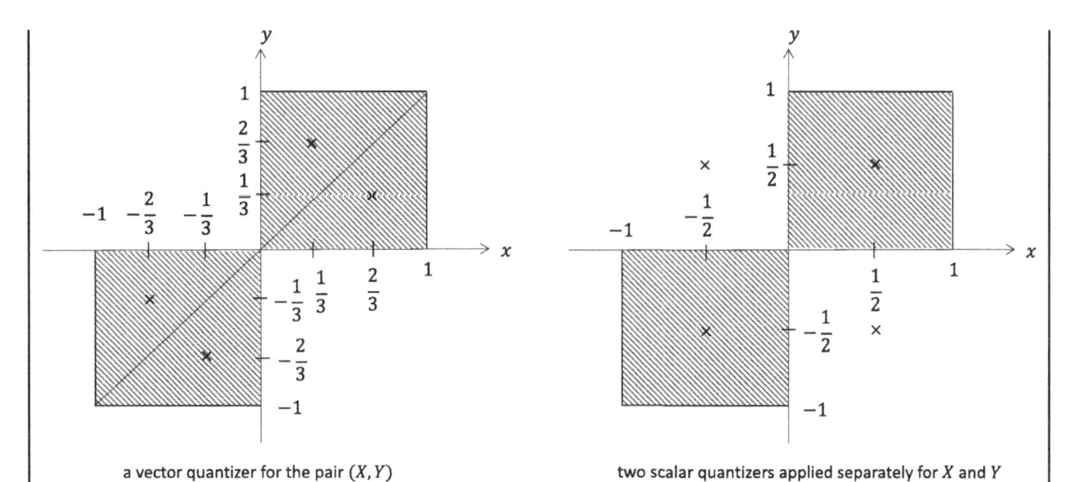

Figure 3.20 The support of the joint PDF, the quantization regions, and reconstruction points (marked with "×"). LHS: the vector quantizer with triangular quantization regions. RHS: two scalar quantizers are used separately for X and Y. The effective quantization regions are separated by the horizontal and vertical axes.

(a) Let us compute the mean squared distortion for X:

$$\mathbb{E}[(X - \hat{X})^2] = \mathbb{E}[(X - \hat{X})^2 | X > 0] \tag{3.105}$$

$$= \int_0^1 \left(x - \frac{1}{2}\right)^2 dx \tag{3.106}$$

$$= \frac{1}{12}, \tag{3.107}$$

where the first line is due to symmetry, and the second line follows since the conditional PDF of X given that it is greater than 0 is uniform on $(0, 1]$.

Since we have complete symmetry, we can also write $\mathbb{E}[(Y - \hat{Y})^2] = 1/12$. Therefore, the overall distortion becomes

$$D = \frac{1}{2}\left[(X - \hat{X})^2 + (Y - \hat{Y})^2\right] = \frac{1}{12}. \tag{3.108}$$

For the two-level quantization of each random variable separately, selecting the quantization region boundary as 0 and the reconstruction levels at $\pm\frac{1}{2}$ is optimal. Therefore, this is the smallest distortion two separate scalar quantizers can provide for this pair of random variables with a total of four quantization levels (for the pair (X, Y)).

(b) For the vector quantization case, we can compute the average distortion as follows:

$$D = \frac{1}{2}\left[(X - \hat{X})^2 + (Y - \hat{Y})^2\right] \tag{3.109}$$

$$= \frac{1}{2}\left[(X - \hat{X})^2 + (Y - \hat{Y})^2 \middle| 0 < Y \leq X \leq 1\right] \tag{3.110}$$

$$= \frac{1}{2} \int_0^1 \int_0^x 2\left(\left(x - \frac{2}{3}\right)^2 + \left(y - \frac{1}{3}\right)^2\right) dydx \tag{3.111}$$

$$= \int_0^1 x\left(x - \frac{2}{3}\right)^2 + \frac{1}{3}\left(x - \frac{1}{3}\right)^3 + \frac{1}{81} dx \tag{3.112}$$

$$= \frac{1}{18}, \tag{3.113}$$

where the first line is the definition of distortion, the second line follows from the complete symmetry of the employed quantizer, and the third line is obtained since the conditional PDF of (X, Y) given that the pair is in the triangular region is uniform.

Therefore, the average distortion obtained by the vector quantizer with four quantization levels is smaller than that of the one computed in part (a) by a factor of 3/2.

We close our short discussion on vector quantization by noting that the iterative Lloyd-Max quantization algorithm can be adopted for this case to design (locally) optimal quantizers as well. Similar to the scalar quantizer design, the algorithm will start with an initial guess on the boundaries of the quantization regions and determine the corresponding optimal reconstruction points by computing the conditional expectation of the set of random variables being quantized, given that they fall in a specific quantization region. Then, the corresponding quantization region boundaries with the given reconstruction points will be determined. These will be in the form of *Voronoi cells*, that is, the set of all points closest to each reconstruction point (compared to the others). The algorithm will iterate between these two steps until convergence.

A more detailed coverage of vector quantization is beyond our scope.

3.3 Pulse-Code Modulation

In this section, we describe a *waveform coding* scheme that combines the ideas of sampling and quantization to come up with a digital representation of an analog source, namely, *pulse-code modulation* (PCM). We will identify several different versions: uniform PCM, non-uniform PCM, differential PCM, and related representations: delta modulation and adaptive delta modulation. A practical application of pulse-code modulation (specifically, non-uniform PCM) is in digital telephony.

3.3.1 Uniform PCM

The block diagram of a uniform PCM scheme is depicted in Fig. 3.21. A continuous-time signal $x(t)$ with a bandwidth of W Hz is sampled at a rate f_s

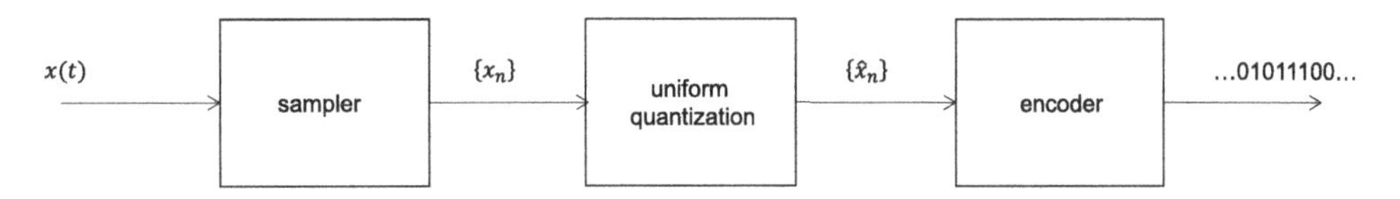

Figure 3.21 Uniform PCM block diagram.

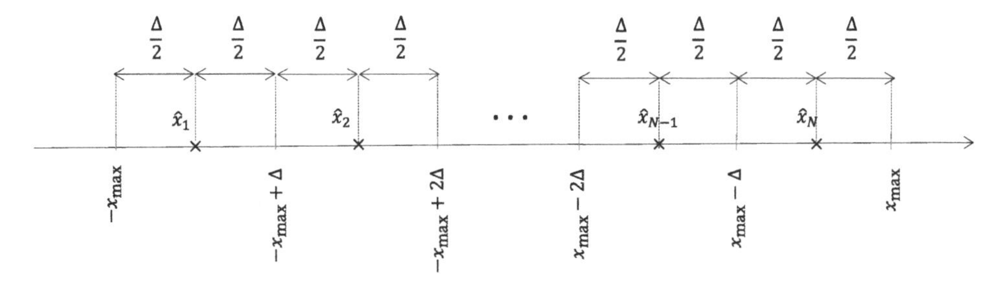

Figure 3.22 Uniform quantizer for the input range $[-x_{\max}, x_{\max}]$.

samples/sec. The samples are passed through a uniform quantizer with N levels, and the quantized values are encoded as a binary sequence. The resulting binary sequence is then transmitted to the destination using digital communication techniques.

To avoid aliasing, the sampling frequency has to be larger than or equal to the Nyquist rate, that is, $f_s \geq 2W$. As long as this condition is satisfied, there is no loss of information because the samples $\{x_n\}$ are sufficient for a perfect reconstruction of the original analog signal. On the other hand, the quantization process is lossy and results in the degradation of signal quality. The number of quantization levels N is selected as a power of 2, that is, $N = 2^\nu$, where ν is a positive integer. Hence, the encoder maps each quantized source sample to ν bits.

Different types of encoders can be used. For instance, natural binary coding maps the quantization levels to ν-bit sequences in lexicographical order. As another example, Gray coding maps adjacent levels to bit sequences that differ in only one position.

PCM is typically used to encode speech signals. For such signals, the analog source samples have a symmetric dynamic range around 0. Therefore, we can think of the quantizer input as being on the interval $[-x_{\max}, x_{\max}]$, where $x_{\max}$ is the largest possible magnitude of the analog source sample. Accordingly, the quantizer divides this input range into $N = 2^\nu$ intervals of equal length Δ. That is,

$$\Delta = \frac{2x_{\max}}{N} = \frac{x_{\max}}{2^{\nu-1}}. \tag{3.114}$$

The reconstruction levels are selected as the midpoints of the quantization regions.

Figure 3.22 illustrates the quantizer structure.

We want to address two main questions: (1) How much channel bandwidth is required for transmission? (2) What is the resulting signal-to-quantization noise ratio?

Bandwidth Requirements. As we are taking f_s samples per second, and representing each sample using v bits, the overall data rate is $f_s v$ bits/sec. We are jumping ahead at this point; however, as we will see in the subsequent chapters, to transmit at a rate of R bits/sec with binary modulation, a minimum of $R/2$ Hz of channel bandwidth is needed. Therefore, the PCM scheme needs a minimum bandwidth of $\frac{vf_s}{2}$ Hz. If we select the sampling rate as the Nyquist rate, that is, with $f_s = 2W$, the minimum bandwidth needed for transmission will be vW Hz.

As intuitively expected, the bandwidth required in transmission increases linearly with the signal bandwidth and the number of bits used in the quantization process.

Signal Quality Analysis. We are interested in computing the SQNR of the resulting representation as a measure of signal quality. Assuming that there are no transmission errors, this will be a suitable assessment of the reproduction quality of the analog signal at the destination.

To accomplish the above goal, we model the source samples as realizations of a random variable denoted by X, taking on values in the interval $[-x_{\max}, x_{\max}]$. Let us denote the probability density function of the source as $f_X(x)$. Define the quantization error as $\tilde{X} = X - Q(X)$, where $Q(\cdot)$ is the quantization function.

Our first observation is that if the source PDF is smooth (as in typical speech signals), for large N, the quantization error $\tilde{X}$ is approximately a uniform random variable on the interval $[-\Delta/2, \Delta/2]$, namely, its PDF is given by $f_{\tilde{X}}(x) = 1/\Delta$ if $-\Delta/2 \leq x \leq \Delta/2$, and 0 otherwise.

To explain the above approximation further, let us consider the (relatively smooth) source PDF illustrated in Fig. 3.23. The figure also shows the conditional PDF of the source sample X and the quantization error $\tilde{X}$, given that the sample falls within the ith quantization interval. It should be evident from the illustration that as long as the source PDF is sufficiently smooth and the quantization intervals are short (i.e., the number of quantization levels is large), the conditional PDF of the quantization error is approximately uniform on the interval $[-\Delta/2, \Delta/2]$. This is true regardless of the quantization interval under consideration, hence averaging the conditional PDFs of $\tilde{X}$ over all possibilities; we argue that the quantization error is almost uniform on $[-\Delta/2, \Delta/2]$.

With the above approximation, the average distortion is given by

$$\mathbb{E}[\tilde{X}^2] = \int_{-\Delta/2}^{\Delta/2} \frac{1}{\Delta} x^2 dx = \frac{x_{\max}^2}{3 \cdot 4^v}. \tag{3.115}$$

Therefore, we can write the resulting SQNR as

$$\text{SQNR} = \frac{P_X}{\mathbb{E}[\tilde{X}^2]} = \frac{3 \cdot 4^v \cdot P_X}{x_{\max}^2}, \tag{3.116}$$

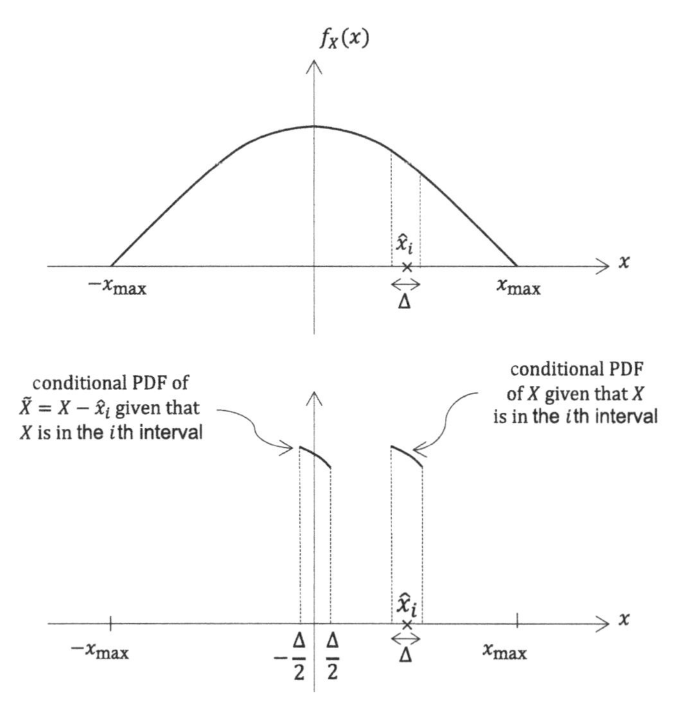

Figure 3.23 Illustration of the PDF of the quantization error given that the source output is in the ith quantization interval. We observe that the PDF of the quantization error can be approximated as uniform on the interval $(-\frac{\Delta}{2}, \frac{\Delta}{2}]$.

where P_X denotes the average signal power, that is,

$$P_X = \mathbb{E}[X^2] = \int_{-x_{\max}}^{x_{\max}} x^2 f_X(x)dx. \tag{3.117}$$

It is customary to express the SQNR in decibel form, namely,

$$\text{SQNR}_{\text{in dB}} = 10\log_{10}\left(\frac{P_X}{x_{\max}^2}\right) + 6.02v + 4.8. \tag{3.118}$$

Note that the ratio $\frac{P_X}{x_{\max}^2}$ is nothing but the average power of the normalized source output $\frac{X}{x_{\max}}$.

This result illustrates that each additional bit in quantization (i.e., doubling the number of quantization levels) increases the SQNR by about 6 dB.

For typical speech sources, the normalized signal power is about 0.1. Thus, for speech representation, with $v = 8$-bit quantization, the resulting SQNR will approximately be 42.8 dB. If $v = 10$ is employed instead, the SQNR will improve to about 54.8 dB.

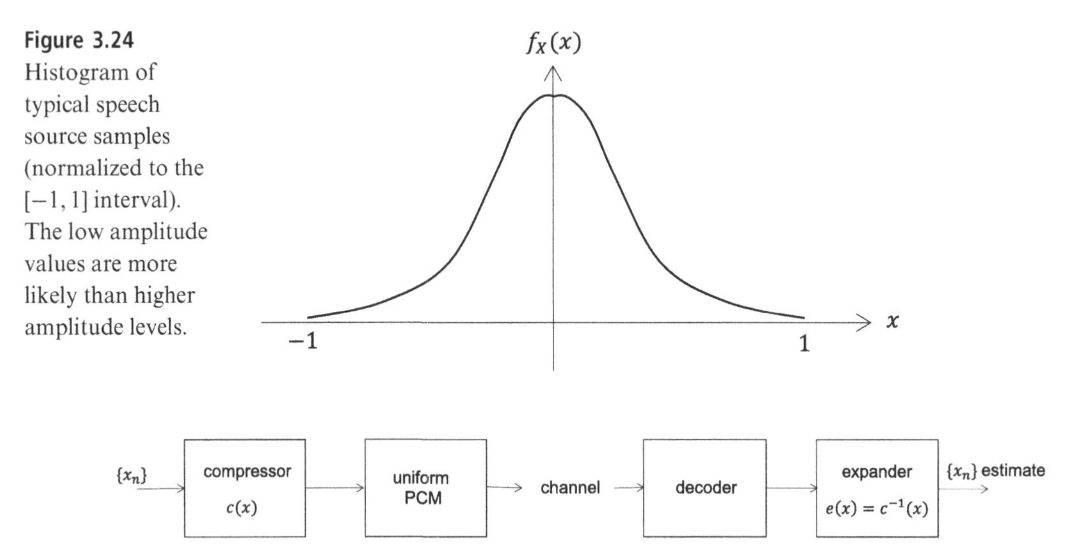

Figure 3.24 Histogram of typical speech source samples (normalized to the $[-1, 1]$ interval). The low amplitude values are more likely than higher amplitude levels.

Figure 3.25 Block diagram of a non-uniform PCM system.

3.3.2 Non-uniform PCM

PCM is conventionally employed in digital telephony to transmit speech signals. Hence, examining these signals to understand their statistics in detail is essential. Figure 3.24 depicts the normalized histogram of a typical speech signal's samples. We can consider this histogram as an approximation of the PDF of source samples.

We deduce from the sample PDF of speech signals that the low amplitude levels are much more likely than the large amplitude levels. With a uniform PCM scheme, we employ a uniform quantizer and keep the lengths of the quantization intervals identical for the entire signal range. Since the low signal amplitude values are more likely, to improve the average SQNR of the system, we can pick smaller quantization intervals for small amplitude levels and employ a coarser quantizer for the higher amplitude values (keeping the total number of quantization levels the same). In other words, we can use a non-uniform quantizer to improve the overall performance. With a non-uniform quantizer, the reconstruction errors for the more frequent samples with low amplitudes are reduced, while they are increased for the seldom occurring samples with high amplitude levels, resulting in a reduced average mean squared error, hence increasing the SQNR.

A practical way to implement non-uniform PCM is through the use of a memoryless non-linear device called a *compressor* at the transmitter side (denoted by $c(x)$), and its inverse called an *expander* at the receiver side (denoted by $e(x) = c^{-1}(x)$). Using a compressor and an expander is referred to as a *compander* for short. The corresponding non-uniform PCM block diagram is depicted in Fig. 3.25.

As the low amplitude values are more likely for speech signals, compressors and expanders matched to this behavior are utilized in practice. Specifically, in the United States and Canada, a μ-law compander (with $\mu = 255$) is employed, while in Europe, an A-law compander is utilized for digital telephony. The compressors for these two systems are given as

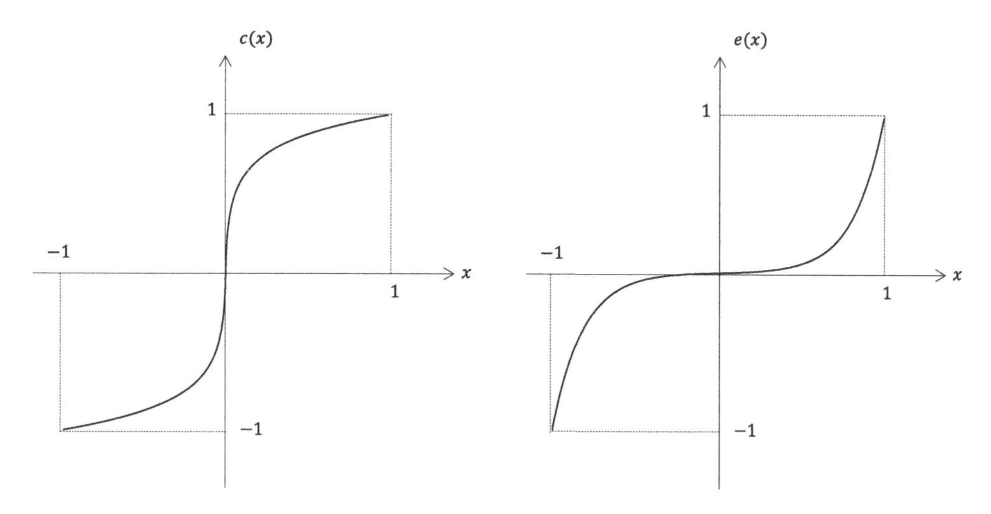

Figure 3.26 Compressor and expander for a μ-law compander with $\mu = 255$.

$$c_{\mu\text{-law}}(x) = \operatorname{sgn}(x)\frac{\ln(1+\mu|x|)}{\ln(1+\mu)}, \qquad \text{for } |x| \le 1 \tag{3.119}$$

and

$$c_{A\text{-law}}(x) = \begin{cases} \frac{Ax}{1+\ln A}, & |x| < 1/A, \\ \operatorname{sgn}(x)\frac{1+\ln(A|x|)}{1+\ln A}, & 1/A \le |x| \le 1. \end{cases} \tag{3.120}$$

The μ-law compressor for $\mu = 255$ and the corresponding expander are depicted in Fig. 3.26. It should be evident that, with this non-linear mapping, the number of quantization levels used to quantize the lower amplitude levels is higher than the number for the high amplitude levels. For instance, for a 10-bit (i.e., $N = 1024$-level quantizer), 730 quantization levels are used for quantizing the source values in the range $[-0.2, 0.2]$, while the remaining 294 levels are used to quantize the source samples whose magnitudes are in the range $[0.2, 1]$.

Example 3.6

Consider a source with PDF as shown in Fig. 3.27. Determine the resulting SQNR for uniform PCM with $v = 10$ bits.

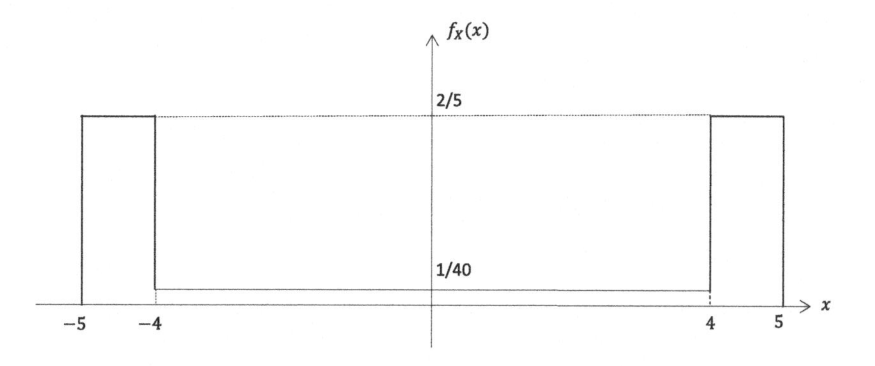

Figure 3.27 PDF of the source for the example.

Solution

We immediately note that the maximum value of (the magnitude of) the source samples is $x_{\max} = 5$ units. Hence, we only need to compute the average power of the source (i.e., $\mathbb{E}[X^2]$); we can determine the signal-to-quantization noise ratio for the uniform PCM case.

We have

$$\mathbb{E}[X^2] = 2\left(\int_0^4 x^2 \frac{1}{40} dx + \int_4^5 x^2 \frac{2}{5} dx\right) \tag{3.121}$$

$$= \frac{1}{60} x^3 \Big|_0^4 + \frac{4}{15} x^3 \Big|_4^5 \tag{3.122}$$

$$= 17.33 \text{ units.} \tag{3.123}$$

Hence, the resulting SQNR with uniform PCM with $v = 10$-bits is given by

$$\text{SQNR} = 10\log_{10}\left(\frac{\mathbb{E}[X^2]}{x_{\max}^2}\right) + 6.02v + 4.8 \tag{3.124}$$

$$= 63.4 \text{ dB.} \tag{3.125}$$

Example 3.7

Consider the same source given in the previous example. Assume that a non-uniform PCM with the compressor shown in Fig. 3.28 is employed instead of a uniform PCM. Observe that this compressor allows for a finer quantization for the more important samples (those with magnitudes between 4 and 5). Note also that while this could be a suitable selection, it is not necessarily optimal.

Determine the resulting SQNR with $v = 10$-bit quantization.

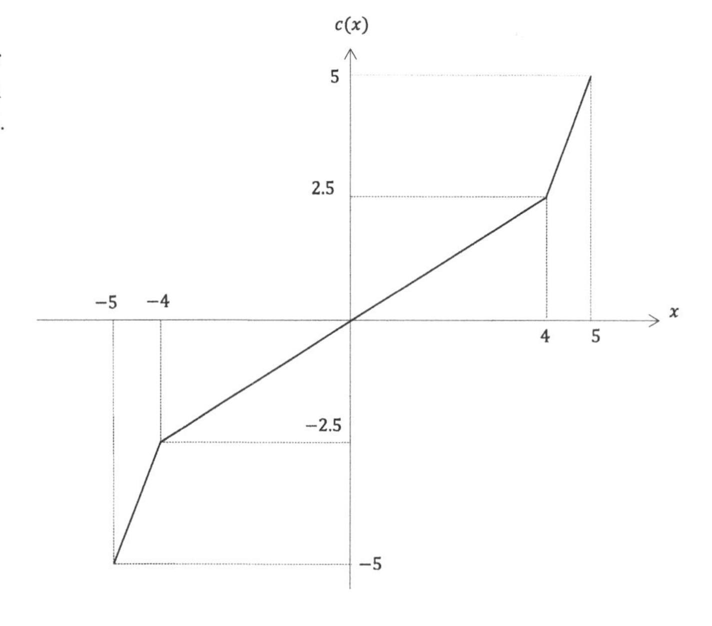

Figure 3.28 Compressor employed in the example.

Solution

With the given compressor, the effective number of quantization levels that are dedicated to the samples with $|X| < 4$ is $\approx \frac{5}{10}2^{\nu}$, that is, 512 levels, and the remaining 512 levels are dedicated for input samples with $|X| \in [4,5]$. Hence, the quantization error for $|X| < 4$ is uniform on $[-\Delta_1/2, \Delta_1/2]$ with $\Delta_1 = \frac{8}{512} = 0.0156$, and that for $|X| \in [4,5]$ is uniform on $[-\Delta_2/2, \Delta_2/2]$ with $\Delta_2 = \frac{2}{512} = 0.0039$. We can determine the average squared error as

$$D = \mathbb{P}(|X| < 4)\mathbb{E}\left[(X - \hat{X})^2 \middle| |X| < 4\right] + \mathbb{P}(4 \le |X| \le 5)\mathbb{E}\left[(X - \hat{X})^2 \middle| 4 \le |X| \le 5\right] \tag{3.126}$$

$$= \frac{1}{5}\frac{\Delta_1^2}{12} + \frac{4}{5}\frac{\Delta_2^2}{12} \tag{3.127}$$

$$\approx 5.10^{-6}. \tag{3.128}$$

The resulting signal-to-quantization ratio is then

$$\text{SQNR} = 10\log_{10}\frac{\mathbb{E}[X^2]}{D} = 10\log_{10}\frac{17.33}{5 \times 10^{-6}} = 65.4\text{ dB}. \tag{3.129}$$

There is an improvement with respect to the uniform PCM case by about 2 dB.

3.3.3 Differential PCM (DPCM)

Observing that when quantizing a typical (speech) source signal, the consecutive samples have values near each other, we can obtain a more efficient scheme compared to PCM by quantizing their differences instead of the sample values directly. Since the differences will have a small dynamic range, we can effectively obtain the same mean squared error performance with a significantly lower number of quantization levels. Or, with the same number of quantization levels, we can use shorter quantization intervals, hence obtaining smaller errors in the reconstruction of the source samples.

The resulting scheme, which quantizes the differences between the consecutive source samples, is referred to as differential PCM (DPCM). A simple implementation of DPCM is shown in Fig. 3.29.

The direct implementation of DPCM suffers from a significant drawback as the quantization errors from consecutive samples accumulate. Hence, it is not

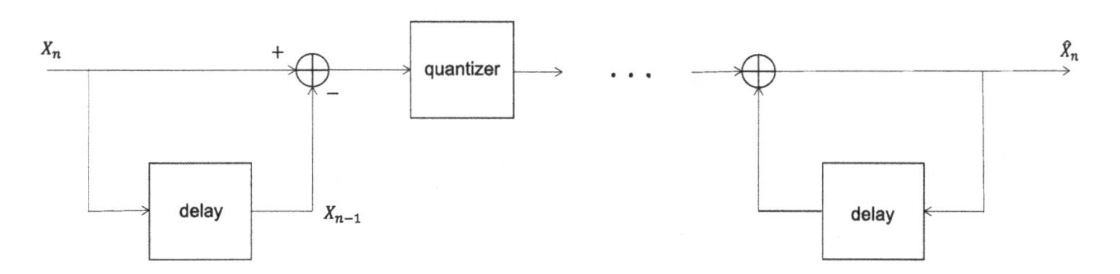

Figure 3.29 Differential PCM (naive implementation).

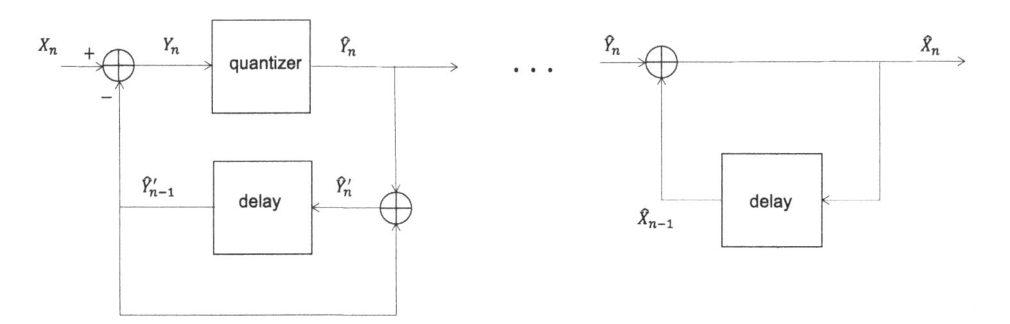

Figure 3.30 Alternate implementation of DPCM with no error propagation.

preferable. To remedy the error accumulation problem, we can utilize an alternate implementation depicted in Fig. 3.30.

Let us explain why the second implementation does not suffer from error propagation. We notice that the encoder side has all the information the decoder has. Therefore, the encoder can accurately track the estimates on the decoder side. Indeed, with the DPCM implementation in Fig 3.30, the encoder produces the estimates at the receiver side as it can be shown that $\hat{X}_n = \hat{Y}'_n$, and it quantizes the difference between the present sample and the estimate at the decoder side for the previous sample. To see this more explicitly, we write

$$\hat{Y}'_n = \hat{Y}_n + \hat{Y}'_{n-1} \tag{3.130}$$

as a simple node equation at the encoder side, and

$$\hat{X}_n = \hat{Y}_n + \hat{X}_{n-1} \tag{3.131}$$

at the decoder side. We notice that these are identical recursions, and with the same initial conditions (e.g., with $\hat{X}_0 = \hat{Y}'_0$), they become identical for all n. We can also write

$$Y_n = X_n - \hat{Y}'_{n-1}. \tag{3.132}$$

By adding this side by side with (3.130), we obtain

$$Y_n + \hat{Y}'_n = X_n + \hat{Y}_n, \tag{3.133}$$

and using $\hat{X}_n = \hat{Y}'_n$, we obtain

$$X_n - \hat{X}_n = Y_n - \hat{Y}_n. \tag{3.134}$$

In other words, the error at the decoder side (which is the left-hand-side term) is simply the error obtained by a single use of the quantizer (with input Y_n and output $\hat{Y}_n$), and hence there is no error accumulation.

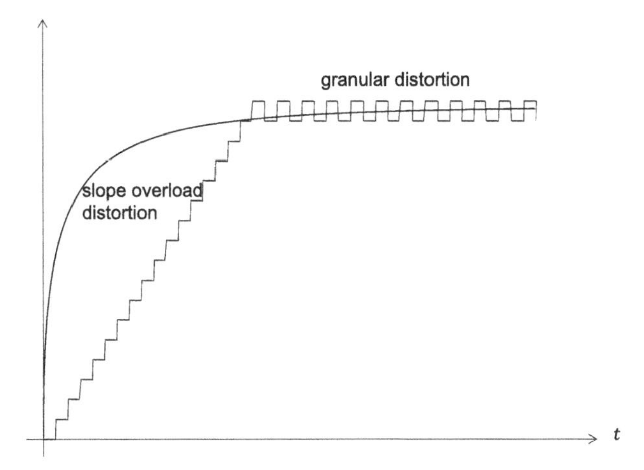

Figure 3.31 An example signal and its delta-modulated version. We observe that there are slope overload and granular distortions depending on the behavior of the signal.

3.3.4 Delta Modulation

A differential PCM scheme with two-level quantization (i.e., 1-bit quantization) is referred to as *delta modulation*. That is, for each new sample, we either increase the signal level by a fixed amount or reduce it by the same amount in an effort to approximate the analog signal. An illustrative example is given in Fig. 3.31.

With delta modulation, we only need 1 bit per source sample; hence, the bandwidth required is very low. However, the signal quality suffers. Specifically, there are two sources of noise: (1) granular noise, namely, the fluctuations in the signal level when the original signal is almost constant; and (2) slope overload distortion, which is because it may take a long time to follow the original signal when its slope is too high.

To remedy the problems with signal quality, an improved version of delta modulation called *adaptive delta modulation* can be used. The idea is to change the step size in a controlled manner in such a way that both the granular distortion and the slope overload distortion are reduced. For instance, if there are two consecutive increases or decreases in the sample values, which signals a higher rate of change in the analog signal, the step size can be increased, while it can be reduced otherwise (which signals a relatively flat behavior). Note that the adaptive delta modulation requires only 1 bit per source sample, the same as the delta modulation, as the decision on the increase or decrease of the step size is based on the reconstructed signal only, which is also available on the decoder side.

An illustration of adaptive delta modulation is depicted in Fig. 3.32. It is clear from the figure that when the signal changes rapidly, the step size increases rapidly. Hence, the signal can be tracked more closely in a shorter time, resulting in a smaller slope-overload distortion. On the other hand, when the signal level is

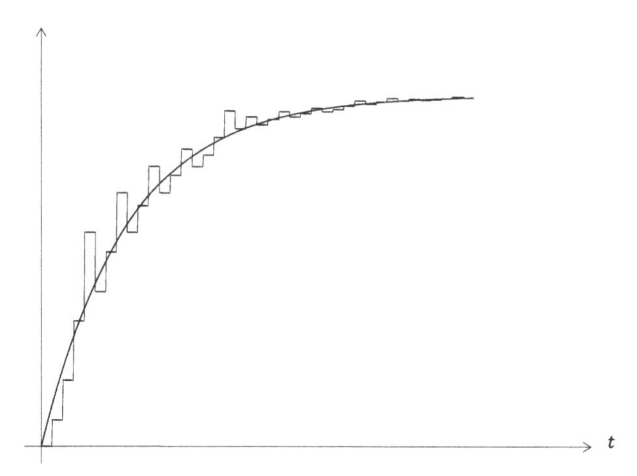

Figure 3.32 The same signal in Fig. 3.31 is being represented by adaptive delta modulation. We observe significant reductions in both slope overload distortion and granular distortion.

relatively flat, the step size becomes smaller and smaller, reducing the fluctuations in the reconstructed signal, and, hence, the granular distortion is alleviated.

3.4 Digital versus Analog Transmission

We close this chapter by highlighting the differences and similarities between analog and digital transmission techniques. With analog transmission, the message signal is a continuous-time waveform (typically, a lowpass signal), which is then embedded into a carrier signal either in the amplitude or in the phase (or frequency) to produce the modulated waveform to be transmitted over the channel. The receiver estimates the message signal as closely as possible by demodulating the attenuated and noisy received signal observed through the channel. On the other hand, in digital transmission, the message signal is a sequence of bits (or, more generally, symbols). Also, note that the message signals in digital communications may be natively digital, as in text messages or computer data, or can be a representation of an analog signal obtained through sampling and quantization.

Being a sequence of symbols, digital messages are not directly in a form that can be transmitted over a communication channel. We need to represent them using signals suitable for the specific communication medium. We need to convert the sequence of symbols into a waveform representation and transmit the resulting waveform. The waveform could be a sound wave for underwater acoustic communications; an optical signal for communication over the air, underwater, or over fiber-optic cables; or a radio wave for transmission over wireless channels. In any case, the underlying digital message needs to be converted to a suitable form for transmission over the channel (which is nothing but the process of digital modulation).

In short, as in analog transmission, in digital communications, the signal that needs to be transmitted over a communication medium is a continuous-time waveform. Namely, the message bits (or symbols) are converted to a continuous-time signal representing the digital information sequence; the channel corrupts the transmitted waveform by noise and, possibly, by other impairments. Observing the corrupted continuous-time signal, the receiver tries to recover the original message sequence. All these details will be covered in the following chapters, and the different components of a digital communication system will be studied in depth.

3.5 Chapter Summary

The first topic we covered in this chapter was a brief review of analog transmission, including amplitude and angle modulation techniques. We only described the fundamental principles, while omitting many details, as the primary goal was to distinguish analog and digital modulation ideas. We then addressed how sampling and quantization can convert an analog signal into a digital form. We determined that as long as the signal is bandlimited, sampling at the Nyquist rate or higher will allow for perfect reconstruction. On the other hand, the quantization process is lossy. We also covered the problem of quantizer design and determined the quality of the quantized signal in terms of the statistics of the source and the quantizer parameters. Finally, we have studied waveform coding techniques, which allow for a digital representation of an analog signal, including pulse-code modulation, differential pulse-code modulation, and delta modulation.

From this point on, with the understanding that a source is either initially digital or can be converted into a digital form by sampling and quantization, we will only consider digital information sources and their transmission via digital modulation techniques.

PROBLEMS

3.1 Consider a message signal given by $m(t) = \cos(2000\pi t) + \sin(4000\pi t)$. Assume that conventional AM is employed to transmit it, where the modulated waveform is given by $x(t) = 5(1 + \frac{1}{3}m(t))\sin(2\pi 200kt)$.
   (a) Find the Fourier transform of $x(t)$.
   (b) Determine the lower sideband of $x(t)$ in the time domain.
   (c) Calculate the average power of the lower sideband of $x(t)$.
   (d) What is the modulation index?

3.2 Assume that conventional AM is used to transmit the message signal $m(t) = \frac{1}{2}\cos(4000\pi t) + \frac{1}{2}\sin(2000\pi t)$, where the modulated signal is given by $x(t) = 10(1 + m(t))\cos(400k\pi t)$.
   (a) Plot the Fourier transforms of $m(t)$ and $x(t)$.
   (b) Determine the ratio of the average power in the sidebands to the overall average power of $x(t)$.

(c) Determine the modulation index. Is the signal overmodulated?

3.3 Let a message signal be given by $m(t) = \sin(2\pi f_m t)$.

(a) Plot the spectrum of the message signal.

(b) Plot the spectrum of the modulated signal if DSB-SC AM with $x(t) = m(t)\sin(2\pi f_c t)$ is employed. Assume that $f_c \gg f_m$.

(c) Plot the spectrum of the modulated signal if conventional AM with $x(t) = (1 + 0.5m(t))\cos(2\pi f_c t)$ is employed. Assume that $f_c \gg f_m$.

3.4 A conventional AM-modulated signal is given by

$$x(t) = 4\sin(2\pi 100kt) + \sin(2\pi 99kt) + \sin(2\pi 101kt) + \cos(2\pi 99kt) - \cos(2\pi 101kt).$$

The carrier frequency is $f_c = 100$ kHz, and the amplitude sensitivity constant is $k_a = 1$.

(a) Determine the power content in the sidebands, and the ratio of the sideband power to the total power of the modulated signal.

(b) Determine the message signal and the modulation index.

3.5 The message signal $m(t) = \frac{1}{2}(\sin(40\pi t) + \cos(60\pi t))$ is being transmitted using conventional AM. The modulated signal is $x(t) = 4(1 + m(t))\sin(2\pi f_c t)$.

(a) What is the modulation index? Is the signal overmodulated?

(b) Plot the Fourier transform of $x(t)$.

(c) What is the ratio of the power in the sidebands of $x(t)$ to its total power?

3.6 A message signal given by $m(t) = \sin(2\pi 10kt) + 2\cos(2\pi 20kt)$ is modulated using DSB-SC AM with a carrier signal $c(t) = 10\cos(2\pi 1000kt + \pi/3)$. The resulting modulated signal is denoted by $x(t)$.

(a) Find and plot the Fourier transform of the modulated signal.

(b) Determine the lower and upper sidebands of the modulated signal in the time domain.

(c) The signal $x(t)$ is demodulated using a carrier signal with an incorrect phase (i.e., $\cos(2\pi 1000kt + \pi/6)$). Determine the ratio of the resulting output signal power to that obtained using the correct carrier phase.

3.7 The message signal $m(t) = \sin(20\pi t) + \cos(40\pi t)$ is transmitted using DSB-SC AM where the modulated signal is given by $x(t) = m(t)\sin(2000\pi t)$.

(a) Plot the Fourier transform of the message signal.

(b) Plot the Fourier transform of the modulated signal and indicate its upper and lower sidebands.

3.8 Let the message signal $m(t)$ be given by $m(t) = \frac{1}{2}(\text{sinc}(1000t) + \text{sinc}^2(2000t))$.

(a) What is the Fourier transform of the message signal? What is its bandwidth?

(b) Assume that conventional AM is used to transmit the message signal $m(t)$, where the modulated signal is given by $x(t) = (1 + m(t))\sin(2\pi f_c t)$.

(i) Find and plot the Fourier transform of the modulated signal.

(ii) What is the modulation index? Is this signal overmodulated?

(c) Assume that DSB-SC AM is used to transmit $m(t)$, with the modulated signal given by $x(t) = m(t)\sin(2\pi f_c t)$. Find and plot the Fourier transform of the modulated signal. Indicate the upper sideband of $x(t)$ in the frequency domain.

3.9 A message signal given by $m(t) = \sin(2\pi f_m t) + \frac{1}{2}\cos(4\pi f_m t)$ is transmitted using conventional AM. The modulated signal is $x(t) = 8(1 + \frac{1}{2}m(t))\cos(2\pi f_c t)$, where $f_c \gg f_m$.

(a) What is the Fourier transform of $m(t)$? Plot its magnitude and phase separately.
(b) What is the Fourier transform of $x(t)$? Giving a carefully labeled plot is sufficient.
(c) What is the bandwidth required to transmit $x(t)$?
(d) What is the modulation index?
(e) What is the upper sideband of $x(t)$ (in the time domain)?
(f) What is the ratio of the power in the sidebands to the total average power of $x(t)$?

3.10 A conventional AM signal is given by

$$x(t) = 2\left(1 + \frac{1}{2}\sin(200\pi t) + \frac{1}{4}\cos(400\pi t)\right)\cos(2\pi 100kt).$$

The carrier signal is $c(t) = 2\cos(2\pi 100kt)$.

(a) Plot the Fourier transform of $x(t)$.
(b) Determine the upper-side-band (USB) signal in the time domain.
(c) Determine the ratio of the average power in the sidebands to the total average power of $x(t)$.
(d) Determine the modulation index.

3.11 A message signal $m(t)$ is being transmitted using DSB-SC AM. Assume that $x(t) = m(t)\cos(2\pi f_c t)$ is the modulated signal, where the carrier frequency f_c is much larger than the bandwidth of the message signal.

We want to recover the message signal from the modulated waveform using a local oscillator, mixers (multipliers), and a lowpass filter. Unfortunately, the local oscillator can generate sinusoidal signals with a frequency of at most $\frac{f_c}{2}$. How can we design the receiver to recover the message signal from the modulated signal? You may ignore any phase offset between the local oscillator and the modulated signal.

3.12 Consider the block diagram of a system used to generate a DSB AM signal shown in Fig. 3.33.

Let $m(t)$ denote the message signal with bandwidth W. Assume that $p(t)$ is periodic with Fourier series expansion

$$p(t) = \sum_{n=-\infty}^{\infty} c_n\, e^{j2\pi f_c n t},$$

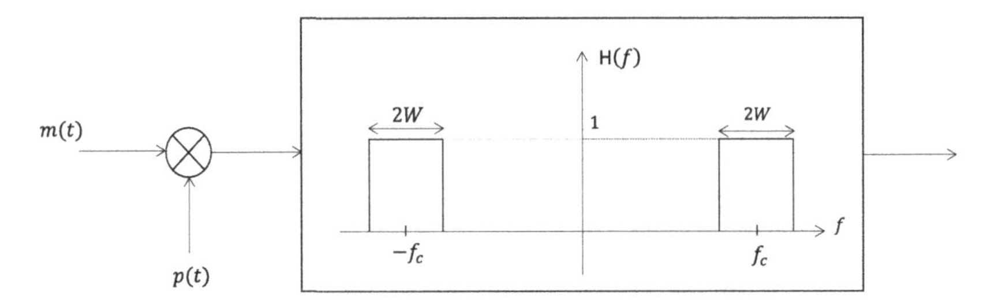

Figure 3.33 Block diagram of the communication system in Problem 3.12.

where the period of $p(t)$ is $1/f_c$. Assume also that $c_1 = j$ and $c_{-1} = -j$.

(a) Plot the Fourier transform of $p(t)$.

(b) Find and plot the Fourier transform of the signal at the input to the bandpass filter.

(c) What is the Fourier transform of the signal at the output of the bandpass filter?

(d) Determine the bandpass filter output in the time domain and argue that this is a DSB-SC AM signal.

3.13 A message signal $m(t)$ is given by

$$m(t) = \text{sinc}^2(t-2) + \text{sinc}^2(t+2).$$

(a) Find the Fourier transform of $m(t)$ and give an approximate plot.

(b) Let $c(t) = 2\cos(2\pi f_c t)$ and $x(t) = m(t)c(t)$. Find and plot the Fourier transform of $x(t)$.

3.14 The message signal $m(t) = \sin(2\pi 12kt) - 5\cos(2\pi 10kt)$ is modulated using SSB AM. The carrier signal is given by $2\cos(2\pi 1000kt)$ and the lower sideband is kept.

(a) Determine and plot the Fourier transform of the modulated signal. What is its bandwidth?

(b) Assume that the carrier phase is known at the receiver. Hence, the receiver can construct the carrier signal $\cos(2\pi 1000kt)$, and a coherent demodulator is employed to recover the message signal. Depict the Fourier transform of the product of the carrier signal (generated at the receiver) with the modulated signal and the lowpass filter output, and demonstrate that the message signal can be recovered.

(c) The SSB AM signal is demodulated using the carrier signal $\cos(2\pi 1000kt + \pi/6)$, namely, using a carrier signal with incorrect phase. Determine the demodulator output. Is the message signal obtained at the demodulator output correctly? Comment on your result.

3.15 Determine the Hilbert transforms of the following signals.

(a) $\cos(2\pi f_0 t)$.

(b) $4\sin(2\pi f_0 t)$.

(c) $2\sin(2\pi 100t + \pi/3) - 5\cos(2\pi 200t + \pi/6)$.

(d) $\text{sinc}(2t)$.

3.16 Consider the message signal given by

$$m(t) = 2\sin(2\pi 20kt) - 3\cos(2\pi 30kt).$$

(a) Determine the Hilbert transform of $m(t)$, denoted by $\hat{m}(t)$.
(b) An SSB AM-modulated signal is given by

$$x(t) = m(t)\cos(2\pi f_c t) + \hat{m}(t)\sin(2\pi f_c t).$$

Determine and plot the Fourier transform of the modulated signal $X(f)$, assuming that $f_c = 1$ MHz. There is no need for separate phase and magnitude plots; a single, carefully labeled plot is sufficient.
(c) Determine the average power of $x(t)$.

3.17 One way to transmit two different signals using DSB-SC AM within the same frequency band is to employ two orthogonal carrier signals. The resulting scheme is called orthogonal carrier multiplexing. In this scheme, given two message signals of bandwidth W, denoted by $m_1(t)$ and $m_2(t)$, respectively, the transmitted signal is given by

$$x(t) = m_1(t)\cos(2\pi f_c t + \theta) + m_2(t)\sin(2\pi f_c t + \theta).$$

(a) Assuming that the carrier phase θ is perfectly known at the receiver, show that the first message signal can be recovered by mixing the received signal with $2\cos(2\pi f_c t + \theta)$ and lowpass filtering the mixer output. Also, show that the second message signal can be recovered by mixing the received signal with $2\sin(2\pi f_c t + \theta)$ and lowpass filtering the mixer output.
(b) Assume that the carrier phase is not perfectly known, and instead, only an estimate $\hat{\theta}$ is available at the receiver. Determine the output of the receiver branch for recovering the ith message signal, $i = 1, 2$. Comment on the result.

3.18 An SSB AM-modulated signal is given by

$$x(t) = m(t)\cos(2\pi f_c t) - \hat{m}(t)\sin(2\pi f_c t),$$

where $m(t)$ is the message signal and $\hat{m}(t)$ is its Hilbert transform. A coherent demodulator with the carrier phase ϕ, namely, $\cos(2\pi f_c t + \phi)$, is employed for demodulation. That is, the carrier signal employed in the demodulation process has a phase offset of ϕ with respect to the correct carrier phase.

Determine the demodulator output. Is the message signal recovered perfectly?

3.19 An analog-modulated signal is given by

$$x(t) = m(t)\cos(2\pi f_c t) - \hat{m}(t)\sin(2\pi f_c t),$$

where the message signal is $m(t)$, and its Hilbert transform is $\hat{m}(t)$.

Assume that the Fourier transform of $m(t)$ is a triangular function, that is, $M(f) = \Lambda(f/10k)$ (with f given in Hz), and $f_c = 1$ MHz. Determine the modulated signal's Fourier transform and identify the modulation type used.

Also, determine the channel bandwidth required to transmit the modulated signal.

3.20 Assume that $m(t) = \cos(2\pi 10kt)$.

(a) Determine the Hilbert transform of $m(t)$, that is, $\hat{m}(t)$.

(b) Let the modulated signal be given by $x(t) = m(t)\cos(2\pi f_c t) + \hat{m}(t)\sin(2\pi f_c t)$, where $f_c = 1$ MHz. Assume that we demodulate this signal using the carrier signal $c(t) = 2\cos(2\pi f_c t + \phi)$ at the receiver (i.e., there is a phase estimation error). We form $v(t) = x(t)c(t)$, and lowpass filter the resulting signal (using an ideal lowpass filter with unit gain and bandwidth of 20 kHz) to obtain an estimate of the transmitted message signal.

Determine the demodulator output (i.e., the message estimate) and comment on your result.

3.21 A message signal given by $m(t) = \cos(2\pi 10kt) - 2\sin(2\pi 35kt)$ is modulated using VSB AM with a bandpass filter whose frequency response is given in Fig. 3.34. Assume that the carrier signal is given by $c(t) = 2\cos(2\pi 760kt)$. Notice that the upper sideband and a vestige of the lower sideband are transmitted.

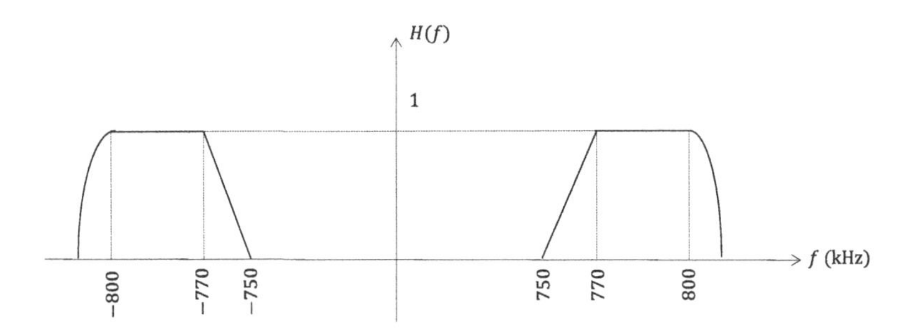

Figure 3.34 Frequency response of the bandpass filter in Problem 3.21.

(a) Determine the Fourier transform of the modulated signal.

(b) Determine the modulated signal in the time domain.

(c) Assume that a coherent demodulator is employed to recover the transmitted signal using the carrier signal $\cos(2\pi 760kt)$ at the receiver (i.e., the carrier phase is perfectly known at the receiver). Plot the Fourier transform of the product of the modulated signal and the carrier signal, as well as the Fourier transform of the lowpass filter output at the receiver. Is the message signal correctly recovered?

(d) Repeat the previous part assuming that the carrier signal used at the receiver is $\cos(2\pi 760kt + \pi/10)$; that is, there is a phase error. Is the message signal recovered correctly in this case?

3.22 Consider a message signal given by

$$m(t) = 2u(t) - 3u(t-1) + u(t-3),$$

where $u(t)$ is the unit step function. Sketch the approximate modulated signals with

(a) frequency modulation,

(b) phase modulation.

3.23 An angle-modulated signal by a pure sinusoid is given by

$$x(t) = A_c \cos(2\pi f_c t + \beta \sin(2\pi f_m t)).$$

Observe that this can be phase modulation of a sine wave or frequency modulation of a cosine wave. Observe that the modulated signal is periodic.

Determine the Fourier series expansion of the modulated signal $x(t)$ and its Fourier transform. (*Note*: The result will be in terms of the Bessel function of the first kind.)

3.24 Let $x(t)$ be a signal with Fourier transform $X(f)$, given in Fig. 3.35.

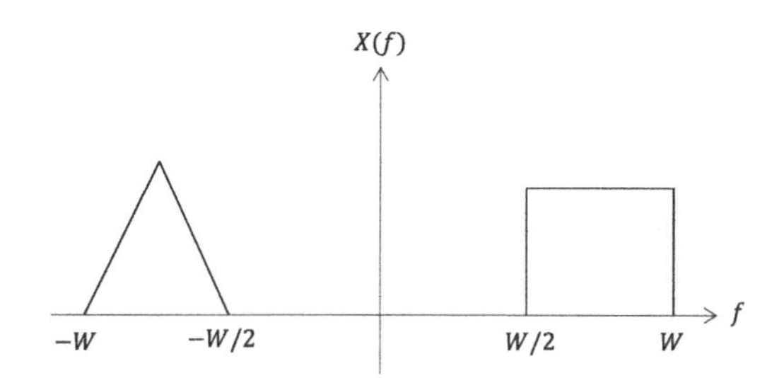

Figure 3.35 Fourier transform of the signal in Problem 3.24.

(a) Is $x(t)$ a real signal?

(b) We know that we can sample $x(t)$ with a sampling period of $T_s \leq \frac{1}{2W}$, and recover the continuous-time signal $x(t)$ from its samples.

Is it possible to sample $x(t)$ at a slower rate and still have perfect reconstruction? What rate is appropriate? How can you recover the continuous-time signal in this case?

3.25 The signal $s(t) = \cos(1500\pi t)$ is sampled at a rate $f_s = 1000$ Hz (which is below the Nyquist rate for this signal), resulting in the discrete-time signal $s[n]$. The resulting samples are impulse modulated, and

$$x(t) = \sum_{n=-\infty}^{\infty} s[n]\delta(t - nT_s)$$

is formed, where T_s is the sampling period ($T_s = 1/f_s$). We pass the signal $x(t)$ through an ideal lowpass filter with bandwidth $W = 500$ Hz.

Determine the frequency of the sinusoid observed at the lowpass filter output.

3.26 The signal $s(t) = 2\sin(5000\pi t) - \cos(2200\pi t)$ is sampled at a rate $f_s = 2000$ Hz. The resulting samples $s[n]$ are impulse modulated, and

$$x(t) = \sum_{n=-\infty}^{\infty} s[n]\delta(t - nT_s)$$

is formed, where T_s is the sampling period ($T_s = 1/f_s$). We pass the signal $x(t)$ through an ideal lowpass filter with a bandwidth of $W = 1000$ Hz. Determine the resulting signal.

3.27 A source output X, which is modeled as a random variable with a probability density function

$$f_X(x) = \begin{cases} \frac{2-|x|}{4}, & \text{if } -2 < x < 2, \\ 0, & \text{else,} \end{cases}$$

is being quantized using a four-level quantizer with quantization regions $[-2, -1]$, $(-1, 0]$, $(0, 1]$, and $(1, 2]$.

(a) Determine the optimal reconstruction levels for the four quantization regions that minimize the mean squared error $D = \mathbb{E}[(X - Q(X))^2]$ (where $Q(\cdot)$ denotes the quantization function).

(b) In this part, assume that a new distortion function $D' = \mathbb{E}[(X - Q(X))^4]$ is being used instead of the mean squared error distortion. Find an equation whose solution is the optimal reconstruction level for the quantization interval $(0, 1]$. Determine the optimal reconstruction level numerically.

3.28 A source output, modeled as a random variable X with a probability density function

$$f_X(x) = c\Lambda\left(\frac{t}{3}\right),$$

is quantized using a six-level quantizer (over the input range $[-3, 3]$). (Determine the constant c.)

(a) Determine the signal-to-quantization ratio assuming that the quantizer is uniform.

(b) Assume that the quantization region boundaries are the same as those of the uniform quantizer in part (a). Determine the corresponding optimal reconstruction levels that minimize the mean squared error.

(c) Determine the signal-to-quantization ratio for the quantizer in part (b).

3.29 A source output X, which is modeled as a random variable with the probability density function shown in Fig. 3.36, is being quantized using a two-level quantizer with quantization regions $[-2, 0]$ and $(0, 2]$.

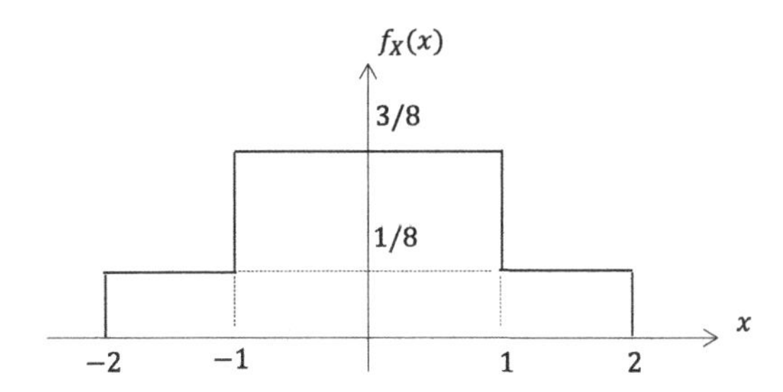

Figure 3.36 Source PDF for Problem 3.29.

(a) Determine the optimal reconstruction levels for the two quantization regions that minimize the mean squared error $D = \mathbb{E}[(X - \hat{X})^2]$ (where $\hat{X}$ is the quantizer output).

(b) Determine the optimal reconstruction levels if the mean absolute value distortion defined as $D' = \mathbb{E}[|X - \hat{X}|]$ is to be minimized instead of the mean squared error distortion.

3.30 Consider a memoryless source that emits X_n $(n = 1, 2, \ldots)$. The source outputs (X_ns) are quantized using an eight-level (scalar) uniform quantizer on the interval $(-4, 4)$. The resulting quantizer output sequence is denoted by $\hat{X}_n$ $(n = 1, 2 \ldots)$, as illustrated on the left-hand side of Fig. 3.37.

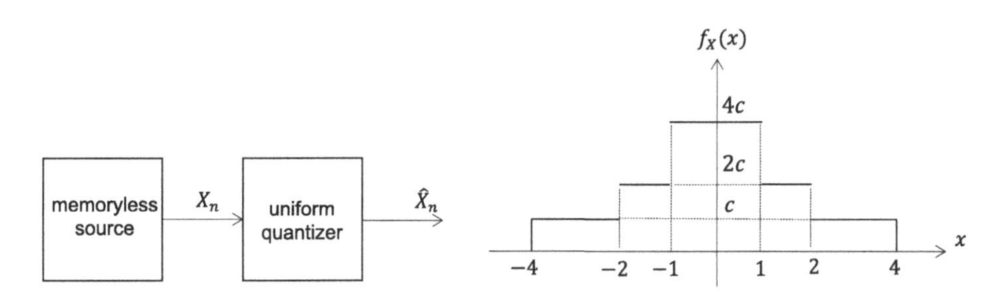

Figure 3.37 Figure for Problem 3.30.

The source outputs are independent and identically distributed random variables with PDF $f_X(x)$ as given on the right-hand side of the figure. Note that c is a constant that needs to be determined.

(a) Determine the PDF of the quantization error $X - \hat{X}$.

(b) Compute the mean squared error $\mathbb{E}[(X - \hat{X})^2]$. What is the signal-to-quantization noise ratio in decibels?

3.31 We are interested in quantizing a source output modeled as a random variable X with probability density function $f_X(x)$. Assume that the range of X is $[-3, 3]$.

(a) The reconstruction levels are given as $\hat{x}_1 = -2,\ \hat{x}_2 = 0,\ \hat{x}_3 = 1,\ \hat{x}_4 = 2$. What should be the corresponding quantization regions to minimize the mean squared error?

(b) Assume that $f_X(x) = cx^2$ for $x \in [-3, 3]$ (where c is a constant), and 0 otherwise. Assume also that a particular quantization region is $(-1, 2)$. Determine the corresponding reconstruction level (for this quantization region only) to minimize the mean squared error.

3.32 Determine the optimal uniform and non-uniform quantizers with seven levels for $X \sim N(-2, 16)$ and the corresponding mean squared errors using the results in Tables 3.2 and 3.3.

3.33 The power spectral density of a wide-sense stationary random process $X(t)$ is given by

$$S_X(f) = 10^{-3}(1 + \cos(\pi f/1000))$$

for $|f| \leq 1000$, and 0 otherwise. Assume that the samples of the random process take on values in the interval $(-20, 20)$.

We represent samples of this signal using uniform PCM. Determine the minimum number of bits v that should be used in quantization so that a signal-to-quantization noise ratio of at least 65 dB is obtained at the output.

3.34 Assume that a discrete memoryless source X with probability density function $f_X(x)$ given in Fig. 3.38 is transmitted using pulse-code modulation. The number of quantization levels is 1024.

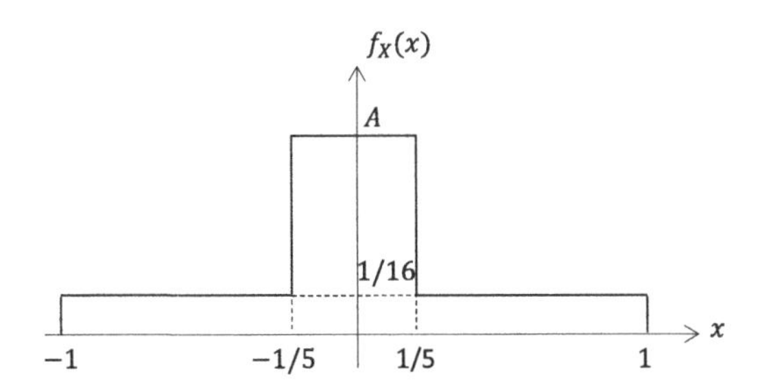

Figure 3.38 Source PDF for Problem 3.34.

(a) Determine the average power of the source $\mathbb{E}[X^2]$. (You will also need to determine A.)

(b) Assuming that uniform PCM is used, determine the output signal-to-quantization noise ratio in decibels.

(c) We decide to transmit the same source with non-uniform PCM using a compander. For this purpose, we identify a compressor of the form given in Fig. 3.39.

Determine the parameter b that maximizes the average SQNR. Also, find the resulting improvement in the SQNR (expressed in dB) compared to the uniform PCM scheme in the previous part.

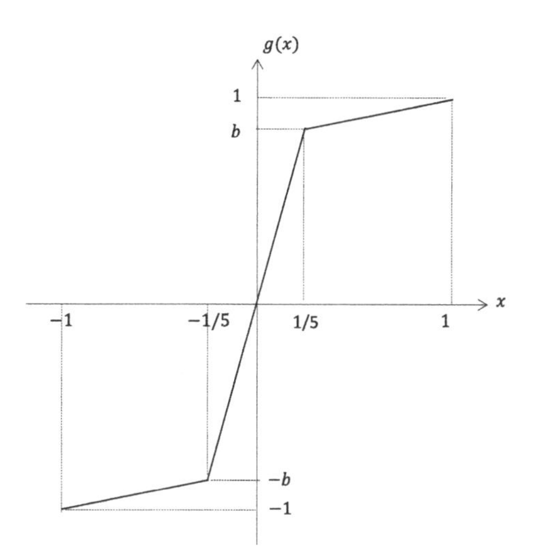

Figure 3.39 Compressor used in Problem 3.34.

3.35 Consider a memoryless source that emits X_n $(n = 1, 2, \dots)$. The source outputs (X_n) are independent and identically distributed random variables with PDF $f_X(x)$ as illustrated in Fig. 3.40. Assume that a uniform quantizer with four levels is used to quantize the source samples; that is, the quantization regions are $[-4, -2]$, $(-2, 0]$, $(0, 2]$, and $(2, 4]$, and the reconstruction levels are the midpoints of these intervals.

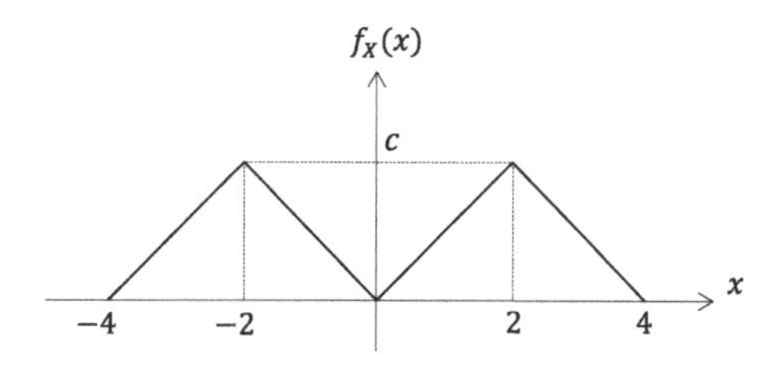

Figure 3.40 Source PDF for Problem 3.35.

(a) Determine the constant c.

(b) Determine the PDF of the quantization error defined as $X - \hat{X}$, given that X falls in the interval $(-2, 0]$.

(c) Determine the resulting MSE distortion, $\mathbb{E}[(X - \hat{X})^2]$.

(d) We decide to use non-uniform PCM to transmit the outputs of this source. What kind of compressor at the transmitter and expander at the receiver would you use to improve the resulting SQNR? Give an approximate plot for both.

3.36 Assume that a waveform whose samples are modeled as random variables with probability density function

$$f_X(x) = \frac{1}{8}\Pi\left(\frac{t}{4}\right) + \frac{1}{4}\Pi\left(\frac{t}{2}\right)$$

is transmitted with uniform PCM. The sampling rate is 10,000 samples per second, and the number of bits per sample is $v = 10$.

(a) Determine the minimum bandwidth required to transmit the PCM signal.

(b) Determine the resulting SQNR (in dB), assuming that there are no bit errors in transmission.

(c) Assume that natural binary mapping is employed and that the resulting bits are transmitted through a noisy communication channel. Assume that the probability of the least significant bit being received in error is 10^{-3}, while there is no error in transmission of the other bits.

(i) Determine the (conditional) PDF of the quantization error given that there is a bit error in the transmission of a sample.

(ii) Compute the resulting SQNR of the PCM scheme with channel errors (express your answer in dB).

3.37 A source output X, modeled as a uniform random variable on the interval $[0, 5]$, is input to a quantizer with the following description:

$$Q(x) = \begin{cases} 1, & \text{if } -\infty < x \leq 2, \\ 3, & \text{if } 2 < x \leq 4, \\ 4.5, & \text{if } 4 < x < \infty. \end{cases}$$

Determine the probability density function of the quantization error $X - Q(X)$.

3.38 Assume that a source with PDF

$$f_X(x) = \begin{cases} 1/3, & \text{for } -1 \leq x < -1/2, \\ 2/3, & \text{for } -1/2 \leq x < 1/2, \\ 1/3, & \text{for } 1/2 \leq x \leq 1, \\ 0, & \text{else} \end{cases}$$

produces 1000 samples per second. We want to transmit the source outputs using uniform PCM with 7 bits per sample.

(a) What is the resulting SQNR (in dB) of the system assuming that there are no bit errors in the transmission?

(b) What is the minimum bandwidth required to transmit the PCM signal?

(c) Assume that the bit error probability in the transmission is 10^{-6}, and natural binary coding is used to map the quantization levels to bit sequences. What is the resulting SQNR (in dB)? How much loss is observed compared to the result in part (a)?

3.39 A source X generates outputs according to the following probability density function:

$$f_X(x) = \begin{cases} x/3, & \text{for } 0 \le x \le 1, \\ 1/3, & \text{for } 1 < x \le 3, \\ (4-x)/3, & \text{for } 3 < x \le 4, \\ 0, & \text{else.} \end{cases}$$

This source is quantized using the following four-level quantizer:

$$Q(x) = \begin{cases} 0.5, & \text{for } 0 \le x \le 1, \\ 1.5, & \text{for } 1 < x \le 2, \\ 2.5, & \text{for } 2 < x \le 3, \\ 3.5, & \text{for } 3 < x \le 4. \end{cases}$$

(a) Calculate the mean absolute error, which is defined as $\mathbb{E}\{|X - Q(X)|\}$.
(b) Calculate the probability that $Q(X)$ is larger than 1, that is, $\mathbb{P}(Q(X) > 1)$.
(c) Suppose that the source produces 20,000 samples per second, and we would like to transmit these samples using uniform PCM with 2 bits per sample, as specified by the quantizer above. What is the minimum bandwidth required to transmit the PCM signal?
(d) Propose a compander to increase the SQNR of this system. Namely, plot or define a function $c(x)$ corresponding to the compressor at the transmitter. Specify the domain, range, and all the critical values of the function $c(x)$. Why do you think the proposed compander will increase the SQNR? (There is no need to compute the resulting SQNR, and the compander does not need to be optimal.)

3.40 The outputs of a stationary source are distributed according to the probability density function

$$f_X(x) = \begin{cases} 1/3, & \text{for } -1 \le x < -1/2, \\ 2/3, & \text{for } -1/2 \le x < 1/2, \\ 1/3, & \text{for } 1/2 \le x \le 1, \\ 0, & \text{else.} \end{cases}$$

(a) Determine the resulting SQNR (in dB) if we use a four-level quantizer given by

$$Q(x) = \begin{cases} -3/4, & \text{for } -1 \le x < -1/2, \\ -1/4, & \text{for } -1/2 \le x < 0, \\ 1/4, & \text{for } 0 \le x < 1/2, \\ 3/4, & \text{for } 1/2 \le x \le 1, \end{cases}$$

to quantize the outputs of this source.
(b) We employ a compander to improve the SQNR of the system. What kind of compressor would you use? Give an approximate form for the compressor and the corresponding expander.

3.41 Consider a source with the probability density function given in Fig. 3.41.

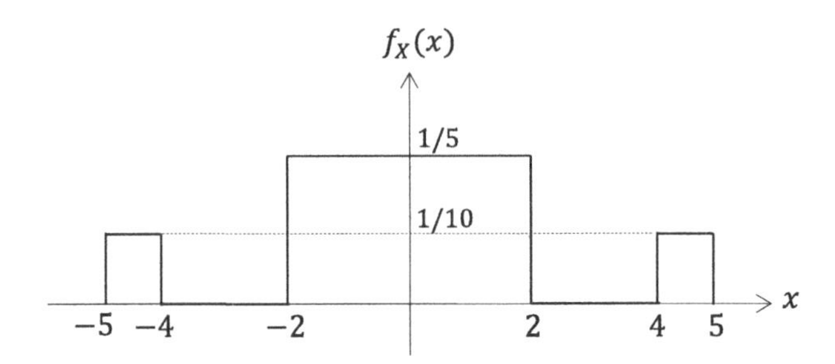

Figure 3.41 Compressor used in Problem 3.41.

Suppose that the source produces 20,000 samples per second. We want to transmit the source outputs using uniform PCM with 10 bits per sample.

(a) What is the resulting SQNR (in dB) of the system?

(b) What is the minimum bandwidth required to transmit the PCM signal (assuming that binary modulation is used)?

(c) To improve the system performance, we use a compander with the compressor at the transmitter given by

$$c(x) = \begin{cases} 2x, & \text{if } |x| \leq 2, \\ 4, & \text{if } 2 < x < 4, \\ -4, & \text{if } -4 < x < -2, \\ x, & \text{if } 4 \leq |x| \leq 5. \end{cases}$$

Determine the expander to be used at the receiver corresponding to $c(x)$, and explain why this compander may be a good choice to improve the SQNR.

Compute the resulting SQNR of this non-uniform PCM scheme. Compare this with the value you computed in part (a).

3.42 A continuous-time source is encoded using pulse-code modulation. The sampling rate is $f_s = 20{,}000$ samples/sec, and each sample is quantized using $N = 1024$ levels. The samples have a probability density function given by $f_X(x) = \Lambda(x)$. Clearly, the range of source outputs is $[-1, 1]$.

(a) Assume that uniform PCM is used and determine the signal-to-quantization noise ratio. Also, determine the minimum channel bandwidth needed for transmission (with binary modulation).

(b) Consider part (a); however, assume that while the quantization regions are of equal length, the reconstruction levels are not at the mid-points of the intervals. Instead, for each interval of the form $[a, b)$, the reconstruction level is at $(a + 3b)/4$, that is, the reconstruction level is closer to the upper boundary of the quantization interval.

Determine the resulting SQNR.

(c) Assume that we perform non-uniform PCM where the equivalent non-uniform quantizer is described as follows (with the input to the quantizer being denoted by x):

– for $x \in [0, 1/4)$, we have 180 quantization regions of equal length;

- for $x \in [1/4, 1/2)$, we have 140 quantization regions of equal length;
- for $x \in [1/2, 3/4)$, we have 116 quantization regions of equal length;
- for $x \in [3/4, 1)$, we have 76 quantization regions of equal length;
- for negative values of x, the quantizer is symmetric; for example, for $x \in [-3/4, -1/2)$, we have 116 quantization regions of equal length;
- for all the quantization regions, the reconstruction levels are at the mid-points of the intervals.

Since there are a large number of quantization levels, we can approximate the quantization errors as uniform.

Determine the resulting SQNR.

COMPUTER PROBLEMS

3.43 A message signal is given by

$$m(t) = \begin{cases} t, & \text{for } 0.1 \leq t < 1, \\ -t+2, & \text{for } 1 \leq t < 1.9, \\ 0.1, & \text{else,} \end{cases}$$

on the interval $[0, 2]$. This signal is modulated using DSB-SC AM with a carrier of frequency of 1 kHz and amplitude 1. Consider sampled versions of the message and modulated signals with a sampling frequency of 3 kHz.

(a) Plot the message signal and the modulated signal in the time domain.

(b) Plot the magnitude of the Fourier transform of $m(t)$ and that of the modulated signal.

Use the fast Fourier transform for part (b) (with the MATLAB command "fft"). Comment on your results.

3.44 The message signal $m(t)$ is given by

$$m(t) = \begin{cases} -1, & \text{for } -2 \leq t < -1, \\ 2t-1, & \text{for } -1 \leq t < 0, \\ -2t+1, & \text{for } 0 \leq t < 1, \\ 1, & \text{for } 1 \leq t < 2, \end{cases}$$

in the interval $[-2, 2]$.

(a) Using MATLAB, plot the magnitude spectrum of the message signal. To represent the continuous-time signal in discrete time, use a sampling period of 0.0001.

(b) Let $x_{AM}(t)$ be the double-sideband suppressed carrier-modulated signal corresponding to $m(t)$. Assume that the carrier frequency is $f_c = 500$ Hz and $A_c = 1$. Plot the magnitude spectrum of $x_{AM}(t)$ by using the fft command in MATLAB.

(c) Let $x_{PM}(t)$ be the phase-modulated signal corresponding to $m(t)$. Assume that the carrier frequency is $f_c = 500$ Hz and $A_c = 1$. Plot the magnitude spectrum of $x_{PM}(t)$ for $k_p = 0.1, 10, 100, 200$, by using the fft command in MATLAB.

(d) Compare the bandwidths of the signals in parts (b) and (c). What do you observe?

3.45 Assume that we use a quantizer to quantize the outputs of a Gaussian source. The source outputs are i.i.d. random variables with zero mean and unit variance. The (optimal) eight-level uniform quantizer is given in Table 3.2. Specifically, $a_1 = -3\Delta$, $a_2 = -2\Delta, \ldots, a_7 = 3\Delta$, and the quantization levels are $\hat{x}_1 = -3.5\Delta$, $\hat{x}_2 = -2.5\Delta, \ldots, \hat{x}_8 = 3.5\Delta$, where $\Delta = 0.586$.

To estimate the expected distortion using Monte Carlo techniques: generate 100,000 source outputs, quantize them individually, determine corresponding realizations of the quantization errors, and estimate the resulting squared error average. What do you observe? Is this result consistent with the theoretical expectations?

Repeat this simulation for the (optimal) eight-level non-uniform quantizer given in Table 3.3. Comment on your results.

3.46 Assume that a continuous-time source given by

$$x(t) = \frac{1}{1+t^2}\sin(2\pi t) + e^{-t}\cos(4\pi t),$$

for $t \in [0,4]$, is to be quantized using delta modulation. Assume that the sampling period is $T_s = 0.1$.

(a) Assume that a constant step size (Δ) is being used. Write a MATLAB code to plot the delta-modulated signal along with the continuous-time signal for $\Delta = 0.3$, $\Delta = 0.1$, and $\Delta = 0.025$. Indicate the regions where granular noise or slope-overload distortion is observed.

(b) Assume that the step size is being adaptively changed as follows: if there are two consecutive increases or decreases, multiply the current step size by α; otherwise, divide it by α. Write a MATLAB code to plot the delta-modulated signal along with the continuous-time signal with an initial step size of $\Delta = 0.4$, and $\alpha = 1.5$. Compare your results with the results of part (a).

(c) Repeat part (b) for a different initial value of Δ and a different scaling factor α.

3.47 Consider a discrete memoryless source whose outputs are standard Gaussian random variables (i.e., with zero mean and unit variance). Assume the source outputs are truncated to the interval $[-7, 7]$, and a uniform quantizer with N levels is employed. Note that with this specification, the quantizer is completely specified.

Generate a long sequence of source samples and their quantized versions. You can take 1 million realizations.

(a) Estimate the source power using the specific sequence of realizations (by averaging the samples' squares). Is this consistent with your expectations from the theory?

(b) Consider $N = 64$ and obtain the sequence of the quantization errors corresponding to the source samples generated. Estimate their average power directly from these samples, and plot the normalized histograms of the quantization errors (to estimate the probability density function). What do you observe? Are the results consistent with your theoretical expectations? How can you approximate the PDF of the quantization errors? What is the estimated signal-to-quantization noise ratio (in dB)? Is this consistent with the theoretical expectations?

(c) Consider the use of a compander to perform non-uniform quantization with $N = 64$ using a μ-law or an A-law compander. Select the parameter of the compander to improve the resulting SQNR, and report on your results. What is the resulting (estimated) mean square error? What is the resulting SQNR? Comment on your results.

(d) Repeat parts (b) and (c) for a different selection of the number of quantization levels (e.g., $N = 256$). Compare your results with those of the previous parts.

3.48 Record a piece of speech in MATLAB with a specific sampling frequency (higher than the Nyquist rate) and form a vector with the resulting samples. Take about 1 million samples. Normalize the vector to ensure the samples are in the interval $[-1, 1]$.

(a) Plot the normalized histogram of the samples (to estimate their probability density function). What do you observe?

(b) Estimate the source power using the samples of the speech signal.

(c) Assume that a uniform quantizer with $N = 64$ is used to quantize the source outputs. Obtain the quantized vector of source outputs, and estimate the PDF of the resulting quantization errors. Also, estimate the resulting SQNR. Compare the result with the theoretical expectations.

(d) Repeat the previous part with a non-uniform quantizer with $N = 64$ using the μ-law compander for different values of μ, and compare the resulting SQNRs.

3.49 Record a piece of speech in MATLAB with a specific sampling frequency (higher than the Nyquist rate) and form a vector with the resulting samples. Take about 1 million samples. Normalize the vector to ensure the samples are in the interval $[-1, 1]$.

Estimate the joint PDF of two consecutive samples of the speech signal you have recorded. To do this, take samples 1 and 2, 3 and 4, 5 and 6, ..., together, and provide a two-dimensional histogram (adequately normalized) to estimate their joint PDF.

What do you observe? Are the consecutive source samples correlated? Can you think of suitable quantizer structures that could be well suited to quantizing two samples at a time? Do you expect significant performance improvements with vector quantization over scalar quantizers?

3.50 Design a 10-level non-uniform quantizer for a zero mean and unit variance Gaussian source using the Lloyd-Max algorithm. Determine the boundaries of the quantization regions and the corresponding reconstruction levels. Also, determine the resulting distortion.

3.51 Consider a discrete memoryless source whose samples have a PDF

$$f_X(x) = \begin{cases} \frac{2+x}{3}, & \text{if } -2 < x \leq -1, \\ \frac{1}{3}, & \text{if } -1 < x \leq 1, \\ \frac{2-x}{3}, & \text{if } 1 < x \leq 2, \\ 0, & \text{else.} \end{cases}$$

(a) Assume that U is a uniform random variable on the unit interval, that is, $[0, 1]$. The MATLAB command "rand" generates realizations of such a random variable. Determine a function $g(\cdot)$ such that $X = g(U)$ has the PDF $f_X(x)$ given above.

(b) Generate 100,000 independent realizations of a uniform random variable on the unit interval and apply the function in the previous part to generate as many samples of the given source. Plot a normalized histogram and verify that the PDF is as expected.

(c) Consider a uniform quantizer with eight levels (on the interval $[-2, 2]$) used to quantize the source samples. Perform quantization of the 100,000 samples generated in the previous part. Estimate the resulting mean squared error and the signal-to-quantization noise ratio. Also, estimate the PDF of the quantization error. Are the results consistent with each other?

(d) Repeat the previous part using 128 quantization levels.

(e) Design a Lloyd-Max quantizer for the given source (using the iterative approach described in the text) with eight levels.

(f) Quantize the 100,000 samples generated earlier using the optimal Lloyd-Max quantizer in the previous part and estimate the corresponding MSE and SQNR. Also, estimate the PDF of the quantization errors. Comment on your results.

3.52 This problem is intended to study PCM and its variations as a waveform coding technique. In particular, we will consider uniform PCM, non-uniform PCM, differential PCM, delta modulation, and adaptive delta modulation.

First, record a piece of speech in MATLAB with a specific sampling frequency (higher than the Nyquist rate) and form a vector with the resulting samples. Make sure to take a sufficiently long speech signal for statistical analysis.

Normalize the vector so that the samples are in the interval $[-1, 1]$.

(a) Estimate the probability density function of the source samples. To do this, remove the silence periods so that there is no (incorrect) "spike" at $x = 0$. Comment on the nature of these samples.

(b) Assuming that uniform PCM is used, estimate the resulting SQNR based on your estimate of the source sample PDF in the first part. Obtain your result as a function of the number of quantization bits (v).

(c) Consider the original vector you have recorded, and apply uniform PCM to obtain a digital representation. Provide samples of (zoomed-in versions) of the encoded version and the original vector. Estimate the mean squared error, the signal power (directly from the vector – by averaging across time), and the resulting SQNR. Do this for two different values of the number of quantization bits. Compare your answer with those of the previous part.

(d) Use an A-law or μ-law compander (no need for both) to perform non-uniform PCM (with $\nu = 8$ bits) and study the resulting signal quality. Vary the parameter of the compander to optimize the quality of representation. What is your optimized value? How much improvement do you observe in the resulting SQNR over uniform PCM?

(e) Consider the first (direct) implementation of differential PCM with 2-bit quantization and encode your recorded speech vector. Decide on the exact quantizer to use (recall that the new dynamic range for the difference between the consecutive samples is small), and report on the quality of the resulting encoding (in terms of SQNR).

Do this for two different sampling rates. (You can use the resample command of MATLAB to generate versions of the same signal with different sampling rates.) Comment on your results.

(f) Repeat the previous part with the alternate implementation of differential PCM, which does not result in error accumulation.

(g) Use delta modulation to encode the speech signal. Try different step sizes and report on the best one (providing the smallest mean squared error). Do this for two different sampling rates. Comment on your results.

(h) Lastly, use adaptive delta modulation to encode the speech signal. Assume that the step size is increased according to $\alpha\Delta$, and reduced according to Δ/α, and optimize the value of α (by estimating the MSE for your example).

4 Digital Information Sources

The basic objective of a communication system is to transmit the output of a source to a destination. Mathematically speaking, the source outputs are realizations of a random process, which could be continuous or discrete time and continuous or discrete valued. If the source outputs are continuous time and continuous valued, the source is analog. That is, analog signals are nothing but continuous-time waveforms such as speech signals. On the other hand, if the source outputs are discrete time and discrete valued, then the source is referred to as digital. Digital information sources produce sequences of symbols such as email messages, text messages, digital images or video, and computer data.

While it is not necessary, it is common to represent the outputs of a digital information source as a sequence of binary digits called bits. Typically, we represent the source symbols using 0s and 1s because the binary system is employed internally by (most) computers and computer-based devices. For instance, each ASCII character on a computer is represented by 8 bits, and long text files are nothing but a sequence of 0s and 1s.

We have seen in the previous chapter that an analog signal can be converted into a digital form via sampling and quantization. Provided that the source signal is bandlimited (which is a valid assumption in general), by selecting the sampling rate at or above the Nyquist rate, the sampling process does not introduce any loss. We can approximate the original analog source as a digital one by choosing the number of quantization levels sufficiently high. Recall that pulse-code modulation and its variants (such as differential PCM) achieve precisely this goal. Once an analog source is digitized, it can be transmitted through digital communication techniques.

In this chapter, our objective is to study digital information sources in detail. Specifically, we will go over the concept of redundancy of a source, and for a particular class of digital sources (namely, discrete memoryless sources), we will provide an exact characterization of its information content. We will then give practical source-coding algorithms that can be used to compress digital information sources.

The chapter is organized as follows. We develop an intuition about the information content of a digital source, in Section 4.1, and we describe the notion of redundancy of a source. We focus on a simple but important class of digital information sources, namely, discrete memoryless sources in Section 4.2. We introduce the concept of entropy, which is used to measure their information content, in Section 4.3. We also give several properties of entropy and state the source-coding theorem for a discrete memoryless source. Section 4.4 is devoted to three practical source coding-algorithms, namely, Huffman coding, which is an optimal data compression algorithm when the source statistics are given, and Lempel–Ziv (LZ) and Lempel–Ziv–Welch (LZW) coding schemes, which are universal compression algorithms (not requiring the source statistics). The chapter is concluded in Section 4.5.

4.1 Information Content of a Source

Consider a typical speech signal, converted into a digital form after sampling and quantization, as shown in Fig. 4.1. The signal is lowpass (i.e., its time variations are "slow"), and its consecutive samples have close values. We argue that if some sample values are missing, they can easily be predicted with good precision based on the rest of the signal. In other words, the signal contains redundancy that can be exploited.

Similar observations hold for other types of signals as well. For example, for a typical digital image, the nearby pixel values are similar, and it is possible to predict some of the pixel values from the neighboring ones; in other words, this source also has redundancy. As another example, consider typical video signals produced at a certain number of frames per second. In addition to the nearby pixel values being similar, consecutive frames are also nearly identical, meaning that the video source also contains significant redundancy.

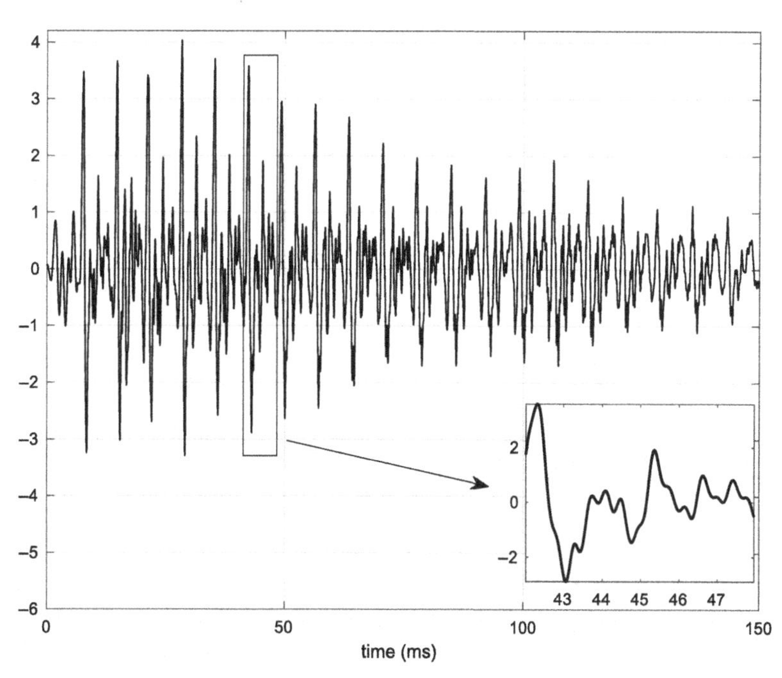

Figure 4.1 A typical 150 ms-long speech signal. With a sampling rate of 44.1 kHz, the waveform shown corresponds to 6615 samples. The zoomed-in portion is approximately 6 ms long, which corresponds to around 265 samples of the signal.

The examples above are sources where different samples of the source (e.g., samples of speech and pixel values of image) are correlated. The signal can contain significant redundancy even when different source samples are independent. For example, consider a digital source that emits independent and identically distributed (i.i.d.) 0s and 1s with the probability of "1" being 0.9 (hence, the probability of "0" being 0.1). A long sequence of bits produced by this source may look like

$$1, 1, 1, 1, 1, 0, 1, 1, 1, 1, 1, 1, 1, 0, 1, 1, \ldots, 1, \tag{4.1}$$

that is, roughly 90% of the bits would be 1s while the remaining (approximately) 10% would be 0s. This follows from the (weak) law of large numbers in probability theory. Hence, without even observing anything about the specific source output, we can guess each bit to be a "1," and we would be correct 90% of the time. Since only the sequences containing about 90% 1s would occur with a very high probability and the rest are unlikely, this source also has significant redundancy.

Let us explain the above point further by considering three binary sequences of the same length containing different amounts of redundancy.

Three Same-Length Binary Sources with Different Information Contents. Assume that the output of the first source is either an all-zero sequence or an all-one sequence with a probability of 1/2 each. In other words, C_1 produces either

$$0, 0, 0, 0, 0, \ldots, 0 \tag{4.2}$$

or

$$1, 1, 1, 1, 1, \ldots, 1 \tag{4.3}$$

with equal probability.

Assume that the second source C_2 results from a sequence of independent, fair coin tosses where "heads" is represented as "0" and "tails" is represented as "1." As the coin toss is fair, the probability of "0" and "1" is $1/2$ each. A typical source output may be

$$0, 1, 1, 1, 0, 1, 0, 0, 1, \ldots, 1. \tag{4.4}$$

Clearly, if the sequence length is n, there are 2^n possible such sequences produced by the source, each with probability $1/2^n$.

Consider equal-length sequences (say n) produced by the two sources above. We notice that while the sequence lengths are the same in both cases, describing the first source's output is much easier. That is, instead of describing the outcome of each coin toss separately, we only need to specify one bit to differentiate between the two possible outputs (namely, the all-zero sequence and the all-one sequence). In other words, these equal-length sequences do not contain the same amount of information; the first is highly redundant and contains little information, while the second is much more uncertain and has a high amount of information.

The source C_1 defined above is quite simple, as only two possible outputs exist. As a more complicated example, consider a third source C_3 whose output results from independent but biased coin tosses (i.e., with unequal probability of "heads" and "tails"). Assume that the probability of a "1" is 0.9, while the probability of a "0" is 0.1. In this case, a specific example of a source output may look like

$$1, 1, 1, 1, 1, 0, 1, 1, 1, 1, 1, 1, 1, 0, 1, 1, \ldots, 1, \tag{4.5}$$

namely, there will be more 1s than 0s in the sequence with a high probability. If the sequence length is n, then there are a total of 2^n possible sequences; however, they are not all equally likely. For instance, the probability of the all-one sequence is 0.9^n, while the probability of a specific sequence with n_1 1s and $n - n_1$ 0s is $0.9^{n_1} 0.1^{n-n_1}$.

Observe that the total number of possible length-n sequences for both $\mathcal{C}_2$ and $\mathcal{C}_3$ is the same (2^n). On the other hand, $\mathcal{C}_3$ contains a higher redundancy than $\mathcal{C}_2$; hence, its information content is lower and it can be represented using significantly fewer bits. This is because not all the sequences of $\mathcal{C}_3$ are equally likely, and it is possible to assign shorter descriptions to the more likely length-n sequences and longer ones for the less likely ones, making the average length of the representation shorter. These ideas will be made precise in the rest of the chapter.

4.2 Discrete Memoryless Sources

A simple mathematical model of a digital information source is a *discrete memoryless source* (DMS), which produces an i.i.d. sequence of symbols from a finite alphabet. While this is too simplistic to model many real-life sources, such as speech or image signals or English text, it will serve our purpose of studying a source's information content or entropy in a tractable manner. Note that it is possible to consider broader classes of source models, for instance, by considering the memory in the source (e.g., using Markov models); however, such generalizations are beyond our scope.

A DMS model is depicted in Fig. 4.2. The source emits an independent sequence of symbols denoted by $\ldots, X_{-2}, X_{-1}, X_0, X_1, X_2, \ldots$. Assuming that the source is M-ary, each sample takes on values from the set $\{x_1, x_2, \ldots, x_M\}$ with probability mass function $P_X(x_i) = P(X = x_i)$ for $i = 1, 2, \ldots, M$.

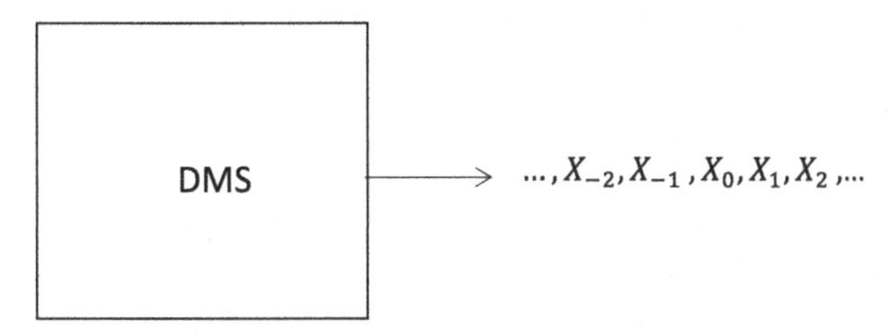

Figure 4.2 Representation of a discrete memoryless source. The sequence of random variables X_i is independent and identically distributed.

Example 4.1

A simple example of a DMS is the binary memoryless source, which is nothing but a sequence of i.i.d. binary digits (bits) usually represented as 0s and 1s. That is, the source samples X_i are i.i.d. Bernoulli random variables taking on 0 and 1 with the PMF

$$\mathbb{P}(X_i = 1) = p \text{ and } \mathbb{P}(X_i = 0) = 1 - p, \tag{4.6}$$

where $0 \leq p \leq 1$.

If $p = 1/2$, the resulting source is called a *binary symmetric source.*

Example 4.2

Assume that X_i takes on values from the set $\{-2, 1, 0, 1, 2\}$ with

$$\mathbb{P}(X_i = k) = \begin{cases} 1/8, & \text{if } k \in \{-2, 2\}, \\ 1/4, & \text{if } k \in \{-1, 0, 1\}, \end{cases} \tag{4.7}$$

and they are i.i.d. Then, the resulting source is a DMS.

4.3 Entropy of a Discrete Memoryless Source

We are interested in quantifying the amount of information a source contains. Consider an M-ary discrete memoryless source with a PMF given by $\{p_1, p_2, \ldots, p_M\}$ (i.e., p_i is the probability that the source takes on the ith value). Denote the source outputs by the random variable X. We define the entropy of the DMS as

$$H(X) = -\sum_{i=1}^{M} p_i \log p_i, \tag{4.8}$$

where we take $0 \log 0 = 0$. If the base of the logarithm is taken as 2, $H(X)$ is measured in *bits* per source output. If the base is taken as e (i.e., natural logarithm), then the unit is in *nats* per source output.

Note that there is an abuse of notation in the above definition: $H(X)$ is not a function of the random variable X (which would be a random variable itself); instead, it is a function of the PMF of the random variable X, and hence it is deterministic.

Intuition Behind This Definition. To develop an intuition for the above definition and why it could be useful, let us return to our earlier discussion on the sequence of independent Bernoulli trials. Assume that the probability of "1" is p and the probability of "0" is $1 - p$ in each trial. As we have argued previously, we will observe nearly np 1s and $n(1 - p)$ 0s in a sequence of source outputs of length n. Here, we assume that np and $n(1 - p)$ are integers for simplicity of notation.[1]

[1] It is possible to make these arguments precise using the weak law of large numbers for any value of p; however, a formal coverage is beyond our scope.

If we ignore all the other outputs and focus on sequences with np 1s and $n(1-p)$ 0s, we can argue that there are nearly $\binom{n}{np}$ sequences that will be observed at the source output (for large n). It is also easy to see that each of these n-length sequences is equally likely (each will be observed with probability $p^{np}(1-p)^{n(1-p)}$). We can then argue that we can represent all of these (important) n-length source output sequences using $\log_2 \binom{n}{np}$ bits. Using Stirling's approximation, which states that $\log(n!) \approx n \log n - n \log e$ for large n, we can write

$$\binom{n}{np} \approx 2^{-np\log_2 p - n(1-p)\log_2(1-p)}. \tag{4.9}$$

Therefore, we can argue that the number of bits needed to represent each of these output sequences uniquely is approximately $nH(X)$, where

$$H(X) = -p\log_2 p - (1-p)\log_2(1-p), \tag{4.10}$$

which is nothing but the entropy definition for a binary source with PMF $\{p, 1-p\}$. In other words, $H(X)$ bits per source output will suffice to represent the binary memoryless source outputs.

Note that the same line of argument can be applied with a similar interpretation for the more general case of an M-ary discrete memoryless source.

Let us give several examples.

Example 4.3

The entropy of the binary memoryless source with PMF $\{p, 1-p\}$ (see Example 4.1) is given by

$$H_b(p) = -p\log p - (1-p)\log(1-p), \tag{4.11}$$

which is referred to as the binary entropy function. It is plotted in Fig. 4.3.

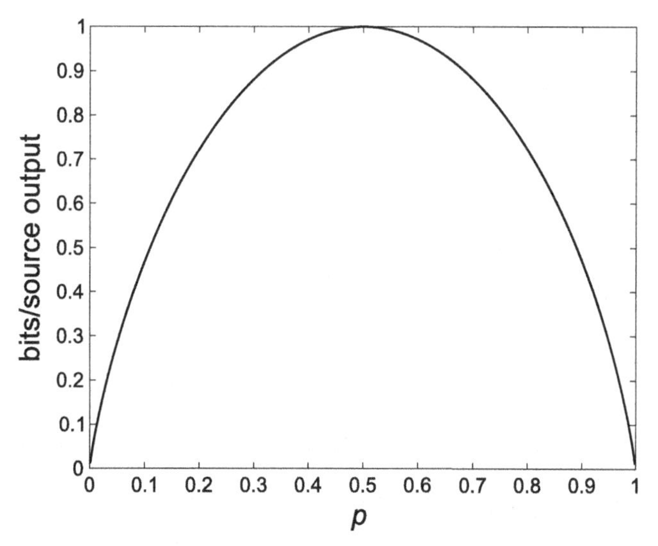

Figure 4.3 Binary entropy function $H_b(p)$. It is maximized for $p = 1/2$, that is, when the binary random variable is the most uncertain. It is zero for $p = 0$ or 1, when the random variable is deterministic.

If $p = 0$ or 1, the entropy is 0 (i.e., the source contains no information and is fully predictable). Meanwhile for $p = 1/2$, it is 1 bit per source output; the source is the most uncertain. This is consistent with our earlier discussion on the i.i.d. binary sequences, and how easy or difficult it is to represent them.

Example 4.4

The entropy of the source in Example 4.2 is given by

$$H(X) = -\sum_{i=1}^{M} p_i \log p_i \tag{4.12}$$

$$= -2\cdot\frac{1}{8}\log\frac{1}{8} - 3\cdot\frac{1}{4}\log\frac{1}{4} \tag{4.13}$$

$$= \frac{9}{4} \text{ bits/source output.} \tag{4.14}$$

Let us now ask the following question: Consider all possible sources with a finite alphabet; which one contains the most information (i.e., which source is the most uncertain)? In other words, what is the PMF that maximizes the entropy among all possible discrete random variables, taking on at most N possible values? A moment of thought suggests that the uniform source (i.e., the one with PMF $\left\{\frac{1}{N}, \frac{1}{N}, \ldots, \frac{1}{N}\right\}$) should be the most uncertain, since it is the most difficult one to guess the next source output value (in the sense that when we guess the next value as the most likely source output, the probability of making an error is the largest).

The above observation is true, as can be shown as follows. Consider a random variable taking on N different values with PMF $\{p_1, p_2, \ldots, p_N\}$. We can write

$$\sum_{i=1}^{N} p_i \ln\left(\frac{1}{Np_i}\right) \leq \sum_{i=1}^{N} p_i \left(\frac{1}{Np_i} - 1\right) \tag{4.15}$$

$$= \sum_{i=1}^{N} \frac{1}{N} - \sum_{i=1}^{N} p_i \tag{4.16}$$

$$= 0, \tag{4.17}$$

where we have used $\ln(x) \leq x - 1$ (which is valid for all $x > 0$) in the first inequality, and the fact that $\sum_{i=1}^{N} p_i = 1$. In other words, we have

$$\sum_{i=1}^{N} p_i \ln\left(\frac{1}{p_i}\right) + \sum_{i=1}^{N} p_i \ln\left(\frac{1}{N}\right) \leq 0. \tag{4.18}$$

Recalling the definition of entropy and using $\sum_{i=1}^{N} p_i = 1$, we can then write

$$H(X) \leq \ln(N) \tag{4.19}$$

for any PMF. Noting also that the equality is achieved for a uniform random variable with cardinality N, we argue that the uniform source is the most uncertain, as expected.[2]

The concept of entropy has an operational meaning in the compression of discrete memoryless sources, which is discussed next.

Source-Coding Theorem. A DMS with entropy $H(X)$ can be encoded with an arbitrarily small error probability at a rate of R bits per source output for any $R > H$. Note that an alternate statement is also true: there exists a source code with any rate $R > H$ with zero probability of error. The converse is also true: there is no lossless compression algorithm with a rate below the source's entropy, that is, $R < H$ is not possible without introducing any distortion in the description of the source.

We will not provide a formal proof of this result. Nevertheless, we can explain it through a connection to our earlier discussion on discrete memoryless sources, that is, sequences of i.i.d. discrete random variables. By invoking the law of large numbers, we know that as the block length is increased, with a probability arbitrarily close to 1, we will observe an output sequence with $\approx np_i$ many x_is (where n is the length of the sequence, and x_i is the source output with probability p_i). We have argued earlier that there are $\approx 2^{nH(X)}$ such sequences. Therefore, for source sequences of this form, only $\approx nH(X)$ bits are sufficient. We can encode the remaining (unlikely) possible sequences using less than $\approx n \log_2 |\mathcal{X}|$ bits. We can differentiate the sequences from the two sets by appending the encoder outputs with "0" or "1." A straightforward calculation on the average length of the description will show that the number of bits needed is $\approx nH(X)$ bits for n source outputs, thus establishing the existence of a lossless encoding scheme which requires on average $H(X)$ bits per source output.

Note that while it is possible to meet a rate of $R = H$ for some sources, we are only guaranteed to achieve a rate arbitrarily close to this limit in general.

Example 4.5

Consider the source with PMF $\{1/2, 1/4, 1/8, 1/16, 1/16\}$, whose entropy can be computed to be $H(X) = \frac{15}{8}$ bits/source output. Hence, to encode this source (i.e., to compress it) losslessly, we only need $R > \frac{15}{8}$ bits/source output. Notice that an encoding scheme achieving a rate close to this limit cannot simply use the trivial binary representation for the individual source outputs (which would require 3 bits for each).

Note that, even though we restricted the discussion in this section to discrete memoryless sources, the concepts of entropy and the corresponding source-coding theorem also generalize to other (more general) sources. However, such extensions are beyond our scope.

[2] Note also that for all $x \neq 1$, the inequality $\ln(x) < x - 1$ is strict; we can argue that no source with cardinality N other than uniform can attain $H(X) = \ln(N)$.

We close this section by noting that there are also practical compression methods. Three such algorithms are described in the next section.

4.4 Source-Coding Algorithms

So far, we have been concerned with the information content of digital sources, the concepts of entropy, and the source-coding theorem. In particular, we have stated that there are lossless source-coding algorithms that can compress the source at a rate down to the entropy per source output for discrete memoryless sources. While the results of the previous section were on the ultimate limits of compression, in this section, we are concerned with source-coding algorithms that can be implemented in practice. Specifically, we provide three lossless source-coding algorithms: Huffman coding, Lempel–Ziv coding, and Lempel–Ziv–Welch coding. While many other practical source-coding algorithms exist, such as arithmetic coding, we do not go over them in our coverage.

Note that it is also possible to represent digital information sources in an approximate form via lossy compression techniques. For instance, a digital image file in a certain format can be compressed in a different format and resolution as an approximation of the original one. While lossy compression is also very important and used in practice in many applications, we limit our coverage to only three examples of lossless compression techniques.

4.4.1 Huffman Coding

The main idea in Huffman coding is to choose the codeword lengths such that the more probable symbols have shorter representations, hence reducing the number of bits needed to describe the source output. The algorithm assumes that the PMF of the source is available. Therefore, it is not a *universal* source-coding algorithm.

The Huffman coding algorithm is given next.

Algorithm 4.1 Huffman coding algorithm.

(1) Sort the source outputs in decreasing order of probability.
(2) Merge the two least probable outputs into a single output, whose probability is the sum of the two probabilities.
(3) If only two outputs remain, go to Step 4, else go to Step 1.
(4) Arbitrarily assign 0 and 1 to the remaining outputs as codewords.
(5) If an output is a merger of two outputs, append the current codeword with "0" and "1" to obtain the codeword for the preceding outputs. Repeat until no output is preceded by another.

A small note about Step 1 is the following: there is no need to explicitly sort the source outputs at each stage with respect to their probabilities; it is only required to pick the two outputs with the lowest probabilities and combine them (in Step 2).

Note that Huffman coding produces prefix-free codewords, that is, no codeword is a prefix of any other; hence, the codewords are uniquely decodable – as we will demonstrate later. Such codes are called *prefix codes.* Prefix codes are also instantaneously decodable, that is, each output can be identified as soon as its codeword is observed – there is no need to wait until the end of the encoded sequence. This is a highly desirable property in practice.

The implementation details of the Huffman coding algorithm can best be illustrated via examples.

Example 4.6

Design a Huffman code for a DMS with five outputs $\{a_1, a_2, a_3, a_4, a_5\}$ whose probabilities are $1/2, 1/4, 1/8, 1/16$, and $1/16$, respectively.

Solution

Our objective is to assign a codeword to each of the five symbols following the rules of the Huffman coding algorithm. Figure 4.4 depicts explicitly the merging steps and the assignment of codewords. In the merging process, the least likely outputs at each step are combined, and the sum of their probabilities is assigned as the probability for the combined symbol. In the assignment of codewords, in this example, the upper branches are labeled 0 while the lower branches are labeled 1, and the overall codeword is read backward to complete the Huffman code design. Note that the assignment of the codewords is not unique, as we arbitrarily selected the upper and lower branch labels.

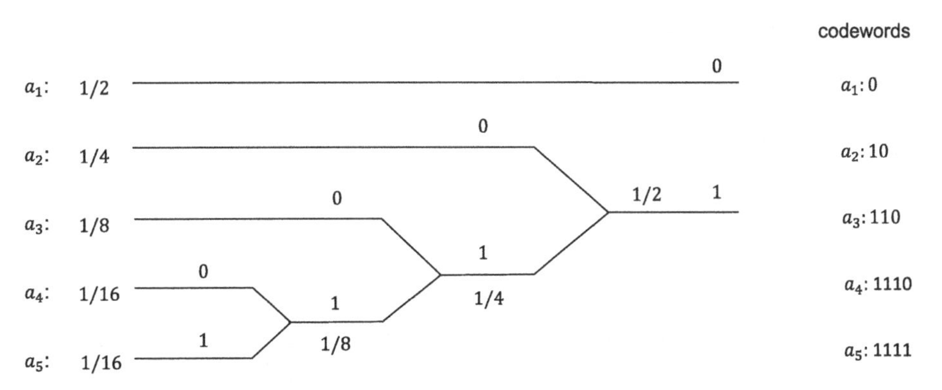

Figure 4.4 Huffman coding of the 5-ary source example. Recall that the codewords are obtained by reading the assigned bits to the branches backward.

Let us now compute the average length of the description per source output. Denoting by $l(a_i)$ the length of the codeword corresponding to the symbol a_i, we obtain

$$\bar{R} = \sum_{i=1}^{5} \mathbb{P}(X = a_i) l(x_i) \tag{4.20}$$

$$= 1 \times \frac{1}{2} + 2 \times \frac{1}{4} + 3 \times \frac{1}{3} + 4 \times \frac{1}{16} + 4 \times \frac{1}{16} \tag{4.21}$$

$$= 15/8 \text{ bits/source output.} \tag{4.22}$$

This means that if we generate a long sequence of source outputs (with the given source statistics) and represent each source output using the designed Huffman code, the length of the encoded sequence is expected to be 15/8 times the number of source samples. Incidentally, for this example, the average length per source output is precisely equal to the source's entropy given in Example 4.5. From the source-coding theorem, we can conclude that this code is optimal and cannot be improved further (i.e., the average length per source output cannot be made smaller). We also notice that no codeword is a prefix of another (i.e., this is a prefix code). Hence, it is a uniquely and instantaneously decodable code. To illustrate this point further, consider the binary sequence

$$1\,1\,0\,0\,1\,0\,1\,1\,1\,0\,0\,1\,0\,1\,1\,1\,1\,0\,1\,1\,0 \tag{4.23}$$

as the encoded version of a sequence of source outputs. Using the codewords for the five symbols shown in Fig. 4.4, the only possible way of parsing this sequence is as follows:

$$1\,1\,0/\ 0/\ 1\,0/\ 1\,1\,1\,0/\ 0/\ 1\,0/\ 1\,1\,1\,1/\ 0/\ 1\,1\,0. \tag{4.24}$$

In other words, the original source samples are obtained as a_3, a_1, a_2, a_4, a_1, a_2, a_5, a_1, a_3. Notice also that each decoded source output is identified as soon as it is observed, without waiting for the rest of the encoded sequence.

Let us give another example.

Example 4.7

Design a Huffman code for the ternary DMS with alphabet $\{a, b, c\}$ with equally likely outputs, that is, $P(X = a) = P(X = b) = P(X = c) = \frac{1}{3}$.

Solution

We apply the Huffman coding algorithm to this source as shown in Fig. 4.5.

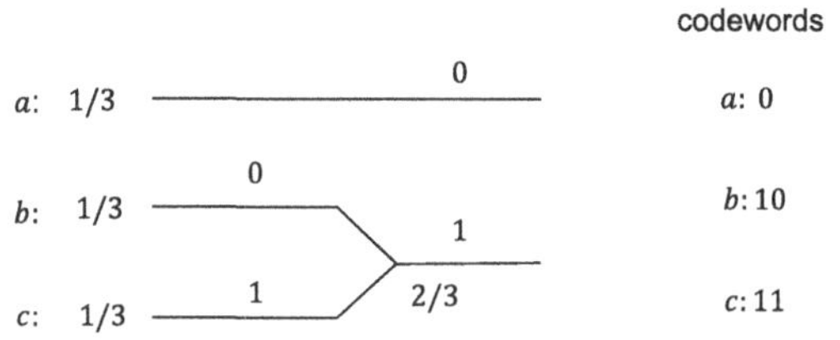

Figure 4.5 Huffman coding of the ternary uniform source.

a: 1/3, b: 1/3, c: 1/3; b and c merged (0, 1) into 2/3; a: 0, 2/3: 1

codewords
a: 0
b: 10
c: 11

The average codeword length is calculated as

$$\bar{R} = 1 \times \frac{1}{3} + 2 \times \frac{1}{3} + 2 \times \frac{1}{3} = \frac{5}{3} \text{ bits/source output.} \tag{4.25}$$

The entropy of the source is given by

$$H(X) = -\sum_{i=1}^{3} \frac{1}{3} \times \log\left(\frac{1}{3}\right) \approx 1.585 \text{ bits/source output,} \tag{4.26}$$

hence $\bar{R} > H$ (as expected from the source-coding theorem).

Since there is a significant difference between the entropy and the average codeword length, it is possible to design another lossless source code with a significantly improved performance for the above DMS, that is, with an average length much closer to thc cntropy.

We can obtain an improved Huffman code for the source in the above example by considering more than one source output at a time. The following example illustrates this point.

Example 4.8

Consider the discrete memoryless source in Example 4.7. Design a Huffman code for it using two source outputs at a time, that is, for the extended source with nine elements *aa*, *ab*, *ac*, *ba*, ..., *cc* with the understanding that the new symbol *xy* means that *x* is the first source output, and *y* is the second.

Solution

Given the original source statistics (and the assumption that the source outputs are independent), we can compute the probability of each of these elements. For instance,

$$\mathbb{P}(X = aa) = \mathbb{P}((X_1, X_2) = (a, a)) = \mathbb{P}(X_1 = a)\mathbb{P}(X_2 = a) = \frac{1}{3} \times \frac{1}{3} = \frac{1}{9}. \quad (4.27)$$

Similarly, all the other combined symbols will have the same probability of $1/9$.

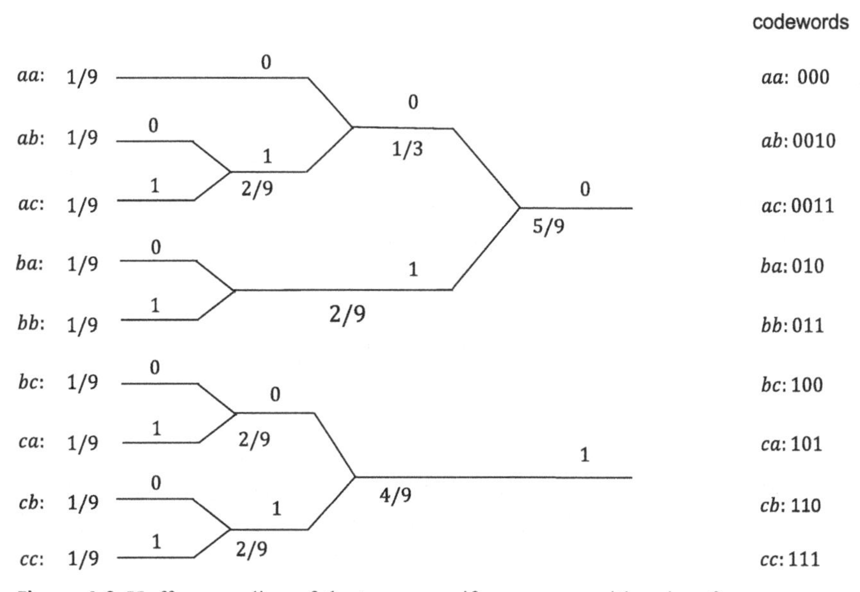

Figure 4.6 Huffman coding of the ternary uniform source with pairs of source output considered at a time.

The Huffman coding algorithm can be applied to the extended source as shown in Fig. 4.6. The average length of the encoded sequence is then

$$\bar{R} = 7 \times 3 \times \frac{1}{9} + 2 \times 4 \times \frac{1}{9} = \frac{29}{9} \text{ bits/extended source output.} \quad (4.28)$$

Since each extended source output contains two original source outputs, the average length is $\bar{R} = \frac{29}{18} \approx 1.611$ bits/source output, which is much closer to the entropy of the source than the average length for the code in Example 4.7. Observe that there is still some room for improvement as the entropy is ≈ 1.585 bits/source output (smaller than the average length of 1.611 bits/source output) by combining more than two source outputs at a time.

We conclude this section by emphasizing some important properties of Huffman coding. Specifically:

- Huffman coding requires the source statistics to be known, that is, it is not a universal source-coding algorithm.
- The resulting code is not unique. Combining at each stage different output pairs (when some outputs have identical probabilities) would result in other Huffman codes. Also, assignments of 0s and 1s at each stage to upper and lower branches are arbitrary, resulting in different codes.
- Huffman coding is a fixed-to-variable-length source-coding algorithm with prefix-free codewords.
- For Huffman coding, the average length of the codewords cannot be far from the source's entropy. It can be shown that the average codeword length is less than $H(X) + 1$ bits/source output, where $H(X)$ is the entropy of the DMS.
- Combining source outputs and applying Huffman coding to multiple source outputs at a time will generally result in an average codeword length closer to entropy (unless it is precisely equal to the entropy, in which case, it will remain the same). Indeed, considering multiple source outputs at a time allows for Huffman coding to exploit the memory in the sequence being encoded further. Considering also the previous property, we can argue that if k outputs are considered together, for the corresponding Huffman code, the average codeword length (per source output) will be at most $1/k$ bits per original source output away from the entropy, that is, by taking k large enough the code is guaranteed to be (near) optimal. Of course, this comes at the cost of increased complexity (in terms of the code construction and the size of the encoding/decoding tables).

4.4.2 Lempel–Ziv Coding

Lempel–Ziv coding is another practical source-coding algorithm. Contrary to the Huffman coding algorithm described in the previous section, it does not require source statistics; hence, it is a universal compression algorithm. Effectively, the algorithm can learn the source statistics from a long sequence of source outputs. Also, it is a variable-to-fixed-length source-coding algorithm, that is, phrases of different lengths are mapped to fixed-length codewords – different from Huffman coding, which is a fixed-to-variable-length encoding scheme.

The main idea is as follows:

- Start with an initial dictionary and parse the input sequence uniquely into phrases of varying lengths. Identify the phrases of the smallest lengths that did not appear when parsing and update the dictionary by including them.
- Use the dictionary for the variable-to-fixed-length encoding.

Let us explain the process using a specific sequence. Consider the Lempel–Ziv encoding of

$$001011\ldots. \tag{4.29}$$

Initially, the only dictionary entry is the empty string (at the 0th location). When parsing, the first (new) phrase we encounter is "0," and this element is placed into the dictionary as the next element and the corresponding encoder output is the address of the dictionary entry (i.e., 0) and the new bit encountered, which is "0." This phrase is added to the dictionary as the next element. Continuing the parsing operation, we encounter the sequence "01" as the next sequence, which has not yet appeared. Therefore, the encoder output is the address of the dictionary entry corresponding to "0" along with the new bit (which is "1"), and the new dictionary element is "01." We next observe the sequence "011" and encode it as the dictionary location of "01" along with the next bit "1." This new phrase "011" is added to the dictionary as the next element. The process continues in the same manner until the end of the sequence is reached.

The decoding process is trivial as the decoder can obtain the dictionary contents from its observed encoded sequence. Note also that the symbols being encoded need not be binary.

Let us show the entire encoding process for a longer sequence in the following example.

Example 4.9

Encode the binary sequence

$$01\,\underbrace{000\cdots0}_{21\text{ many}}\,\underbrace{111\cdots1}_{15\text{ many}}\,\underbrace{000\cdots0}_{24\text{ many}}\,\underbrace{111\cdots1}_{30\text{ many}}\,\underbrace{000\cdots0}_{60\text{ many}} \tag{4.30}$$

using LZ coding.

Solution

Using the steps described above, we parse the sequence as follows:

$$\begin{aligned}&0/1/00/000/0000/00000/000000/01/11/111/1111/11111/0000000/00000000/\\&000000000/111111/1111111/11111111/111111111/0000000000/00000000000/\\&000000000000/0000000000000/00000000000000.\end{aligned} \tag{4.31}$$

As per the LZ algorithm, these phrases are placed in a dictionary, and the dictionary elements' address and the new bit at each step comprise the encoded sequence. This process is completed in Table 4.1 (note that the dictionary location "0" contains the null string).

Table 4.1 LZ encoding process for the given sequence.

Dictionary location	Content	Codeword (location + new bit)
1	0	0 0
2	1	0 1
3	00	1 0
4	000	3 0
5	0000	4 0
6	00000	5 0
7	000000	6 0
8	01	1 1
9	11	2 1
10	111	9 1
11	1111	10 1
12	11111	11 1
13	0000000	7 0
14	00000000	13 0
15	000000000	14 0
16	111111	12 1
17	1111111	16 1
18	11111111	17 1
19	111111111	18 1
20	0000000000	15 0
21	00000000000	20 0
22	000000000000	21 0
23	0000000000000	22 0
24	00000000000000	23 0

In this example, we have a total of 25 dictionary entries (including the null string); hence, we need 5 bits to specify their locations. Note that the dictionary itself is not part of the encoded sequence. Therefore, each codeword (corresponding to each phrase) consists of $5 + 1 = 6$ bits. Hence, the overall encoded sequence for this example is $24 \times 6 = 144$ bits long, as there are a total of 24 phrases identified in the sequence. The original sequence is 152 bits; hence, there is some compression. If we encode longer sequences, the compression rate will be increased. Indeed, asymptotically, the average length will achieve the theoretical lower limit (the *entropy rate* of the source), under some conditions on the source statistics.

Given the encoded sequence (i.e., the third column in the table), we can easily determine the original sequence. This works because the dictionary can be reconstructed at the receiver side during decoding. For instance, when the decoder sees the dictionary location along with the new bit as "0 0," it recovers the null string appended with the new bit "0" and hence declares "0" as the decoded phrase, and obtains the new dictionary entry (at location "1") as "0". This process follows in exactly the same way until the decoding process is complete. Notice also that we do not need to wait until the end of the entire sequence to run the decoding process.

An important property is that the LZ algorithm can exploit the memory in the sequence being compressed in a natural manner. The phrases that appear are placed in the dictionary so that when they are re-encountered, their description is very simple. For example, an English text about the subject of communications technology may contain many instances of the word "information"; hence, this word will be placed in the dictionary (at some point) and, in all subsequent instances, will only require a few bits to represent.

4.4.3 Lempel–Ziv–Welch Coding

A variation of Lempel–Ziv coding is the Lempel–Ziv–Welch algorithm. In this case, the encoder and decoder first agree on an initial dictionary. At any given step, the encoder continues parsing the text by using the longest sequence currently in the dictionary, and it employs the specific dictionary entry as the next element of the encoded version. Also, a new dictionary entry is added at each step by appending the current phrase with the next character in the text.

Let us give a detailed example.

Example 4.10

Assume that we have a 4-ary alphabet consisting of the letters a, b, c, and d. Encode the following sequence using LZW coding:

$$\underbrace{aaa\cdots a}_{15\text{ times}}\underbrace{bbb\cdots b}_{15\text{ times}}\underbrace{ccc\cdots c}_{15\text{ times}}\underbrace{aaa\cdots a}_{15\text{ times}}\underbrace{ddd\cdots d}_{15\text{ times}}\#. \tag{4.32}$$

Assume also that the initial dictionary is $a \rightarrow 1$, $b \rightarrow 2$, $c \rightarrow 3$, and $d \rightarrow 4$, and the end of file symbol is "#," which is the 0th dictionary element.

Solution

The encoder first reads "a," the longest phrase currently in the dictionary, and declares "1" as the next phrase. At the same time, it appends the next character, also "a," to the presently encoded phrase to obtain the new dictionary element "aa" as "5." Continuing with the parsing operation, the encoder next sees the phrase "aa" as the longest piece of text in the (extended) dictionary. Therefore, it encodes this phrase as "5" and obtains the following dictionary entry "aaa" as "6." The process proceeds until the end of file symbol is reached. The parsing operation, as well as the creation of the new dictionary entries for the first few characters of the text, is given below:

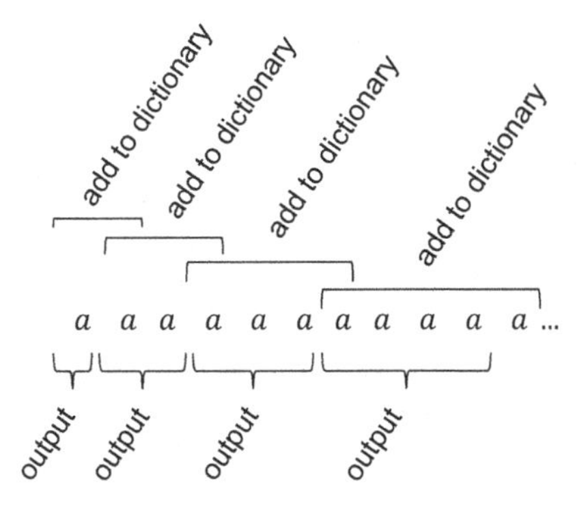

Table 4.2 presents the entire encoding process. The encoder output is given by

$$1,5,6,7,8,2,10,11,12,13,3,15,16,17,18,8,20,7,4,23,24,25,26,0. \quad (4.33)$$

Table 4.2 LZW encoding process.

Current sequence	Next character	Output	Extended dictionary
a	a	1	$aa : 5$
aa	a	5	$aaa : 6$
aaa	a	6	$aaaa : 7$
$aaaa$	a	7	$aaaaa : 8$
$aaaaa$	b	8	$aaaaab : 9$
b	b	2	$bb : 10$
bb	b	10	$bbb : 11$
bbb	b	11	$bbbb : 12$
$bbbb$	b	12	$bbbbb : 13$
$bbbbb$	c	13	$bbbbbc : 14$
c	c	3	$cc : 15$
cc	c	15	$ccc : 16$
ccc	c	16	$cccc : 17$
$cccc$	c	17	$ccccc : 18$
$ccccc$	a	18	$ccccca : 19$
$aaaaa$	a	8	$aaaaaa : 20$
$aaaaaa$	a	20	$aaaaaaa : 21$
$aaaa$	d	7	$aaaad : 22$
d	d	4	$dd : 23$
dd	d	23	$ddd : 24$
ddd	d	24	$dddd : 25$
$dddd$	d	25	$ddddd : 26$
$ddddd$	#	26	
#	—	0	

Using 5 bits to represent each of the dictionary entries (as there are 27 entries in the extended dictionary, this is sufficient), we need a total of $24 \times 5 = 120$ bits to represent the LZW encoded version of the text. This is in contrast to the 150 bits needed by the original source (with 2 bits per source symbol ignoring the end of file symbol), that is, we achieve a 20% compression.

We observe from the example that when a new phrase (of the shortest length) is encountered, it is placed in the dictionary, and when it is re-encountered, it becomes extremely simple to encode. For instance, the phrase "$aaaaaa$" is encoded by a single dictionary entry (using only 5 bits), while it normally would have required 12 bits to represent without any encoding.

We emphasize that we do not need the source statistics in LZW coding. The algorithm learns the repetitive patterns from the text itself, and hence, it can exploit the memory in the text to achieve a significant compression performance as in LZ coding.

The decoder observes the encoded sequence, and along with the knowledge of the initial dictionary (which is already available), it decompresses the encoded sequence and obtains the original text. This procedure is illustrated in the following example.

Example 4.11

Consider the encoded sequence in Example 4.10, that is,

$$1, 5, 6, 7, 8, 2, 10, 11, 12, 13, 3, 15, 16, 17, 18, 8, 20, 7, 4, 23, 24, 25, 26, 0. \tag{4.34}$$

Given that the initial dictionary is $a \to 1$, $b \to 2$, $c \to 3$, $d \to 4$, and $\# \to 0$, show the steps of the decoding process and obtain the original text.

Solution

The decoder maps the elements of the encoded text back to the original text. At the same time, it reconstructs the extended dictionary obtained at the encoder during the compression process. The decoding process is given in Table 4.3.

Table 4.3 LZW decoding process.

		New dictionary entry	
Input	Output	Full	Conjectured
1	a	—	$5 : a?$
5	aa	$5 : aa$	$6 : aa?$
6	aaa	$6 : aaa$	$7 : aaa?$
7	$aaaa$	$7 : aaaa$	$8 : aaaa?$
8	$aaaaa$	$8 : aaaaa$	$9 : aaaaa?$
2	b	$9 : aaaaab$	$10 : b?$
10	bb	$10 : bb$	$11 : bb?$
11	bbb	$11 : bbb$	$12 : bbb?$
12	$bbbb$	$12 : bbbb$	$13 : bbbb?$
13	$bbbbb$	$13 : bbbbb$	$14 : bbbbb?$
3	c	$14 : bbbbbc$	$15 : c?$
15	cc	$15 : cc$	$16 : cc?$
16	ccc	$16 : ccc$	$17 : ccc?$
17	$cccc$	$17 : cccc$	$18 : cccc?$
18	$ccccc$	$18 : ccccc$	$19 : ccccc?$
8	$aaaaa$	$19 : cccccca$	$20 : aaaaa?$
20	$aaaaaa$	$20 : aaaaaa$	$21 : aaaaaa?$
7	$aaaa$	$21 : aaaaaaa$	$22 : aaaa?$
4	d	$22 : aaaad$	$23 : d?$
23	dd	$23 : dd$	$24 : dd?$
24	ddd	$24 : ddd$	$25 : ddd?$
25	$dddd$	$25 : dddd$	$26 : dddd?$
26	$ddddd$	$26 : ddddd$	—
0	#		

While the reconstruction of the extended dictionary is mostly straightforward, there is one point that needs clarification. For certain steps, the dictionary entry is not fully known when encountered in the encoded sequence. This requires resolving the dictionary entry along with the next phrase in the text. For instance, in the second line of Table 4.3, "5" is encountered when it is not fully known what phrase it represents. Noting that such a case can only occur if a newly created dictionary element is immediately encountered in the original text, the only possibility is that the question mark in the conjectured entry corresponding to "5" (in the first line of the table) is its first character. Hence, the phrase and the dictionary entry are resolved as "*aa* : 5." This situation is repeatedly encountered in this example; all are resolved similarly.

4.4.4 Some Remarks

Lempel–Ziv coding and its variants (such as LZW coding) are universal source-coding algorithms, that is, they do not require the source statistics. In addition, they are highly effective at exploiting the memory in the source. Specifically, it can be shown that they are optimal for important classes of stationary and ergodic sources (whose proof or detailed discussion is beyond our scope). That is, the average length per source output by this compression technique becomes the entropy rate of the source as the sequence length goes to infinity.

Note also that LZ coding is a variable-to-fixed-length coding algorithm (as opposed to fixed-to-variable-length encoding, as in Huffman coding).

When encoding long sequences, any fixed dictionary size will become insufficient at some point. To address this problem, the encoder must purge some dictionary entries when the dictionary is full. This can be done in different ways, with the main idea being to purge the entries that are less likely to be re-encountered. This should be done in a deterministic manner in such a way that the decoder can implement it in a precise way during the decompression process.

Due to its attractive properties, LZ coding and its variants have found many practical applications. For instance, the compression algorithms used in different computer operating systems implement these ideas.

4.5 Chapter Summary

This chapter was concerned with basic digital information source modeling, the information content of a source, and practical data compression techniques. We stated Shannon's celebrated source-coding theorem for discrete memoryless sources and detailed several specific lossless source-coding algorithms: Huffman, Lempel–Ziv, and Lempel–Ziv–Welch coding. This is the only chapter that deals exclusively with digital information sources; in subsequent chapters, we will be concerned with transmitting digital information (i.e., a sequence of symbols) over noisy channels and related concepts.

PROBLEMS

4.1 The output of a discrete memoryless source is modeled as a geometric random variable with parameter p. Determine the entropy of the source. What is the minimum (average) number of bits per source output required to represent the source outputs in a lossless fashion?

4.2 Let X be a discrete random variable taking values on non-zero integers with

$$\mathbb{P}(X = k) = \mathbb{P}(X = -k) = p_k,$$

for $k = 1, 2, 3, \ldots$. Define another random variable as $Y = |X|$.
Determine the entropy of Y as a function of the entropy of X. Comment on your result.

4.3 Let the output of a discrete memoryless source be a binomial random variable with parameters $n = 5$ and $p = 0.2$. Determine the entropy of the source in bits/source output.

4.4 Let X be a positive integer-valued random variable with mean $\mathbb{E}[X] = \alpha$. Determine the largest possible value of the entropy of X and the corresponding probability mass function.

4.5 Can the entropy of a discrete random variable be unbounded? (*Hint*: Is a probability mass function of the form $1/k \log k$ for $k \in \mathbb{Z}^+$ a valid one?)

4.6 Consider an 8-ary discrete memoryless source with output alphabet $\{a, b, c, d, e, f, g, h\}$ and PMF $\{1/4, 1/4, 1/6, 1/8, 1/8, 1/36, 1/36, 1/36\}$.
(a) What is its entropy in bits/source output?
(b) Design a binary Huffman code for the source.
(c) Determine the average length of the code in the previous part in bits/source output. Can a different (lossless) source code have a smaller average length? Why or why not?

4.7 Design a binary Huffman code for a DMS with PMF $\{1/5, 1/5, 1/5, 1/10, 1/10, 1/10, 1/10\}$. Determine the average length of the code and compare it with the source entropy. Is the code optimal in terms of average length?

4.8 Consider a DMS with a PMF $\{1/2, 1/3, 1/6\}$.
(a) Design a binary Huffman code for the source, compute the average codeword length, and compare it with the source entropy.
(b) Design a binary Huffman code considering pairs of source outputs. What is the average codeword length (per source output)? How does the result compare with that of the previous part?

4.9 Consider a discrete memoryless source with alphabet $\{a_1, a_2, \ldots, a_8\}$ with PMF

$$\mathbb{P}(a_1) = 0.25,\ \mathbb{P}(a_2) = 0.2,\ \mathbb{P}(a_3) = \mathbb{P}(a_4) = 0.15,\ \mathbb{P}(a_5) = 0.1,$$
$$\mathbb{P}(a_6) = \mathbb{P}(a_7) = \mathbb{P}(a_8) = 0.05.$$

(a) Determine the entropy of the source in bits per source output.
(b) Design a Huffman code for the source and determine the average number of bits per source output.

4.10 Consider an 8-ary discrete memoryless source with output alphabet $\{a, b, c, d, e, f, g, h\}$ and PMF

$$\{1/2, 1/4, 1/8, 1/16, 1/64, 1/64, 1/64, 1/64\}.$$

(a) What is the entropy of the source in bits/source output?
(b) Design a Huffman code for the source.
(c) Determine the average length of the code in the previous part in bits/source output. Can a different (lossless) source code have a smaller average length? Why or why not?

4.11 Consider a discrete memoryless source $\mathcal{S}_1$ with outputs $X_1, X_2, \ldots$. Assume that the source samples take on the values $a_1, a_2, \ldots, a_6$ with probabilities $1/4, 1/4, 1/4, 1/8, 1/16, 1/16$, respectively.

(a) Design a Huffman code for $\mathcal{S}_1$ assuming that each source sample is encoded separately. Compute the source's entropy and determine whether the code can be improved in terms of its average length.
(b) Consider a second discrete memoryless source $\mathcal{S}_2$ with outputs $Y_1, Y_2, \ldots$. Assume that the samples take on the values b_1, b_2, b_3. Also assume that the ith source output Y_i is correlated with the ith source output X_i of the first DMS $\mathcal{S}_1$ (but it is independent of all other source samples of $\mathcal{S}_1$). The conditional PMF of Y_i given X_i is as follows:

$$\begin{aligned}
&\mathbb{P}(Y = b_1 | X = a_1) = 1, \\
&\mathbb{P}(Y = b_2 | X = a_2) = 1/2, \quad \mathbb{P}(Y = b_3 | X = a_2) = 1/2, \\
&\mathbb{P}(Y = b_2 | X = a_3) = 1/2, \quad \mathbb{P}(Y = b_3 | X = a_3) = 1/2, \\
&\mathbb{P}(Y = b_1 | X = a_k) = 1, \quad \text{for } k = 4, 5, 6.
\end{aligned}$$

Determine the PMF of Y_i. Design a Huffman code for $\mathcal{C}_2$, assuming the samples are encoded separately. Can the code be improved in terms of average length or not?

(c) We now encode the outputs of the two sources $\mathcal{S}_1$ and $\mathcal{S}_2$ together, that is, we encode the pairs (X_i, Y_i) jointly. Determine the possible (X_i, Y_i) values and the respective probabilities, and design a Huffman code (taking one such pair at a time). What is the average length of the code? Is this lower than the sum of the entropies you have obtained in the previous two parts? If so, how can you explain this? Can the code be improved in terms of the average length by applying Huffman coding to multiple pairs of source outputs together? Why or why not?

4.12 Design a Huffman code for a discrete memoryless source with PMF

$$\{1/16, 1/16, 1/16, 1/16, 1/8, 1/8, 1/8, 1/6, 5/24\}.$$

Determine the entropy of the source and the average length of the code. Can this code be improved in terms of its average length?

4.13 Determine whether or not the given set of codewords can be a Huffman code. Justify your reasoning.

(a) 00, 10, 11, 01, 101, 100.
(b) 10, 11, 01, 001, 0000, 00001.
(c) 000, 001, 010, 011, 100, 101, 110, 1110, 1111.

4.14 Consider a discrete memoryless source (call it DMS-1) consisting of three symbols $\{s_1, s_2, s_3\}$ in its alphabet with probabilities 0.2, 0.5, and 0.3, respectively. Consider another discrete memoryless source (call it DMS-2) consisting of two symbols $\{\tilde{s}_1, \tilde{s}_2\}$ in its alphabet with probabilities 0.6 and 0.4, respectively. Suppose that DMS-1 and DMS-2 generate symbols at the same rate, and their outputs are independent. We want to consider two symbols simultaneously, where the first symbol is from DMS-1 and the second is from DMS-2, and perform Huffman coding. (For example, $s_1\tilde{s}_1$ is one such symbol pair.)

Perform Huffman coding for this scenario and list the codewords. Also, calculate the average codeword length.

4.15 Consider a discrete memoryless source with alphabet $\{a, b, c\}$. The PMF of the source outputs X is given by $\mathbb{P}(X = a) = 2/3$, $\mathbb{P}(X = b) = 1/6$, $\mathbb{P}(X = c) = 1/6$.

Apply Huffman coding considering two source samples at a time and determine the resulting average length of the encoder output in bits/source sample.

Is it possible to improve this average length further, for example, by considering a larger number of source outputs at a time?

4.16 Encode the following binary sequence using Lempel–Ziv coding:

$$1101\ \underbrace{000\cdots0}_{10\text{ many}}\ \underbrace{111\cdots1}_{20\text{ many}}\ \underbrace{000\cdots0}_{12\text{ many}}\ \underbrace{111\cdots1}_{20\text{ many}}\ \underbrace{000\cdots0}_{20\text{ many}}. \tag{4.35}$$

How do the original sequence and its encoded version compare in terms of length?

4.17 Encode the sequence in the previous problem using LZW coding with the initial dictionary of $0 \to 1$ and $1 \to 2$. How do the original sequence and its encoded version compare in terms of length?

4.18 Assume that the codeword lengths for a binary prefix code are $l_1, l_2, l_3, \ldots, l_N$, where N is the cardinality of the DMS. Prove that $\sum_{i=1}^{N} 2^{-l_i} \leq 1$.

4.19 Assume that a binary sequence has been compressed using Lempel–Ziv–Welch coding where the initial dictionary is simply $0 \to 1$ and $1 \to 2$. The end of file character # is denoted by the 0th dictionary entry.

Determine the original text (sequence of 0s and 1s) if the encoded version is given by

$$1, 3, 1, 2, 4, 7, 4, 6, 8, 9, 11, 3, 10, 13, 0. \tag{4.36}$$

4.20 Assume that LZW coding is used to encode a particular text written using an alphabet of only three distinct characters: a, b, and c. Assume also that the initial dictionary is $a \to 1$, $b \to 2$, and $c \to 3$.

Determine the message if the encoder output is

$$1, 3, 2, 4, 6, 5, 7, 10, 9, 12, 8, 14, 11, 15, 13, 16, 19, 18, 21, 17, 23.$$

What is the resulting compression ratio (assuming that the original text would be represented using two bits per character)?

4.21 Consider a source with alphabet $\{a, b, c, d\}$. Use Lempel–Ziv–Welch coding to encode the sequence

$$\underbrace{aaaaaaaaaa}_{10 \text{ many}}\underbrace{bbbbbbbbbb}_{10 \text{ many}}\underbrace{cccccccccc}_{10 \text{ many}}. \tag{4.37}$$

Assume that the initial dictionary is $a \to 1$, $b \to 2$, $c \to 3$, and $d \to 4$.

Compare the number of bits required to encode the original text and the compressed versions.

4.22 Encode the sequence in the previous problem using LZ coding and compare the number of bits required to encode the original text and the compressed versions.

4.23 Encode the binary sequence

$$1\,0\underbrace{1\,1\,1\cdots1}_{120 \text{ times}}$$

using:

(a) Lempel–Ziv coding;

(b) Lempel–Ziv–Welch coding (with initial dictionary $0 \to 1$, $1 \to 2$).

COMPUTER PROBLEMS

4.24 Consider a discrete memoryless source with alphabet $\{1, 2, 3, 4, 5\}$ and PMF $\{1/2, 1/4, 1/12, 1/12, 1/12\}$.

(a) Generate three different realizations of this source, each containing 100,000 samples independently of each other. Plot the normalized histograms of the number of realizations from each particular symbol for the three cases and compare the result with the source PMF. Are the empirical PMF estimates (i.e., the normalized histograms) close to the true PMF?

(b) Repeat the previous part with three other realizations but with fewer samples, say 1000. Comment on your results.

4.25 Design a binary Huffman code for a discrete memoryless source with alphabet $\{1, 2, 3, 4, 5\}$ and PMF $\{2/5, 1/5, 1/5, 1/10, 1/10\}$. Using this Huffman code, encode a realization of the source with 10,000 samples. Determine the number of bits needed to encode this specific sequence and compare it with the expected length and with the source entropy.

4.26 Design a binary Huffman code for the source in the previous problem by taking two source outputs at a time. Using this Huffman code, encode the same realization of 10,000 samples. What is the total number of bits needed to encode the sequence? How does this compare with the result of the previous problem?

4.26 Design a Huffman code for a source with PMF $\{1/2, 1/3, 1/18, 1/18\ 1/18\}$ (call it PMF-1). Let us call the resulting code as Code 1. This code is used to encode the output sequence of a discrete memoryless source with PMF $\{1/2, 1/4, 1/12, 1/12, 1/12\}$ (call it PMF-2). Note that Code 1 is not specifically designed for this PMF; hence, there is a mismatch between the code and the statistics of the source being encoded.

(a) Compute the entropy of the source with PMF-2.

(b) Generate 10,000 realizations of the source (with PMF-2) and encode the resulting sequence using Code 1. Determine the number of bits needed to encode this sequence and compare this with the entropy computed in part (a). Are the results close to each other?

4.27 Write a MATLAB code that implements Lempel–Ziv coding with the binary alphabet and test it on a sequence of independent Bernoulli random variables with a probability of 0 of 0.9 and a probability of 1 of 0.1.

(a) Generate 10,000 realizations and use the Lempel–Ziv code to encode the source outputs. How many bits are needed to represent the given sequence? Is there a compression? By how much?

(b) Repeat part (a) with 100,000 realizations.

4.28 Repeat the previous problem with Lempel–Ziv–Welch coding.

5 Digital Modulation – Fundamentals

In this chapter, we consider transmission of messages via digital communication techniques. Specifically, we explain the fundamental principles of digital modulation for a channel model commonly encountered in practice: the additive white Gaussian noise (AWGN) channel.

The main idea is as follows. Each message is represented by a specific waveform (of a given duration, called the symbol period), and the messages are transmitted over the communication medium through these waveforms. Assuming the AWGN channel model, the effect of a communication channel is to add noise; hence, the received signal is the transmitted waveform corrupted by additive noise. For most of our coverage throughout the chapter, we assume that there are no timing problems (i.e., the symbol boundaries are known precisely). The receiver observes the noisy waveform corresponding to the transmitted message, and its job is to decide which symbol is transmitted in such a way that the probability of error is minimized.

We will rely on the signal space concepts and a mathematical description of the channel effects for a detailed treatment of digital communication fundamentals. For a given modulation scheme, we identify the signal space and a basis for it, and express the transmitted signals in a vector form. Similarly, the received signal through the noisy channel is also written in a vector form by projecting it onto the signal space. Uncovering the statistics of the received signal vector given the transmitted signals, we identify the optimal decision rule that minimizes the symbol error probability, whose implementation can be accomplished through a correlation-type or a matched filter-type receiver. To assess the system performance, we compute the average error probability as a function of the signal-to-noise ratio at the receiver.

The chapter is organized as follows. We introduce the fundamental problem of digital transmission through a simplified but critical example of antipodal signaling in Section 5.1. In the same section, we introduce the concept of white Gaussian noise to explain the channel effects and give the basic additive white Gaussian channel model. In Section 5.2, we present the concept of signal space, highlighting that the transmitted signals can be written in a vector form for easy mathematical manipulation. We also show that many important calculations involving signals are greatly simplified by using an orthogonal and normalized set of basis functions. The Gram–Schmidt orthonormalization procedure, which is a systematic method to obtain an orthogonal and normalized basis for a given set of signals, is also described. Section 5.3 provides a detailed study of binary antipodal signaling. We derive the maximum a posteriori decision rule and the maximum likelihood decision rule, and we discuss the implementation of correlation-type and matched filter-type receivers. We show that the probability of error analysis for this scheme is tractable and derive the average bit error probability as a function of the signal-to-noise ratio. As matched filtering is a fundamental concept also finding widespread use in other applications, we pay special attention to it and review some of its properties. We extend the coverage to the general case of M-ary signaling in Section 5.4. As obtaining a closed-form average probability of error expression for the general case is not always feasible, we introduce the union bound as a simple technique and argue that it is highly beneficial, particularly for higher signal-to-noise ratios. Furthermore, through a simplified (asymptotic) analysis, we show how to compare different signaling techniques in terms of their error rate performance. In Sections 5.5 and 5.6, we provide as examples the details of two important digital modulation schemes: pulse amplitude modulation and orthogonal signaling, respectively. A brief discussion of timing recovery techniques is given in Section 5.7, and the chapter is concluded in Section 5.8.

5.1 Brief Introduction to Digital Transmission Over Noisy Channels

A basic digital communication block diagram is shown in Fig. 5.1, where we only focus on the operations of the digital modulator/demodulator and the effects of the channel. We do not consider the other elements of an overall digital communication system (generation of the source, source encoding/decoding, and channel encoding/decoding blocks) as they are treated in other chapters.

As illustrated in Fig. 5.1, the digital modulator's role is to generate a waveform to be transmitted over the channel to convey the information being transmitted to the receiver. For concreteness and ease of exposition, let us start with the transmission of binary messages (bits represented as sequences of 0s and 1s) and explain the basic operations at the transmitter and the receiver sides for this setting. We select two distinct signals of duration T_b to represent the two bits, where T_b is referred to as the bit period. Assume that the signal $s_1(t)$ is used to transmit the bit 1 and the signal $s_2(t)$ is used to send the bit 0. Figure 5.2 illustrates an example where a rectangular waveform is used for transmission.

In this (binary) communication system example, the main problem is transmitting a sequence of bits. The time interval $[(i-1)T_b, iT_b]$ is used to transmit the ith

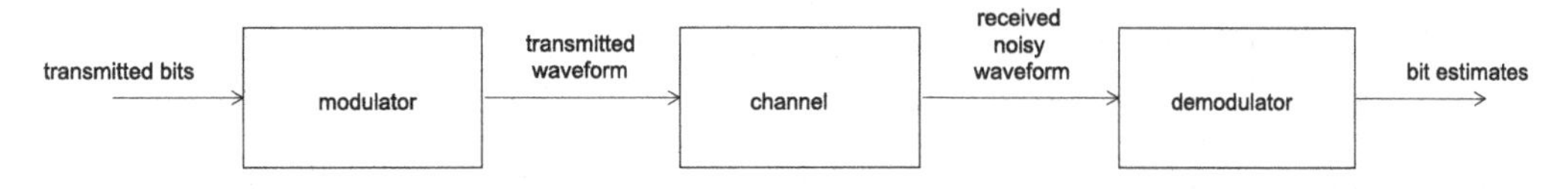

Figure 5.1 Illustration of digital modulation and demodulation over a generic noisy channel.

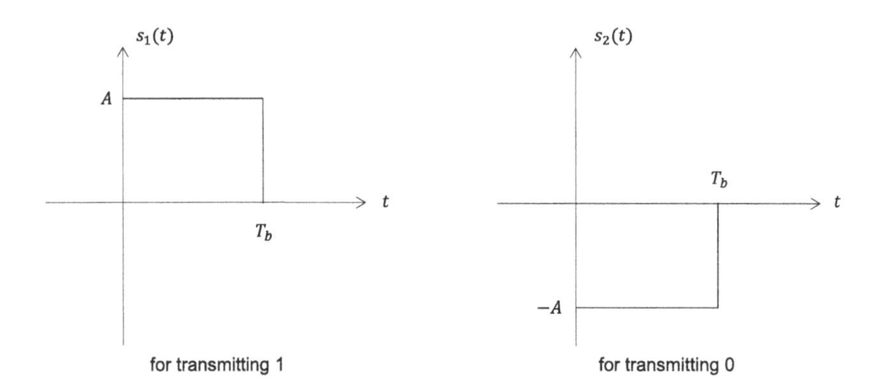

Figure 5.2 Waveform examples for representing different bits.

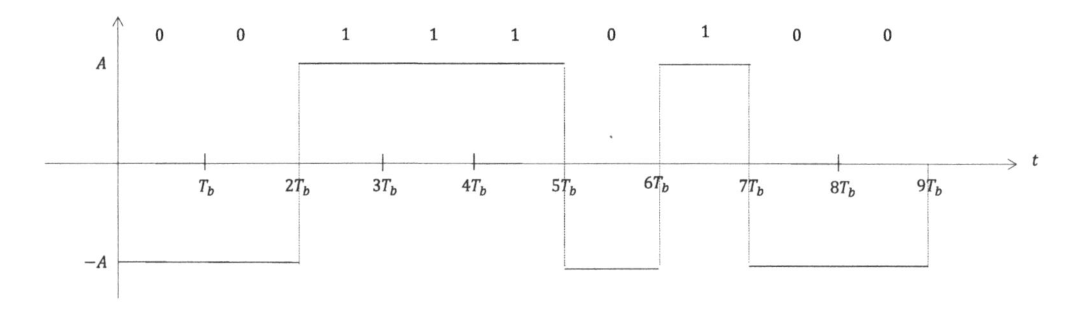

Figure 5.3 Sample transmitted signal.

bit in the sequence (using the pre-specified waveforms), for $i = 1, 2, 3, \ldots$. For instance, assuming that the two waveforms in Fig. 5.2 are utilized, the transmitted signal corresponding to the sequence of bits

$$0, 0, 1, 1, 1, 0, 1, 0, 0 \tag{5.1}$$

is as shown in Fig. 5.3.

Let us assume that this signal is transmitted over an additive noise channel; hence, the received signal is simply the transmitted signal (or, more precisely, an attenuated version of the transmitted signal) corrupted with additive noise (modeled as a random process). This is an idealistic assumption since the transmitted signal is not distorted by the channel, regardless of its bandwidth. In the subsequent chapters, we will study extensions of this model which also account for other channel impairments, including the effects of a bandlimited channel on a communication pulse. An illustrative example of a received signal is shown in Fig. 5.4.

The receiver knows the waveforms used in the transmission, the symbol period, the amount of channel attenuation (if any), and the statistics of the noise process. For simplicity, we also assume that the receiver has perfect timing information, that is, it knows the individual symbol boundaries and can extract the part of the received signal corresponding to any transmitted bit. We will briefly review the problem of timing recovery at the end of this chapter. On the other hand, the receiver does not know which bit is transmitted in different bit intervals, and its

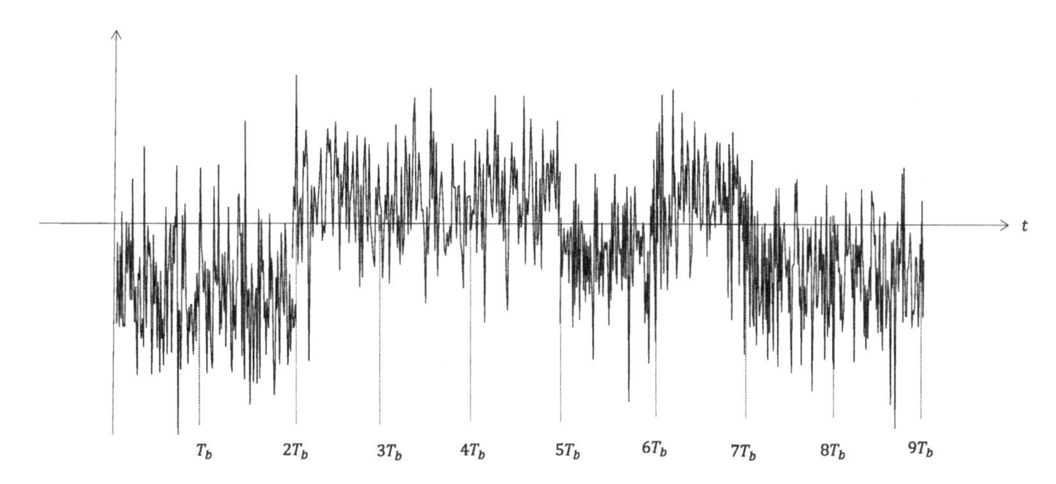

Figure 5.4 Sample received signal.

objective is to decide on the sequence of transmitted bits using the (noisy) signal it observes to minimize the average error probability of the system.

Under the perfect synchronization assumption, the transmission and detection of different bits over an additive noise channel can be considered separately. In other words, for the ith bit, we only need to extract the received signal on the interval $[(i-1)T_b, iT_b)$ and use it to decide on the bit. Therefore, without loss of generality, we will take the interval as $[0, T_b)$ and describe the receiver's operation for the first bit only – with the understanding that the process repeats for the detection of remaining bits. At this stage, we are unconcerned with imperfect synchronization or other non-idealities that the noisy communication medium may introduce. We only note that different bits/symbols may affect each other for such scenarios, complicating the receiver design and degrading the system performance.

5.1.1 White Gaussian Noise Processes

The noise process at an ideal receiver is due to the random movement of electrons with thermal agitation. As there are many of them, their aggregate effect is modeled as a Gaussian random process by applying the central limit theorem. Refer to Chapter 2 for a review of random processes, and specifically, Section 2.5 for Gaussian random processes.

For an ideal receiver, the thermal noise has a power spectral density given by

$$S_n(f) = \frac{h|f|}{2\left(e^{h|f|/kT} - 1\right)}, \tag{5.2}$$

where $h = 6.6 \times 10^{-34}$ J sec is Plank's constant, $k = 1.38 \times 10^{-23}$ J/K is Boltzman's constant, and T is temperature in Kelvin. Taking T as the room temperature, we see that for frequencies of interest in typical communication systems, we have $\left|\frac{hf}{kT}\right| \ll 1$. Therefore, we can employ the approximation $e^{hf/kT} \approx 1 + \frac{hf}{kT}$, and obtain

$$S_n(f) \approx \frac{kT}{2} \tag{5.3}$$

for the range of frequencies of interest.

As the PSD of the noise process is a constant for the range of frequencies relevant to communication systems, we define an idealized noise process called *white noise* as a zero-mean wide-sense stationary random process, denoted by $W(t)$. Its PSD is given by

$$S_W(f) = \frac{N_0}{2}, \tag{5.4}$$

where $N_0 = kT$ is referred to as the one-sided PSD of the noise process. Taking the inverse Fourier transform of $S_W(f)$, we write the autocorrelation function of the white noise as

$$R_W(\tau) = \frac{N_0}{2}\delta(\tau). \tag{5.5}$$

The power spectral density and the autocorrelation function of the white noise are illustrated in Fig. 5.5. This idealized noise process is called *white* since it includes all the frequencies similar to white light (which includes all the frequencies in the visible band).

Note that this is an idealization as the power content of the white noise is not bounded, that is,

$$P_W = \int_{-\infty}^{\infty} \frac{N_0}{2} df = \infty. \tag{5.6}$$

While it is not physically realizable, white noise serves as a simple and useful model for noise when the bandwidth of the noise at the system input is much greater than the communication system bandwidth, which is normally the case.

If the white noise is also Gaussian, it is called a white Gaussian noise process. In this case, the samples of the process at different time instances are independent. This is because

$$\mathbb{E}[W(t_1)W(t_2)] = R_W(t_1 - t_2) = 0 \text{ if } t_1 \neq t_2, \tag{5.7}$$

and since the process is zero mean, $\mathbb{E}[W(t_1)]\mathbb{E}[W(t_2)] = 0$. That is, $W(t_1)$ and $W(t_2)$ are uncorrelated if $t_1 \neq t_2$, and being jointly Gaussian, they are independent.

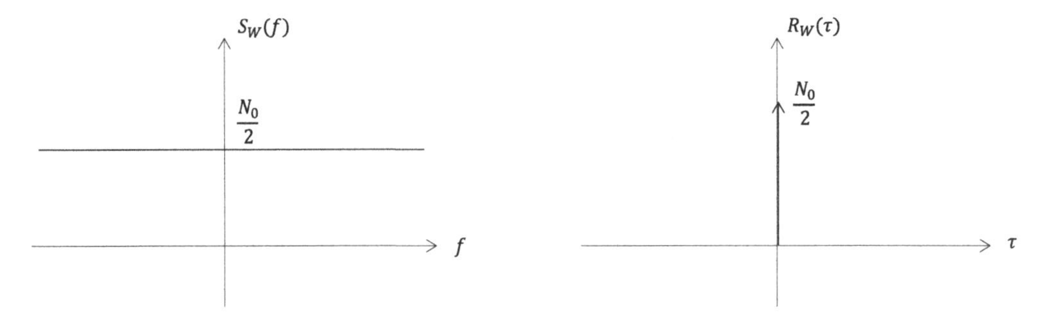

Figure 5.5 Power spectral density and autocorrelation function of white noise.

To summarize, the thermal noise is commonly modeled as a zero-mean, stationary, additive white Gaussian noise process.

Let us go through two important examples involving white noise.

Example 5.1

A zero-mean white noise process with PSD $S_W(f) = \frac{N_0}{2}$ is input to a lowpass filter with cut-off frequency B, as shown in Fig. 5.6. Determine the PSD and the autocorrelation function of the noise process at the output (denoted by $N(t)$).

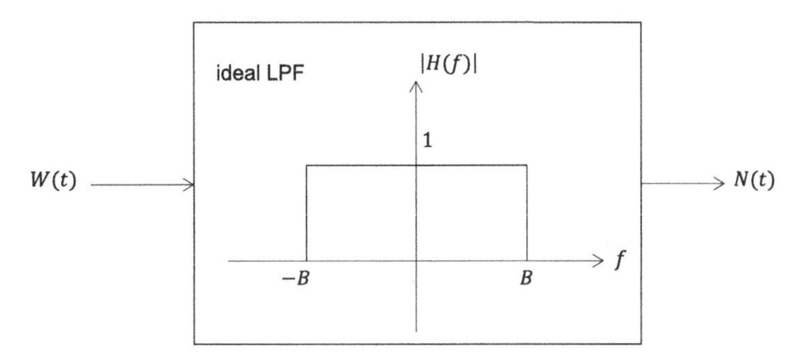

Figure 5.6 White noise input to a lowpass filter.

Solution

Recalling from Chapter 2, we can argue that the output process is also zero mean, and it is wide-sense stationary (as it is the output of an LTI system to a WSS input process). The power spectral density of $N(t)$ is given by

$$S_N(f) = S_W(f)|H(f)|^2 = \begin{cases} \frac{N_0}{2}, & \text{for } |f| \leq B, \\ 0, & \text{else.} \end{cases} \tag{5.8}$$

Taking the inverse Fourier transform, we obtain the autocorrelation function of $N(t)$ as

$$R_N(\tau) = N_0 B \text{sinc}(2B\tau). \tag{5.9}$$

$S_N(f)$ and $R_N(\tau)$ are depicted in Fig. 5.7. Note that the noise samples taken $m/2B$ time units apart, that is, $N(t)$ and $N(t + \frac{m}{2B})$, are uncorrelated for

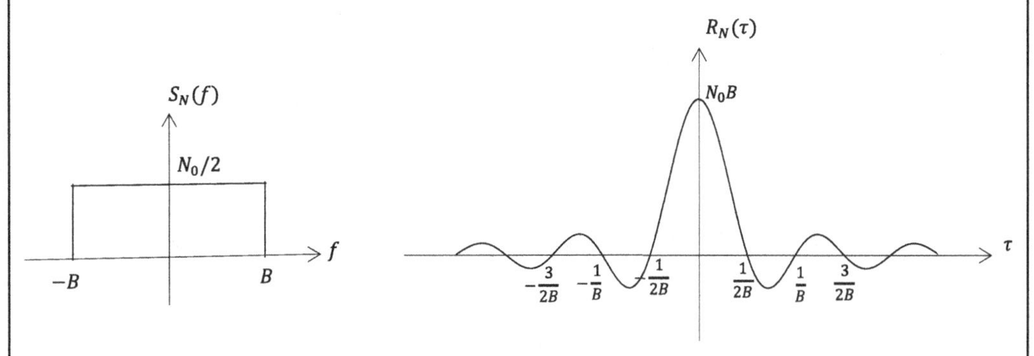

Figure 5.7 Power spectral density and autocorrelation function of the process at the output of an ideal lowpass filter with white noise at its input.

$m \in \mathbb{Z} - \{0\}$. Furthermore, if $W(t)$ is a Gaussian random process, then $N(t)$ is also a Gaussian random process, and its samples are jointly Gaussian. Being uncorrelated, we can conclude that the samples of $N(t)$ taken at a rate of $2B$ samples per second are independent.

Example 5.2

Consider the random variable obtained by correlating a zero-mean white noise process with PSD $S_W(f) = \frac{N_0}{2}$ with a deterministic function $h(t)$, that is,

$$Y = \int_0^T W(t)h(t)dt, \tag{5.10}$$

for some T. Note that Y can also be thought of as the output of an LTI system evaluated at a specific time instant when its input is a white noise process. Such operations will be encountered frequently in our study of digital communication systems. Compute the mean and variance of Y.

Solution

We can write

$$\mathbb{E}[Y] = \mathbb{E}\left[\int_{-\infty}^{\infty} W(t)h(t)dt\right] \tag{5.11}$$

$$= \int_{-\infty}^{\infty} \mathbb{E}[W(t)]h(t)dt \tag{5.12}$$

$$= 0, \tag{5.13}$$

and since $\text{Var}(Y) = \mathbb{E}[Y^2] - (\mathbb{E}[Y])^2$, with $\mathbb{E}[Y] = 0$, we have $\text{Var}(Y) = \mathbb{E}[Y^2]$, which can be computed as

$$\text{Var}(Y) = \mathbb{E}\left[\int_0^T W(t_1)h(t_1)dt_1 \int_0^T W(t_2)h(t_2)dt_2\right] \tag{5.14}$$

$$= \mathbb{E}\left[\int_0^T \int_0^T W(t_1)W(t_2)h(t_1)h(t_2)dt_1dt_2\right] \tag{5.15}$$

$$= \int_0^T \int_0^T \mathbb{E}[W(t_1)W(t_2)]h(t_1)h(t_2)dt_1dt_2 \tag{5.16}$$

$$= \int_0^T h(t_1)\left[\int_0^T \frac{N_0}{2}\delta(t_2 - t_1)h(t_2)dt_2\right]dt_1 \tag{5.17}$$

$$= \frac{N_0}{2}\int_0^T h^2(t)dt, \tag{5.18}$$

where the first line follows by writing Y in two different ways using two different integrals (with different integration variables), the second line follows by converting the product of two integrals into a single two-dimensional integral, the third line comes from exchanging the order of integral and expectation, and the final result is by treating t_1 as a constant and evaluating the inner integral.

Furthermore, if $W(t)$ is a Gaussian process, then Y is a Gaussian random variable, that is,

$$Y \sim \mathcal{N}\left(0, \frac{N_0}{2}\int_0^T h^2(t)dt\right). \tag{5.19}$$

In other words, if we correlate a Gaussian random process with a deterministic signal (multiply the random process with the signal and integrate over some time window), the result is a Gaussian random variable whose mean and variance are easily computed using the PSD level and the LTI system's impulse response. This is an important result that we will encounter frequently.

5.1.2 Additive White Gaussian Noise Channel

We noted that due to the thermal agitation of electrons, the received signal experiences additive noise that is well modeled as a zero mean, stationary, white Gaussian random process. That is, as a fundamental channel model, we have the following characterization: for a given transmitted signal $s(t)$, the received signal becomes

$$r(t) = \alpha s(t) + W(t), \tag{5.20}$$

where α is a deterministic attenuation factor and $W(t)$ is an additive white Gaussian noise process. This channel model is referred to as an AWGN channel. Despite its simplicity, this model is applicable in a wide variety of scenarios of practical significance, and it is commonly adopted. The AWGN channel model will be used throughout this chapter (even the entire book) as the basic communication channel model.

5.1.3 A Comment on the Receiver Structure

Let us close this section with a further comment on the receiver structure. The details will be covered later in the chapter; however, following up with the running example of this section, one may speculate that the decision block can integrate the received signal over the bit interval and decide as "1" if the result is positive and "0" if the result is negative, as depicted in Fig. 5.8.

Indeed, this specific receiver operation minimizes the probability of error for the given example of a rectangular waveform and its negative when the bits are equally likely to be transmitted over an AWGN channel (while it is not necessarily optimal in general). This point will be discussed rigorously in the rest of this chapter. The optimal receiver structure for digital modulation over an AWGN channel will be derived, and the corresponding performance will be analyzed.

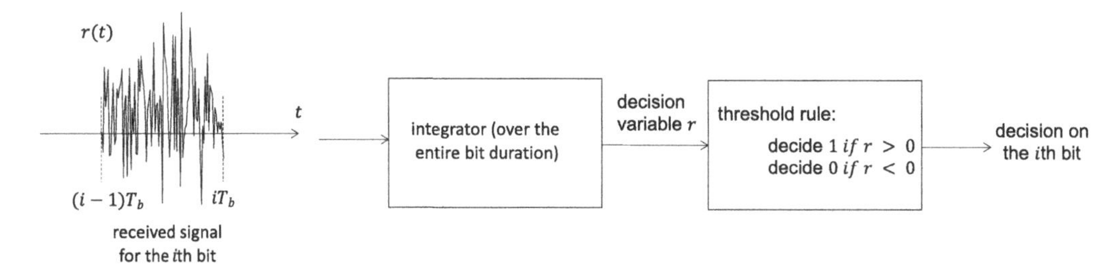

Figure 5.8 Receiver operation for the binary signaling example using rectangular waveforms.

5.2 Signal Space Concepts

We are ready to tackle the fundamental problem of digital modulation over an AWGN channel. We consider the scenario of M-ary signaling for a general exposition. We select M distinct waveforms of duration T_s to represent the M symbols. We denote the M messages by $m_1, m_2, \ldots, m_M$, and the corresponding waveforms employed for transmission as $s_1(t), s_2(t), \ldots, s_M(t)$, $t \in [0, T_s)$, respectively.

We consider the set of M signals of duration T_s as vectors. And the space they span, that is, the set of signals that can be written as their linear combinations, as the signal space. Here, the (vector) addition is nothing but the standard addition of the signals of time variable t, and the multiplication of a scalar with the signal (i.e., vector) is defined in the usual way. We also need an inner product of two signals (vectors) $f(t)$ and $g(t)$, which is defined as follows:

$$\langle f(t), g(t) \rangle = \int_0^{T_s} f(t)g(t)dt, \tag{5.21}$$

under the assumption that the signals are real.

Throughout this chapter, we will deal with real signals. Therefore, considering the vector space over the field of real numbers is sufficient, so is the inner product definition above. However, we will also need to consider complex signals as vectors over the field of complex numbers. In this case, we need a slightly extended version of the inner product definition. For two complex signals on $t \in [0, T_s)$, the inner product is defined as

$$\langle f(t), g(t) \rangle = \int_0^{T_s} f(t)g^*(t)dt. \tag{5.22}$$

The set of M signals of duration T_s used to transmit the M symbols define a finite-dimensional vector space, which we refer to as the *signal space*. Our first objective is to understand the signal space's properties and obtain a simple (vector) representation of different signals being used for transmission.

Assume that a basis for the signal space spanned by M real signals $s_1(t), s_2(t), \ldots, s_M(t)$ is given by the set of functions $\psi_1(t), \psi_2(t), \ldots, \psi_N(t)$ defined on the interval $[0, T_s)$, where the following two conditions hold.

- Orthogonality: $\langle \psi_i(t), \psi_j(t) \rangle = \int_0^{T_s} \psi_i(t)\psi_j(t)dt = 0, \quad \text{if } i \neq j.$
- Normalization: $\langle \psi_i(t), \psi_i(t) \rangle = \int_0^{T_s} \psi_i^2(t)dt = 1.$

Since $\psi_1(t), \psi_2(t), \ldots, \psi_N(t)$ are orthogonal to each other, and they are normalized, we call such a basis an *orthonormal* basis. The dimensionality of the signal set is less than or equal to the number of distinct signals used in transmission, that is, $N \leq M$.

Since the set of signals $\{\psi_i(t)\}_{i=1}^N$ form a basis, any signal in the original signal set can be written as a linear combination of them. Namely, we have

$$s_i(t) = \sum_{i=1}^{N} s_{ij}\psi_j(t), \quad i = 1, 2, \ldots, M, \tag{5.23}$$

where s_{ij} is easily obtained as the inner product of the signal $s_i(t)$ with $\psi_j(t)$. To see this, we note that

$$\langle s_i(t), \psi_j(t) \rangle = \int_0^{T_s} s_i(t)\psi_j(t)dt \tag{5.24}$$

$$= \sum_{k=1}^{N} s_{ik} \int_0^{T_s} \psi_k(t)\psi_j(t)dt \tag{5.25}$$

$$= \sum_{k=1}^{N} s_{ik}\delta_{kj} \tag{5.26}$$

$$= s_{ij}. \tag{5.27}$$

Here, δ_{kj} is the Kronecker delta, defined as $\delta_{kj} = 1$ if $k = j$, and 0 otherwise.

By stacking the coefficients s_{ij} into an $N \times 1$ vector, we obtain the vector representation for the signal $s_i(t)$, denoted by $\boldsymbol{s}_i$. Namely,

$$\boldsymbol{s}_i = \begin{bmatrix} s_{i1} \\ s_{i2} \\ \vdots \\ s_{iN} \end{bmatrix}. \tag{5.28}$$

The two representations for the ith signal, $s_i(t)$ as a time-domain function and $\boldsymbol{s}_i$ as an N-dimensional vector, are equivalent.

For notational convenience, let us define the *dot product* of two length-N vectors $\boldsymbol{x}$ and $\boldsymbol{y}$,

$$\boldsymbol{x} = \begin{bmatrix} x_1 \\ x_2 \\ \vdots \\ x_N \end{bmatrix} \text{ and } \boldsymbol{y} = \begin{bmatrix} y_1 \\ y_2 \\ \vdots \\ y_N \end{bmatrix}, \tag{5.29}$$

as

$$\boldsymbol{x} \cdot \boldsymbol{y} = \sum_{i=1}^{N} x_i y_i. \tag{5.30}$$

Any computations with the time-domain waveforms involving inner products (as defined above) can be carried out using the vector representations and vice versa. For instance, the energy of the signal $s_i(t)$ can be computed as

$$E_{s_i} = \langle s_i(t), s_i(t) \rangle \tag{5.31}$$

$$= \boldsymbol{s}_i \cdot \boldsymbol{s}_i \tag{5.32}$$

$$= \|\boldsymbol{s}_i\|^2 \tag{5.33}$$

$$= \boldsymbol{s}_i^T \boldsymbol{s}_i, \tag{5.34}$$

for $i = 1, 2, \ldots, M$.

To see this, we write

$$\langle s_i(t), s_i(t) \rangle = \int_0^{T_s} s_i(t) s_i(t) dt \tag{5.35}$$

$$= \int_0^{T_s} \sum_{k=1}^{N} s_{ik} \psi_k(t) \sum_{l=1}^{N} s_{il} \psi_l(t) dt \tag{5.36}$$

$$= \sum_{k=1}^{N} s_{ik} \sum_{l=1}^{N} s_{il} \int_0^{T_s} \psi_k(t) \psi_l(t) dt \tag{5.37}$$

$$= \sum_{k=1}^{N} s_{ik} \sum_{l=1}^{N} s_{il} \delta_{kl} \tag{5.38}$$

$$= \sum_{k=1}^{N} s_{ik}^2 \tag{5.39}$$

$$= \boldsymbol{s}_i^T \boldsymbol{s}_i \tag{5.40}$$

$$= \|\boldsymbol{s}_i\|^2. \tag{5.41}$$

Similarly, we can show that

$$\langle s_i(t), s_j(t) \rangle = \boldsymbol{s}_i \cdot \boldsymbol{s}_j \tag{5.42}$$

$$= \boldsymbol{s}_i^T \boldsymbol{s}_j \tag{5.43}$$

for $i, j = 1, 2, \ldots, M$.

Another useful calculation involves the squared Euclidean distance between two signals, namely,

$$d_{ij}^2 = \langle s_i(t) - s_j(t), s_i(t) - s_j(t) \rangle = \int_0^{T_s} (s_i(t) - s_j(t))^2 dt. \tag{5.44}$$

We can write

$$d_{ij}^2 = \langle s_i(t) - s_j(t), s_i(t) - s_j(t) \rangle \tag{5.45}$$
$$= (\boldsymbol{s}_i - \boldsymbol{s}_j) \cdot (\boldsymbol{s}_i - \boldsymbol{s}_j) \tag{5.46}$$
$$= \left\| \boldsymbol{s}_i - \boldsymbol{s}_j \right\|^2. \tag{5.47}$$

In other words, the squared Euclidean distance between two signals can also be computed straightforwardly using their vector representations.

Each vector $\boldsymbol{s}_i$ is referred to as a *constellation point*, and the set of M constellation points make up the *signal constellation*.

We usually represent the signal vectors $\boldsymbol{s}_i,\ i = 1, 2, \ldots, M$, in the N-dimensional signal space geometrically; namely, we plot the signal constellation as a way to specify the particular digital modulation scheme.

Example 5.3

A 4-ary modulation scheme uses the four signals shown in Fig. 5.9. Determine an orthonormal basis for this signal set and express the signals in vector form.

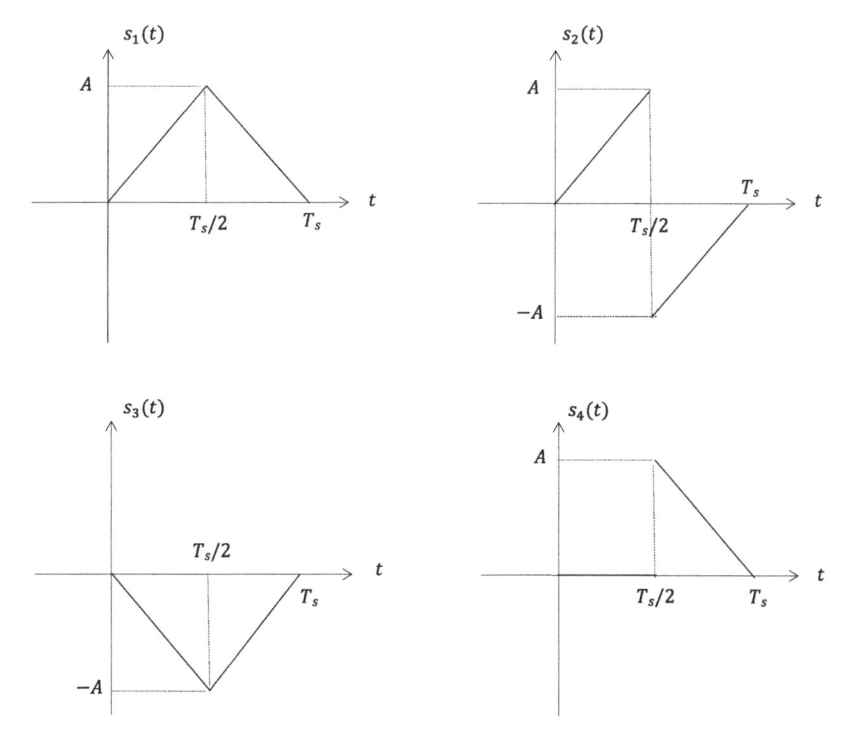

Figure 5.9 Signals used for the 4-ary communication system example.

Solution

It is clear that this signal set is not one-dimensional as the signals are not scaled versions of a single waveform. Indeed, it is easy to see that the signal set is two-dimensional, and all four signals given in the figure can be written as linear combinations of the two triangular functions:

$$\psi_1(t) = \begin{cases} c_1 t, & \text{for } t \in [0, T_s/2), \\ 0, & \text{else;} \end{cases} \tag{5.48}$$

$$\psi_2(t) = \begin{cases} c_2(T_s - t), & \text{for } t \in [T_s/2, T_s), \\ 0, & \text{else,} \end{cases} \tag{5.49}$$

for some constants c_1 and c_2. It is obvious that the two triangular functions shown are orthogonal since, whenever one signal is non-zero, the other one is, making their product simply zero for all t, and integrating over the symbol duration results in 0. Furthermore, the coefficients c_1 and c_2 can be computed by normalizing them to form an orthonormal basis. For instance, for the first basis function, we need $E_{\psi_1} = \langle \psi_1(t), \psi_1(t) \rangle = 1$. Since

$$E_{\psi_1} = \int_0^{T_s} \psi_1^2(t) dt \tag{5.50}$$

$$= \int_0^{T_s/2} c_1^2 t^2 dt \tag{5.51}$$

$$= \frac{T_s^3}{24} c_1^2, \tag{5.52}$$

we obtain $c_1 = \sqrt{\frac{24}{T_s^3}}$. Similarly, $c_2 = \sqrt{\frac{24}{T_s^3}}$ also results. In other words, we have

$$\psi_1(t) = \begin{cases} \sqrt{\frac{24}{T_s^3}} t, & \text{for } t \in [0, T_s/2), \\ 0, & \text{else,} \end{cases} \tag{5.53}$$

$$\psi_2(t) = \begin{cases} \sqrt{\frac{24}{T_s^3}} (T_s - t), & \text{for } t \in [T_s/2, T_s), \\ 0, & \text{else.} \end{cases} \tag{5.54}$$

A plot of these orthonormal basis functions is depicted in Fig. 5.10.

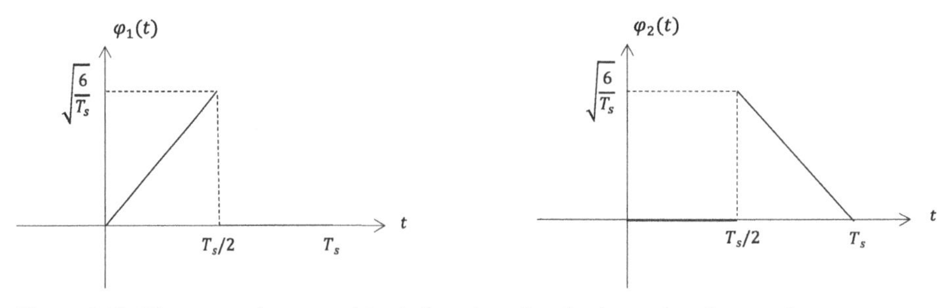

Figure 5.10 The two orthonormal basis functions for the 4-ary signal set under consideration.

Note that this selection of the basis functions is not unique. For instance, another orthonormal basis can be obtained by adequately scaling the two signals $s_1(t)$ and $s_2(t)$.

With the above selection of the orthonormal basis functions, we obtain the components of the signal $s_1(t)$ in the directions of $\psi_1(t)$ and $\psi_2(t)$ as follows:

$$s_{11} = \langle s_1(t), \psi_1(t) \rangle \tag{5.55}$$

$$= \int_0^{T_s} s_1(t)\psi_1(t)dt \tag{5.56}$$

$$= \int_0^{T_s/2} \frac{2A}{T_s}\sqrt{\frac{24}{T_s^3}}t^2 dt \tag{5.57}$$

$$= A\sqrt{\frac{T_s}{6}} \tag{5.58}$$

and

$$s_{12} = \langle s_1(t), \psi_2(t) \rangle \tag{5.59}$$

$$= \int_0^{T_s} s_1(t)\psi_2(t)dt \tag{5.60}$$

$$= \int_{T_s/2}^{T_s} \frac{2A}{T_s}\sqrt{\frac{24}{T_s^3}}(T_s - t)^2 dt \tag{5.61}$$

$$= A\sqrt{\frac{T_s}{6}}. \tag{5.62}$$

Hence, the vector representation for the first signal becomes

$$\mathbf{s}_1 = \begin{bmatrix} A\sqrt{\frac{T_s}{6}} \\ A\sqrt{\frac{T_s}{6}} \end{bmatrix}. \tag{5.63}$$

Similar computations yield the vector representation for the other three signals as

$$\mathbf{s}_2 = \begin{bmatrix} A\sqrt{\frac{T_s}{6}} \\ -A\sqrt{\frac{T_s}{6}} \end{bmatrix}, \ \mathbf{s}_3 = \begin{bmatrix} -A\sqrt{\frac{T_s}{6}} \\ -A\sqrt{\frac{T_s}{6}} \end{bmatrix}, \ \mathbf{s}_4 = \begin{bmatrix} 0 \\ A\sqrt{\frac{T_s}{6}} \end{bmatrix}. \tag{5.64}$$

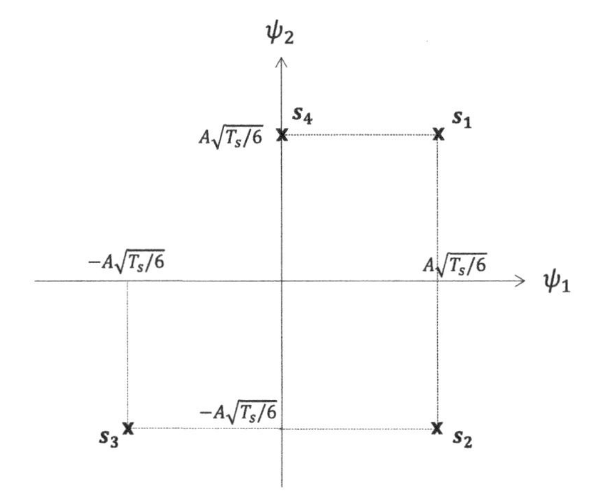

Figure 5.11 Signal constellation in the 4-ary signaling example.

Figure 5.11 depicts the resulting 4-ary signal constellation.

We next give an example illustrating that calculations involving inner products (namely, the multiplication of two signals integrated over the signal duration) can readily be performed using the vector representations of the signals.

Example 5.4

Let us consider Example 5.3 and compute the inner product of the signals $s_1(t)$ and $s_3(t)$ in two different ways. In the first method, we use the time-domain signals directly, that is,

$$\langle s_1(t), s_3(t)\rangle = \int_0^{T_s} s_1(t)s_3(t)dt \tag{5.65}$$

$$= \int_0^{T_s/2} (2A/T_s)t \cdot (-2A/T_s)tdt$$
$$+ \int_{T_s/2}^{T_s} (2A/T_s)(T_s - t) \cdot (-2A/T_s)(T_s - t)dt \tag{5.66}$$

$$= -\frac{A^2 T_s}{3}. \tag{5.67}$$

In the second method, we use the vector representations obtained in Example 5.3 to get

$$\langle \mathbf{s}_1, \mathbf{s}_3\rangle = \left\langle \begin{bmatrix} A\sqrt{T_s/6} \\ A\sqrt{T_s/6} \end{bmatrix}, \begin{bmatrix} -A\sqrt{T_s/6} \\ -A\sqrt{T_s/6} \end{bmatrix} \right\rangle \tag{5.68}$$

$$= -\frac{A^2 T_s}{3}, \tag{5.69}$$

which is the same as the result of the first method, as expected.

Example 5.3 demonstrates that, in some cases, it is possible to figure out an orthonormal basis for a given signal set using inspection. However, this is not always possible or straightforward, and there is a need for a systematic procedure to accomplish this – as described next.

Gram–Schmidt Orthonormalization. We can systematically determine an orthonormal basis for the space spanned by the M signals $s_1(t), s_2(t), \ldots, s_M(t)$ using the Gram–Schmidt orthonormalization procedure. In the rest of this subsection, we describe this procedure in detail and provide a step-by-step example.

Gram–Schmidt Orthonormalization Algorithm

(1) Assuming that $s_1(t) \neq 0$, obtain the first basis function by a simple normalization, that is,

$$\phi_1(t) = \frac{s_1(t)}{\sqrt{E_{s1}}}, \tag{5.70}$$

with $E_{s1} = \langle s_1(t), s_1(t)\rangle$ being the energy of the signal. (If $s_1(t) = 0$, move to the next signal.)

(2) Subtract the projection of the signal $s_2(t)$ in the direction of the first basis function from itself, to obtain

$$d_2(t) = s_2(t) - \langle s_2(t), \phi_1(t)\rangle \phi_1(t), \tag{5.71}$$

which is orthogonal to $\phi_1(t)$. This signal, along with the first basis function ($\phi_1(t)$), spans the same signal space as the first two signals $s_1(t)$ and $s_2(t)$. If $d_2(t) \neq 0$, normalize it to obtain the second basis function with unit norm. Namely,

$$\phi_2(t) = \frac{d_2(t)}{\sqrt{\langle d_2(t), d_2(t)\rangle}}. \tag{5.72}$$

If $d_2(t) = 0$ for $t \in [0, T_s)$, there is no new basis function; move to the next step.

For $k = 3, 4, \ldots, M$:

(k) Assume that the orthonormal basis functions generated until this step are $\phi_1(t), \phi_2(t), \ldots, \phi_{k'-1}(t)$, which span the signal space spanned by $s_1(t), s_2(t), \ldots, s_{k-1}(t)$. Compute the projection of $s_k(t)$ onto the basis functions already determined and subtract them from the signal $s_k(t)$ to determine the k'th difference signal

$$d_k(t) = s_k(t) - \sum_{i=1}^{k'-1} \langle s_k(t), \phi_i(t)\rangle \phi_i(t). \tag{5.73}$$

The difference function $d_k(t)$, along with the previous basis functions, spans the space of the signals $s_1(t), s_2(t), \ldots, s_k(t)$. Also, $d_k(t)$ is orthogonal to all the basis functions determined in the previous steps. If $d_k(t) \neq 0$, normalize it to obtain the next basis function, that is,

$$\phi_{k'}(t) = \frac{d_k(t)}{\sqrt{\langle d_k(t), d_k(t) \rangle}}. \tag{5.74}$$

If $d_k(t) = 0$ for $t \in [0, T_s)$, then there is no new basis function, so move to the next step.

Let us illustrate the Gram–Schmidt orthonormalization procedure via an example.

Example 5.5

Determine an orthonormal basis for the four signals given in Fig. 5.12 using the Gram–Schmidt orthogonalization procedure. (T denotes the signal duration.)

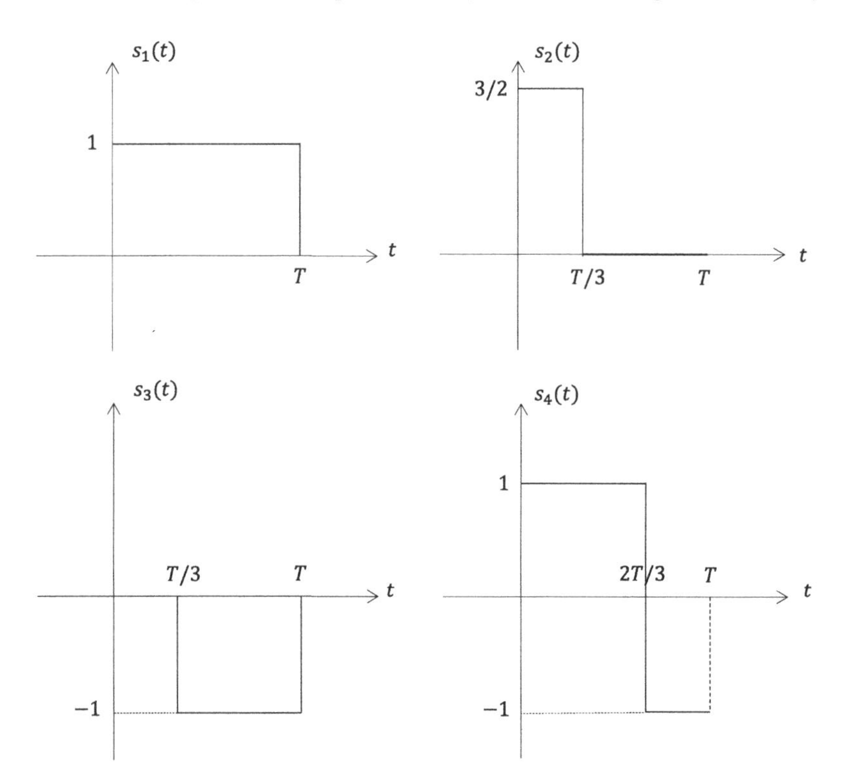

Figure 5.12 The four signals used in the example.

Solution

We apply the following procedure.

Step 1: Obtain the first basis function by normalizing $s_1(t)$ by its norm. Since

$$\langle s_1(t), s_1(t) \rangle = \int_0^T s_1^2(t)dt = \int_0^T 1dt = T, \tag{5.75}$$

we obtain $\phi_1(t) = 1/\sqrt{T}$ for $t \in [0, T)$.

Step 2: The inner product of the second signal with the first basis function is

$$\langle s_2(t), \phi_1(t)\rangle = \int_0^T s_2(t)\phi_1(t)dt = \int_0^{T/3} \frac{3}{2\sqrt{T}}dt = \frac{\sqrt{T}}{2}. \tag{5.76}$$

Hence, the difference signal is

$$d_2(t) = s_2(t) - \langle s_2(t), \phi_1(t)\rangle\phi_1(t) \tag{5.77}$$

$$= s_2(t) - \frac{\sqrt{T}}{2}\phi_1(t) \tag{5.78}$$

$$= \begin{cases} 1, & \text{for } t \in \left[0, \frac{T}{3}\right), \\ -1/2, & \text{for } t \in \left[\frac{T}{3}, T\right). \end{cases} \tag{5.79}$$

We need to normalize this signal to obtain the second basis function. Since

$$\langle d_2(t), d_2(t)\rangle = \int_0^T d_2^2(t)dt = \frac{T}{2}, \tag{5.80}$$

we obtain

$$\phi_2(t) = \begin{cases} \sqrt{\frac{2}{T}}, & \text{for } t \in \left[0, \frac{T}{3}\right), \\ -\sqrt{\frac{1}{2T}}, & \text{for } t \in \left[\frac{T}{3}, T\right). \end{cases} \tag{5.81}$$

Step 3: The inner products of the third signal $s_3(t)$ with the basis functions already determined are given by

$$\langle s_3(t), \phi_1(t)\rangle = -\int_{T/3}^{T} \frac{1}{\sqrt{T}}dt = -\frac{2\sqrt{T}}{3} \tag{5.82}$$

and

$$\langle s_3(t), \phi_2(t)\rangle = \int_{T/3}^{T} \sqrt{\frac{1}{2T}}dt = \frac{\sqrt{2T}}{3}. \tag{5.83}$$

Hence,

$$d_3(t) = s_3(t) + \frac{2\sqrt{T}}{3}\phi_1(t) - \frac{\sqrt{2T}}{3}\phi_2(t) \tag{5.84}$$

$$= 0. \tag{5.85}$$

In other words, $s_3(t)$ is a linear combination of the first two basis signals, and no new basis function is introduced in this step.

Step 4: At this point, we have two basis functions; thus, we compute

$$\langle s_4(t), \phi_1(t)\rangle = \int_0^{2T/3} \frac{1}{\sqrt{T}}dt - \int_{2T/3}^{T} \frac{1}{\sqrt{T}}dt = \frac{\sqrt{T}}{3} \tag{5.86}$$

and

$$\langle s_4(t), \phi_2(t)\rangle = \int_0^{T/3} \sqrt{\frac{2}{T}}dt - \int_{T/3}^{2T/3} \sqrt{\frac{1}{2T}}dt + \int_{2T/3}^{T} \sqrt{\frac{1}{2T}}dt = \frac{\sqrt{2T}}{3}. \tag{5.87}$$

Hence, we obtain

$$d_4(t) = s_4(t) - \frac{\sqrt{T}}{3}\phi_1(t) - \frac{\sqrt{2T}}{3}\phi_2(t) \tag{5.88}$$

$$= \begin{cases} 0, & \text{for, } t \in \left[0, \frac{T}{3}\right), \\ 1, & \text{for, } t \in \left[\frac{T}{3}, \frac{2T}{3}\right), \\ -1, & \text{for, } t \in \left[\frac{2T}{3}, T\right). \end{cases} \tag{5.89}$$

With $E_{d4} = \langle d_4(t), d_4(t)\rangle = 2T/3$, we normalize $d_4(t)$ by $\sqrt{E_{d4}}$ to obtain the third basis function as

$$\phi_3(t) = \begin{cases} 0, & \text{for } t \in \left[0, \frac{T}{3}\right), \\ \sqrt{\frac{3}{2T}}, & \text{for } t \in \left[\frac{T}{3}, \frac{2T}{3}\right), \\ -\sqrt{\frac{3}{2T}}, & \text{for } t \in \left[\frac{2T}{3}, T\right). \end{cases} \tag{5.90}$$

The three orthonormal basis functions are depicted in Fig. 5.13.

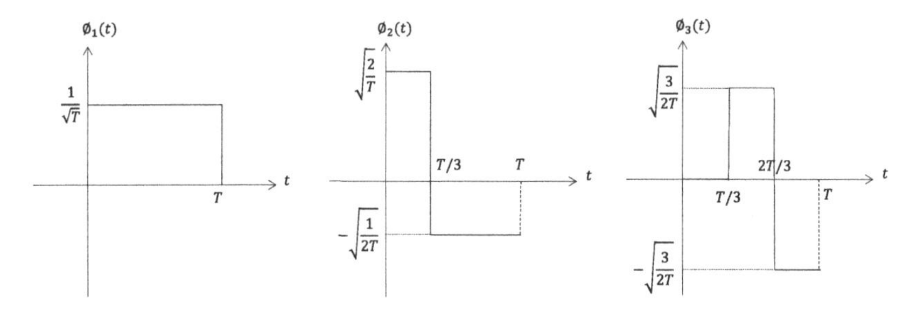

Figure 5.13 The resulting orthonormal basis functions in the example.

Using these basis functions, we can also write the vector forms of the four signals as follows:

$$s_1 = \begin{bmatrix} \sqrt{T} \\ 0 \\ 0 \end{bmatrix}, \quad s_2 = \begin{bmatrix} \frac{\sqrt{T}}{2} \\ \sqrt{\frac{T}{2}} \\ 0 \end{bmatrix}, \quad s_3 = \begin{bmatrix} -\frac{2\sqrt{T}}{3} \\ \frac{\sqrt{2T}}{3} \\ 0 \end{bmatrix}, \quad s_4 = \begin{bmatrix} \frac{\sqrt{T}}{3} \\ \frac{\sqrt{2T}}{3} \\ \sqrt{\frac{2T}{3}} \end{bmatrix}. \tag{5.91}$$

Note that we could have also determined an orthonormal basis for this signal set using inspection. For instance, we could have picked three non-overlapping pulses of duration $T/3$, adequately normalized, to represent the signal set. Furthermore, we could have also applied the Gram–Schmidt orthonormalization procedure using a different order of signals, resulting in a different set of basis functions.

5.3 Binary Antipodal Modulation

In this section, we study a specific binary modulation scheme called *antipodal signaling* over an AWGN channel and describe the transmitter and receiver operations in detail. This is perhaps the most straightforward yet interesting scenario, consisting of only two symbols for transmission. We will extend this discussion to the case of M-ary modulation in Section 5.4.

Let the bit 1 be transmitted by $s_1(t) = p(t)$ and the bit 0 be sent by $s_0(t) = -p(t)$, where $p(t)$ is an arbitrary signal (or pulse) defined on the interval $[0, T_b)$, with T_b denoting the bit duration. Since the transmission of individual symbols (in this case, bits) can be studied separately from the other symbols without loss of generality, we focus only on the bit interval from 0 to T_b. Since the same pulse is used with different polarities to represent the two bits, this modulation scheme is called binary antipodal signaling.

Binary antipodal signaling utilizes a one-dimensional signal set for which a normalized basis function is given by

$$\psi(t) = \frac{p(t)}{\sqrt{E_p}}, \tag{5.92}$$

where E_p is the energy of the pulse, that is,

$$E_p = \int_0^{T_b} p^2(t)dt. \tag{5.93}$$

Hence, the two signals used for transmission are represented as scalars, namely, $s_0 = -\sqrt{E_p}$ and $s_1 = \sqrt{E_p}$. Figure 5.14 shows the resulting signal constellation.

For transmission over an AWGN channel, the received signal is given by

$$r(t) = s_i(t) + n(t), \tag{5.94}$$

where $s_i(t)$ is the transmitted signal and $n(t)$ is a white Gaussian noise process with power spectral density $\frac{N_0}{2}$.

The receiver observes the signal $r(t)$ for $t \in [0, T_b)$, and its job is to decide on the transmitted bit in such a way that the probability of error is minimized.

Let us denote the transmitted bit by the random variable B. We assume that the prior probabilities are $\mathbb{P}(B = 1) = p$ and $\mathbb{P}(B = 0) = 1 - p$ for some $0 \leq p \leq 1$. The case of $p = 1/2$ corresponds to the transmission of equally likely bits.

Projection of the Received Signal Onto the Signal Space. The realization of the additive noise $n(t)$ does not lie in the signal space (which is only one-dimensional for

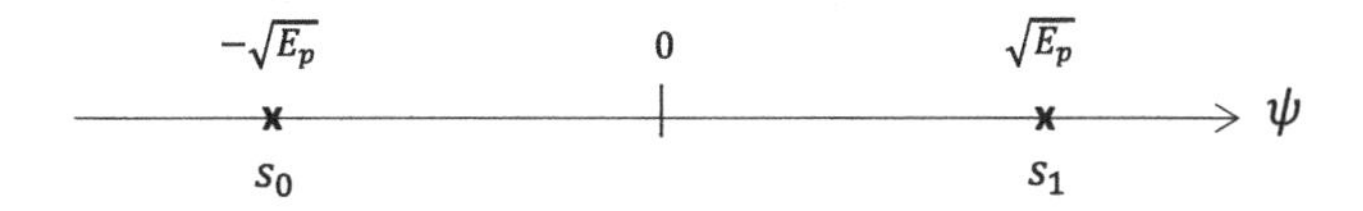

Figure 5.14 Binary antipodal signal constellation.

binary antipodal signaling). Therefore, the received signal $r(t)$ does not lie in the signal space either. To proceed further, we first project the received signal $r(t)$ onto the signal space and obtain

$$r'(t) = r\psi(t), \tag{5.95}$$

where

$$r = \langle r(t), \psi(t) \rangle \tag{5.96}$$

$$= \int_0^{T_b} r(t)\psi(t)dt. \tag{5.97}$$

Clearly,

$$r = s_i + n, \tag{5.98}$$

where

$$n = \langle n(t), \psi(t) \rangle \tag{5.99}$$

$$= \int_0^{T_b} n(t)\psi(t)dt, \tag{5.100}$$

namely, $n\psi(t)$ is the projection of the white Gaussian noise term onto the signal space. We can write

$$r(t) - r'(t) = s_i(t) + n(t) - r\psi(t) \tag{5.101}$$

$$= s_i(t) + n(t) - s_i\psi(t) - n\psi(t) \tag{5.102}$$

$$= n(t) - n\psi(t). \tag{5.103}$$

From this expression, we conclude that the part of the received signal component (not represented by the coefficient "r") contains only noise (with no useful signal component). We also argue that what remains is a Gaussian random process uncorrelated with the random variable r, since

$$\mathbb{E}[r(r(t) - r'(t))] = \mathbb{E}[r(n(t) - n\psi(t))] \tag{5.104}$$

$$= \mathbb{E}[(s_i + n)(n(t) - n\psi(t))] \tag{5.105}$$

$$= s_i\mathbb{E}[n(t)] - s_i\mathbb{E}[n]\psi(t) + \mathbb{E}\left[\left(\int_0^{T_b} n(\tau)\psi(\tau)d\tau\right) n(t)\right] - \mathbb{E}[n^2]\psi(t) \tag{5.106}$$

$$= \int_0^{T_b} \mathbb{E}[n(\tau)n(t)]\psi(\tau)d\tau - \frac{N_0}{2}\psi(t) \tag{5.107}$$

$$= \int_0^{T_b} \frac{N_0}{2}\delta(\tau - t)\psi(\tau)d\tau - \frac{N_0}{2}\psi(t) \tag{5.108}$$

$$= \frac{N_0}{2}\psi(t) - \frac{N_0}{2}\psi(t) \tag{5.109}$$

$$= 0. \tag{5.110}$$

Since the $r(t) - r'(t)$ and r are jointly Gaussian, and $\mathbb{E}[r(t) - r'(t)] = 0$, the random variable r and $r(t) - r'(t)$ are independent. Hence, we can conclude that $r(t) - r'(t)$ does not contain any information that can help in our decision-making process, and the random variable r forms *sufficient statistics* for the receiver operation.

An Equivalent Problem. At this point, we have a representation of the received signal as a realization of a random variable (denoted by r), and the two possible transmitted symbols are the scalars $\sqrt{E_p}$ and $-\sqrt{E_p}$. Therefore, the receiver design problem boils down to the following: observe the realization of the random variable r and decide on the transmitted bit in such a way that the probability of error is minimized. In the following, we tackle this problem.

Maximum A Posteriori (MAP) Decision Rule. Intuitively, given the received signal r, we can compute the posterior probabilities of the transmitted bit being 0 or 1, namely, $\mathbb{P}(0|r)$ and $\mathbb{P}(1|r)$, and pick the bit which gives the larger value as our decision on what was transmitted. Namely,

$$\hat{m} = \arg\max_{m=0,1} \mathbb{P}(m|r), \tag{5.111}$$

This is called the MAP decision rule, and it minimizes the probability of error, as will be justified shortly. Using Bayes' rule, we can equivalently write

$$\hat{m} = \arg\max_{m=0,1} \frac{f(r|m)\mathbb{P}(m)}{f(r)} \tag{5.112}$$

where $f(r)$ is the PDF of the received signal, and $f(r|m)$ is the conditional PDF of r given that the bit m is transmitted. Noting that $f(r)$ is positive for any r, and it does not depend on which bit is being transmitted, we can drop the denominator and obtain

$$\hat{m} = \arg\max_{m=0,1} \mathbb{P}(m)f(r|m) \tag{5.113}$$

as an equivalent form of the MAP decision rule.

MAP Rule Minimizes the Probability of Error. A (deterministic) decision rule partitions the space of received signals (in this case, the set of real numbers) into distinct decision regions. Let us denote the decision region for the bit 0 as $\mathcal{D}_0$, and for the bit 1 as $\mathcal{D}_1$. Minimizing the error probability is the same as maximizing the correct decision probability, which can be written as

$$P_c = \mathbb{P}(0)\mathbb{P}(r \in \mathcal{D}_0|0 \text{ sent}) + \mathbb{P}(1)\mathbb{P}(r \in \mathcal{D}_1|1 \text{ sent}) \tag{5.114}$$

$$= \int_{\mathcal{D}_0} \mathbb{P}(0)f(r|0)dr + \int_{\mathcal{D}_1} \mathbb{P}(1)f(r|1)dr. \tag{5.115}$$

The decision rule assigns each possible received signal $r \in \mathbb{R}$ to either $\mathcal{D}_0$ or to $\mathcal{D}_1$. That is, a specific r contributes either to the first integral or to the second one. Therefore, to maximize the correct decision probability, we need to pick

$$r \in \mathcal{D}_0 \quad \text{if} \quad \mathbb{P}(0)f(r|0) > \mathbb{P}(1)f(r|1) \tag{5.116}$$

and

$$r \in \mathcal{D}_1 \quad \text{if} \quad \mathbb{P}(1)f(r|1) > \mathbb{P}(0)f(r|0). \tag{5.117}$$

Ties can be broken arbitrarily. Since this is nothing but the MAP decision rule we described earlier, we conclude that the MAP rule minimizes the average probability of error.

MAP Rule for the AWGN Channel. Let us now specialize the MAP decision rule to the case of an AWGN channel with noise power spectral density $N_0/2$. First, we notice that the noise term n is a Gaussian random variable since it is obtained by integrating the product of a Gaussian random process with a deterministic function over the symbol period. Its mean is

$$\mathbb{E}[n] = \mathbb{E}\left[\int_0^{T_b} n(t)\psi(t)dt\right] \tag{5.118}$$

$$= \int_0^{T_b} \mathbb{E}[n(t)]\psi(t)dt \tag{5.119}$$

$$= 0, \tag{5.120}$$

and its variance is given by

$$\text{Var}(n) = \mathbb{E}[n^2] \tag{5.121}$$

$$= \mathbb{E}\left[\int_0^{T_b} n(t)\psi(t)dt \int_0^{T_b} n(\tau)\psi(\tau)d\tau\right] \tag{5.122}$$

$$= \int_0^{T_b}\int_0^{T_b} \mathbb{E}[n(t)n(\tau)]\psi(t)\psi(\tau)dtd\tau, \tag{5.123}$$

where the second line follows by writing the noise random variable n in two different ways using different integration variables, and the second line is obtained by changing the integration order. As $n(t)$ is a white noise process with power spectral density $\frac{N_0}{2}$, we can write

$$\text{Var}(n) = \int_0^{T_b}\int_0^{T_b} \frac{N_0}{2}\delta(t-\tau)\psi(t)\psi(\tau)dtd\tau \tag{5.124}$$

$$= \int_0^{T_b} \psi(\tau) \left\{ \int_0^{T_b} \frac{N_0}{2} \delta(t-\tau)\psi(t)dt \right\} d\tau \tag{5.125}$$

$$= \frac{N_0}{2} \int_0^{T_b} \psi^2(\tau)d\tau \tag{5.126}$$

$$= \frac{N_0}{2}. \tag{5.127}$$

Here, the second line follows by changing the order of the integrals, and the third line is obtained by evaluating the inner integral (considering $\tau \in [0, T_b)$ as fixed), and the final result follows as the basis function $\psi(t)$ is normalized. In other words, the noise term n is a Gaussian random variable with mean 0 and variance $\frac{N_0}{2}$, that is, $n \sim \mathcal{N}(0, N_0/2)$.

When the bit 0 is transmitted, the signal component is nothing but $s_i = -\sqrt{E_p}$, and the received signal is given by

$$r = -\sqrt{E_p} + n. \tag{5.128}$$

Hence, the conditional PDF of (the mathematically equivalent received signal) r given that $\{B = 0\}$ is a Gaussian random variable with mean $-\sqrt{E_p}$ and variance $\frac{N_0}{2}$, that is, $\sim \mathcal{N}(-\sqrt{E_p}, N_0/2)$. More explicitly,

$$f(r|0) = \frac{1}{\sqrt{\pi N_0}} e^{-\left(r+\sqrt{E_p}\right)^2/N_0}, \quad r \in \mathbb{R}. \tag{5.129}$$

Similarly, when the bit 1 is transmitted, the received signal r is a Gaussian random variable with mean $\sqrt{E_p}$ and variance $N_0/2$, that is, $r \sim \mathcal{N}(\sqrt{E_p}, N_0/2)$. In other words,

$$f(r|1) = \frac{1}{\sqrt{\pi N_0}} e^{-\left(r-\sqrt{E_p}\right)^2/N_0}, \quad r \in \mathbb{R}. \tag{5.130}$$

These two conditional PDFs are depicted in Fig. 5.15.

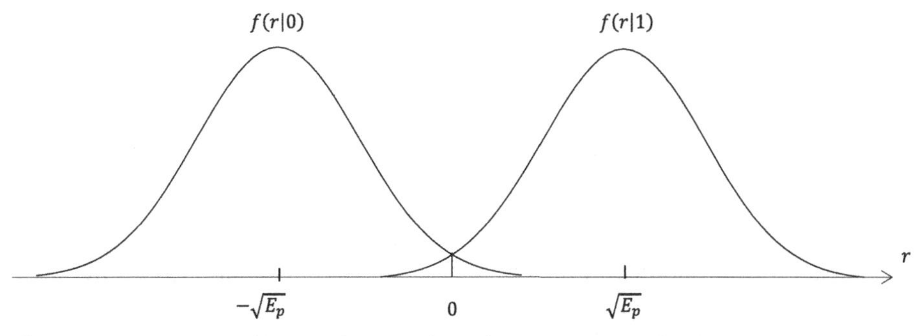

Figure 5.15 Conditional PDFs of the received signal given that bit 1 is sent ($f(r|1)$) and bit 0 is sent ($f(r|0)$).

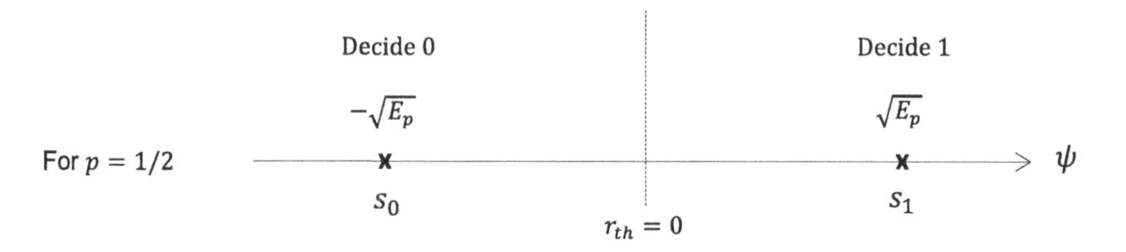

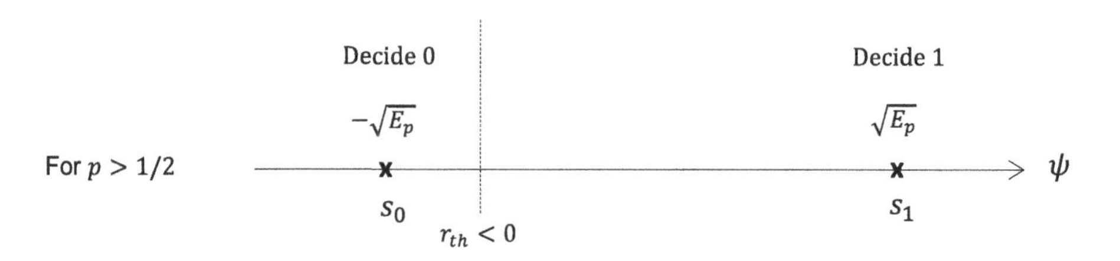

Figure 5.16 Decision regions for binary antipodal signaling for different prior probabilities.

Substituting these conditional PDFs into the generic MAP rule, with $\mathbb{P}(B = 1) = p$ and $\mathbb{P}(B = 0) = 1 - p$, we obtain the MAP decision rule for an AWGN channel as follows:

$$p \frac{1}{\sqrt{\pi N_0}} e^{-\left(r-\sqrt{E_p}\right)^2/N_0} \underset{0}{\overset{1}{\gtrless}} (1-p) \frac{1}{\sqrt{\pi N_0}} e^{-\left(r+\sqrt{E_p}\right)^2/N_0}. \tag{5.131}$$

This expression can be simplified to

$$\hat{m} = \begin{cases} 1, & \text{if } r > r_{th}, \\ 0, & \text{if } r < r_{th}, \end{cases} \tag{5.132}$$

where the threshold value is

$$r_{th} = \frac{N_0}{4\sqrt{E_p}} \ln\left(\frac{1-p}{p}\right). \tag{5.133}$$

Since the probability of the received signal r being precisely equal to the threshold is zero (as r is a continuous random variable), this case does not need to be considered explicitly.

This decision rule is illustrated in Fig. 5.16 for three different cases: $p = 1/2$, $p < 1/2$, and $p > 1/2$. If the bits being transmitted are equally likely, then the threshold is at the midpoint of the two constellation points, namely, $r_{th} = 0$. On the other hand, if $p > 1/2$, that is, the bit 1 is more likely than the bit 0, then the decision boundary becomes negative ($r_{th} < 0$). If $p < 1/2$, the decision boundary becomes positive ($r_{th} > 0$). Note that for some prior probabilities, the decision region for a particular bit can even include the other constellation point.

Maximum Likelihood (ML) Decision Rule. The optimal decision rule simplifies further if the transmitted bits are equally likely. This is a critical case that deserves a detailed study. This is because, in most communication systems, there is a source encoder removing the redundancy in the data being transmitted, and the output of an (effective) source encoder consists of independent and uniformly distributed bits (otherwise, the source can be compressed further). Also, the assumption of equally likely transmitted bits is highly reasonable when the exact source statistics are not available.

If the transmitted bits are equally likely, that is, $p = 1/2$, the MAP decision rule in (5.113) simplifies to the ML decision rule, yielding

$$\hat{m} = \arg\max_{m=0,1} f(r|m). \tag{5.134}$$

The conditional PDF $f(r|m)$ is called the likelihood function. Since, logarithm is an increasing function, an equivalent form for the ML rule maximizes the log-likelihood function defined as $\ln f(r|m)$, namely,

$$\hat{m} = \arg\max_{m=0,1} \ln f(r|m). \tag{5.135}$$

ML Decision Rule for an AWGN Channel. The ML decision rule for antipodal signaling over an AWGN channel is obtained from (5.132) and (5.133) by substituting $p = 1/2$, resulting in

$$\hat{m} = \begin{cases} 1, & \text{if } r > 0, \\ 0, & \text{if } r < 0. \end{cases} \tag{5.136}$$

In other words, the ML decision rule for antipodal signaling over an AWGN channel is straightforward: obtain the projection of the received signal onto the basis function, and compare it with the threshold of 0 to make a decision.

Overall Receiver Structure. The complete receiver structure for binary antipodal signaling with equally likely bits over an AWGN channel is given in Fig. 5.17. The decision variable is obtained from the received signal $r(t)$ using a mixer (multiplier) and an integrator over the bit duration, that is, $r = \langle r(t), \psi(t) \rangle$ is computed, and then it is fed to an ML decision block implementing a simple threshold rule: if $r > 0$, declare "1," if $r < 0$, declare "0" as the receiver output.

Recall that we have described a simple communication scheme using a rectangular pulse and its negative at the beginning of the chapter. For this example, we speculated that a receiver could integrate the received signal over the symbol period and compare the result with a threshold of 0. The scenario considered was, in fact, antipodal signaling with a rectangular pulse; therefore, our development

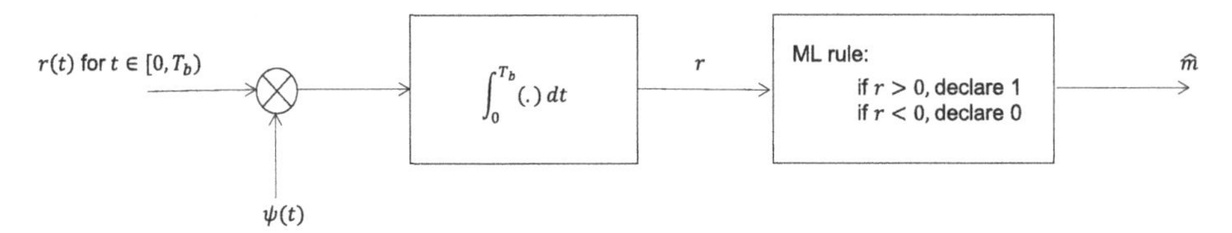

Figure 5.17 Correlation-type receiver for binary antipodal signaling. The receiver structure is explicitly shown for only a single bit period (arbitrarily taken as $[0, T_b)$).

in this section demonstrates that the speculated receiver structure in Fig. 5.8 was indeed optimal (assuming equally likely transmitted bits), that is, it minimizes the average error probability when the additive noise is a zero-mean white Gaussian random process.

Finally, we note that if the transmitted bits are not equally likely, then the receiver structure implementing the optimal detection rule (MAP rule) to minimize the error probability is almost the same as that of the ML receiver, the only difference being the comparison with a different threshold value r_{th} calculated in (5.133) instead of 0.

5.3.1 Matched Filter-Type versus Correlation-Type Receiver

We considered a simple binary digital modulation scheme in the previous subsection and explained the overall receiver structure. A main ingredient in the receiver operation is the computation of the decision variable r, that is, the projection of the received signal onto the basis function $\psi(t)$:

$$r = \int_0^{T_b} r(t)\psi(t)dt. \tag{5.137}$$

The receiver structure depicted in Fig. 5.17, which computes the projection of $r(t)$ onto the signal space, is referred to as a *correlation-type* receiver, since the decision variable r is directly obtained using a mixer and an integrator. An alternate – and often preferable – implementation employs a matched filter. To see how this can be accomplished, consider a linear time-invariant filter whose impulse response is given by $h(t) = \psi(T_b - t)$. If the received signal is input to the filter, then the output is given by the convolution integral

$$y(t) = r(t) * h(t) \tag{5.138}$$

$$= \int_0^{T_b} r(\tau)h(t-\tau)d\tau \tag{5.139}$$

$$= \int_0^{T_b} r(\tau)\psi(T_b - t + \tau)d\tau. \tag{5.140}$$

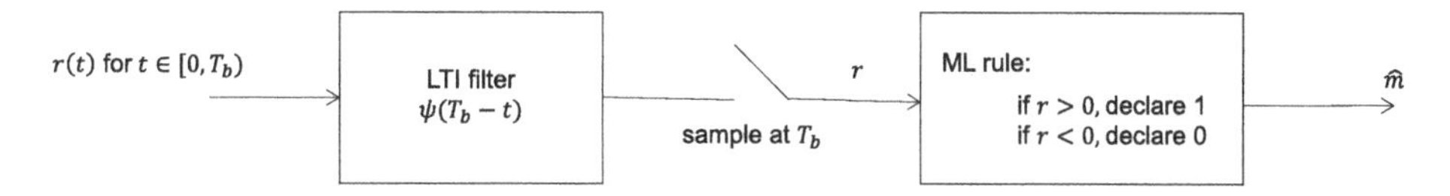

Figure 5.18 Matched filter-type receiver for binary antipodal signaling.

We notice that the decision variable needed in the receiver implementation r can be obtained by sampling the matched filter output at $t = T_b$. That is,

$$r = y(T_b) = \int_0^{T_b} r(\tau)\psi(\tau)d\tau. \tag{5.141}$$

In other words, we can replace the mixer and integrator in the correlation-type receiver with a matched filter and sampler. The resulting receiver is called a *matched filter-type* receiver. This is depicted in Fig. 5.18 for the case of binary antipodal signaling.

Example 5.6

Take the symbol period as T_b and consider a basis function given by $\psi(t) = 2\left(1 - \frac{t}{T_b}\right)$ for $t \in [0, T_b)$. Then, the impulse response of the filter matched to $\psi(t)$ is given by

$$\psi(T_b - t) = \frac{2t}{T_b}. \tag{5.142}$$

The function $\psi(t)$ and the impulse response of its matched filter are depicted in Fig. 5.19.

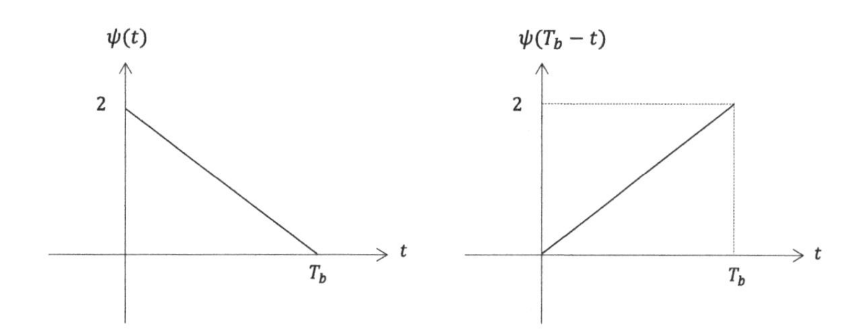

Figure 5.19 An exemplary signal and its matched filter.

Matched Filter Maximizes the Output Signal-to-Noise Ratio. Matched filtering is a widely used operation as it is the optimal solution for some problems encountered in signal processing and communications. For example, consider the basic problem of detecting a known signal in additive white Gaussian noise. The optimal solution to minimize the error probability is to pass the noisy signal through a matched filter, sample the output signal, and apply a threshold test. While we will not go through

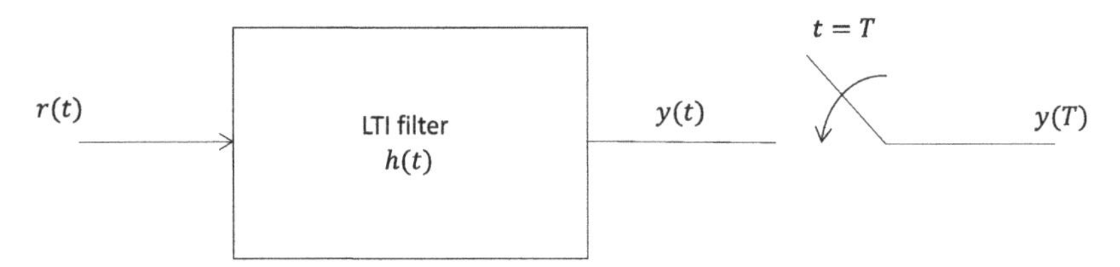

Figure 5.20 Among all possible choices for $h(t)$, the matched filter maximizes the output signal-to-noise ratio in this setup.

the details of the detection problem itself, we will illustrate an important property of the matched filter in the following.

Let

$$r(t) = s(t) + n(t) \quad \text{for} \quad t \in [0, T), \tag{5.143}$$

where $s(t)$ is a deterministic signal, and $n(t)$ is a white Gaussian noise process with power spectral density $N_0/2$. We are interested in filtering this signal using an LTI filter with impulse response $h(t)$ and sampling the filter output at time instant $t = T$. The corresponding block diagram is shown in Fig. 5.20. The filter whose impulse response is matched to $s(t)$, that is, $h(t) = s(T - t)$, maximizes the signal-to-noise ratio of the output sample.

Let us prove the above statement. Using the convolution integral, we can write the filter output as

$$y(t) = \int_0^T r(\tau)h(t - \tau)d\tau, \tag{5.144}$$

and the output sample as

$$y(T) = \int_0^T r(\tau)h(T - \tau)d\tau \tag{5.145}$$

$$= y_s(T) + y_n(T), \tag{5.146}$$

where

$$y_s(T) = \int_0^T s(\tau)h(T - \tau)d\tau \tag{5.147}$$

is the signal sample, while

$$y_n(T) = \int_0^T n(\tau)h(T - \tau)d\tau \tag{5.148}$$

is the noise sample at the filter output.

Noting that the signal power at the output is $y_s^2(T)$ (recall that the signal is deterministic) and that the average noise power in the output sample is $\mathbb{E}[y_n^2(T)]$, we define the output signal-to-noise ratio (SNR_{out}) as

$$\mathrm{SNR}_{\mathrm{out}} = \frac{y_s^2(T)}{\mathbb{E}[y_n^2(T)]}. \tag{5.149}$$

The average noise power can be computed as follows:

$$\mathbb{E}[y_n^2(T)] = \mathbb{E}\left[\int_0^T n(\tau)h(T-\tau)d\tau \int_0^T n(t)h(T-t)dt\right] \tag{5.150}$$

$$= \int_0^T\int_0^T \mathbb{E}[n(t)n(\tau)]h(T-t)h(T-\tau)dtd\tau \tag{5.151}$$

$$= \int_0^T h(T-\tau)\left(\int_0^T \frac{N_0}{2}\delta(t-\tau)h(T-t)dt\right)d\tau \tag{5.152}$$

$$= \frac{N_0}{2}\int_0^T h^2(T-\tau)d\tau. \tag{5.153}$$

Therefore, we can upper bound $\mathrm{SNR}_{\mathrm{out}}$ as

$$\mathrm{SNR}_{\mathrm{out}} = \frac{\left(\int_0^T s(\tau)h(T-\tau)d\tau\right)^2}{\frac{N_0}{2}\int_0^T h^2(T-\tau)d\tau} \tag{5.154}$$

$$\leq \frac{2}{N_0}\frac{\int_0^T s^2(\tau)d\tau \int_0^T h^2(T-\tau)d\tau}{\int_0^T h^2(T-\tau)d\tau} \tag{5.155}$$

$$= \frac{2}{N_0}\int_0^T s^2(\tau)d\tau, \tag{5.156}$$

where the bound follows from the Cauchy–Schwarz inequality (see the footnote in Section 2.3.5). Noting that the bound is attained with equality if and only if $h(T-\tau) = \kappa s(\tau)$ where κ is a constant, that is, $h(t) = \kappa s(T-t)$, we argue that $\mathrm{SNR}_{\mathrm{out}}$ is maximized if $h(t)$ is selected as the filter matched to $s(t)$, resulting in the largest possible signal-to-noise ratio of

$$\frac{2}{N_0}\int_0^T s^2(\tau)d\tau. \tag{5.157}$$

5.3.2 Probability of Error for Binary Antipodal Signaling

We determine the average symbol (or bit) error probability as a figure of merit to assess the performance of a digital transmission system. In this section, we carry out this calculation for binary antipodal signaling over an AWGN channel.

Denoting the transmitted bit by B, and using the total probability theorem, we can write the average probability of error as

$$P_b = \mathbb{P}(B = 1)P_{b,1} + \mathbb{P}(B = 0)P_{b,0} \tag{5.158}$$

$$= pP_{b,1} + (1 - p)P_{b,0}, \tag{5.159}$$

where $P_{b,m}$ is the conditional probability of error given that m is transmitted ($m = 0, 1$). We have already determined the PDF of the received signal r given each transmitted bit and the decision rule for use at the receiver. Armed with this knowledge, we can write

$$P_{b,1} = \mathbb{P}(r < r_{th}|m = 1 \text{ is sent}) \tag{5.160}$$

$$= \mathbb{P}\left(\sqrt{E_p} + n < r_{th}\right) \tag{5.161}$$

$$= \mathbb{P}\left(n < r_{th} - \sqrt{E_p}\right) \tag{5.162}$$

$$= \Phi\left(\frac{r_{th} - \sqrt{E_p}}{\sqrt{N_0/2}}\right) \tag{5.163}$$

$$= Q\left(\frac{\sqrt{E_p} - r_{th}}{\sqrt{N_0/2}}\right), \tag{5.164}$$

where r_{th} is as given in (5.133). To arrive at this result, we have used the following: (1) when $m = 1$ is sent, the received signal is given by $r = \sqrt{E_p} + n$ and we make an error if the received signal is below the optimal threshold value r_{th}; (2) the noise PDF is given by $n \sim \mathcal{N}(0, N_0/2)$; and (3) $\Phi(x) = Q(-x)$.

Similarly, we write

$$P_{b,0} = \mathbb{P}(r > r_{th}|m = 0 \text{ is sent}) \tag{5.165}$$

$$= \mathbb{P}\left(-\sqrt{E_p} + n > r_{th}\right) \tag{5.166}$$

$$= \mathbb{P}\left(n > r_{th} + \sqrt{E_p}\right) \tag{5.167}$$

$$= Q\left(\frac{\sqrt{E_p} + r_{th}}{\sqrt{N_0/2}}\right). \tag{5.168}$$

We then obtain the average probability of error for binary antipodal signaling with the MAP detector as

$$P_b = p\,Q\left(\frac{\sqrt{E_p} - r_{th}}{\sqrt{N_0/2}}\right) + (1 - p)\,Q\left(\frac{\sqrt{E_p} + r_{th}}{\sqrt{N_0/2}}\right). \tag{5.169}$$

We have already determined the optimal threshold r_{th} for an AWGN channel using the MAP rule directly. Note that we could have calculated the optimal threshold using (5.169) as well. Noticing that the optimal decision rule should be a thresholding rule on the received signal r, the average bit error probability for an arbitrary threshold value is written as (5.169). To determine the optimal threshold value that minimizes P_b, we can differentiate this expression with respect to r_{th}, and set it to zero. The result would be exactly the same as that in (5.133).

For the average bit error probability computation of the ML receiver (which is optimal when the prior probabilities are identical), we set $r_{th} = 0$ and obtain

$$P_b = Q\left(\sqrt{\frac{2E_p}{N_0}}\right). \tag{5.170}$$

Notice that we have complete symmetry for ML detection (since the threshold is set at 0). In this case, both conditional probability terms $P_{b,0}$ and $P_{b,1}$ are identical as the noise PDF is symmetric around the threshold. Therefore, we could compute one of these terms and obtain the overall error probability. While it is not very important for the simple case of binary antipodal signaling, it is generally helpful to recognize such symmetries to avoid unnecessary computations.

Finally, we note that it is common practice to give the error probability expression in terms of average signal-to-noise ratios. For the case of antipodal signaling, the transmission of either bit requires the same energy E_p. Hence, we have an energy per bit of $E_b = E_p$. Defining the SNR per bit as $\gamma_b = E_b/N_0$, for ML decoding, we obtain

$$P_b = Q\left(\sqrt{2\gamma_b}\right). \tag{5.171}$$

We plot the bit error probability P_b versus γ_b in Fig. 5.21. Note that it is customary to use a log scale for the error probability axis and to express γ_b in decibel scale (i.e.,

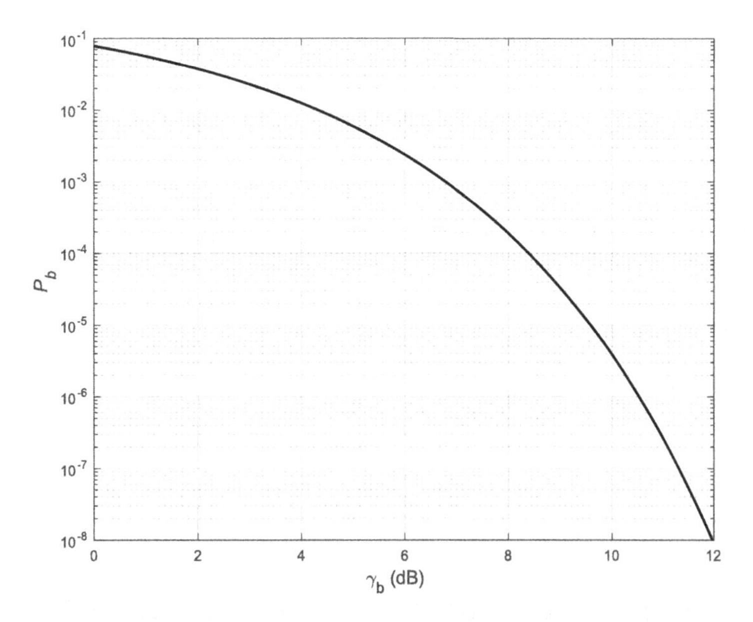

Figure 5.21 Bit error probability for antipodal signaling.

γ_b in dB $= 10\log_{10}\gamma_b$), namely, both axes are shown in log scale. This is because, in this way, we can read very low error probability values clearly and assess the system performance from the error probability plot. For example, reading directly from the figure, we obtain an error probability of 10^{-5} at around $\gamma_b = 9.7$ dB. To reduce the error probability by an order of magnitude to 10^{-6}, we need an additional 0.9 dB, corresponding to increasing the signal-to-noise ratio γ_b by a factor of approximately 1.23.

We emphasize that the dependence of the error probability on $\gamma_b = \frac{E_b}{N_0}$ is an exponential decay. This can be seen from the simple upper bound on the Q-function

$$Q(x) < \exp\left(-\frac{x^2}{2}\right), \tag{5.172}$$

which results in an upper bound on the probability of error for antipodal signaling over an AWGN channel as

$$P_b < \exp(-\gamma_b). \tag{5.173}$$

Note that this upper bound on the Q-function is reasonably tight for large values of its argument; hence, this simple expression well approximates the error probability of antipodal signaling when $\gamma_b \gg 1$.

5.3.3 Transmission of a Sequence of Symbols

So far, we have focused on the transmission of a single bit in isolation from the rest of the data. Let us now consider the transmission of a sequence of bits (using antipodal signaling) with a bit period of T_b seconds. In this case, the transmitted signal is a long waveform generated from the sequence of transmitted bits, obtained by concatenating the pulses used to represent each. The receiver observes a noisy version of this signal corrupted by AWGN, and it decides on the sequence of transmitted bits, that is, it produces a bit decision for every T_b seconds.

The receiver's operation is illustrated in Fig. 5.22 for the case of a matched filter-type receiver. The received signal is fed (continuously) to a matched filter, and its

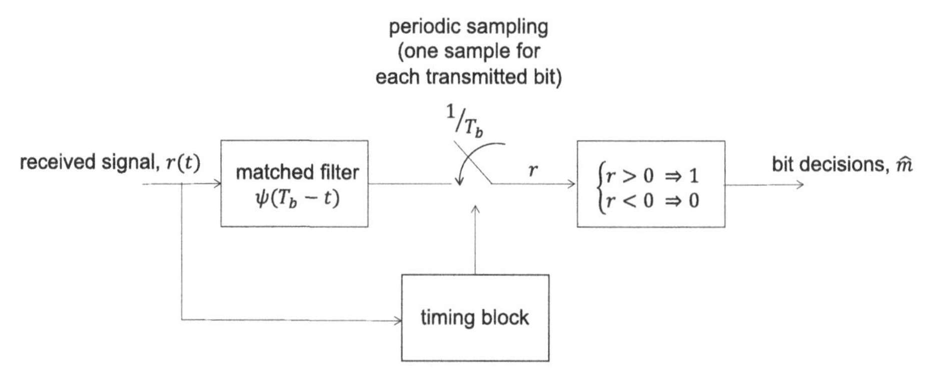

Figure 5.22 Receiver structure for transmission of a sequence of bits. The timing block determines the bit boundaries in the received signal, hence the sampling instances of the matched filter output.

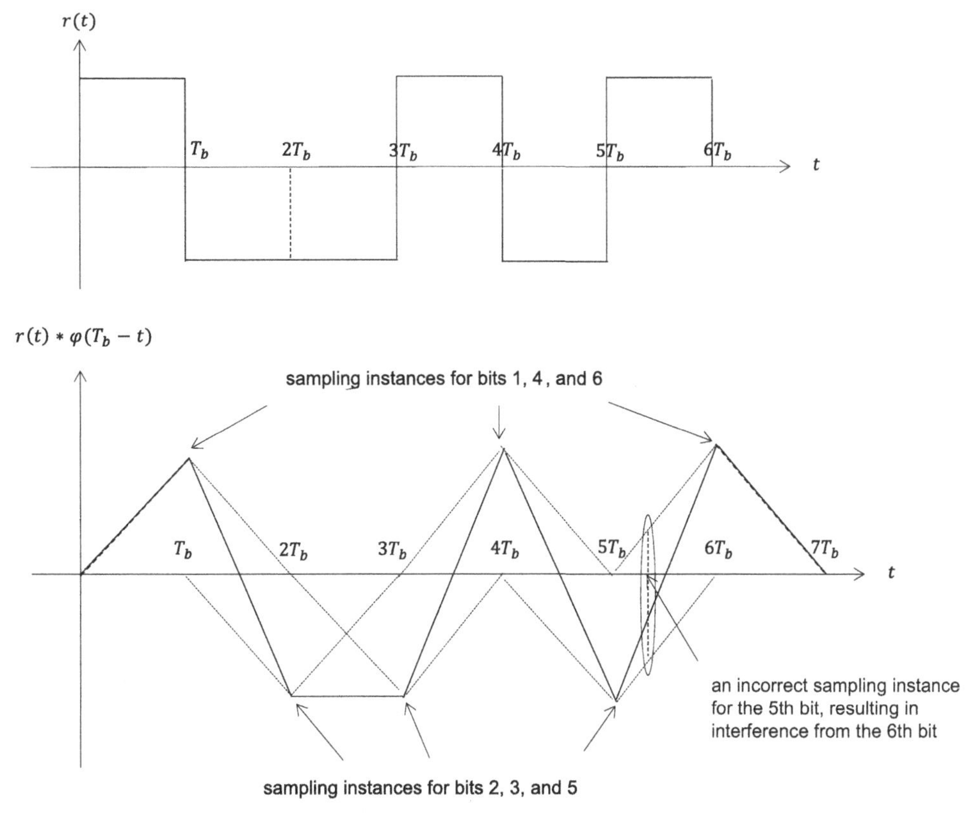

Figure 5.23 Received signal for a sequence of bits ignoring noise (when a rectangular pulse is employed) as well as the matched filter output and the sampling instances for making transmitted bit decisions. Also depicted is an incorrect sampling instance for the fifth bit's decision, resulting in interference from the next bit.

output is sampled at a rate of one sample per T_b seconds. Bit decisions are made periodically by comparing these samples with the threshold of "0." The exact sampling locations are determined through a timing recovery algorithm – as will be discussed at the end of the chapter.

It is also instructive to give an exemplary received signal for a sequence of bits, the corresponding matched filter outputs, and the sampling instances. Consider a rectangular pulse with a positive amplitude as $p(t)$, used for transmitting the bit 1, and its negative used for transmitting the bit 0. Figure 5.23 depicts the received waveform (in the absence of noise) for transmission of the sequence $1, 0, 0, 1, 0, 1$. In the same figure, we also depict the corresponding matched filter output. The contributions of the individual pulses corresponding to different bits are also shown (using dashed lines). Also depicted are the sampling instances of the matched filter output.

When the pulses used (for representing each bit) are confined to the bit interval, the channel is ideal (i.e., it does not distort the transmitted signal), and the receiver has the perfect timing information (i.e., it knows the exact sampling instances for each bit), the bit decisions do not interfere with each other. This should be clear from Fig. 5.23 for the case of a rectangular pulse. On the other hand, if the timing information is not perfect, that is, when the sampling instances are not precisely at multiples of T_b, there will also be a contribution from adjacent signals being transmitted. Such an incorrect sampling instance for the fifth bit's decision is also marked on the figure. Owing to the incorrect timing, there is a reduction in the desired sample value and interference due to the following bit, that is, intersymbol interference. Throughout this chapter and the rest of the book, we will assume that the timing information is perfect; hence, there is no such impairment. In other words, in our study, we will consider *one-shot transmission*, namely, individual bit/symbol transmissions in isolation without considering the effects of others.

5.4 *M*-ary Modulation

In the previous section, we studied the simplified digital modulation scheme of binary antipodal signaling over an AWGN channel. We now consider the general case: M-ary digital modulation. We have M distinct symbols $1, 2, \ldots, M$ transmitted via the pulses $s_1(t), s_2(t), \ldots, s_M(t)$, respectively. Denote the prior probability of the ith signal being transmitted by p_i. For the case of equally likely symbols, $p_i = 1/M$. The M signals are defined on the interval $[0, T_s)$, where T_s is the symbol period. As discussed previously for the binary case, while a long sequence of symbols drawn from the M-ary alphabet is transmitted continuously, different transmitted symbols do not interfere with each other (under ideal channel conditions and the perfect synchronization assumption). Hence, we can focus on an isolated symbol transmission on the interval $[0, T_s)$ for receiver design. The receiver operation will then be periodically repeated to demodulate the other symbols.

Recalling the signal space concepts, we identify an orthonormal basis for the set of signals spanned by the M pulses. Let us assume that the signal dimensionality is N (clearly, $N \leq M$) and the orthonormal basis signals are denoted by $\psi_1(t), \psi_2(t), \ldots, \psi_N(t)$. We then have the vector representation of the M signals as

$$\mathbf{s}_i = \begin{bmatrix} s_{i1} \\ s_{i2} \\ \vdots \\ s_{iN} \end{bmatrix} \tag{5.174}$$

for $i = 1, 2, \ldots, M$.

Assuming that the ith symbol represented by the pulse $s_i(t)$ is transmitted over an AWGN channel, the received signal is given by

$$r(t) = s_i(t) + n(t), \quad \text{for} \quad t \in [0, T_s), \tag{5.175}$$

where $n(t)$ is white Gaussian noise with power spectral density $\frac{N_0}{2}$.

The receiver needs to decide on the transmitted symbol using the received signal $r(t)$. To accomplish this, as in binary antipodal signaling, we first project the received signal onto the signal space. Since, in this case, the signal space is N-dimensional, the resulting projection will be an N-dimensional vector (as opposed to a scalar as in binary antipodal signaling). The components of the received signal vector are obtained by the inner products:

$$r_j = \langle r(t), \psi_j(t) \rangle \tag{5.176}$$

$$= \int_0^{T_s} r(t)\psi_j(t)dt \tag{5.177}$$

for $j = 1, 2, \ldots, N$, and the received signal is represented as

$$\boldsymbol{r} = \begin{bmatrix} r_1 \\ r_2 \\ \vdots \\ r_N \end{bmatrix}. \tag{5.178}$$

While this vector representation of the received signal carries all the relevant information for the receiver's operation over an AWGN channel, it does not fully represent the received signal (which is infinite-dimensional). As discussed in the previous section, what is left out is orthogonal to the signal space. Since the transmission is over an AWGN channel, the projection of the received signal onto the signal space ($\boldsymbol{r}$) and the remainder are jointly Gaussian, and they are uncorrelated with each other; hence, what is left out is independent of all the received signal components in the signal space. Therefore, it is irrelevant in making a decision about the transmitted symbol and can be omitted.

We now have a mathematically equivalent model as follows: to transmit the symbols $1, 2, \ldots, M$, we use the vectors $\boldsymbol{s}_1, \boldsymbol{s}_2, \ldots, \boldsymbol{s}_M$, respectively. When the ith symbol is transmitted, the received signal vector is given by

$$\boldsymbol{r} = \boldsymbol{s}_i + \boldsymbol{n}, \tag{5.179}$$

where

$$\boldsymbol{n} = \begin{bmatrix} n_1 \\ n_2 \\ \vdots \\ n_N \end{bmatrix}, \tag{5.180}$$

with $n_j = \langle n(t), \psi_j(t) \rangle$ being the jth component of the N-dimensional AWGN noise vector.

In the following, we will determine the optimal receiver structure to minimize the average symbol error probability.

5.4.1 MAP and ML Decision Rules

The receiver's job is to decide on the transmitted symbol (out of M possibilities) given the received signal in vector form $\boldsymbol{r}$. The MAP decision rule that we have previously introduced for binary antipodal signaling, which picks the symbol with the largest posterior probability, becomes

$$\hat{m} = \underset{m=1,2,\ldots,M}{\arg\max}\ \mathbb{P}(m|\boldsymbol{r}). \tag{5.181}$$

As we have shown for the case of binary antipodal signaling, this decision rule is optimal in the sense that it minimizes the average probability of error. Noting Bayes' rule

$$\mathbb{P}(m|\boldsymbol{r}) = \frac{\mathbb{P}(m)f(\boldsymbol{r}|m)}{f(\boldsymbol{r})}, \tag{5.182}$$

and that the joint PDF of the received signal vector $f(\boldsymbol{r})$ does not depend on the transmitted symbol m, the MAP decision rule can also be written as

$$\hat{m} = \underset{m=1,2,\ldots,M}{\arg\max}\ \mathbb{P}(m)f(\boldsymbol{r}|m). \tag{5.183}$$

For the important case of M-ary equally likely signaling $\left(\text{i.e., when } p_i = \frac{1}{M}\right)$, the above rule boils down to the ML decision rule which maximizes the likelihood function $f(\boldsymbol{r}|m)$:

$$\hat{m} = \underset{m=1,2,\ldots,M}{\arg\max}\ f(\boldsymbol{r}|m), \tag{5.184}$$

or equivalently, the log-likelihood function $\ln f(\boldsymbol{r}|m)$:

$$\hat{m} = \underset{m=1,2,\ldots,M}{\arg\max}\ \ln f(\boldsymbol{r}|m). \tag{5.185}$$

5.4.2 ML Receiver for the AWGN Channel

Let us now determine the ML decision rule for the case of an AWGN channel to obtain the overall receiver structure. To accomplish this, we need to determine the conditional (joint) PDF of the received signal vector given a particular symbol transmission, namely, $f(\boldsymbol{r}|m)$. As an intermediate step, let us first determine the joint PDF of the noise vector obtained by projecting the zero mean white Gaussian noise process with power spectral density $N_0/2$, $n(t)$, onto the signal space (for which an orthonormal basis is given by the functions $\psi_j(t), j = 1, 2, \ldots, N$).

We have already computed the marginal PDF of the jth component of the noise vector

$$n_j = \langle n(t), \psi_j(t) \rangle \tag{5.186}$$

when we studied binary antipodal signaling in the previous section. To recall, as the noise process is Gaussian, its inner product with a (normalized) basis function becomes a Gaussian random variable with mean

$$\mathbb{E}[n_j] = \mathbb{E}\left[\int_0^{T_s} n(t)\psi_j(t)dt\right] \tag{5.187}$$

$$= \int_0^{T_s} \mathbb{E}[n(t)]\psi_j(t)dt \tag{5.188}$$

$$= 0, \tag{5.189}$$

and variance

$$\mathrm{Var}(n_j) = \mathbb{E}[n_j^2] \tag{5.190}$$

$$= \mathbb{E}\left[\int_0^{T_s} n(t)\psi_j(t)dt \int_0^{T_s} n_j(\tau)\psi_j(\tau)d\tau\right] \tag{5.191}$$

$$= \int_0^{T_s}\int_0^{T_s} \mathbb{E}[n(t)n(\tau)]\psi_j(t)\psi_j(\tau)dtd\tau \tag{5.192}$$

$$= \int_0^{T_s}\left(\int_0^{T_s} \frac{N_0}{2}\delta(t-\tau)\psi_j(t)dt\right)\psi_j(\tau)d\tau \tag{5.193}$$

$$= \frac{N_0}{2}\int_0^{T_s} \psi_j^2(\tau)d\tau \tag{5.194}$$

$$= \frac{N_0}{2}. \tag{5.195}$$

That is, each component of the noise vector $\boldsymbol{n}$ is a Gaussian random variable with zero mean and variance $N_0/2$: $n_j \sim \mathcal{N}(0, N_0/2)$.

We also need to determine the interdependency of the components of the noise vector. Fortunately, in the case of an AWGN channel, this is a straightforward task. We first recall from the properties of Gaussian processes that the N random variables $n_1, n_2, \ldots, n_N$ all derived from the same Gaussian random process (by mixing with deterministic signals and integrating) are jointly Gaussian. Hence, knowing their means (which are zero), we only need to determine their covariance matrix to characterize the joint distribution of the noise vector. Towards this end, let us compute $\mathbb{E}[n_j n_l]$ for $j \neq l$:

$$\mathbb{E}[n_j n_l] = \mathbb{E}\left[\int_0^{T_s} n(t)\psi_j(t)dt \int_0^{T_s} n_l(\tau)\psi_j(\tau)d\tau\right] \tag{5.196}$$

$$= \int_0^{T_s}\int_0^{T_s} \mathbb{E}[n(t)n(\tau)]\psi_j(t)\psi_l(\tau)dtd\tau \tag{5.197}$$

$$= \int_0^{T_s}\int_0^{T_s} \frac{N_0}{2}\delta(t-\tau)\psi_j(t)\psi_l(\tau)dtd\tau \tag{5.198}$$

$$= \frac{N_0}{2} \int_0^{T_s} \psi_j(t)\psi_l(t)dt \tag{5.199}$$

$$= 0, \tag{5.200}$$

where the last line follows from the orthogonality of the basis functions.

In other words, the different components of the noise vector are uncorrelated, that is, the covariance matrix is diagonal, with all the elements on the diagonal being $N_0/2$. Being jointly Gaussian, we can also conclude that the noise terms are independent.

At this point, we have a complete description of the noise vector as

$$\boldsymbol{n} \sim \mathcal{N}\left(\boldsymbol{0}, \frac{N_0}{2}\boldsymbol{I}\right), \tag{5.201}$$

where $\boldsymbol{0}$ is an $N \times 1$ vector of zeros and $\boldsymbol{I}$ is an $N \times N$ identity matrix. More explicitly, we can write

$$f(\boldsymbol{n}) = f(n_1, n_2, \ldots, n_N) \tag{5.202}$$

$$= f(n_1)f(n_2)\ldots f(n_N) \tag{5.203}$$

$$= \prod_{j=1}^{N} \frac{1}{\sqrt{\pi N_0}} \exp\left(-\frac{n_j^2}{N_0}\right) \tag{5.204}$$

$$= \left(\frac{1}{\pi N_0}\right)^{N/2} \exp\left(-\frac{\|\boldsymbol{n}\|^2}{N_0}\right). \tag{5.205}$$

Having obtained the joint probability density function of the noise vector, it is a simple matter to determine the desired conditional PDFs, $f(\boldsymbol{r}|m)$, for $m = 1, 2, \ldots, M$. Given that the symbol m is transmitted, the received signal is written as $\boldsymbol{r} = \boldsymbol{s}_m + \boldsymbol{n}$. Therefore, the received signal is a Gaussian random vector with mean $\boldsymbol{s}_m$ and covariance matrix $\frac{N_0}{2}\boldsymbol{I}$. More explicitly, we have

$$f(\boldsymbol{r}|m) = \prod_{j=1}^{N} \frac{1}{\sqrt{\pi N_0}} \exp\left(-\frac{(r_j - s_{mj})^2}{N_0}\right)$$

$$= \left(\frac{1}{\pi N_0}\right)^{N/2} \exp\left(-\frac{\|\boldsymbol{r} - \boldsymbol{s}_m\|^2}{N_0}\right). \tag{5.206}$$

Hence, given that a particular symbol is transmitted, the components of the received signal vector are independent Gaussian random variables with means equal to the corresponding elements of the transmitted signal vector and variance $N_0/2$. We also write

$$\boldsymbol{r}|m \sim \mathcal{N}\left(\boldsymbol{s}_m, \frac{N_0}{2}\boldsymbol{I}\right). \tag{5.207}$$

An exemplary joint PDF of the received signal given that a particular symbol is transmitted is depicted in Fig. 5.24 for the case of two-dimensional signaling

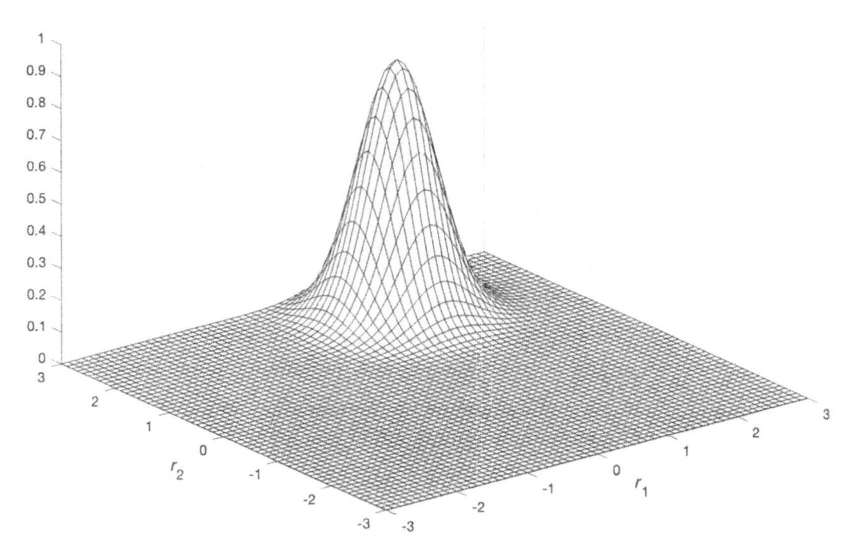

Figure 5.24 Joint PDF of the received signal conditioned on a specific symbol transmission (with $s_{11} = 1$ and $s_{12} = 2$) over an AWGN channel.

($N = 2$). It is instructive to compare this with the previously obtained received signal PDFs for the one-dimensional case in Fig. 5.15.

At this point, we are in a position to simplify the ML detector for M-ary modulation over an AWGN channel. From (5.184), we know that we are to maximize the conditional PDF $f(\boldsymbol{r}|m)$ over all possible symbols. Consider (5.206); as the coefficient of the exponential term is the same for all m, the exponential function is an increasing function, and the term in the exponent is negative, the ML receiver operation boils down to minimizing the squared Euclidean distance between $\boldsymbol{r}$ and $\boldsymbol{s}_m$. In other words, the ML decision rule over an AWGN channel becomes

$$\hat{m} = \underset{m=1,2,\ldots,M}{\arg\min} \; \|\boldsymbol{r} - \boldsymbol{s}_m\|^2. \tag{5.208}$$

Note that, since $\|\boldsymbol{r} - \boldsymbol{s}_m\|^2 = \|\boldsymbol{r}\|^2 - 2\langle \boldsymbol{r}, \boldsymbol{s}_m\rangle + \|\boldsymbol{s}_m\|^2$ and the $\|\boldsymbol{r}\|^2$ term is the same for all the terms being minimized, the ML rule can also be written as

$$\hat{m} = \underset{m=1,2,\ldots,M}{\arg\min} \; \langle \boldsymbol{r}, \boldsymbol{s}_m\rangle - \frac{1}{2}\|\boldsymbol{s}_m\|^2, \tag{5.209}$$

which is nothing but maximization of correlation (along with a correction term for the energy of each constellation point). For constant energy signal constellations, the second term in the argument is the same for all m, and the rule is further simplified to

$$\hat{m} = \underset{m=1,2,\ldots,M}{\arg\min} \; \langle \boldsymbol{r}, \boldsymbol{s}_m\rangle. \tag{5.210}$$

It is remarkable that the optimal receiver (for equally likely signaling), that is, the ML receiver, is quite simple. We project the received signal onto the signal space, obtain an N-dimensional vector representation, and compare the resulting received vector with the M constellation points to minimize the squared Euclidean distance or to maximize the correlation metric so as to decide which symbol is the most likely one. The ML receiver divides the N-dimensional signal space (on which $\boldsymbol{r}$ belongs) into M non-overlapping Voronoi cells (denoted as $\mathcal{D}_m$, $m = 1, 2, \ldots, M$), where each cell is the decision region for a particular symbol. In other words, when $\boldsymbol{r} \in \mathcal{D}_{m'}$ for a particular symbol m', we decide in favor of that symbol. An illustration of the decision regions for a two-dimensional signal constellation is given in Fig. 5.25.

The decision regions may become quite simple when the signal constellation possesses certain symmetry. We will see many examples of such signal constellations. We have already studied one instance (binary antipodal signaling), for which the signal space is one-dimensional and the decision regions for equally likely signaling are the sets of positive and negative real numbers, respectively.

Figure 5.26 depicts the block diagram of the overall correlation-type ML receiver for M-ary modulation over an AWGN channel. Figure 5.27 illustrates the matched filter-type implementation.

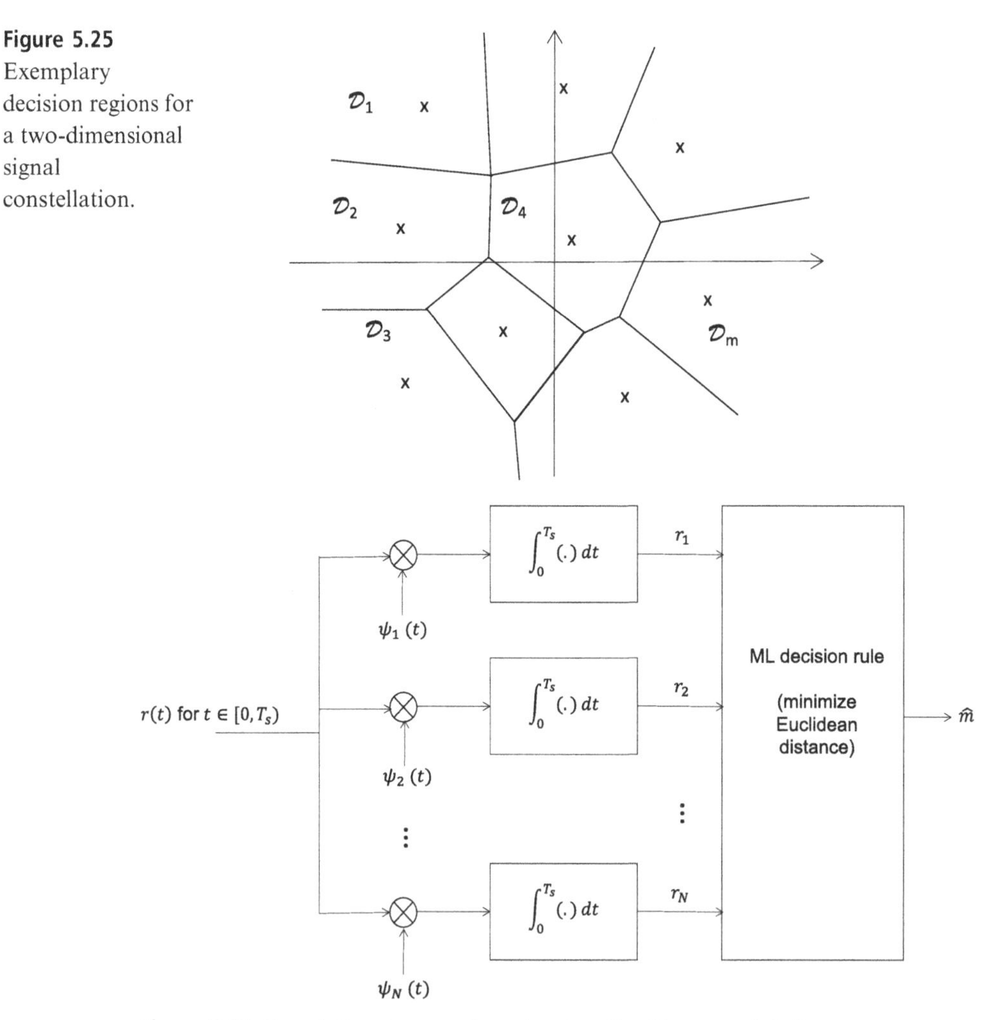

Figure 5.25 Exemplary decision regions for a two-dimensional signal constellation.

Figure 5.26 Correlation-type receiver structure for M-ary modulation over an AWGN channel.

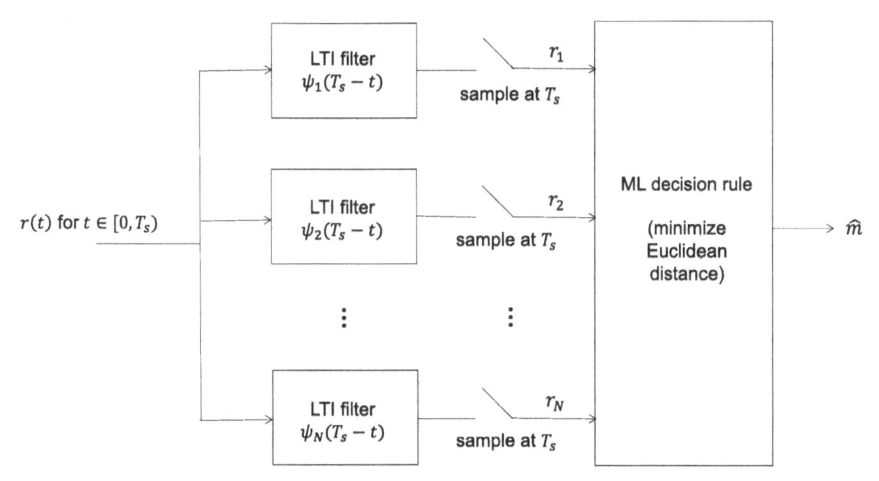

Figure 5.27 Matched filter-type receiver structure for M-ary modulation over an AWGN channel.

Example 5.7

Consider two-dimensional 4-ary modulation with the signal constellation given in Fig. 5.28 for transmission over an AWGN channel. Denote the symbol period by T_s and the two orthonormal basis functions by $\phi_1(t)$ and $\phi_2(t)$. Assuming equally likely signaling, determine the ML receiver structure.

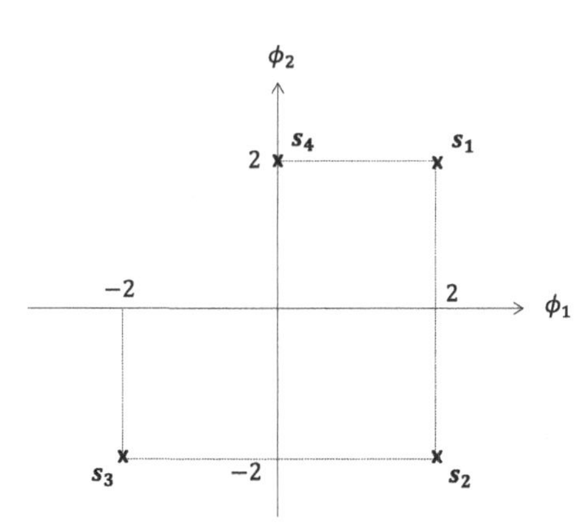

Figure 5.28 The 4-ary signal constellation for the example.

Solution

The received signal $r(t)$ is first projected onto the two basis functions using a correlation-type or a matched filter-type detector, as described earlier in the section, to obtain the received signal in vector form $\boldsymbol{r} = \begin{bmatrix} r_1 \\ r_2 \end{bmatrix}$.

All we need to do is determine the decision region for each of the four signals. Recalling that for an AWGN channel, the ML receiver minimizes the Euclidean distance between the received signal vector $\boldsymbol{r}$ and the transmitted signals, the mth decision region is given by

$$\mathcal{D}_m = \{\boldsymbol{r}: \|\boldsymbol{r} - \boldsymbol{s}_m\| < \|\boldsymbol{r} - \boldsymbol{s}_k\|, \text{ for } k \neq m\}. \tag{5.211}$$

Note that we do not worry about the case where the received signal is at an equal distance from two different constellation points, as this will occur with probability 0. Let us explicitly carry out the necessary computation for the second signal to determine $\mathcal{D}_2$. We need to compare

$$\|\boldsymbol{r} - \boldsymbol{s}_2\|^2 = (r_1 - 2)^2 + (r_2 + 2)^2 \tag{5.212}$$

with the three squared distances

$$\|\boldsymbol{r} - \boldsymbol{s}_1\|^2 = (r_1 - 2)^2 + (r_2 - 2)^2, \tag{5.213}$$

$$\|\boldsymbol{r} - \boldsymbol{s}_3\|^2 = (r_1 + 2)^2 + (r_2 + 2)^2, \tag{5.214}$$

$$\|\boldsymbol{r} - \boldsymbol{s}_4\|^2 = r_1^2 + (r_2 - 2)^2. \tag{5.215}$$

Namely,

$$\mathcal{D}_2 = \{(r_1, r_2) : (r_1 - 2)^2 + (r_2 + 2)^2 < (r_1 - 2)^2 + (r_2 - 2)^2, \tag{5.216}$$

$$(r_1 - 2)^2 + (r_2 + 2)^2 < (r_1 + 2)^2 + (r_2 + 2)^2, \tag{5.217}$$

$$(r_1 - 2)^2 + (r_2 + 2)^2 < r_1^2 + (r_2 - 2)^2\}, \tag{5.218}$$

which can be simplified as

$$\mathcal{D}_2 = \{(r_1, r_2): r_1 > 0,\ r_2 < 0,\ r_1 - 2r_2 > 1\}. \tag{5.219}$$

Similarly, the other decision regions are obtained as follows:

$$\mathcal{D}_1 = \{(r_1, r_2): r_1 > 1,\ r_2 > 0\}, \tag{5.220}$$

$$\mathcal{D}_3 = \{(r_1, r_2): r_1 < 0,\ r_1 + 2r_2 < -1\}, \tag{5.221}$$

$$\mathcal{D}_4 = \{(r_1, r_2): r_1 < 1,\ r_1 + 2r_2 > -1,\ r_1 - 2r_2 < 1\}. \tag{5.222}$$

These four decision regions are illustrated in Fig. 5.29.

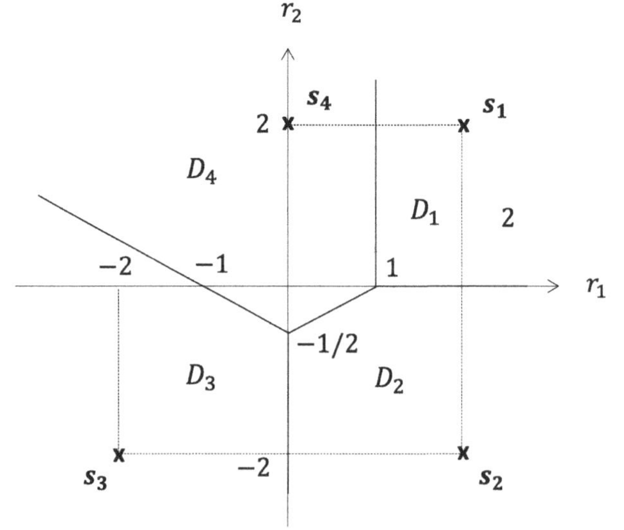

Figure 5.29 Decision regions for the four symbols.

To summarize: the receiver finds the vector representation of the received signal (corresponding to the symbol period), for instance, by using matched filters as in Fig. 5.27, and then applies the decision rule

$$\hat{m} = \begin{cases} 1, & \text{if } r_1 > 1,\ r_2 > 0, \\ 2, & \text{if } r_1 > 0,\ r_2 < 0,\ r_1 - 2r_2 > 1, \\ 3, & \text{if } r_1 < 0,\ r_1 + 2r_2 < -1, \\ 4, & \text{if } r_1 < 1,\ r_1 + 2r_2 > -1,\ r_1 - 2r_2 < 1. \end{cases} \tag{5.223}$$

5.4.3 Probability of Error for *M*-ary Modulation and the Union Bound

We are now in a position to examine the performance of M-ary signaling over an AWGN channel in terms of its average probability of error. We focus on equally likely signaling and ML detection.

We can write the conditional error probability given that the mth symbol is transmitted as

$$P_{e,m} = \mathbb{P}(\boldsymbol{r} \notin \mathcal{D}_m | m \text{ is sent}). \tag{5.224}$$

Hence, by averaging over the possible transmitted symbols (which are assumed equally likely), we obtain the average error probability for M-ary transmission as

$$P_e = \frac{1}{M} \sum_{m=1}^{M} P_{e,m} \tag{5.225}$$

$$= \frac{1}{M} \sum_{m=1}^{M} \mathbb{P}(\boldsymbol{r} \notin \mathcal{D}_m | m \text{ is sent}). \tag{5.226}$$

While writing the general expression for the exact error probability is straightforward, its evaluation for an AWGN channel may be challenging, particularly for higher-dimensional signaling. This is because, for an exact evaluation, one needs to integrate multivariate Gaussian PDFs over Voronoi cells (as depicted in Fig. 5.25), which becomes intractable when there is not sufficient symmetry in the signal constellation. To proceed, we can resort to a union bound on the error probability.

Take two constellation points at a time ($\boldsymbol{s}_m$ and $\boldsymbol{s}_k$) and define the pairwise error event $A_{m,k}$ as the event that the received signal $\boldsymbol{r}$ is closer to $\boldsymbol{s}_k$ (in the Euclidean distance sense) given that $\boldsymbol{s}_m$ is transmitted. The pairwise error probability

$$\mathbb{P}(A_{m,k}) = \mathbb{P}(\|\boldsymbol{r} - \boldsymbol{s}_k\| < \|\boldsymbol{r} - \boldsymbol{s}_m\| \, | \boldsymbol{s}_m \text{ is transmitted}) \tag{5.227}$$

is easy to compute as the condition $\{\|\boldsymbol{r} - \boldsymbol{s}_k\| < \|\boldsymbol{r} - \boldsymbol{s}_i\|\}$ divides the N-dimensional signal space into two half-spaces, and the required integration of the multivariate Gaussian PDF becomes tractable. We will illustrate this point in detail shortly. The error event given that the mth symbol is transmitted is the union of the pairwise error events $A_{m,1}, A_{m,2}, \ldots, A_{m,m-1}, A_{m,m+1}, \ldots, A_{m,M}$. See Fig. 5.30 for a 4-ary signal constellation as an illustration. We can write

$$P_{e,m} = \mathbb{P}(A_{m,1} \cup A_{m,2} \cup \cdots \cup A_{m,m-1} \cup A_{m,m+1} \cup \cdots \cup A_{m,M}). \tag{5.228}$$

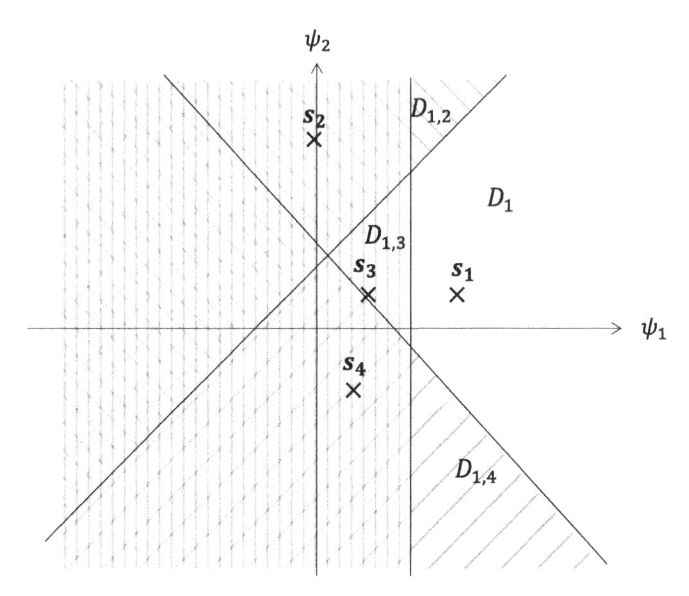

Figure 5.30 Illustration of the error region for a particular signal constellation and its representation as the union of pairwise error regions ($\mathcal{D}_{1,k} = \{\boldsymbol{r}: \|\boldsymbol{r} - \boldsymbol{s}_k\| < \|\boldsymbol{r} - \boldsymbol{s}_1\|\}$, $k = 2, 3, 4$).

Employing the union bound, we obtain

$$P_{e,m} \leq \sum_{\substack{k=1 \\ k \neq m}}^{M} \mathbb{P}(A_{m,k}). \tag{5.229}$$

In other words, we can write an upper bound on the conditional error probability given that $\boldsymbol{s}_m$ is transmitted in terms of the pairwise error probabilities involving the mth symbol.

Let us now calculate the individual pairwise error probabilities. When $\boldsymbol{s}_m$ is transmitted, the received signal is of the form $\boldsymbol{r} = \boldsymbol{s}_m + \boldsymbol{n}$, hence we can write

$$\mathbb{P}(A_{m,k}) = \mathbb{P}(\|\boldsymbol{r} - \boldsymbol{s}_k\|^2 < \|\boldsymbol{r} - \boldsymbol{s}_m\|^2 \,|\boldsymbol{s}_m \text{ is transmitted}) \tag{5.230}$$

$$= \mathbb{P}(\|\boldsymbol{s}_m + \boldsymbol{n} - \boldsymbol{s}_k\|^2 < \|\boldsymbol{n}\|^2) \tag{5.231}$$

$$= \mathbb{P}(\|\boldsymbol{n}\|^2 + \|\boldsymbol{s}_k - \boldsymbol{s}_m\|^2 - 2\,\boldsymbol{n} \cdot (\boldsymbol{s}_k - \boldsymbol{s}_m) < \|\boldsymbol{n}\|^2) \tag{5.232}$$

$$= \mathbb{P}\left(\boldsymbol{n} \cdot (\boldsymbol{s}_k - \boldsymbol{s}_m) > \frac{\|\boldsymbol{s}_m - \boldsymbol{s}_k\|^2}{2}\right). \tag{5.233}$$

Since the dot product of $\boldsymbol{n}$ and $\boldsymbol{s}_k - \boldsymbol{s}_m$ is nothing but the sum of scaled noise components, that is,

$$\boldsymbol{n} \cdot (\boldsymbol{s}_m - \boldsymbol{s}_m) = \sum_{j=1}^{N} (s_{kj} - s_{mj}) n_j, \tag{5.234}$$

and the noise terms n_js are independent Gaussian random variables with zero mean and variance $\frac{N_0}{2}$, the dot product becomes a Gaussian random variable with zero mean and variance

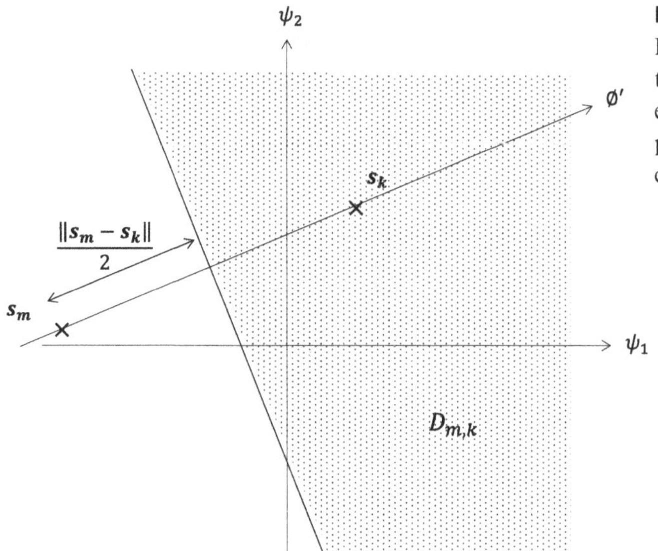

Figure 5.31 Illustration of the pairwise error probability calculation.

$$\sum_{j=1}^{N}(s_{kj} - s_{mj})^2 \frac{N_0}{2} = \|s_m - s_k\|^2 \frac{N_0}{2}. \tag{5.235}$$

Hence, we obtain

$$\mathbb{P}(A_{m,k}) = Q\left(\frac{\|s_m - s_k\|}{\sqrt{2N_0}}\right). \tag{5.236}$$

Let us also give an alternate derivation of the pairwise error probability expression. Consider the two constellation points (in an N-dimensional signal space) as illustrated in Fig. 5.31. We can use a different set of basis functions for the signal space, for which one of the (normalized) basis functions is that in the direction of a vector from s_m to s_k, denoted by $\phi'(t)$. The only component of the noise vector that can move a received signal point closer to s_k compared to s_m is the projection of the $n(t)$ onto $\phi'(t)$, that is, $n' = \langle n(t), \phi'(t) \rangle$. This is because all the other noise components are orthogonal to this direction, and they will not have an effect on how close the received signal vector is to the wrong constellation point s_k versus the correct one s_m. The pairwise error event will occur if the noise component n' is greater than half the Euclidean distance between the two constellation points. Therefore, we can readily write

$$\mathbb{P}(A_{m,k}) = \mathbb{P}\left(n' > \frac{\|s_m - s_k\|}{2}\right) \tag{5.237}$$

$$= Q\left(\frac{\|s_m - s_k\|}{\sqrt{2N_0}}\right), \tag{5.238}$$

where the second line follows using $n' \sim \mathcal{N}(0, N_0/2)$. As expected, this is the same result as that obtained in (5.236).

We highlight that the pairwise error probability term depends only on the Euclidean distance between the pair of constellation points being considered and the variance of the Gaussian noise. Substituting this expression into the union bound for the error probability of the mth symbol, we can write

$$P_{e,m} \leq \sum_{\substack{k=1 \\ k \neq m}}^{M} Q\left(\frac{\|\boldsymbol{s}_m - \boldsymbol{s}_k\|}{\sqrt{2N_0}}\right), \tag{5.239}$$

and hence, we obtain the union bound on the error probability of the M-ary signaling over an AWGN channel as

$$P_e \leq \frac{1}{M} \sum_{m=1}^{M} \sum_{\substack{k=1 \\ k \neq m}}^{M} Q\left(\frac{\|\boldsymbol{s}_m - \boldsymbol{s}_k\|}{\sqrt{2N_0}}\right). \tag{5.240}$$

It should be evident that we can easily compute this expression for arbitrary signal constellations by simply enumerating the pairwise Euclidean distances among all the constellation points.

We can simplify the union bound further by upper bounding the right-hand side of (5.240). Since the Q-function is a decreasing function of its argument, if we replace each Euclidean distance term with a smaller value, the expression will become larger. Defining the minimum Euclidean distance of the signal constellation as

$$d_{\text{min}} = \min_{\substack{m,k \\ m \neq k}} \|\boldsymbol{s}_m - \boldsymbol{s}_k\|, \tag{5.241}$$

and replacing $\|\boldsymbol{s}_m - \boldsymbol{s}_k\|$ with d_{min} in the union bound expression, we can obtain a looser version as

$$P_e \leq \frac{1}{M} \sum_{m=1}^{M} \sum_{\substack{k=1 \\ k \neq m}}^{M} Q\left(\frac{d_{\text{min}}}{\sqrt{2N_0}}\right) \tag{5.242}$$

$$\leq (M-1) Q\left(\frac{d_{\text{min}}}{\sqrt{2N_0}}\right), \tag{5.243}$$

which depends only on the minimum distance of the signal constellation and the noise variance, and hence is very easy to compute.

We close this section by noting that despite its simplicity, the union bound offers an instrumental performance analysis tool for studying M-ary modulation. Both the original version and the loose form are quite tight for large signal-to-noise ratios (i.e., for low noise levels). This is because the Q-function terms decay exponentially in the signal-to-noise ratio; hence, the terms with the smallest pairwise distances dominate the expression.

To clarify the above point further, let us also obtain a lower bound on the error probability. It is easy to see that P_e cannot be smaller than $\frac{1}{M} P_{e,m}$ (the scaled version of any of the conditional error probabilities), and at least one of the $P_{e,m}$ terms is larger than (or equal to) $Q\left(\frac{d_{\text{min}}}{\sqrt{2N_0}}\right)$ (since all the pairwise error regions for the symbol m are subsets of the overall error region for the mth symbol). In other words, the probability of error is sandwiched as follows:

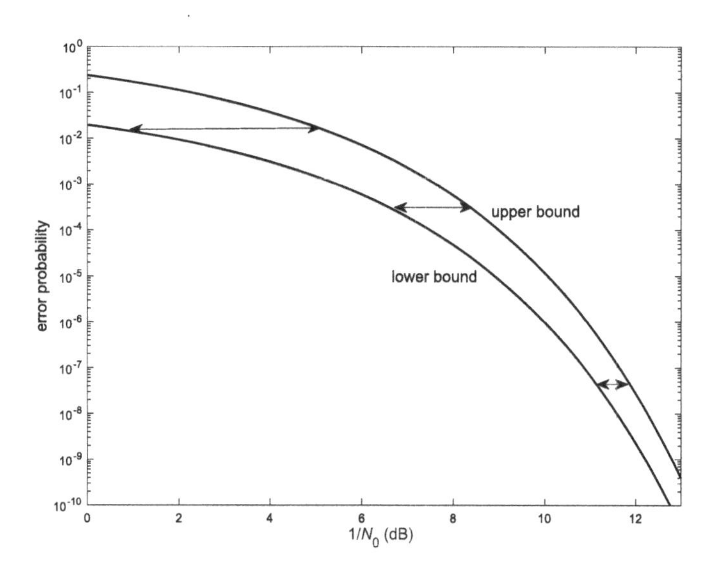

Figure 5.32 Illustration of the upper and lower bounds on the probability of error in (5.244) for a constellation with $M = 4$ and $d_{\min} = 2$ as a function of $1/N_0$.

$$\frac{1}{M}Q\left(\frac{d_{\min}}{\sqrt{2N_0}}\right) \leq P_e \leq (M-1)Q\left(\frac{d_{\min}}{\sqrt{2N_0}}\right), \tag{5.244}$$

where the right-hand side is the loose form of the union bound. Since the Q-function decays exponentially in the square of its argument, the above upper and lower bounds decay exponentially in $1/N_0$. Since the factors $\frac{1}{M}$ and $(M-1)$ are simply constants, the right-hand and left-hand sides behave similarly for large signal-to-noise ratios. Therefore, the union bound cannot be far away from the exact error probability when the noise variance is small (i.e., for large signal-to-noise ratios).

As an example, Fig. 5.32 depicts the upper and lower bounds in (5.244) for $M = 4$ and $d_{\min} = 2$ as a function of $1/N_0$. As expected, the gap between the lower and upper bounds shrinks for small values of the noise variance.

Example 5.8

Let us continue with the setup in Example 5.7. Assume that the white Gaussian noise power spectral density is $N_0/2$. Calculate the conditional error probability given that the first symbol is transmitted, namely, $P_{e,1}$.

Solution

We can write

$$P_{e,1} = \mathbb{P}(\boldsymbol{r} \neq \mathcal{D}_1 | \boldsymbol{s}_1 \text{ is sent}) \tag{5.245}$$

$$= 1 - \mathbb{P}(\boldsymbol{r} \in \mathcal{D}_1 | \boldsymbol{s}_1 \text{ is sent}) \tag{5.246}$$

$$= 1 - \mathbb{P}(r_1 > 1 \text{ and } r_2 > 0 | \boldsymbol{s}_1 \text{ is sent}). \tag{5.247}$$

Conditioned on $\boldsymbol{s}_1$ being transmitted, we have $\boldsymbol{r} = \boldsymbol{s}_1 + \boldsymbol{n}$, that is, $r_1 = 2 + n_1$ and $r_2 = 2 + n_2$. Therefore,

$$P_{e,1} = 1 - \mathbb{P}(2 + n_1 > 1 \text{ and } 2 + n_2 > 0) \tag{5.248}$$
$$= 1 - \mathbb{P}(n_1 > -1)\mathbb{P}(n_2 > -2), \tag{5.249}$$

which follows since the two noise terms (in orthogonal directions) n_1 and n_2 are independent. We also recall that $n_1, n_2 \sim \mathcal{N}(0, N_0/2)$. Hence,

$$P_{e,1} = 1 - Q\left(-\frac{1}{\sqrt{N_0/2}}\right) Q\left(-\frac{2}{\sqrt{N_0/2}}\right) \tag{5.250}$$
$$= 1 - \left(1 - Q\left(\sqrt{\frac{2}{N_0}}\right)\right)\left(1 - Q\left(\sqrt{\frac{8}{N_0}}\right)\right) \tag{5.251}$$
$$= Q\left(\sqrt{\frac{2}{N_0}}\right) + Q\left(\sqrt{\frac{8}{N_0}}\right) - Q\left(\sqrt{\frac{2}{N_0}}\right) Q\left(\sqrt{\frac{8}{N_0}}\right). \tag{5.252}$$

Let us now express the result in terms of the average signal-to-noise ratio $\bar{\gamma}_s = \bar{E}_s/N_0$, where $\bar{E}_s$ is the average energy of the signal constellation given by

$$\bar{E}_s = \frac{1}{4}\|s_1\|^2 + \frac{1}{4}\|s_2\|^2 + \frac{1}{4}\|s_3\|^2 + \frac{1}{4}\|s_4\|^2 \tag{5.253}$$
$$= \frac{1}{4}(8 + 8 + 8 + 4) \tag{5.254}$$
$$= 7 \text{ units.} \tag{5.255}$$

Therefore,

$$P_{e,1} = Q\left(\sqrt{\frac{2}{7}\bar{\gamma}_s}\right) + Q\left(\sqrt{\frac{8}{7}\bar{\gamma}_s}\right) - Q\left(\sqrt{\frac{2}{7}\bar{\gamma}_s}\right) Q\left(\sqrt{\frac{8}{7}\bar{\gamma}_s}\right). \tag{5.256}$$

For large average signal-to-noise ratios ($\bar{\gamma}_s \gg 1$), noting that the Q-function terms decay exponentially in the square of their arguments, the first term dominates, and we have

$$P_{e,1} \approx Q\left(\sqrt{\frac{2}{7}\bar{\gamma}_s}\right). \tag{5.257}$$

As there is not sufficient symmetry, we cannot calculate the conditional error probability for the other symbols in closed form.

Example 5.9

Continue with the previous example and compute the union bound on the average symbol error probability.

Solution

The union bound is given by

$$P_e \leq \frac{1}{M}\sum_{m=1}^{M}\sum_{\substack{k=1 \\ k\neq m}}^{M} Q\left(\frac{\|s_m - s_k\|}{\sqrt{2N_0}}\right). \tag{5.258}$$

Therefore, we can write

$$P_e \leq \frac{1}{2}Q\left(\frac{2}{\sqrt{2N_0}}\right) + Q\left(\frac{4}{\sqrt{2N_0}}\right) + Q\left(\frac{2\sqrt{5}}{\sqrt{2N_0}}\right) + \frac{1}{2}Q\left(\frac{4\sqrt{2}}{\sqrt{2N_0}}\right) \tag{5.259}$$

$$= \frac{1}{2}Q\left(\sqrt{\frac{2}{7}\gamma_s}\right) + Q\left(\sqrt{\frac{8}{7}\bar{\gamma}_s}\right) + Q\left(\sqrt{\frac{10}{7}\bar{\gamma}_s}\right) + \frac{1}{2}Q\left(\sqrt{\frac{16}{7}\bar{\gamma}_s}\right). \tag{5.260}$$

For $\bar{\gamma}_s \gg 1$, this bound is approximated by the first term, that is,

$$P_e \approx \frac{1}{2}Q\left(\sqrt{\frac{2}{7}\bar{\gamma}_s}\right). \tag{5.261}$$

Note also that $d_{\min} = 2$ units, hence the loose version of the union bound immediately follows as

$$P_e \leq 3Q\left(\frac{2}{\sqrt{2N_0}}\right) = 3Q\left(\sqrt{\frac{2}{7}\bar{\gamma}_s}\right). \tag{5.262}$$

5.4.4 Comparison of Different Signal Constellations

We can use the performance analysis tool developed in the previous section to compare different M-ary signal constellations in terms of their average error probabilities (for high signal-to-noise ratios). We have a (tight) bound on the error probability of an M-ary modulation scheme with equally likely symbols with a specific signal constellation as

$$P_e \leq (M-1)Q\left(\frac{d_{\min}}{\sqrt{2N_0}}\right). \tag{5.263}$$

Assume that the average energy of the signal constellation is $\bar{E}_s$, and define the average signal-to-noise ratio as $\bar{\gamma}_s = \frac{\bar{E}_s}{N_0}$. We can write

$$P_e \leq (M-1)Q\left(\sqrt{\frac{d^2_{\min}}{2\bar{E}_s}\bar{\gamma}_s}\right). \tag{5.264}$$

We want to compare different M-ary constellations in terms of their error probabilities at the same signal-to-noise ratio $\bar{\gamma}_s$. From the union bound expression in (5.264), we identify that a metric that can be used is the ratio of the minimum squared Euclidean distance to the average energy of the constellation, that is, $d^2_{\min}/\bar{E}_s$. If this quantity is larger for a signal constellation, its average error probability (at high signal-to-noise ratios) is lower; hence, it has a better error probability performance.

We can also assess the approximate performance difference among different signal constellations. Consider two signal constellations with minimum Euclidean distances $d_{\min,1}$ and $d_{\min,2}$, and average energies $\bar{E}_{s,1}$ and $\bar{E}_{s,2}$, respectively. The two performance metrics are

$$\frac{d^2_{\min,1}}{\bar{E}_{s,1}} \quad \text{and} \quad \frac{d^2_{\min,2}}{\bar{E}_{s,2}}. \tag{5.265}$$

Suppose, for instance, that the first metric is larger than the second, that is,

$$\frac{d^2_{\min,1}}{\bar{E}_{s,1}} \geq \frac{d^2_{\min,2}}{\bar{E}_{s,2}}. \tag{5.266}$$

Consider a specific average error probability for the first signal constellation obtained at a (high) average signal-to-noise ratio of $\bar{\gamma}_{s,1} = \gamma$. To obtain the same error rate using the second signal constellation, we need an average signal-to-noise ratio of

$$\bar{\gamma}_{s,2} \approx \frac{d^2_{\min,1}/\bar{E}_{s,1}}{d^2_{\min,2}/\bar{E}_{s,2}}\gamma. \tag{5.267}$$

Expressed in decibels, we need

$$\bar{\gamma}_{s,2}(\text{dB}) \approx \gamma(\text{dB}) + 10\log_{10}\left(\frac{d^2_{\min,1}/\bar{E}_{s,1}}{d^2_{\min,2}/\bar{E}_{s,2}}\right), \tag{5.268}$$

that is, the second constellation is worse than the first one by about $10\log_{10}\left(\frac{d^2_{\min,1}/\bar{E}_{s,1}}{d^2_{\min,2}/\bar{E}_{s,2}}\right)$ dB. To obtain the same error rate, transmission with the second signal constellation would require about $10\log_{10}\left(\frac{d^2_{\min,1}/\bar{E}_{s,1}}{d^2_{\min,2}/\bar{E}_{s,2}}\right)$ dB higher average signal-to-noise ratio.

Example 5.10

Compare the two 8-ary signal constellations in Fig. 5.33 in terms of their error probabilities assuming equally likely symbols and transmission over an AWGN channel.

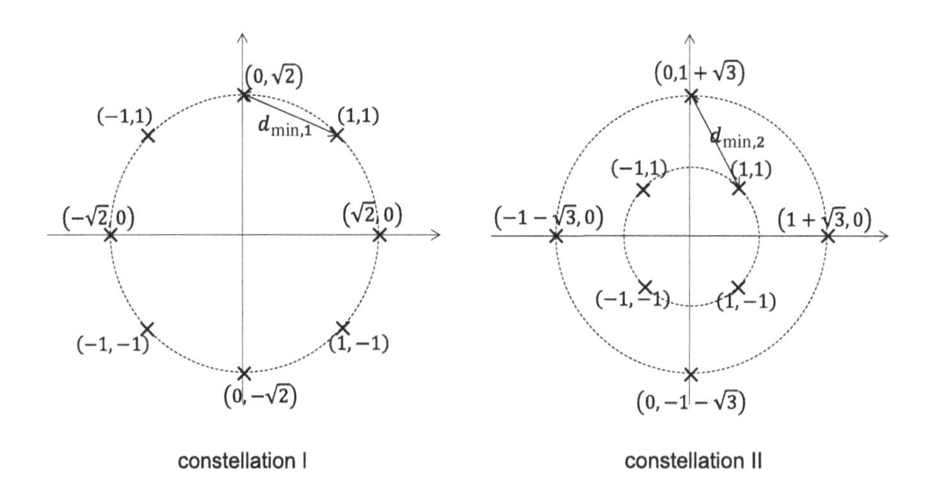

Figure 5.33 Two 8-ary signal constellations to be compared.

Solution

We compute the average energies and the minimum distances of the two constellations. That is,

$$\bar{E}_{s,1} = 2 \text{ and } \bar{E}_{s,2} = \frac{1}{2}\left(2 + \left(1 + \sqrt{3}\right)^2\right) = 3 + \sqrt{3}, \tag{5.269}$$

and

$$d^2_{\min,1} = \left(\sqrt{2} - 1\right)^2 + 1 = 4 - 2\sqrt{2} \text{ and } d^2_{\min,2} = 4. \tag{5.270}$$

The normalized (squared) minimum distances are

$$\frac{d^2_{\min,1}}{\bar{E}_{s,1}} = \frac{4 - 2\sqrt{2}}{2} = 2 - \sqrt{2} \approx 0.59 \text{ and } \frac{d^2_{\min,2}}{\bar{E}_{s,2}} = \frac{4}{3 + \sqrt{3}} \approx 0.845. \tag{5.271}$$

Hence, the second constellation is better than the first one in terms of its error performance by about

$$10 \log_{10}\left(\frac{0.845}{0.59}\right) = 1.56 \text{ dB} \tag{5.272}$$

at high signal-to-noise ratios.

5.5 *M*-ary Pulse Amplitude Modulation

In this section, we study M-ary pulse amplitude modulation (M-PAM) as an important example of a digital modulation scheme.

For M-PAM, information is carried by the amplitude of a basic pulse, namely,

$$s_i(t) = A_i p(t) \quad \text{for } t \in [0, T_s), \tag{5.273}$$

where $i = 1, 2, \ldots, M$ and $p(t)$ is an arbitrary signal defined on the symbol interval $[0, T_s)$. This is a one-dimensional signal set with the basis function

$$\psi(t) = \frac{p(t)}{\sqrt{E_p}} \quad \text{for } t \in [0, T_s), \tag{5.274}$$

where $E_p = \int_0^{T_s} p^2(t)dt$. Hence, the vector forms of the signals are given by

$$s_i = A_i\sqrt{E_p}, \quad \text{for } i = 1, 2, \ldots, M, \tag{5.275}$$

which are, indeed, scalars.

Typically, we select

$$A_i = (2i - 1 - M)A \tag{5.276}$$

for $i = 1, 2, \ldots, M$ and some constant A. This is because the error probability performance of a modulation scheme over an AWGN channel is well characterized by the minimum distance of the signal constellation; hence, selecting the consecutive constellation points of M-PAM with equal separation is advantageous. Furthermore, placing the center of the signal constellation at the origin minimizes the average energy without changing the distance properties. Hence, this selection of the amplitude levels minimizes the average error probability with equally likely signaling over an AWGN channel for a fixed average constellation energy.

For simplicity of notation, let us define $d = A\sqrt{E_p}$. Hence, $s_i = (2i - 1 - M)d$ and the resulting signal constellation is as depicted in Fig. 5.34.

Consider transmission of M-ary PAM signals over an AWGN channel. The receiver projects the received signal $r(t)$ for $t \in [0, T_s)$ onto the signal space, obtaining the mathematically equivalent received signal, that is, the scalar decision variable, as

$$r = \int_0^{T_s} r(t)\psi(t)dt. \tag{5.277}$$

Assuming that the symbols are equally likely, the receiver applies the ML decision rule, which minimizes the Euclidean distance of the received signal r with the constellation points, resulting in a threshold-based rule described as

$$\hat{m} = \begin{cases} 1, & \text{if } r < -(M-2)d, \\ i, & \text{if } (2i-2-M)d < r < (2i-M)d, i = 2, 3, \ldots, M-1 \\ M, & \text{if } r > (M-2)d. \end{cases} \tag{5.278}$$

The correlation-type receiver structure for M-PAM signaling is depicted in Fig. 5.35.

Let us now compute the resulting average symbol error probability P_e. We have already considered the case of $M = 2$ as it is nothing but the binary antipodal signaling scheme studied earlier in the chapter. Consider the case with $M > 2$. We can write

$$P_e = \frac{1}{M} \sum_{m=1}^{M} P_{e,i}, \tag{5.279}$$

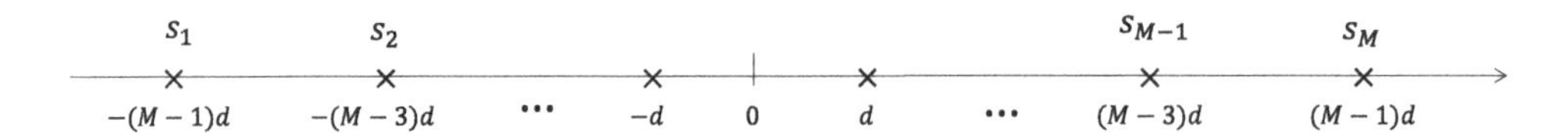

Figure 5.34 An M-PAM signal constellation.

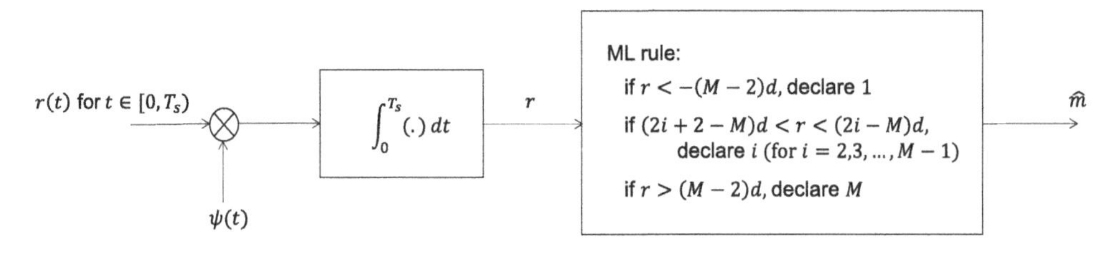

Figure 5.35 Correlation-type M-PAM receiver structure.

where $P_{e,i}$ is the conditional error probability given that the symbol i is sent. It is clear from the symmetry of the signal constellation that $P_{e,1} = P_{e,M}$, and that $P_{e,2} = P_{e,3} = \cdots = P_{e,M-1}$. We write

$$P_{e,1} = \mathbb{P}\big(\hat{m} \neq m_1 | m_1 \text{ sent}\big) \tag{5.280}$$

$$= \mathbb{P}\big(r > -(M-2)d \,\big|\, m_1 \text{ sent}\big) \tag{5.281}$$

$$= \mathbb{P}\Big(-(M-1)d + n > -(M-2)d\Big), \tag{5.282}$$

where $n \sim \mathcal{N}\left(0, \frac{N_0}{2}\right)$ is the additive noise term. Hence, we obtain

$$P_{e,1} = \mathbb{P}(n > d) \tag{5.283}$$

$$= Q\left(\frac{d}{\sqrt{N_0/2}}\right) \tag{5.284}$$

$$= Q\left(\sqrt{\frac{2d^2}{N_0}}\right). \tag{5.285}$$

Similarly,

$$P_{e,2} = \mathbb{P}\big(\hat{m} \neq m_2 \,\big|\, m_2 \text{ sent}\big) \tag{5.286}$$

$$= \mathbb{P}\Big(r < -(M-2)d \text{ or } r > -(M-4)d \,\Big|\, m_2 \text{ sent}\Big) \tag{5.287}$$

$$= \mathbb{P}\Big(-(M-3)d + n < -(M-2)d \text{ or } -(M-3)d + n > -(M-4)d\Big) \tag{5.288}$$

$$= \mathbb{P}\Big(n < -d \text{ or } n > d\Big) \tag{5.289}$$

$$= 2Q\left(\sqrt{\frac{2d^2}{N_0}}\right). \tag{5.290}$$

Therefore, we obtain the average error probability as

$$P_e = \frac{2}{M}P_{e,1} + \frac{M-2}{M}P_{m,2} \tag{5.291}$$

$$= \frac{2M-2}{M}Q\left(\sqrt{\frac{2d^2}{N_0}}\right). \tag{5.292}$$

To express the result in terms of the average signal-to-noise ratio $\bar{\gamma}_s = \frac{\bar{E}_s}{N_0}$, we compute the average energy of the signal constellation as

$$\bar{E}_s = \frac{1}{M}\sum_{i=1}^{M} s_i^2 = \frac{1}{M}\sum_{i=1}^{M}(2i-1-M)^2 d^2. \tag{5.293}$$

Using the identity

$$1^2 + 3^2 + \cdots + (2j-1)^2 = \frac{j(2j-1)(2j+1)}{3}, \tag{5.294}$$

we obtain

$$\bar{E}_s = \frac{(M^2-1)d^2}{3}. \tag{5.295}$$

Therefore,

$$P_e = \frac{2(M-1)}{M} Q\left(\sqrt{\frac{2}{N_0}\frac{3\bar{E}_s}{M^2-1}}\right) \tag{5.296}$$

$$= \frac{2(M-1)}{M} Q\left(\sqrt{\frac{6\bar{\gamma}_s}{M^2-1}}\right). \tag{5.297}$$

Notice that, for $M = 2$, this formula boils down to the error probability expression obtained for binary antipodal signaling calculated earlier, as expected.

Let us also express the symbol error probability in terms of the average signal-to-noise ratio per bit. With $M = 2^k$, each symbol carries k bits of information. Therefore, the average symbol energy is $\bar{E}_s = k\bar{E}_b = (\log_2 M)\bar{E}_b$, where $\bar{E}_b$ is the energy per bit. With this observation, we can write the average signal-to-noise ratio per bit $\bar{\gamma}_b$ as

$$\bar{\gamma}_b = \frac{\bar{\gamma}_s}{\log_2 M}. \tag{5.298}$$

Hence, we obtain the average symbol error probability as

$$P_e = \frac{2(M-1)}{M} Q\left(\sqrt{\frac{6\log_2 M}{M^2-1}\bar{\gamma}_b}\right). \tag{5.299}$$

The average symbol error probability is depicted in Fig. 5.36 for different values of M. Notice that while we transmit more information with an increasing constellation size, there is a trade-off, as the average error probability worsens (i.e., the signaling scheme becomes less power efficient). In subsequent chapters, we will discuss the power and bandwidth efficiency trade-offs in digital communication systems in detail.

To close this section, let us also comment on the bit error probability. Assume that $M = 2^k$ for some $k \in \mathbb{Z}^+$. To determine the average bit error probability with M-PAM, we also need to specify the mapping of bits to symbols. Once the rule is established, it is possible to compute the exact bit error probability considering all possible symbol transmissions and calculating the probability of deciding in favor of distinct symbols (and taking into account the number of bit errors for each case).

Let us consider the important special case of Gray mapping, for which the length-k bit sequences corresponding to adjacent symbols differ by only one bit. For instance, for $M = 4$, the assignment

$$00 \to 1, \ \ 01 \to 2, \ \ 11 \to 3, \quad 10 \to 4$$

is an example of Gray mapping. For large average signal-to-noise ratios ($\bar{\gamma}_b \gg 1$), we can deduce that most errors will be made in favor of the closest constellation

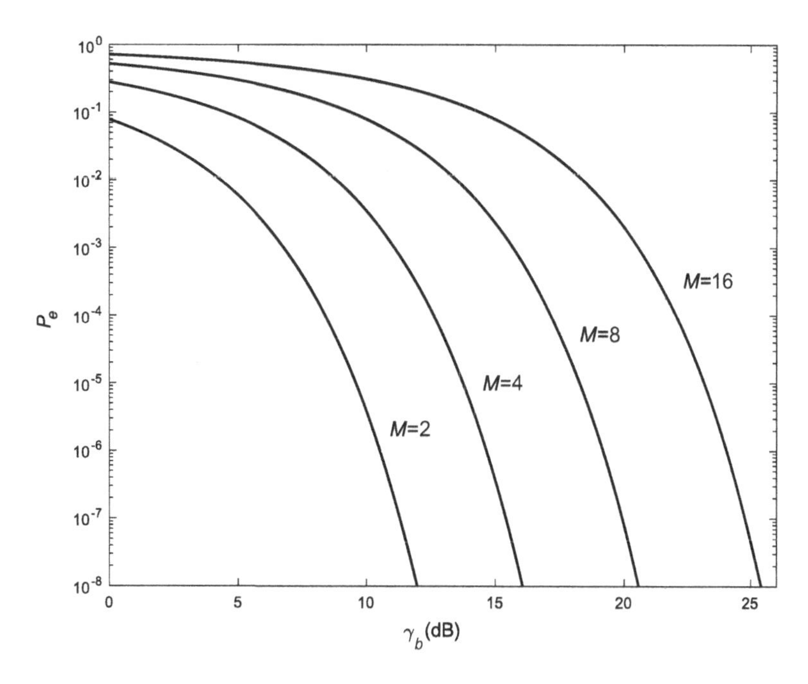

Figure 5.36 M-PAM error probability as a function of $\overline{\gamma}_b = \overline{E}_b/N_0$.

points; hence, each symbol error will correspond to a 1-bit error (out of $k = \log_2 M$ transmitted bits) with high probability. Therefore, we can approximate the average bit error probability of M-PAM signaling with Gray mapping as

$$P_b \approx \frac{2(M-1)}{M \log_2 M} Q\left(\sqrt{\frac{6 \log_2 M}{M^2 - 1} \bar{\gamma}_b}\right). \tag{5.300}$$

An exact calculation would show that this is an excellent approximation of the bit error probability, particularly for $\bar{\gamma}_b \gg 1$.

5.6 *M*-ary Orthogonal Signals and Some Variants

In the previous section, we studied M-PAM as an example of a one-dimensional signaling scheme. In this section, we present another exemplary modulation scheme, which is higher dimensional. Specifically, we consider another example where the dimensionality of the signal set is equal to the number of distinct symbols being transmitted, namely, M-ary orthogonal signaling and some of its variants.

Assume that M different symbols are transmitted using M (non-zero) signals $s_1(t), s_2(t), \ldots, s_M(t)$ that satisfy

$$\int_0^{T_s} s_i(t) s_j(t) dt = 0, \tag{5.301}$$

for $i \neq j$, $i,j \in \{1,2,\ldots,M\}$, where T_s is the symbol period. Hence, the signal space is M-dimensional and a set of orthonormal basis functions is given by $\psi_1(t), \psi_2(t), \ldots, \psi_M(t)$, where

$$\psi_i(t) = \frac{s_i(t)}{\sqrt{E_{s_i}}}, \quad i = 1,2,\ldots,M, \tag{5.302}$$

with E_{s_i} being the energy of the signal $s_i(t)$. We call such a modulation scheme M-ary orthogonal signaling.

Example 5.11

The signals given in Fig. 5.37 form a 4-ary orthogonal signal set.

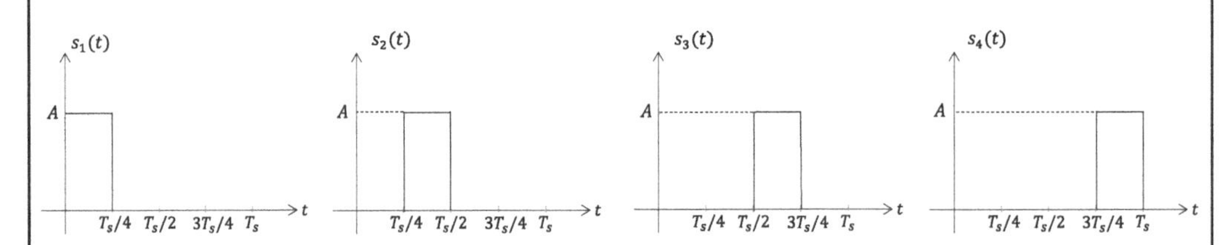

Figure 5.37 4-ary pulse position modulation signal set.

This signal set is called pulse position modulation (PPM), as the digital information is embedded in the position of a basic pulse within the symbol duration.

Example 5.12

Another example of an orthogonal signal set is given in Fig. 5.38.

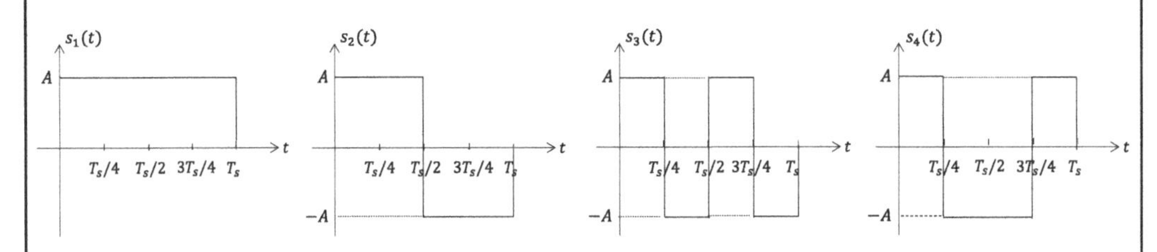

Figure 5.38 An example of an orthogonal signal set.

Note that this signal set and the one in Example 5.11 span the same signal space.

Biorthogonal Signals. A set of M-ary biorthogonal signals is constructed from $M/2$ orthogonal signals by also including their negatives in the signal set. That is, given $M/2$ orthogonal signals

$$s_1(t), s_2(t), \ldots, s_{M/2}(t), \tag{5.303}$$

the M biorthogonal signal set constructed from them is given by

$$s_1(t), s_2(t), \ldots, s_{M/2}(t), -s_1(t), -s_2(t), \ldots, -s_{M/2}(t). \tag{5.304}$$

Example 5.13

The four signals shown in Fig. 5.39 form an example of a biorthogonal signal set.

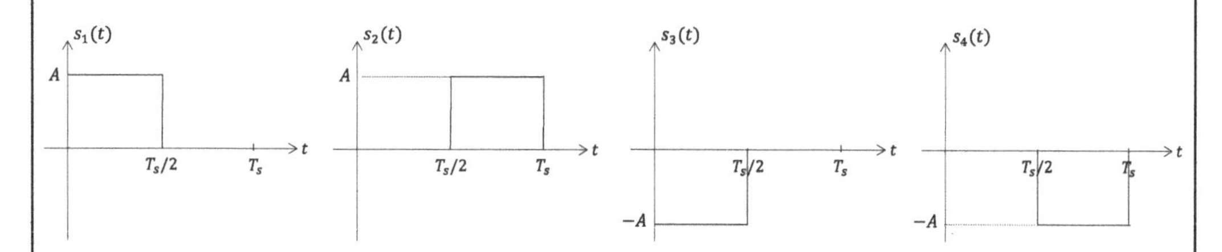

Figure 5.39 A biorthogonal signal set.

This 4-ary biorthogonal signal set is constructed from the two orthogonal signals $s_1(t)$ and $s_2(t)$.

It should be clear that the minimum distance of the signal constellation does not reduce by introducing negatives of the $M/2$ orthogonal signals into the signal set. Also, the average energy of the constellation is the same. Therefore, the error rate performance remains unchanged even though we transmit twice as many symbols (i.e., one more bit of information with each symbol) using the biorthogonal signal set. On the other hand, the receiver implementation may be more complicated as the overall signal set is no longer orthogonal.

Simplex Signals. We observe that the average of M orthogonal signals is non-zero; in other words, these signals are not centered at the origin. We have discussed the merits of placing the center of signal constellations at the origin when discussing M-PAM. To recall, by centering the signal constellation at the origin, we will reduce the average signal energy without changing the structure of the signal constellation (or its distance properties), hence without increasing the average error probability.

Simplex signals are obtained by subtracting the average of the M orthogonal signals from each. Namely, given the orthogonal signals $s_1(t), s_2(t), \ldots, s_M(t)$, for $t \in [0, T_s)$, each with energy E_s, a set of simplex signals is given by

$$\tilde{s}_i(t) = s_i(t) - \frac{1}{M}\sum_{k=1}^{M} s_k(t), \tag{5.305}$$

for $i = 1, 2, \ldots, M$. Note that the resulting signal space is $(M-1)$-dimensional (as the sum of all the signals is 0).

We can compute the energy of $\tilde{s}_i(t)$ as

$$E_{\tilde{s}} = \int_0^{T_s} \left(s_i(t) - \frac{1}{M} \sum_{k=1}^{M} s_k(t) \right)^2 dt \tag{5.306}$$

$$= \int_0^{T_s} s_i^2(t) dt - \frac{2}{M} \sum_{k=1}^{M} \int_0^{T_s} s_i(t) s_k(t) dt + \frac{1}{M^2} \sum_{k=1}^{M} \sum_{l=1}^{M} \int_0^{T_s} s_k(t) s_l(t) dt \tag{5.307}$$

$$= E_s - \frac{2}{M} E_s + \frac{1}{M^2} M E_s \tag{5.308}$$

$$= \left(1 - \frac{1}{M}\right) E_s, \tag{5.309}$$

which is smaller than E_s. Since the distance properties of the signal constellation remain the same with this translation, the average error probability is identical to the original orthogonal signaling and the corresponding simplex signal set. Since the latter has lower average energy, we conclude that the simplex signal set is better in terms of the error probability over an AWGN channel. The amount of improvement can be quantified as

$$10 \log_{10} \left(\frac{M}{M-1} \right) \text{ dB.} \tag{5.310}$$

As an example, for a 4-ary signal constellation, the improvement is about 1.25 dB. For $M = 8$, the improvement reduces to 0.58 dB. The improvement reduces even further with increasing M.

Note that we can also compute the correlation of two signals in the simplex signal set as

$$\int_0^{T_s} \tilde{s}_i(t) \tilde{s}_j(t) dt = -\frac{E_s}{M-1}, \tag{5.311}$$

for $i \neq j$, $i,j \in \{1, 2, \ldots, M\}$, that is, the simplex signal set is not orthogonal. Hence, implementing the receiver may be more difficult compared to an orthogonal signaling scheme.

5.7 Timing Recovery

Throughout this chapter, we studied the transmission of a sequence of symbols via digital modulation and the optimal receiver structure in the presence of AWGN. We identified that, with a matched filter-type implementation, the receiver passes the received signal through a bank of LTI filters whose impulse responses are matched to the basis functions, and the filter outputs are periodically sampled at the symbol rate to produce the vector form of the received signal, which is then used to make decisions on the transmitted symbol. The operation is similar for a correlation-type receiver as well. Namely, the timing information of the different bits or symbols is used in forming the product of the received signal with the basis functions, and the integrations over the symbol period are carried out.

For a proper operation, whether a matched filter type or a correlation type, the receiver must determine the symbol boundaries to produce the vector form of the received signal for each symbol interval. As illustrated earlier in Section 5.3, if this process is not carried out with good precision, the receiver experiences a significant reduction in the signal amplitude (hence, a reduction in the signal-to-noise ratio) as well as intersymbol interference, deteriorating the system performance. Refer back to Fig. 5.23 for an illustration of the effect of incorrect timing information at the matched filter output.

The process of obtaining the correct symbol boundaries, or equivalently, the correct instances for sampling the outputs of the matched filters (to the basis functions), is referred to as *timing recovery.* While it is possible to obtain the timing information by employing a master clock, the most effective and widely used methods are based on deriving it directly from the received signal, also referred to as *self-synchronization.*

There are different ways of timing recovery from the received digitally modulated signal. These methods include both decision-directed methods, which employ the symbol decisions being made, and non-decision-directed methods, which do not. Since our objective is to only give the general ideas behind timing recovery, we only describe a commonly used non-decision-directed timing recovery method, called *early–late gate synchronization*, as an exemplary approach.

Early–Late Gate Synchronization. This timing recovery method relies on a basic property of the matched filter: the output of a matched filter with impulse response matched to the signal at its input is symmetric around $t = T$ where T is the symbol period, with its peak at $t = T$. This was detailed previously in this chapter. Two examples are given in Fig. 5.40; one for a rectangular pulse, the other for a triangular pulse.

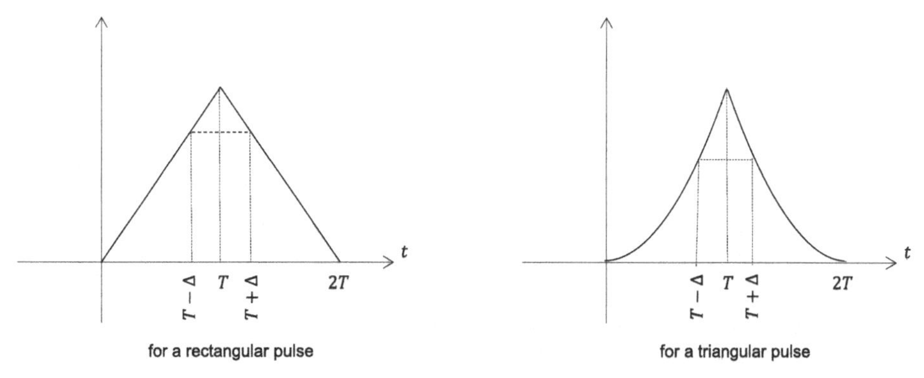

Figure 5.40 Matched filter output corresponding to LHS: a rectangular input to an LTI filter with a rectangular impulse response, RHS: a triangular input to an LTI filter with a an impulse response matched to it.

The basic observation is as follows. The optimal sampling point is when the matched filter (or correlator) output is the largest in magnitude. We can take two samples to achieve this: one intended to be slightly before the optimal sampling point $t = T$, the other slightly after. Hence we can take an early sample, ideally at time $t = T - \Delta$, and a late sample at $t = T + \Delta$. If these samples are taken at the intended time instances, then the difference between the signal samples' magnitudes (or their squares) at the early and late sampling instances will be zero. If this difference is positive, we observe that the samples are being taken late; hence, we need to advance the sampling instances. If this difference is negative, then the samples are being taken early, and we need to retard them. See Fig. 5.41 for an illustration of this point for a binary antipodal modulation scheme using a triangular pulse for transmission.

Given a received signal corresponding to a sequence of symbol transmissions, the process of advancing/retarding the sampling instances to find the optimal one and tracking it can be accomplished using the basic block diagram depicted in Fig. 5.42.

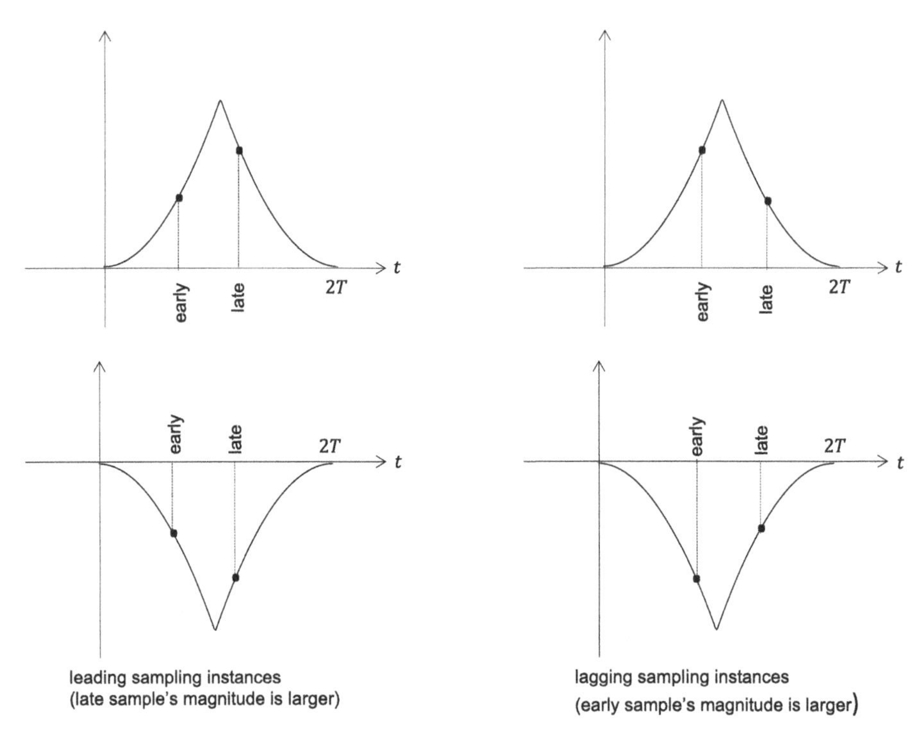

Figure 5.41 Examples of early and late samples taken at the matched filter output for antipodal signaling (with $+1$ and -1 transmissions). On the left-hand side, the samples are leading: the magnitude of the late sample is larger than that of the early sample. Hence, the sampling points should be retarded. On the right-hand side, the samples are lagging: the magnitude of the late sample is smaller than that of the early sample. Hence, the sampling points should be advanced.

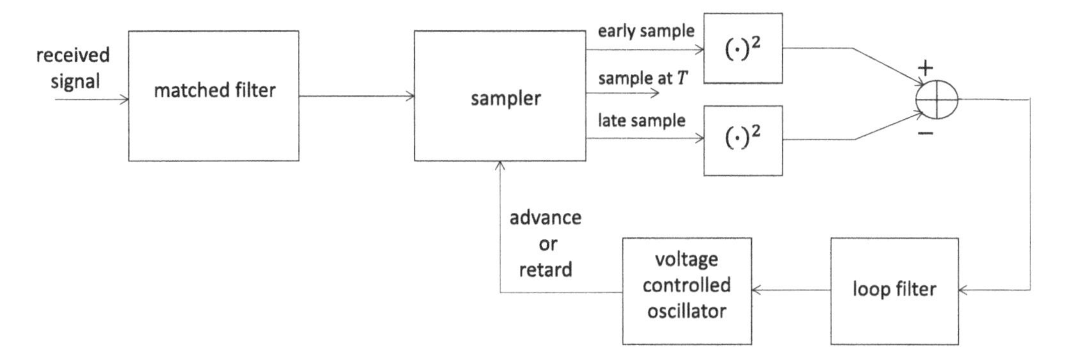

Figure 5.42 A high-level block diagram of an early–late gate timing recovery.

Since the operation is repeated for each symbol (at a rate of $1/T$), the contributions of the noise and the other symbols will average out, and the sampling instance will be driven towards the optimal one, hence establishing (and then tracking) the correct symbol boundaries.

5.8 Chapter Summary

This chapter covered digital modulation techniques for transmission over additive white Gaussian noise channels. We started with the basic AWGN model widely encountered in various communication systems and established the signal-space representation of transmitted signals. We then derived the MAP and ML optimal receiver structures for AWGN channels. We also covered the matched filter, which provides a convenient demodulation method, and summarized some of its properties. Furthermore, we identified the probability of error analysis as the primary tool to determine the performance of a digital communication system. We showed that the performance assessment is greatly simplified by using a union bound on the error probability. Based on this analysis, we gave a simple performance assessment method for a specific digital modulation scheme and, accordingly, a way of comparing different signal constellations in terms of their performance over an AWGN channel. As specific examples of digital modulation schemes, we covered pulse amplitude modulation and orthogonal signaling, along with some of its variants. The entire coverage relied on the assumption that there is perfect timing recovery, and hence, the symbol boundaries are precisely known. With this motivation and to close the loop, finally, we briefly covered timing recovery methods and described the early–late gate synchronization method.

This chapter is the core of digital modulation, and as such, it is the longest and most detailed. Subsequent chapters are heavily founded on this material.

PROBLEMS

5.1 Consider a binary modulation scheme for which the bit 1 is transmitted by $p(t)$ and the bit 0 is transmitted by $-p(t)$, with $p(t)$ given by

$$p(t) = A \sin\left(\frac{\pi t}{T_b}\right),$$

for $t \in [0, T_b)$. Here A is the pulse amplitude and T_b is the bit duration.

(a) Plot the transmitted signal corresponding to the bit sequence

$$0, 0, 1, 0, 1, 1, 0, 0, 1.$$

(b) Repeat the previous part assuming that on–off signaling is employed, that is, the bit 1 is transmitted by $p(t)$ while, 0 is transmitted by a zero pulse.

5.2 A zero-mean white Gaussian noise process $X(t)$ with power spectral density $N_0/2$ is input to an LTI filter with impulse response

$$h(t) = \delta(t) - \frac{1}{2}\delta(t-1) - \frac{1}{2}\delta(t+1).$$

Denote the filter output by $Y(t)$.

(a) Is $Y(t)$ a Gaussian random process? Why or why not?

(b) Is $Y(t)$ a wide-sense stationary process? Is it strict-sense stationary? Why or why not?

5.3 Consider a binary communication system using the two signals $s_1(t)$ and $s_2(t)$ shown in Fig. 5.43. The transmitted bits are equally likely; the bit duration is $T = 2$, and the transmission is over an AWGN channel with power spectral density $N_0/2$.

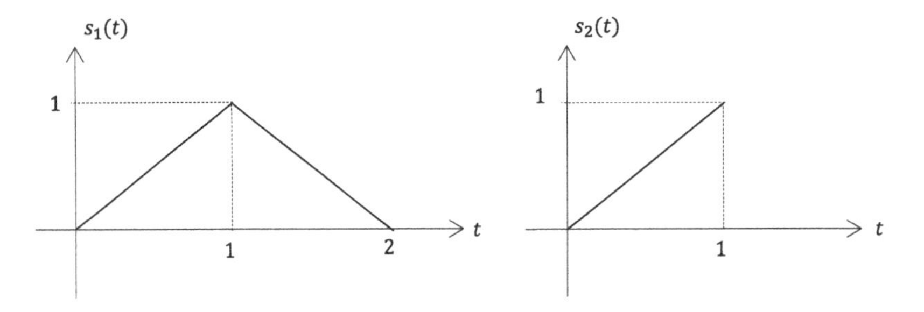

Figure 5.43 The two signals used in Problem 5.3.

(a) Find an orthonormal basis for the signal space and show the signal constellation.

(b) Show the impulse responses matched to the basis functions and describe the optimal matched filter-type receiver. Is it possible to implement the optimal receiver using only one matched filter? If so, how? Be specific, and provide a block diagram as necessary.

(c) What is the average error probability? Express your answer in terms of the average SNR per bit $\bar{\gamma}_b$.

5.4 A binary communication system uses two pulses

$$s_1(t) = \cos\left(\frac{2\pi t}{T_b}\right) \text{ and } s_2(t) = \sin\left(\frac{2\pi t}{T_b}\right), t \in [0, T_b),$$

to transmit 0 and 1, respectively, where T_b is the bit duration.

(a) Find an orthonormal basis for the signal space and show the signal constellation.

(b) Plot the transmitted signal for the sequence of bits 0, 0, 1, 1, 0, 1, 0, 1, 1, 1.

(c) Assume that the matched filter(s) to the basis function(s) in part (a) are employed at the receiver. Approximately sketch the matched filter output corresponding to the transmitted signal in part (b). Also, specify the sampling times for deciding on each bit and determine the matched filter output values at the sampling instances.

5.5 Consider the binary communication system in Problem 5.4 and further assume that the transmitted bits are equally likely and the transmission is over an AWGN channel with power spectral density $N_0/2$.

(a) Determine the optimal correlation-type receiver structure. Give a block diagram and be specific.

(b) What is the average error probability? Express your answer in terms of the average SNR per bit $\bar{\gamma}_b$.

(c) How would your answers to parts (a) and (b) change if the bits are not equally likely, that is, if 1 is transmitted with probability p and 0 is transmitted with probability $1 - p$?

5.6 Consider an antipodal signaling scheme using $p(t)$ and $-p(t)$ to transmit 1 and 0, respectively, over an AWGN channel with power spectral density $N_0/2$. Assume that $p(t) = \cos(\pi t/T_b) + \sin(\pi t/T_b)$, for $t \in [0, T_b)$, where T_b is the bit duration. The transmitted bits are equally likely.

Assume that a receiver multiplies the received signal (AWGN channel output) with a rectangular pulse of amplitude A (where $A > 0$) and integrates the result over the bit period to produce a decision variable r, and then it uses a threshold detector to decide on the transmitted bit. Namely, the receiver declares 1 as its output if $r > r_{th}$ and outputs 0 otherwise.

(a) Determine the optimal value of the threshold r_{th} to minimize the average bit error probability.

(b) Take $r_{th} = 0$ and compute the average error probability. Express the result in terms of E_b/N_0.

(c) The receiver structure in part (b) is not optimal. Determine the optimal receiver structure and compute the resulting average error probability. Compare the result with that of part (b).

5.7 A 4-ary communication system uses the following four signals (represented in vector form with respect to an orthonormal basis):

$$\mathbf{s}_1 = \begin{bmatrix} \sqrt{E/2} \\ \sqrt{E/2} \\ 0 \end{bmatrix}, \mathbf{s}_2 \begin{bmatrix} -\sqrt{E/2} \\ -\sqrt{E/2} \\ 0 \end{bmatrix}, \mathbf{s}_3 = \begin{bmatrix} 0 \\ \sqrt{E} \\ 0 \end{bmatrix}, \mathbf{s}_4 = \begin{bmatrix} 0 \\ 0 \\ \sqrt{E} \end{bmatrix},$$

where $E \in \mathbb{R}^+$. All the signals are equally likely, and the channel is AWGN with noise power spectral density $\frac{N_0}{2}$.

(a) What is the average energy per symbol?

(b) Determine the union bound on the conditional probability of symbol error $\mathbb{P}(\text{error}|\mathbf{s}_i \text{ is sent})$, for $i = 1, 2, 3, 4$.

(c) Determine the union bound on the average symbol error probability.

5.8 Consider the three signal constellations obtained by two orthonormal basis functions, depicted in Fig. 5.44. Assume that all the $M = 6$ signals transmitted are equally likely.

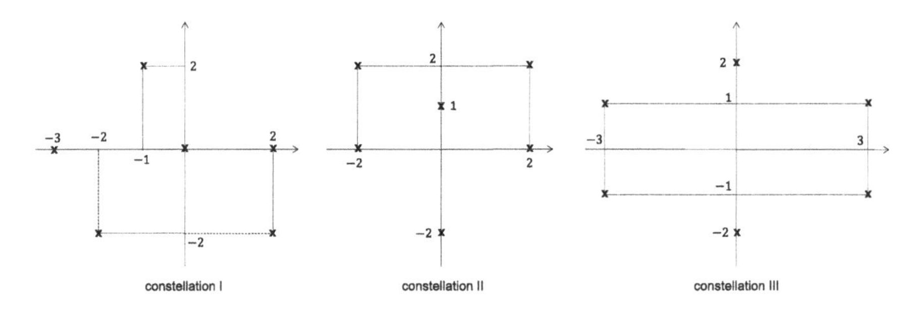

Figure 5.44 The signal constellations for Problem 5.8.

(a) Determine the average energy of the transmitted signal for each constellation.

(b) Assuming that they are used over an AWGN channel and considering high signal-to-noise ratios, order these constellations in terms of their error probability performance and indicate their performance differences in decibels (i.e., the differences in the required SNRs for the same average error probability for high SNR values).

5.9 The four signals shown in Fig. 5.45 are used to transmit four different messages (where the symbol period is $T = 2$). Assume that the symbols are equally likely.

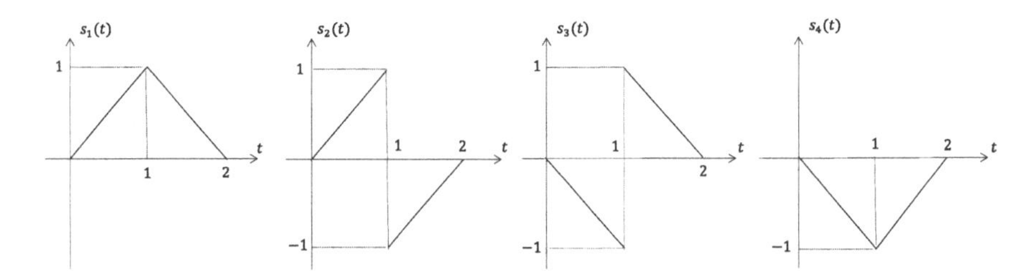

Figure 5.45 The four signals used in Problem 5.9.

(a) Find an orthonormal set of basis functions for the signal set.

(b) Plot the signal constellation.

(c) What are the impulse responses of the filters matched to the basis functions in part (a)?

(d) This system is used over an AWGN channel. Describe the structure of the optimal receiver that minimizes the average error probability.

(e) Assume that mapping from two bits to the signals is as follows: 00 is transmitted by $s_1(t)$, 01 by $s_2(t)$, 10 by $s_3(t)$, and 11 by $s_4(t)$. Also, assume that communication is over an AWGN channel with a power spectral density of $\frac{N_0}{2}$. What is the resulting (exact) bit error probability? Express your answer in terms of the average signal-to-noise ratio per bit, $\bar{\gamma}_b = \frac{\bar{E}_b}{N_0}$.

5.10 Four equally likely symbols are transmitted using the signals in Fig. 5.46. (Assume that the symbol period is $T = 2$ units.)

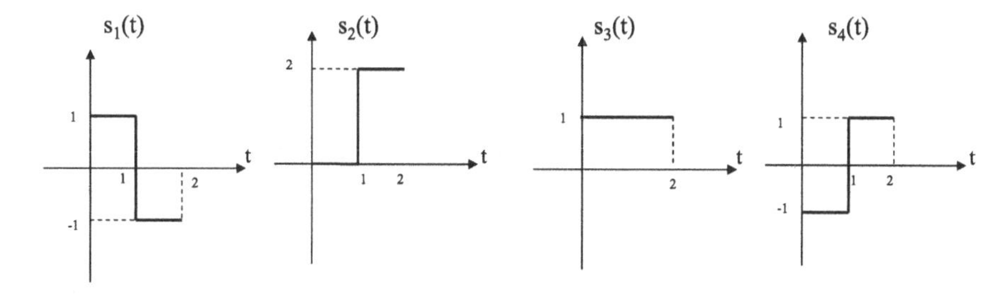

Figure 5.46 The four signals used in Problem 5.10.

(a) Find an orthonormal basis for this signal set. You can use inspection.

(b) Express the transmitted signals in vector form and show the signal constellation.

(c) Show the impulse responses of the filters matched to $s_1(t)$, $s_2(t)$, $s_3(t)$, and $s_4(t)$.

(d) What is the optimal receiver for an AWGN channel? Be specific and simplify as much as you can.

(e) Compute the conditional error probability given that $s_2(t)$ is transmitted (assuming that the AWGN power spectral density is $N_0/2$). Express your result in terms of average signal-to-noise ratio $\bar{E}_s/N_0$.

5.11 Assume that four equally likely signals are being transmitted over an additive white Gaussian noise (AWGN) channel with power spectral density $N_0/2$. Assume that the vector representations of the signals (used for sending the four symbols), with respect to an orthonormal basis of dimension 3, are given by

$$s_1 = \begin{bmatrix} 2 \\ 0 \\ 1 \end{bmatrix}, s_2 = \begin{bmatrix} 0 \\ 0 \\ 1 \end{bmatrix}, s_3 = \begin{bmatrix} 2 \\ -2 \\ 1 \end{bmatrix}, s_4 = \begin{bmatrix} 2 \\ 1 \\ -1 \end{bmatrix}.$$

(a) Let the received signal in vector form (with respect to the same basis) be given by $\boldsymbol{r} = [r_1 \ r_2 \ r_3]^T$. Determine the decision region of the first symbol (corresponding to $\boldsymbol{s}_1$) with maximum likelihood decoding.

(b) Determine the union bound on the conditional error probability of the first symbol, namely, on $P_{e,1} = \mathbb{P}(\text{error} \mid \text{first symbol is sent})$. (Do not use the loose version of the union bound.) Express your result in terms of the average signal-to-noise ratio $\bar{\gamma}_s = \frac{\bar{E}_s}{N_0}$, where $\bar{E}_s$ is the average energy per symbol.

5.12 Assume that four equally likely signals are being transmitted over an additive white Gaussian noise (AWGN) channel with power spectral density $N_0/2$. Assume that the vector representations of the signals (used for sending the four symbols) with respect to an orthonormal basis of dimension 4 are given by

$$\boldsymbol{s}_1 = \begin{bmatrix} 2 \\ 0 \\ 1 \\ 1 \end{bmatrix}, \boldsymbol{s}_2 = \begin{bmatrix} 1 \\ 2 \\ 0 \\ 1 \end{bmatrix}, \boldsymbol{s}_3 = \begin{bmatrix} 1 \\ 1 \\ 2 \\ 0 \end{bmatrix}, \boldsymbol{s}_4 = \begin{bmatrix} 0 \\ 1 \\ 1 \\ 2 \end{bmatrix}.$$

(a) Let the received signal in vector form (with respect to the same basis) be given by $\boldsymbol{r} = [r_1 \ r_2 \ r_3 \ r_4]^T$. Determine the maximum likelihood decision rule and simplify it as much as possible.

(b) Determine the union bound on the average symbol error probability for the modulation scheme in part (a). Express your answer in terms of the average signal-to-noise ratio $\bar{E}_s/N_0$ (where $\bar{E}_s$ is the average energy of the constellation per symbol). (Do not utilize the loose version of the union bound.)

(c) In this part, consider the binary modulation scheme which uses only the first two signals ($\boldsymbol{s}_1$ and $\boldsymbol{s}_2$). The two bits are equally likely. Determine the average probability of error with ML detection and express it in terms of $\bar{E}_b/N_0$, where $\bar{E}_b$ is the energy per bit.

5.13 Four equally likely symbols are transmitted over an AWGN channel with power spectral density $N_0/2$ using the four pulses

$$s_i(t) = A_i p(t), \quad i = 1, 2, 3, 4,$$

where $A_1 = -3$, $A_2 = 0$, $A_3 = 2$, and $A_4 = 5$, and $p(t)$ is a pulse of duration T with energy E_p.

(a) Describe the optimal receiver structure (that minimizes the probability of symbol error). Give a block diagram as necessary.

(b) Determine the conditional error probability given that the symbol 2 is transmitted. Express your answer in terms of the average signal-to-noise ratio per symbol $\bar{\gamma}_s$.

(c) Determine the average symbol error probability. Express your answer in terms of $\bar{\gamma}_s$.

5.14 The four signals shown in Fig. 5.47 are used to transmit four different messages (where the symbol period is T_s). Assume that the symbols are equally likely.

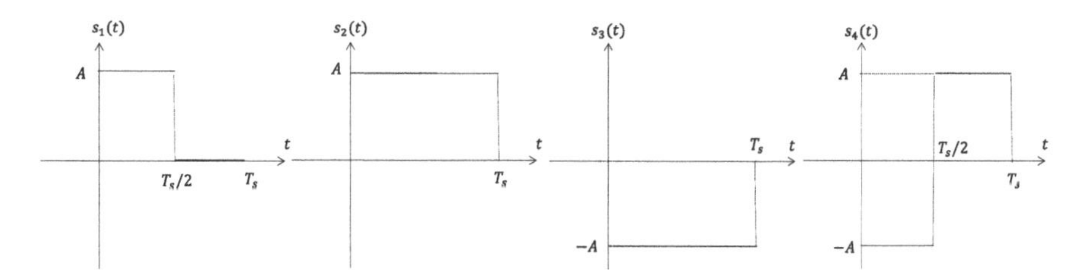

Figure 5.47 The four signals used in Problems 5.14 and 5.15.

(a) Find an orthonormal set of basis functions for this signal set.
(b) Plot the signal constellation.
(c) Plot the impulse responses of the filters matched to the basis functions in part (a)?
(d) This system is used over an AWGN channel with power spectral density $N_0/2$. At the receiver, matched filters to the basis functions are employed, and the received signal is projected onto the signal space. Determine the maximum likelihood decision rule. Show the decision regions on the signal space for clarity.
(e) Determine the union bound on the average symbol error probability. Express your answer in terms of $\bar{\gamma}_s = \frac{\bar{E}_s}{N_0}$, where $\bar{E}_s$ is the average energy of the constellation per symbol.
(f) Determine the exact conditional error probability given that the second symbol is transmitted (using $s_2(t)$). Express your answer in terms of $\bar{\gamma}_s$.

5.15 Apply the Gram–Schmidt procedure to find an orthonormal basis for the signal set shown in Fig. 5.47. Apply the procedure in the following order of signals: $s_4(t), s_3(t), s_2(t)$, and $s_1(t)$.

5.16 The four signals shown in Fig. 5.48 are used for digital modulation with symbol period T. Apply the Gram–Schmidt orthonormalization procedure in the given order of the four signals to determine an orthonormal basis. Also, determine the corresponding vector representations of the transmitted signals.

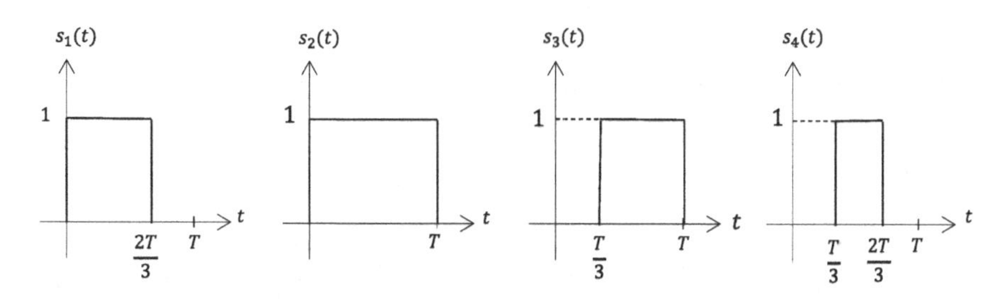

Figure 5.48 The four signals used in Problem 5.16.

5.17 The four signals in Fig. 5.49 are used for digital transmission.

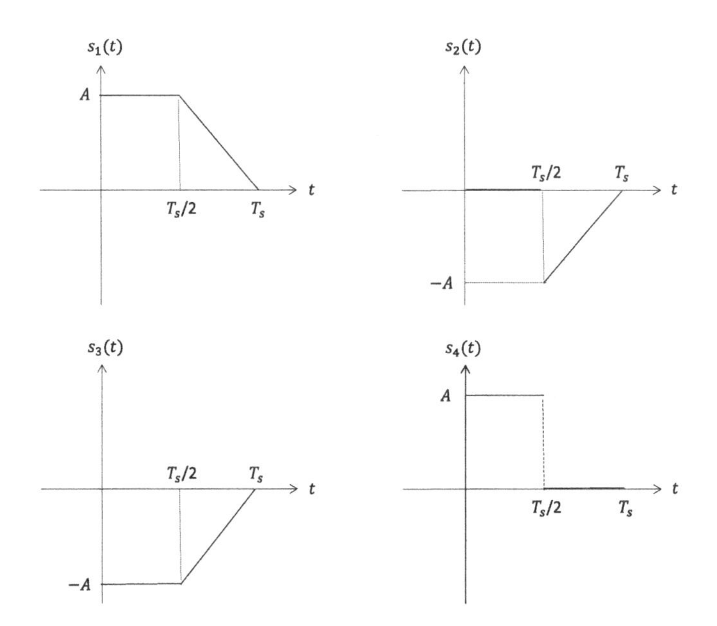

Figure 5.49 The four signals used in Problems 5.17 and 5.18.

(a) Determine an orthonormal basis for the signal space using the Gram–Schmidt orthonormalization procedure with the given order of the four signals.

(b) Determine the vector representation of the signals and plot the signal constellation.

(c) Determine the Euclidean distance between all six pairs of signals using their vector representation.

5.18 Assume that four equally likely symbols are transmitted using the four signals in Fig. 5.49. The channel is AWGN with a power spectral density of $N_0/2$.

(a) Determine an orthonormal basis using inspection, express the signals in vector form, and plot the signal constellation.

(b) Describe the optimal matched filter-type receiver using a block diagram. Be specific and simplify as much as possible.

(c) Compute the union bound on the average symbol error probability and express the result in terms of the average signal-to-noise ratio $\bar{E}_s/N_0$, where $\bar{E}_s$ is the average energy of the constellation per symbol.

5.19 Assume that the following input–output relationship describes an equivalent model for a communication channel:

$$y = \alpha x + \eta,$$

where x is the channel input, y is the channel output, α is a (random) channel gain, and η is a Gaussian random variable with zero mean and variance $\frac{N_0}{2}$. The random channel gain has the probability density function

$$P_\alpha(\alpha) = c\delta(\alpha + 1) + (1 - c)\delta(\alpha - 1),$$

where $c \in [0, 1]$.

Assume that binary antipodal signaling (with $x = -\sqrt{E_b}$ or $x = +\sqrt{E_b}$) is used over this channel, and that the transmitted bits are equally likely. Derive the optimal decision rule and find the probability of error (in terms of c and E_b/N_0) for the following two cases (in parts (a) and (b)).

(a) The realization of the random channel gain α is known at the receiver. That is, the receiver makes a decision based on both the realization of the random variable α and the channel output y.

(b) The realization of the random channel gain α is not known at the receiver. That is, the decision is based only on the channel output y (as the realization of α is not available).

(c) For $c = 0.5$, is antipodal signaling a good choice? Explain.

5.20 Consider the two constellations as defined below used for equally likely signaling over an AWGN channel:

$$\text{Constellation 1: } \mathbf{s}_1 = \begin{bmatrix} 0 \\ 0 \end{bmatrix}, \quad \mathbf{s}_2 = \begin{bmatrix} A \\ A \end{bmatrix}, \quad \mathbf{s}_3 = \begin{bmatrix} -A \\ A \end{bmatrix}, \quad \mathbf{s}_4 = \begin{bmatrix} 0 \\ -2A \end{bmatrix}$$

$$\text{Constellation 2: } \mathbf{s}_1 = \begin{bmatrix} B \\ 0 \end{bmatrix}, \quad \mathbf{s}_2 = \begin{bmatrix} B/2 \\ \sqrt{3}B/2 \end{bmatrix}, \quad \mathbf{s}_3 = \begin{bmatrix} B/2 \\ -\sqrt{3}B/2 \end{bmatrix},$$

$$\mathbf{s}_4 = \begin{bmatrix} -2B/3 \\ 0 \end{bmatrix}$$

where $A > 0$ and $B > 0$.

(a) Calculate the average energy of each constellation.

(b) Find a relation between A and B such that the average error probability is approximately equal for the two constellations at high signal-to-noise ratios.

(c) Considering the relation in part (b), which constellation is more energy efficient at high signal-to-noise ratios?

5.21 Consider a ternary communication system using three signals:

$$s_1(t) = 1 \text{ for } t \in [0, 1/2), \ 0 \text{ otherwise};$$
$$s_2(t) = 1 \text{ for } t \in [1/2, 1], \ 0 \text{ otherwise};$$
$$s_3(t) = 1 \text{ for } t \in [0, 1].$$

The symbol period is $T = 1$. The transmitted symbols are equally likely, and the channel is AWGN with power spectral density $N_0/2$.

(a) Find an orthonormal basis and the corresponding vector representations of the transmitted signals. Also, determine the optimal receiver structure that minimizes the symbol error probability.

(b) What is the (exact) conditional error probability assuming that $s_3(t)$ is transmitted?

(c) Determine the union bound on the overall error probability.

5.22 A 5-ary modulation scheme uses the following four signals (where $s_i(t)$ is used to represent the ith symbol):

$$s_1(t) = -2\sin(\pi t/T) - 2\cos(\pi t/T),$$
$$s_2(t) = -2\sin(\pi t/T) + 2\cos(\pi t/T),$$
$$s_3(t) = 2\sin(\pi t/T) + 2\cos(\pi t/T),$$
$$s_4(t) = 2\sin(\pi t/T) - 2\cos(\pi t/T),$$
$$s_5(t) = 0,$$

for $t \in [0, T_s)$, where T_s is the symbol period. The symbols are equally likely. This modulation scheme is used over an additive white Gaussian noise channel with noise power spectral density $N_0/2$. The received signal is denoted by $r(t)$.

(a) Determine an orthonormal basis for this signal set. Also, determine the vector representation of the signals and plot the signal constellation.

(b) Determine the maximum likelihood receiver structure and simplify as much as possible. Depict the block diagram of the correlation-type receiver.

(c) Determine the union bound on the average symbol error probability. Express your answer in terms of the signal-to-noise ratio defined as $\bar{\gamma}_s = \bar{E}_s/N_0$, where $\bar{E}_s$ is the average energy of the signal constellation.

(d) Determine the exact (conditional) probability of error given that $s_5(t)$ is transmitted. Express your result in terms of $\bar{\gamma}_s$.

5.23 Consider a binary communication system where the prior probabilities are 2/3 (for bit 1) and 1/3 (for bit 0). Assume that when 0 is transmitted, the (equivalent) received signal is an exponential random variable with a mean of 1, and when 1 is sent, the received signal is exponential with a mean of 10. Denote the received signal by (the scalar) r. Determine the optimal decision rule (to minimize the error probability) and the resulting error probability.

5.24 Consider a binary antipodal modulation scheme with the transmit pulse $p(t) = at$ (for $t \in [0, T)$), where T is the bit period and a is a positive constant. The bit 1 is transmitted by $p(t)$, and the bit 0 is transmitted by $-p(t)$. Assume that the transmitted bits are equally likely and that the transmission is over an AWGN channel with power spectral density $N_0/2$.

(a) Determine the optimal receiver structure. Be specific and clearly label any diagrams you use. Also, determine the average bit error probability (in terms of a, T, and N_0).

(b) Assume that another receiver structure is as follows: we integrate the received signal $r(t)$ over the symbol period, that is, we obtain

$$r' = \int_0^T r(t)dt,$$

and decide in favor of 1 if $r' > 0$ and in favor of 0 if $r' < 0$. What is the resulting average bit error probability? How does this compare with the one in part (a)?

(c) Consider a different receiver that integrates the received signal over two overlapping intervals $(0, 2T/3)$ and $(T/3, T)$ and produces

$$r_1 = \int_0^{2T/3} r(t)dt, \quad r_2 = \int_{T/3}^{T} r(t)dt.$$

(i) Determine the (conditional) joint PDF of r_1 and r_2 given that the transmitted bit is 1, that is, $f(r_1, r_2|1)$. Also compute $f(r_1, r_2|0)$.

(ii) The ML receiver (based on the two integrator outputs r_1 and r_2) maximizes $f(r_1, r_2|m)$ over $m = 0$ and 1. Simplify this decision rule as much as you can.

5.25 Consider transmission of four equally likely symbols (1, 2, 3, and 4) using digital modulation over an additive white Gaussian noise channel with noise power spectral density $S_n(f) = \frac{N_0}{2}$. Assume that the signal space is two-dimensional with an orthonormal basis given by $\psi_1(t)$ and $\psi_2(t)$. Also, assume that to transmit the symbol i, the signal $s_i(t)$ with vector representation $\mathbf{s}_i$ for $i = 1, 2, 3, 4$ is used, where

$$s_1 = \begin{bmatrix} 2 \\ 1 \end{bmatrix}, s_2 = \begin{bmatrix} -2 \\ 2 \end{bmatrix}, s_3 = \begin{bmatrix} 2 \\ -2 \end{bmatrix}, s_4 = \begin{bmatrix} -2 \\ -1 \end{bmatrix}.$$

The receiver projects the received signal onto the signal space and obtains its vector representation as $\mathbf{r} = [r_1 \ \ r_2]^T$ (with respect to the same basis $\{\psi_1(t), \psi_2(t)\}$).

(a) Assume that the decision rule adopted at the receiver is as follows: if $r_1, r_2 > 0$, select 1; if $r_1 < 0$, $r_2 > 0$, select 2; if $r_1 > 0, r_2 < 0$, select 3; if $r_1, r_2 < 0$, choose 4. Note that this decision rule is not optimal.

Determine the average symbol error probability. Express your answer in terms of the average signal-to-noise ratio per symbol $\bar{E}_s/N_0$.

(b) Determine the decision region for the first symbol $\mathbf{s}_1$ for the optimal decision rule, that is, the one that minimizes the average symbol error probability.

(c) Determine the union bound on the error probability corresponding to the optimal decision rule (do not use the loose version of the bound), and compare your answer to the error probability computed in part (a) for high signal-to-noise ratios.

5.26 Three symbols (m_1, m_2, and m_3) are transmitted using pulse amplitude modulation over an additive white Gaussian noise channel with power spectral density $N_0/2$. Assume that m_i is transmitted using the signal $s_i(t) = A_i g(t)$, where $g(t)$ is a pulse defined on the interval $[0, T)$ (T being the symbol period), and $A_1 = -A$, $A_2 = 0$, and $A_3 = A$ for some $A > 0$. Also, assume that the prior probabilities of m_1 and m_3 are p, while the prior probability of m_2 is $1 - 2p$ (where $0 \le p \le 1/2$). Denote the energy of the pulse $g(t)$ by E_g, and the received signal by $r(t)$ (for $t \in [0, T)$).

(a) Determine the optimal decision rule that minimizes the average symbol error probability.

(b) Determine the range of values for p (in terms of the parameters A, E_g, N_0) for which the receiver never decides m_2. What is the decision rule in this case?

5.27 Consider a ternary ($M = 3$) communication system employing the three signals

$$s_1(t) = -\sqrt{6}\,t, \quad s_2(t) = 0, \quad s_3(t) = 2\sqrt{6}\,t, \quad \text{for } t \in [0, 2).$$

The symbol i is transmitted by the signal $s_i(t)$, $i = 1, 2, 3$, and the three symbols are equally likely. The symbol period is $T = 2$ units. Observe that this is a one-dimensional signal set, and a basis for it is $\psi(t) = \sqrt{\frac{3}{8}}\,t$ for $t \in [0, 2)$, and 0 otherwise. The transmission is over an additive white Gaussian noise channel where the noise $n(t)$ is zero mean, and it has a power spectral density $S_n(f) = \frac{N_0}{2}$.

At the receiver, we compute the correlation of the received signal $r(t)$ for $t = [0, 2)$ with the basis function and obtain the decision variable $r = \int_0^2 r(t)\psi(t)dt$.

(a) Determine the decision rule (based on r) that minimizes the probability of error and computes the corresponding average symbol error probability. Express your answer in terms of the signal-to-noise ratio per symbol $\bar{\gamma}_s = \frac{\bar{E}_s}{N_0}$ (where $\bar{E}_s$ is the average energy of the constellation).

(b) Assume that the mapping from bits to the three symbols is as follows: $00 \to 1$, $01 \to 2$, $11 \to 3$. Determine the average error probability assuming that the decision rule in the previous part is used. Express your answer in terms of $\bar{\gamma}_s$.

(c) For this part, assume that the receiver correlates the received signal with the function $\psi'(t) = \frac{1}{\sqrt{2}}$ for $t \in [0, 2)$, that is, it computes $r' = \int_0^2 r(t)\psi'(t)dt$ and applies a decision rule based on this (non-optimal) correlator output r' only. Determine the decision rule based on r' that minimizes the probability of error. What is the corresponding average symbol error probability? Express your answer in terms of $\bar{\gamma}_s$ and compare the result with that in part (a).

(d) Assume that the receiver uses the correct basis function as in part (a) and implements the correlation calculation using a matched filter (matched to the correct basis function). However, instead of taking the output sample at time $t = 2$, it takes it at $t = 1.9$. Also, assume that the transmission takes place in isolation, that is, only a single symbol is transmitted with no neighboring symbols – no contribution from any previously transmitted or future symbols to the output sample of the matched filter. Except for the incorrect sampling instance, the receiver is the same as that in part (a). Determine the average symbol error probability in terms of $\bar{\gamma}_s$, and compare your result with that in part (a).

5.28 Rank the following three 4-PAM signal constellations in terms of their power efficiency at high signal-to-noise ratios:

$$\{-3, -1, 1, 3\}, \ \{-5, -3, -1, 1\}, \ \{-2, -1, 1, 3\}.$$

Also quantify their performance difference in terms of the average signal-to-noise ratio (in dB) needed for the same error probability level at high signal-to-noise ratios.

5.29 Consider the following (one-dimensional) signal constellations (obtained with respect to a normalized basis function):

$$\{-3, -1, 1, 3\}, \ \{-4, -2, -1, 1\}, \ \{-2, -1, 0, 5\}.$$

Assume that these signal constellations are used over an additive white Gaussian noise (AWGN) channel with power spectral density $N_0/2$. The symbols are equally likely.

(a) Compute the loose form of the union bound on the symbol error probability for each of the three constellations. Express your answer in terms of the average signal-to-noise ratio per symbol $\bar{\gamma}_s = \bar{E}_s/N_0$, where $\bar{E}_s$ is the average energy per symbol.

(b) Rank the three signal constellations in terms of their power efficiency and quantify their performance differences in terms of the signal-to-noise ratio (in dB) needed for the same error probability level at high signal-to-noise ratios.

5.30 Consider 4-ary PAM signaling with a rectangular pulse of duration T_s and unit energy. Assume that the amplitude levels used for transmitting the symbols $1, 2, 3, 4$ are given by $-3, -1, 1$, and 3, respectively.

(a) Plot the transmitted signal corresponding to the sequence of symbols $2, 3, 1, 1, 2, 4, 1, 2, 3$.

(b) A matched filter-type receiver is employed with a filter matched to the basis function. Plot the matched filter output corresponding to the transmitted signal in part (a). Also, mark the sampling instances for deciding on each transmitted symbol.

5.31 Consider an M-ary simplex signal set obtained from an orthogonal set of signals with equal energies E_s. Show that two different signals $\tilde{s}_i(t)$ and $\tilde{s}_j(t)$ $(i \neq j)$ in the simplex signal set have a correlation of

$$\int_0^{T_s} \tilde{s}_i(t)\tilde{s}_j(t)dt = -\frac{E_s}{M-1}.$$

5.32 Consider transmission of four equally likely symbols (1, 2, 3, and 4) using digital modulation over an additive white Gaussian noise channel with noise power spectral density $S_n(f) = \frac{N_0}{2}$. Assume that the signal space is two-dimensional with an orthonormal basis given by $\psi_1(t)$ and $\psi_2(t)$. Also, assume that $s_i(t)$ with a vector representation $\mathbf{s}_i$ is used to transmit the symbol i, for $i = 1, 2, 3, 4$. We are given that

$$s_1 = \begin{bmatrix} 2 \\ 3 \end{bmatrix}, s_2 = \begin{bmatrix} 0 \\ 2 \end{bmatrix}, s_3 = \begin{bmatrix} 0 \\ -2 \end{bmatrix}, s_4 = \begin{bmatrix} 2 \\ -3 \end{bmatrix}.$$

The receiver projects the received signal onto the signal space and obtains its vector representation as $\mathbf{r} = \begin{bmatrix} r_1 \\ r_2 \end{bmatrix}$ (with respect to the same basis $\{\psi_1(t), \psi_2(t)\}$).

(a) Assume that the decision rule adopted at the receiver is as follows:

$$\hat{m} = \begin{cases} 1, & \text{if } r_1 > 1 \text{ and } r_2 > 0, \\ 2, & \text{if } r_1 < 1 \text{ and } r_2 > 0, \\ 3, & \text{if } r_1 < 1 \text{ and } r_2 < 0, \\ 4, & \text{if } r_1 > 1 \text{ and } r_2 < 0. \end{cases}$$

Determine the average symbol error probability. Express your answer in terms of the average signal-to-noise ratio per symbol $\bar{\gamma}_s = \bar{E}_s/N_0$.

(b) Determine the maximum likelihood decision rule.

(c) Determine the union bound on the error probability of the ML detector (do not use the loose version of the bound) and compare your answer to that in part (a) for $\bar{\gamma}_s \gg 1$, that is, determine the difference in the required $\bar{\gamma}_s$ for the same error probability level (in dB).

5.33 Assume that a signal is transmitted through two parallel channels. The channel input X (which takes on -1 or 1) is transmitted, and two signals Y_1 and Y_2 are received. Assume that -1 and 1 are transmitted with equal probability. In other words, we have

$$Y_1 = a_1 X + \eta_1,$$
$$Y_2 = a_2 X + \eta_2,$$

where a_1, a_2 are positive constants (representing channel gains) and the noise terms η_1 and η_2 are independent Gaussian random variables with zero means and variances of $N_0'/2$ and $N_0''/2$, respectively. The receiver has access to both Y_1 and Y_2.

(a) What is the optimal decision rule that minimizes the error probability?

(b) What is the resulting error probability?

(c) An alternative receiver structure is as follows. The receiver adds Y_1 and Y_2 first, then applies threshold detection to decide if -1 or 1 is transmitted (the threshold is selected at zero). What is the probability of error in this case? Compare this result with your answer in part (b).

5.34 Consider binary antipodal signaling over a Gaussian noise channel. Assume that the channel is used twice to transmit two independent and identically distributed bits. Using the mathematically equivalent model for transmission over the Gaussian channels, the received signals are given by

$$r_1 = I_1 + n_1, \qquad r_2 = I_2 + n_2,$$

where $I_1, I_2 = \pm 1$ are two independent message bits with $\mathbb{P}(I_1 = 1) = \mathbb{P}(I_2 = 1) = 1/2$.

Assume that the noise terms for the two uses of the channel, n_1 and n_2, are correlated, zero-mean Gaussian random variables with variance σ^2. Denote their correlation coefficient by ρ (assume that $0 \le \rho < 1$).

The receiver observes both r_1 and r_2, and using these two observations, it decides on the two transmitted bits I_1 and I_2. Determine the optimal decision rule that minimizes the average probability of error $\mathbb{P}((\hat{I}_1, \hat{I}_2) \neq (I_1, I_2))$, where $\hat{I}_1$ and $\hat{I}_2$ denote the decisions at the receiver output. Simplify the decision rule as much as possible.

5.35 Assume that M equally likely signals are being transmitted over an additive white Gaussian noise channel with power spectral density $N_0/2$. Assume that the vector representation of the ith signal (used for transmitting the ith symbol) (with respect to an orthonormal basis) is given by $\boldsymbol{s}_i = [s_{i1}\ s_{i2}\ \dots\ s_{iM}]^T$, where $s_{ii} = 0$ and $s_{ij} = a$ if $i \neq j$, $i,j = 1, 2, \dots, M$. Assume that $a > 0$.

(a) Let the received signal in vector form (with respect to the same basis) be given by $\boldsymbol{r} = [r_1\ r_2\ \dots\ r_M]^T$. Determine the maximum likelihood decision rule and simplify it as much as possible.

(b) Consider the specific case with $M = 2$. Determine the average probability of error and express it in terms of $\bar{\gamma}_s = \bar{E}_s/N_0$, where $\bar{E}_s$ is the average energy per symbol.

(c) Consider the case of arbitrary M, and determine the union bound on the average symbol error probability. Express your answer in terms of $\bar{\gamma}_s$.

(d) Determine an expression for the exact symbol error probability (in the form of an integral). Do not evaluate it.

5.36 Assume that M equally likely signals are being transmitted over an additive white Gaussian noise channel with power spectral density $N_0/2$. Assume that the vector representation of the ith signal (used for transmitting the ith symbol) (with respect to an orthonormal basis) is given by $\boldsymbol{s}_i = [s_{i1}\ s_{i2}\ \dots\ s_{iM}]^T$, where $s_{ii} = 2a$ and $s_{ij} = a$ if $i \neq j$, $i,j = 1, 2, \dots, M$. Assume that $a > 0$.

(a) Let the received signal in vector form (with respect to the same basis) be given by $\boldsymbol{r} = [r_1\ r_2\ \dots\ r_M]^T$. Determine the maximum likelihood decision rule and simplify it as much as possible.

(b) Determine the union bound on the average symbol error probability. Express your answer in terms of $\bar{E}_s/N_0$, where $\bar{E}_s$ is the average energy of the signal constellation.

5.37 We use the signal $g(t)$ to transmit binary PAM signals (at a rate of $1/T$ bits per second). The receiver employs a matched filter to the pulse $g(t)$ and a symbol rate sampler. However, there is a timing error of ΔT in determining the exact symbol boundaries. Assume that the channel is AWGN with power spectral density $N_0/2$, and the transmitted bits are equally likely. Also, assume that the pulse $g(t) = A$ for $0 \le t \le 4T/5$, and 0 otherwise.

Compute the average bit error rate.

(a) assuming that $0 \le \Delta T \le T/5$;

(b) assuming that $T/5 \le \Delta T \le T/2$.

Express your answers in terms of the SNR E_b/N_0 and the fractional timing error $\alpha = \Delta T/T$.

5.38 We use a rectangular pulse of amplitude A and duration T to transmit binary PAM signals (at a rate of $1/T$ bits per second). There is a timing error of ΔT (less than half a symbol duration) in determining the exact symbol boundaries. Assume that the channel is AWGN with power spectral density $N_0/2$, and the bits are equally likely.

(a) What is the average symbol error rate assuming that ΔT is a fixed quantity. Express your answer in terms of the signal-to-noise ratio $\gamma_b = E_b/N_0$ and fractional timing error $\alpha = \Delta T/T$. How much is the performance degradation for high SNRs due to the mistiming?

(b) Use the approximation of the Q-function given by $Q(x) \approx \frac{1}{2}\exp(-x^2/2)$ to obtain an estimate of the average symbol error probability under the assumption that the timing error is a uniform random variable on the interval $(-\Delta, \Delta)$, where $\Delta = cT$ (c is a constant less than $1/2$).

COMPUTER PROBLEMS

5.39 Write a MATLAB program to simulate the probability of error for binary PAM over an additive white Gaussian noise channel for different signal-to-noise ratios. Estimate the error probability using the simplified mathematically equivalent model with 100,000 transmissions. You can limit the range of signal-to-noise ratios to 2–9 dB. Compare your results with the theoretical expectations.

5.40 Write a MATLAB program to simulate the error probability for on–off signaling over an additive white Gaussian noise channel for different signal-to-noise ratios. As in the previous problem, use the mathematically equivalent model and 100,000 samples to estimate the error rates. You can limit the range of signal-to-noise ratios to 5–12 dB. Compare your results with the theoretical expectations.

5.41 Write a MATLAB program to simulate the symbol error rate for M-level PAM over an additive white Gaussian noise channel with $M = 4, 8, 16$ for different signal-to-noise ratios. Use 100,000 transmissions to estimate the probability of error. You can limit the range of signal-to-noise ratios to 5–13 dB for $M = 4$, 8–17 dB for $M = 8$, and 10–22 dB for $M = 16$. Compare your estimates with the theoretical probability of error expressions for the respective modulation schemes.

5.42 Consider binary PAM transmission with equally likely bits. Select a pulse of duration T (any pulse shape other than rectangular) for transmission. Simulate the transmission of a large number of bits using the selected pulse with binary PAM. That is, generate a large number of bits randomly (e.g., 1000 or more) and accordingly determine the (sampled version) of the transmitted signal in MATLAB. You can pick $T/10$ or a similar value as the sampling period. Obtain the corresponding received signal by simulating a white Gaussian noise process. The objective is to work with this received signal, simulate the operation of the matched filter-type receiver, and estimate the corresponding error rates.

(a) Simulate the implementation of a matched filter-type receiver using MATLAB and make decisions on the transmitted bits. By running a large number of simulations, obtain an error rate estimate. The results should match those in Problem 5.39, which were based on the mathematically equivalent model.

(b) In this part, instead of using the matched filter (to the pulse you have selected), consider the use of a receiver filter with a rectangular impulse response (extending from 0 to T). Estimate the corresponding error rates, and compare your results with the previous part. What do you observe?

(c) Consider the matched filter-type receiver (in part (a)), but assume that the receiver does not have perfect timing information, and instead of sampling the matched filter outputs at the correct time instances, it is off by some ΔT. Pick three different values of ΔT (e.g., $T/10$, $T/5$, and $2T/5$) and repeat the simulations in part (a) to estimate the corresponding error rates as a function of the signal-to-noise ratio. Compare your results with those in part (a), and comment on them.

5.43 Pick arbitrarily a symbol period T and a sampling period significantly smaller (e.g., you may use a sampling frequency of $f_s = 10/T$ or higher). Assume that the transmitted bits are equiprobable and independent of each other.

(a) Generate a long realization of a sample transmitted signal using binary PAM with a rectangular pulse. Plot a portion of this signal (no need to plot the entire vector).

(b) Repeat the previous part with the pulse $\sin(\pi t/T)$.

5.44 Select a pulse of duration T (any pulse shape other than rectangular) and simulate the transmission of a large number of bits with binary PAM using the selected pulse. That is, generate a large number of bits randomly (e.g., 1000 or so) and accordingly determine the (sampled version) of the transmitted signal in MATLAB. For the sampling period, you can pick $T/20$. Obtain the corresponding received signal by simulating a white Gaussian noise process.

(a) Simulate the implementation of a matched filter-type receiver using MATLAB and make decisions on the transmitted bits. By running a large number of simulations, obtain an error rate estimate for several noise levels (signal-to-noise ratios). Plot these results and compare them with the theoretical expectations.

(b) In this case, assume that there is a "jitter," and the receiver cannot sample the matched filter outputs at the required time instances precisely. Instead, the samples are taken at $T + \Delta$, where Δ is a Gaussian random variable with zero mean and variance σ_Δ^2. Plot the simulated error rates for three different σ_Δ^2 values, and comment on your results.

Appendix 5A: Transmission over Non-Gaussian Channels

Since Gaussian noise is commonly observed in digital communication systems, our focus throughout the chapter has been the study of AWGN channels. However, other types of channel impairments can also be observed in practice. While the specific details would differ significantly, the general tools developed in this section (e.g., MAP or ML detection rules) are also suitable for studying such scenarios. We provide a simple example in this appendix.

Consider a binary communication system using light to transmit digital information. Assume that an on–off-type signaling is used, where the bit 1 is transmitted with light emission and the bit 0 is transmitted with no light. The receiver is a photon-counting detector. Assume that the number of photons detected (denoted by N) is a Poisson random variable with parameter λ_0 when 0 is transmitted and parameter λ_1 when 1 is transmitted. Also, assume that the prior probability of bit 1 is p (and correspondingly, the prior probability of bit 0 is $1 - p$). We assume that $\lambda_0 < \lambda_1$, that is, we expect to detect more photons when 1 is transmitted.

We are given the following conditional probabilities for the received signal N (which is a discrete random variable in this example):

$$\Pr\left(N = k \,\middle|\, 0 \text{ is sent}\right) = e^{-\lambda_0}\frac{\lambda_0^k}{k!} \tag{5.312}$$

and

$$\Pr\left(N = k \,\middle|\, 1 \text{ is sent}\right) = e^{-\lambda_1}\frac{\lambda_1^k}{k!}. \tag{5.313}$$

Therefore, with k denoting the specific received signal, the MAP decision rule becomes

$$pe^{-\lambda_1}\frac{\lambda_1^k}{k!} \underset{0}{\overset{1}{\gtrless}} (1-p)e^{-\lambda_0}\frac{\lambda_0^k}{k!}, \tag{5.314}$$

which can be simplified to

$$\left(\frac{\lambda_1}{\lambda_0}\right)^k \underset{0}{\overset{1}{\gtrless}} \frac{1-p}{p}e^{\lambda_1-\lambda_0}, \tag{5.315}$$

or simply to the threshold rule

$$\hat{m} = \begin{cases} 1, & \text{if } k > \frac{\lambda_1-\lambda_0+\ln(1-p/p)}{\ln(\lambda_1)-\ln(\lambda_0)}, \\ 0, & \text{else.} \end{cases} \tag{5.316}$$

If the number of photons detected is above the computed threshold, then we decide that 1 is transmitted, and if it is below, we declare that 0 is transmitted. If this number is exactly equal to the threshold (which may have a non-zero probability), we can decide either way without losing optimality. We also notice that if p increases, the value of the threshold is reduced, enlarging the decision region for bit 1, as expected.

We can obtain the ML decision rule for this problem by substituting $p = 1/2$ into the MAP rule, namely,

$$\hat{m} = \begin{cases} 1, & \text{if } k > \frac{\lambda_1 - \lambda_0}{\ln(\lambda_1) - \ln(\lambda_0)}, \\ 0, & \text{else.} \end{cases} \quad (5.317)$$

It should be clear that, although it is not explored here, it is a straightforward task to calculate the error probability to assess the detector's performance.

6 Single-Carrier Bandpass Transmission

If the signals used in digital modulation (as described in Chapter 5) are lowpass (i.e., if their Fourier transforms occupy a frequency band around 0, or equivalently, their time variations are "slow"), the transmitted signal is lowpass, and hence, the resulting scheme is referred to as baseband transmission. While baseband communication systems have widespread usage (e.g., in storage applications or wireline communications), not all communications occur at low frequencies. In fact, for most practical scenarios (e.g., for communication through radio waves) a frequency band around a high carrier frequency f_c is allocated for the particular application, and the transmission must occur within this band.

In this chapter, our focus is on digital transmission using bandpass signals. While the general methodologies of Chapter 5 are applicable, and the study of such systems can be carried out using the tools we have developed thus far, some critical specific aspects of bandpass transmission require a separate detailed exposition.

Throughout this chapter, as in the previous one, we assume that the communication channel is ideal, meaning that the transmitted signals do not undergo any distortion apart from the (fixed) attenuation and corruption by additive white Gaussian noise. However, unlike the previous chapter, we need to explicitly take the delay between the transmitter and receiver into account for a more precise description of the relevant channel models and detection methods.

The chapter is organized as follows. We introduce the transmission of bandpass signals and the corresponding channel effects in Section 6.1. We argue that the channel will introduce a carrier phase shift due to the propagation delay. We then study basic single-carrier bandpass modulation schemes, namely, bandpass pulse amplitude modulation, phase-shift keying, and quadrature amplitude modulation, in Section 6.2. We focus on the case where the channel phase is estimated and available at the receiver; hence, coherent detection is employed. In Section 6.3, we describe the representation of bandpass signals using their lowpass equivalents and introduce the in-phase and quadrature components of a bandpass signal. Section 6.4 is devoted to the additive white Gaussian noise effects, modeled as a bandpass white noise, and its lowpass-equivalent representation. We address the study of single-carrier bandpass modulation schemes through the lowpass-equivalent signal and noise representations in Section 6.5. Two practically motivated variations of the 4-ary phase-shift keying scheme, $\frac{\pi}{4}$-QPSK and offset QPSK, are presented in Section 6.6. We make a distinction among coherent, non-coherent, and differentially coherent detection schemes in Section 6.7, and study in some detail differential phase-shift keying in Section 6.8. Then, we describe the basic methodology for carrier phase synchronization in Section 6.9 and conclude the chapter in Section 6.10.

6.1 Bandpass Signals and Channel Effects

In single-carrier bandpass communication systems, we utilize signals of the form

$$p(t)\cos(2\pi f_c t), \tag{6.1}$$

where $t \in [0, T_s)$, with T_s denoting the symbol duration; $p(t)$ is a lowpass signal, which is used to control the frequency occupancy of the transmitted signals, as will be made clear in Chapter 7, and f_c is the carrier frequency. Digital information being transmitted is embedded in the carrier signal's amplitude and/or phase. The carrier frequency is typically very large compared to the bandwidth of $p(t)$ and the inverse of the symbol duration, namely, $f_c \gg 1/T_s$. For instance, f_c could be in the gigahertz (GHz) range, while the symbol durations could be in the range of microsecond (μs), meaning that there could be thousands of *cycles* of the sinusoidal signal within each symbol duration.

Consider transmission of the above signal, $p(t)\cos(2\pi f_c t)$, over a communication channel. Assume that the channel is ideal, that is, it does not distort the signal at all; however, there is a propagation delay. For example, suppose radio waves with $f_c = 1$ GHz are employed and the transmitter/receiver separation is 1 km; with the speed of light $c = 3 \times 10^8$ m/s, the resulting delay is around 3.3 μs, corresponding to more than 3000 cycles of the sinusoid.

It is reasonable to assume that the receiver can estimate and compensate for this delay, for example, by using a timing recovery scheme such as an early–late gate synchronizer, as discussed in Chapter 5. However, there will inevitably be an error in the receiver's timing estimate, denoted by τ. This error would be much smaller than the symbol duration for a typical communication system, that is, $\tau \ll T_s$. Therefore, when the signal $p(t)\cos(2\pi f_c t)$ is transmitted, the received signal (ignoring the noise and taking the attenuation factor as unity) will be of the form

$$p(t-\tau)\cos(2\pi f_c(t-\tau)). \tag{6.2}$$

Since the pulse $p(t)$ is a lowpass signal of duration T_s, which is much larger than τ, we can say that

$$p(t) \approx p(t-\tau), \tag{6.3}$$

and hence, the symbol boundaries are identified with sufficient accuracy. On the other hand, even if the residual delay τ is very small, its product with the carrier frequency f_c would typically be large, meaning that there will be a significant phase shift in the received sinusoid. For instance, if the delay estimate has an error in the range of $\tau = 10$ ns (which is highly accurate), for $f_c = 1$ GHz, the phase shift in the signal will correspond to around ten cycles.

With the above discussion, we see that when $p(t)\cos(2\pi f_c t)$ is transmitted over a communication channel, even under ideal conditions, the received signal will be of the form

$$p(t)\cos(2\pi f_c t + \phi) \tag{6.4}$$

(excluding the effects of noise and attenuation), where ϕ is the phase shift introduced by the residual error in the delay estimate (given by $\phi = -2\pi f_c t$). In other words, even an ideal channel will result in a carrier phase shift in the transmitted signal even after a highly accurate timing recovery.

It may be possible to estimate the channel phase shift and compensate for it. We will discuss how this can be accomplished at the end of the chapter when describing phase synchronization.

We say that the receiver is *coherent* if the phase shift introduced by the channel is estimated at the receiver and is utilized for detecting the transmitted symbol. For coherent detection, since the channel phase ϕ can be removed by properly delaying the received signal, we can say that the received signal for an AWGN channel is given by

$$p(t)\cos(2\pi f_c t) + \text{noise}. \tag{6.5}$$

If, however, the channel phase is unknown at the receiver, it cannot be used during the detection process, and the corresponding receiver is referred to as *non-coherent*. In this case, the dependency of the received signal on ϕ must be explicitly shown, that is, the AWGN channel model becomes

$$p(t)\cos(2\pi f_c t + \phi) + \text{noise}, \tag{6.6}$$

where ϕ is a random variable, typically taken as uniform on $[0, 2\pi)$.

In the first part of our detailed study that follows, we assume that coherent detection is employed; however, we will also consider the unavailability of the channel phase at the receiver and discuss possible receiver structures for different modulation techniques later in this chapter.

6.2 Basic Single-Carrier Modulation Schemes

In this section, we focus on three fundamental bandpass digital modulation techniques, namely, bandpass pulse amplitude modulation (PAM), phase-shift keying (PSK), and quadrature amplitude modulation (QAM). Throughout the section, we assume that the communication medium is an AWGN channel and coherent detection is employed; hence, the phase shift introduced by the residual error in the delay estimate is eliminated at the receiver, hence it is not explicitly shown in the received signal model. We will deal with the scenarios where the phase is unknown at the receiver later in the chapter.

6.2.1 Bandpass Pulse Amplitude Modulation

In bandpass (or carrier-modulated) PAM, information is carried by varying the amplitude of a signal of the form

$$g(t) = p(t)\cos(2\pi f_c t), \quad t \in [0, T_s), \tag{6.7}$$

where $p(t)$ is a lowpass signal, f_c is the carrier frequency, and T_s is the symbol duration. The idea is very similar to the (baseband) PAM studied in Chapter 5, with the only difference being the introduction of the sinusoidal signal, whose role is to translate the frequency band of the modulated signal to frequencies around f_c. Another name for this scheme is *amplitude-shift keying (ASK)*.

For M-ary bandpass PAM, as in the case of baseband PAM studied in Chapter 5, we commonly select M different amplitude levels that are equispaced and symmetric around zero. That is, the symbol i is transmitted via

$$s_i(t) = A_i p(t)\cos(2\pi f_c t), \quad t \in [0, T_s), \tag{6.8}$$

where $A_i = (2i - 1 - M)A$, $i = 1, 2, \ldots, M$, and A is a constant. Since all the signals are scaled versions of $g(t) = p(t)\cos(2\pi f_c t)$, the signal space is one-dimensional. Therefore, we only need to normalize this basic pulse to obtain an orthonormal basis. The energy of this pulse (in terms of the energy of $p(t)$, E_p) is calculated as follows:

$$E_g = \int_0^{T_s} g^2(t)dt \tag{6.9}$$

$$= \int_0^{T_s} p^2(t)\cos^2(2\pi f_c t)dt \tag{6.10}$$

$$= \int_0^{T_s} p^2(t)\frac{1}{2}\Big(1 + \cos(4\pi f_c t)\Big)dt, \tag{6.11}$$

where we have used a basic trigonometric identity to arrive at the third line. At this point, we note that $p(t)$ is a lowpass signal (hence a slowly varying signal over the integration window of duration T_s), and $f_c \gg 1/T_s$, that is, there are many cycles of the sinusoid over $[0, T_s)$, hence

$$\int_0^{T_s} p^2(t)\cos(4\pi f_c t)dt \approx 0. \tag{6.12}$$

This point is further illustrated in Fig. 6.1.

If $p(t)$ is a rectangular pulse and f_c is an integer multiple of $1/T_s$, the integral in (6.12) will be exactly zero. However, this is not very realistic, as many communication systems also use non-rectangular pulses, and precision needed to guarantee

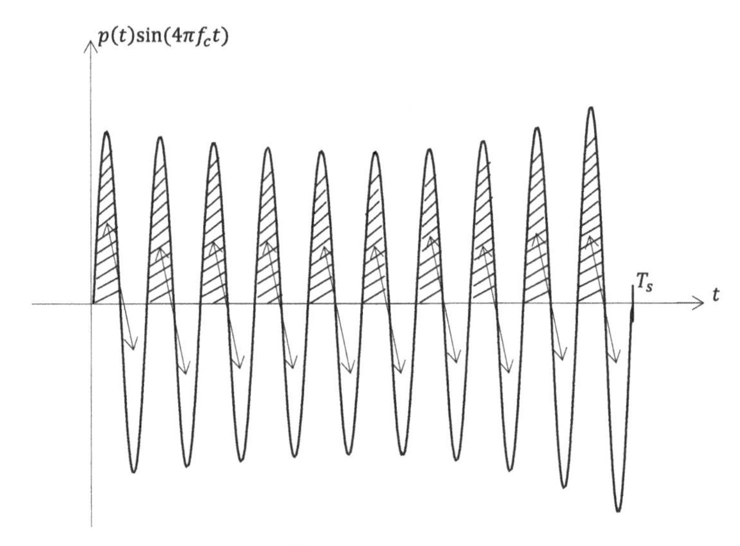

Figure 6.1 Illustration of $p^2(t)\sin(4\pi f_c t)$ over a symbol duration. The shaded and unshaded portions cancel out, making its integral from 0 to T_s (approximately) zero.

that f_c is an integer multiple of $1/T_s$ is impractical. Nevertheless, the approximation above is sufficient for our purposes; hence, we can conclude that

$$E_g = E_p/2. \tag{6.13}$$

We will make similar arguments in the rest of this chapter and the subsequent chapters when dealing with bandpass signals.

To summarize, we have a one-dimensional signal space with a normalized basis function of

$$\psi(t) = \sqrt{\frac{2}{E_p}}p(t)\cos(2\pi f_c t), \quad \text{for } t \in [0, T_s), \tag{6.14}$$

resulting in a vector representation of the transmitted signals (as scalars) of the form

$$s_i = (2i - 1 - M)A\sqrt{\frac{E_p}{2}}. \tag{6.15}$$

Defining the constant $d = A\sqrt{\frac{E_p}{2}}$, we simply have

$$s_i = (2i - 1 - M)d, \tag{6.16}$$

which is identical to the vector representation obtained for baseband PAM in Section 5.5. The signal constellation is depicted in Fig. 6.2.

Noticing that the signal constellation is the same as the one for baseband PAM, which was previously studied in detail, it is clear that the same receiver structure and performance analysis apply. In other words, the ML receiver structure would be the same (except that the basis function is different, and accordingly, the determination

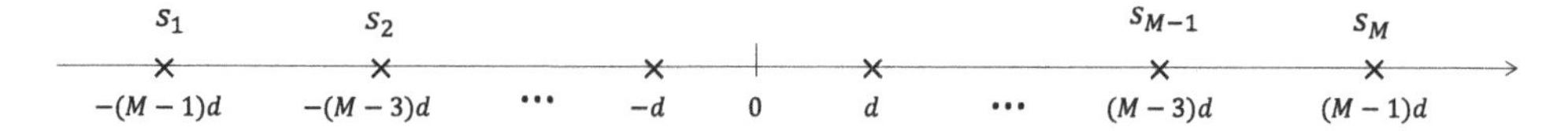

Figure 6.2 M-PAM signal constellation.

of the received signal in vector form is slightly different). The same error probability analysis also applies, resulting in identical error probability expressions in terms of the average signal-to-noise ratio $\bar{\gamma}_s = \bar{E}_s/N_0$ with baseband PAM, which was given in (5.299). Also, given the mapping of the bits to the symbols, the bit error probability can be calculated in a similar fashion.

6.2.2 Phase-Shift Keying

In M-ary PSK signaling, M different symbols (represented as $1, 2, \ldots, M$) are transmitted by M different phases of the sinusoidal carrier signal. Namely,

$$s_i(t) = Ap(t)\cos(2\pi f_c t + \theta_i), \tag{6.17}$$

for $i = 1, 2, \ldots, M$ and $t \in [0, T_s)$. The amplitude of the signal used in transmission does not contain any information, as it is the same for all M symbols. Typically, the phases are selected as

$$\theta_i = \frac{2\pi(i-1)}{M}. \tag{6.18}$$

The reason behind selecting the phases uniformly spaced in the entire range $[0, 2\pi)$ is similar to the uniform selection of the constellation points in M-PAM signaling, that is, to obtain the best error probability performance by making sure that the minimum separation of the constellation points is as large as possible for a given average energy.

Using a trigonometric identity, we can write

$$s_i(t) = Ap(t)\cos\left(2\pi f_c t + \frac{2\pi(i-1)}{M}\right) \tag{6.19}$$

$$= A\cos\left(\frac{2\pi(i-1)}{M}\right)p(t)\cos(2\pi f_c t) - A\sin\left(\frac{2\pi(i-1)}{M}\right)p(t)\sin(2\pi f_c t). \tag{6.20}$$

In other words, all the signals used in transmission can be written as linear combinations of the signals

$$p(t)\cos(2\pi f_c t) \quad \text{and} \quad -p(t)\sin(2\pi f_c t). \tag{6.21}$$

Let us compute the inner product of these two signals:

$$\langle p(t)\cos(2\pi f_c t), -p(t)\sin(2\pi f_c t)\rangle = -\int_0^{T_s} p^2(t)\cos(2\pi f_c t)\sin(2\pi f_c t)dt \tag{6.22}$$

$$= -\frac{1}{2}\int_{0}^{T_s} p^2(t)\sin(4\pi f_c t)dt \tag{6.23}$$

$$\approx 0, \tag{6.24}$$

where the second line is due to a trigonometric identity, while the third line follows by the same line of arguments we made in the previous subsection. Namely, $p^2(t)$ is a lowpass (slowly varying) signal, $f_c \gg 1/T_s$; hence, the integral is approximately zero. (Note that for the case of a rectangular pulse, if f_c is an integer multiple of $1/T_s$, the integral will be exactly zero.)

Since the signals $p(t)\cos(2\pi f_c t)$ and $-p(t)\sin(2\pi f_c t)$ are orthogonal, we obtain an orthonormal basis for PSK signaling by simply normalizing them, resulting in

$$\psi_1(t) = \sqrt{\frac{2}{E_p}}p(t)\cos(2\pi f_c t) \text{ and } \psi_2(t) = -\sqrt{\frac{2}{E_p}}p(t)\sin(2\pi f_c t), \tag{6.25}$$

with E_p denoting the energy of the pulse $p(t)$. The corresponding vector representations of $s_i(t)$ are given by

$$\mathbf{s}_i = \begin{bmatrix} s_{i1} \\ s_{i2} \end{bmatrix} = \begin{bmatrix} A\sqrt{\frac{E_p}{2}}\cos\left(\frac{2\pi(i-1)}{M}\right) \\ A\sqrt{\frac{E_p}{2}}\sin\left(\frac{2\pi(i-1)}{M}\right) \end{bmatrix}, \tag{6.26}$$

for $i = 1, 2, \ldots, M$. Recalling that all the M signals have equal energy

$$E_s = A^2\frac{E_p}{2}, \tag{6.27}$$

we can also write

$$\mathbf{s}_i = \begin{bmatrix} \sqrt{E_s}\cos\left(\frac{2\pi(i-1)}{M}\right) \\ \sqrt{E_s}\sin\left(\frac{2\pi(i-1)}{M}\right) \end{bmatrix}, \tag{6.28}$$

for $i = 1, 2, \ldots, M$.

M-PSK signal constellations for several values of M are depicted in Fig. 6.3. Note that the case of $M = 2$ is the same signaling scheme as binary (bandpass) PAM, as using phases of 0 and π is the same as using the pulse $p(t)\cos(2\pi f_c t)$ and

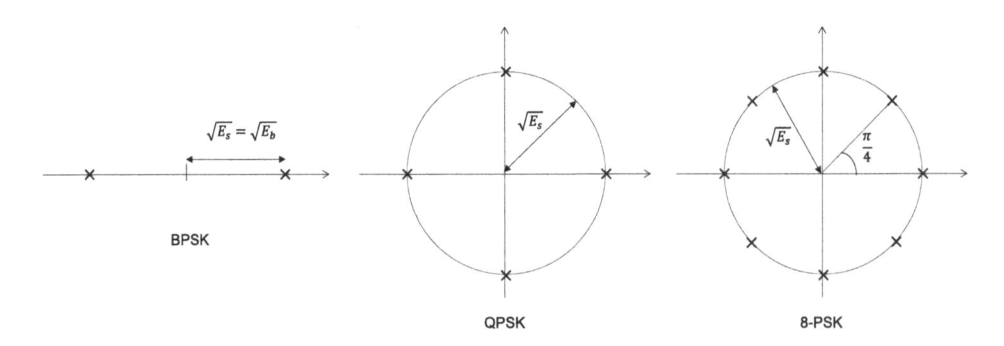

Figure 6.3 M-ary PSK signal constellations for $M = 2$, 4, and 8.

its negative. The case with $M = 2$ is called binary phase-shift keying (BPSK), while the case with $M = 4$ is called quadrature phase-shift keying (QPSK).

Let us determine the ML receiver structure for an AWGN channel, assuming that the symbols are equally likely. Denote the received signal by $r(t)$, that is,

$$r(t) = s_i(t) + n(t), \quad t \in [0, T_s), \tag{6.29}$$

where $s_i(t)$ is the transmitted signal and $n(t)$ is the AWGN with zero mean and power spectral density $\frac{N_0}{2}$. We project $r(t)$ onto the signal space (along the directions of the orthonormal basis functions) and obtain the vector representation $\boldsymbol{r} = \begin{bmatrix} r_1 \\ r_2 \end{bmatrix}$, and use the ML decision rule, which minimizes the Euclidean distance between the received signal and the constellation points. Since the constellation points are uniformly spaced on a circle, the ML rule becomes

$$\hat{m} = i \text{ if } (2i-3)\frac{\pi}{M} < \angle \boldsymbol{r} < (2i-1)\frac{\pi}{M}, \text{ for } i = 1, 2, \ldots, M, \tag{6.30}$$

where $\angle \boldsymbol{r}$ denotes the angle of the complex number $r_1 + jr_2$. In other words, the decision region for the ith symbol is given by

$$\mathcal{D}_i = \left\{ \boldsymbol{r} : (2i-3)\frac{\pi}{M} < \angle \boldsymbol{r} < (2i-1)\frac{\pi}{M} \right\}. \tag{6.31}$$

For the case of QPSK transmission, the decision rule is specialized to

$$\hat{m} = \begin{cases} 1, & \text{if } r_1 > r_2, r_1 > -r_2, \\ 2, & \text{if } r_1 < r_2, r_1 > -r_2, \\ 3, & \text{if } r_1 < r_2, r_1 < -r_2, \\ 4, & \text{if } r_1 > r_2, r_1 < -r_2. \end{cases} \tag{6.32}$$

The receiver structure is illustrated in Fig. 6.4, assuming that a correlation-type implementation is employed. Of course, a matched filter-type implementation can also be readily obtained by replacing each mixer and integrator block with a matched filter and a symbol-rate sampler.

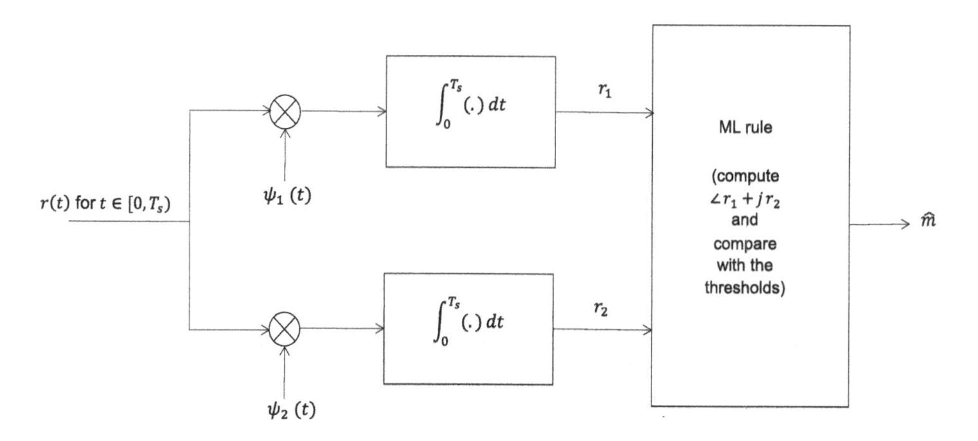

Figure 6.4 M-ary PSK receiver structure.

The decision regions are explicitly shown for the cases $M = 4$ and $M = 8$ in Fig. 6.5. We have omitted the case $M = 2$ as it is identical to binary PAM.

In the above discussion, we have selected the phases utilized in M-PSK signaling as

$$0, \frac{2\pi}{M}, 2\frac{2\pi}{M}, \ldots, (M-1)\frac{2\pi}{M}. \tag{6.33}$$

It should be clear that this is not the only possibility. Indeed, we could have selected any set of uniformly spaced phases on $[0, 2\pi)$ for M-ary PSK transmission, that is, we could rotate the signal constellation by any angle without changing anything. For instance, we could use a $\frac{\pi}{M}$-rotated signal constellation, which employs the phases

$$\frac{\pi}{M}, 3\frac{\pi}{M}, 5\frac{\pi}{M}, \ldots, (2M-1)\frac{\pi}{M}. \tag{6.34}$$

Notice that there is no change in the signal space; the only difference is that a different set of orthonormal basis functions is utilized. This point is illustrated for the case of QPSK signaling in Fig. 6.6.

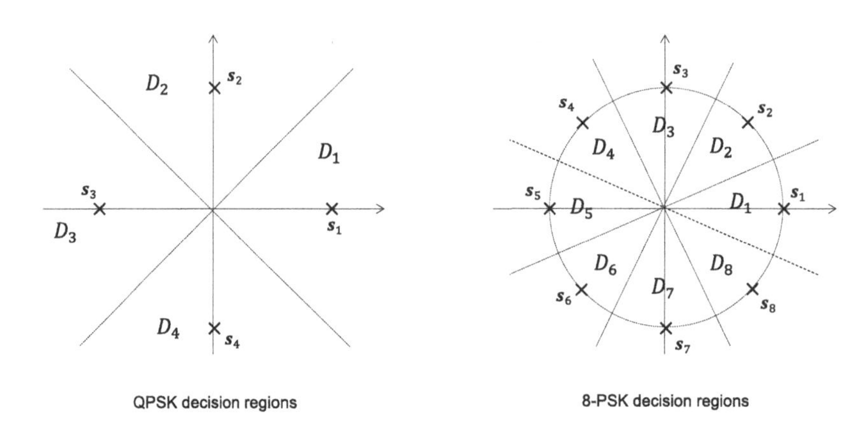

Figure 6.5 Decision regions for QPSK and 8-PSK signaling.

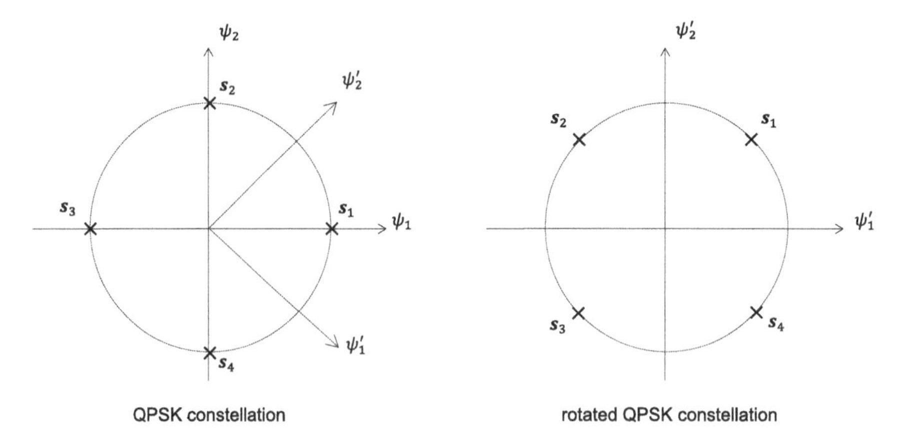

Figure 6.6 QPSK signaling with two different selections for phases.

6.2.2.1 Error Probability Analysis

BPSK Signaling. The binary PSK signaling case is the same as binary PAM. Thus, the same analysis applies, and the average bit error probability with equally likely bits over an AWGN channel is simply

$$P_b = Q\left(\sqrt{2\gamma_b}\right), \tag{6.35}$$

where $\gamma_b = \frac{E_b}{N_0}$ is the signal-to-noise ratio, and E_b is the energy per bit.

QPSK Signaling. Let us determine the average symbol error probability for QPSK signaling. Due to complete symmetry, the conditional error probability for each symbol transmission is the same, that is,

$$P_{e,1} = P_{e,2} = P_{e,3} = P_{e,4}, \tag{6.36}$$

where $P_{e,i} = \mathbb{P}(\text{error}|\boldsymbol{s}_i \text{ is sent})$.

Let us determine $P_{e,1}$ using the rotated signal constellation in Fig. 6.6. Noting that a given $\boldsymbol{s}_1$ is sent, the received signal is $\boldsymbol{r} = \boldsymbol{s}_1 + \boldsymbol{n}$ and we can write

$$P_{e,1} = 1 - \mathbb{P}\left(\boldsymbol{r} \in \mathcal{D}_1 \,\middle|\, \boldsymbol{s}_1 \text{ is sent}\right) \tag{6.37}$$

$$= 1 - \mathbb{P}\left(r_1 > 0, r_2 > 0 \,\middle|\, \boldsymbol{s}_1 \text{ is sent}\right) \tag{6.38}$$

$$= 1 - \mathbb{P}\left(\sqrt{\frac{E_s}{2}} + n_1 > 0 \text{ and } \sqrt{\frac{E_s}{2}} + n_2 > 0\right) \tag{6.39}$$

$$= 1 - \mathbb{P}\left(n_1 > -\sqrt{\frac{E_s}{2}}\right)\mathbb{P}\left(n_2 > -\sqrt{\frac{E_s}{2}}\right), \tag{6.40}$$

where the last line follows from the independence of the AWGN terms in orthogonal directions n_1 and n_2. Recall from our discussion in Chapter 5 that, with the use of orthonormal basis functions, n_1 and n_2 are independent, zero-mean Gaussian random variables with variance $N_0/2$ each. Hence, continuing with the derivation above,

$$P_{e,1} = 1 - \left(Q\left(-\frac{\sqrt{E_s/2}}{\sqrt{N_0/2}}\right)\right)^2 \tag{6.41}$$

$$= 1 - \left(1 - Q\left(\sqrt{\frac{E_s}{N_0}}\right)\right)^2 \tag{6.42}$$

$$= 2Q\left(\sqrt{\frac{E_s}{N_0}}\right) - Q^2\left(\sqrt{\frac{E_s}{N_0}}\right). \tag{6.43}$$

Therefore, the average symbol error probability is given by

$$P_e = 2Q\left(\sqrt{\frac{E_s}{N_0}}\right) - Q^2\left(\sqrt{\frac{E_s}{N_0}}\right). \tag{6.44}$$

Noting that the symbol energies are all E_s, the average symbol energy is also E_s. Hence, defining the signal-to-noise ratio per symbol as $\gamma_s = E_s/N_0$, we can write

$$P_e = 2Q(\sqrt{\gamma_s}) - Q^2(\sqrt{\gamma_s}). \tag{6.45}$$

To express the result in terms of the signal-to-noise ratio per bit γ_b, we note that with each symbol, 2 bits are transmitted, hence $E_s = 2E_b$ (with E_b being the average energy per bit). Therefore, $\gamma_s = 2\gamma_b$. That is,

$$P_e = 2Q\left(\sqrt{2\gamma_b}\right) - Q^2\left(\sqrt{2\gamma_b}\right). \tag{6.46}$$

Note that for typical digital communication systems, we operate at signal-to-noise ratios for which the resulting probability of error is small (e.g., in the range 10^{-4}–10^{-5} or lower). Therefore, the Q-function term is small; thus, the second term with the square of the same Q-function can be neglected. That is,

$$P_e \approx 2Q(\sqrt{\gamma_s}) = 2Q\left(\sqrt{2\gamma_b}\right). \tag{6.47}$$

Note that we could have carried out the average symbol error probability computation using the original signal constellation (with phases $0, \frac{\pi}{2}, \pi, \frac{3\pi}{2}$) as well. The following example illustrates this point.

Example 6.1

Determine the average error probability of QPSK signaling using the original (unrotated) signal constellation in Fig. 6.6.

Solution

Due to symmetry, $P_e = P_{e,1}$, hence

$$P_e = 1 - \mathbb{P}\Big(\text{decide } s_1 \,\Big|\, s_1 \text{is sent}\Big) \tag{6.48}$$

$$= 1 - \mathbb{P}\Big(r_1 > r_2,\ r_1 > -r_2 \,\Big|\, s_1 \text{is sent}\Big) \tag{6.49}$$

$$= 1 - \mathbb{P}\Big(\sqrt{E_s} + n_1 > n_2,\ \sqrt{E_s} + n_1 > -n_2\Big) \tag{6.50}$$

$$= 1 - \mathbb{P}\Big(n_1 - n_2 > -\sqrt{E_s} \text{ and } n_1 + n_2 > -\sqrt{E_s}\Big). \tag{6.51}$$

Since n_1 and n_2 are jointly Gaussian random variables (indeed, they are also independent), their linear combinations are also jointly Gaussian. Namely, $n_1 - n_2$ and $n_1 + n_2$ are jointly Gaussian. Also, $\mathbb{E}[n_1 - n_2] = \mathbb{E}[n_1 + n_2] = 0$, and

$$\mathbb{E}\Big[(n_1 - n_2)(n_1 + n_2)\Big] = \mathbb{E}[n_1^2] - \mathbb{E}[n_2^2] = 0. \tag{6.52}$$

Hence, $n_1 - n_2$ and $n_1 + n_2$ are uncorrelated; being jointly Gaussian, they are also independent. Furthermore,

$$\text{Var}(n_1 + n_2) = \text{Var}(n_1 - n_2) = \text{Var}(n_1) + \text{Var}(n_2) = N_0. \tag{6.53}$$

Therefore, we can write

$$P_e = 1 - \mathbb{P}\left(n_1 - n_2 > -\sqrt{E_s}\right)\mathbb{P}\left(n_1 + n_2 > -\sqrt{E_s}\right) \tag{6.54}$$

$$= 1 - Q^2\left(-\frac{\sqrt{E_s}}{\sqrt{N_0}}\right) \tag{6.55}$$

$$= 2Q(\sqrt{\gamma_s}) - Q^2(\sqrt{\gamma_s}), \tag{6.56}$$

which is the same result we obtained in (6.45).

We now consider the mapping of bits to symbols and determine the average bit error probability. QPSK with Gray mapping, depicted in Fig. 6.7, is particularly interesting. With this mapping, the first bit is affected by n_2 only (as the noise in the direction of ψ_1, i.e., n_1, cannot cause an error for this bit), while for the second bit, only n_1 matters. The error probability for either bit is the same. Hence, we can write the bit error probability of QPSK with Gray mapping as

$$P_b = \mathbb{P}\left(n_1 > \sqrt{\frac{E_s}{2}}\right) \tag{6.57}$$

$$= Q\left(\sqrt{\gamma_s}\right) \tag{6.58}$$

$$= Q\left(\sqrt{2\gamma_b}\right), \tag{6.59}$$

which, incidentally, is the same as the bit error probability of BPSK signaling. In other words, while transmitting at twice the rate of BPSK, we do not lose in terms of the error probability performance. Indeed, what is going on is the following: QPSK with Gray mapping is nothing but two BPSK modulation schemes in orthogonal directions that do not interfere.

M-PSK Error Probability for $M > 4$. We obtained the average error probabilities for BPSK and QPSK signaling over an AWGN channel in closed form. On the other hand, for higher-order PSK constellations, there is no closed-form expression. We can only write an integral that needs to be evaluated numerically.

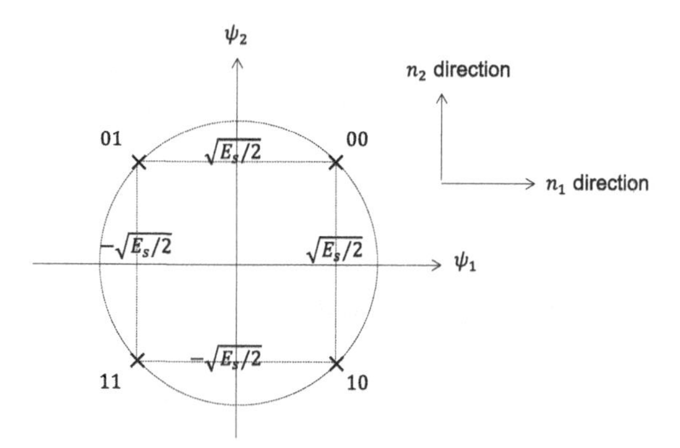

Figure 6.7 QPSK signaling with Gray mapping.

However, we can use the union bound and obtain an upper bound on the symbol error probability for a simple analysis.

Denoting the pairwise error probability of making an error in favor of the kth symbol when s_1 is transmitted by $P(s_1 \to s_k)$, we can write

$$P_e = P_{e,1} \leq \sum_{k=2}^{M} \mathbb{P}(s_1 \to s_k) \tag{6.60}$$

$$= \sum_{k=2}^{M} Q\left(\frac{\|s_1 - s_k\|}{\sqrt{2N_0}}\right). \tag{6.61}$$

We can compute the pairwise distances between s_1 and the other constellation points and compute the right-hand side. However, we will take the easier approach and simply use the fact that $d_{\min} \leq \|s_1 - s_k\|$ to obtain

$$P_e \leq (M-1)Q\left(\frac{d_{\min}}{\sqrt{2N_0}}\right) \tag{6.62}$$

$$= (M-1)Q\left(\frac{2\sqrt{E_s}\sin(\pi/M)}{\sqrt{2N_0}}\right) \tag{6.63}$$

$$= (M-1)Q\left(\sqrt{2\gamma_s}\sin(\pi/M)\right), \tag{6.64}$$

which is nothing but the loose form of the union bound. Note that $d_{\min}$ is the Euclidean distance between any two neighboring constellation points (see Fig. 6.8).

As an alternative to the union bound, we can obtain a tighter result on the symbol error probability by using the pairwise error probabilities with the two nearest neighbors of s_1. Observing that the region $\mathcal{D}_1^c$ is exactly the union of two half-spaces as shown in Fig. 6.9, we can write

$$P_e = P_{e,1} \leq \mathbb{P}(s_1 \to s_2) + P(s_1 \to s_M) \tag{6.65}$$

$$= 2Q\left(\sqrt{2\gamma_s}\sin(\pi/M)\right). \tag{6.66}$$

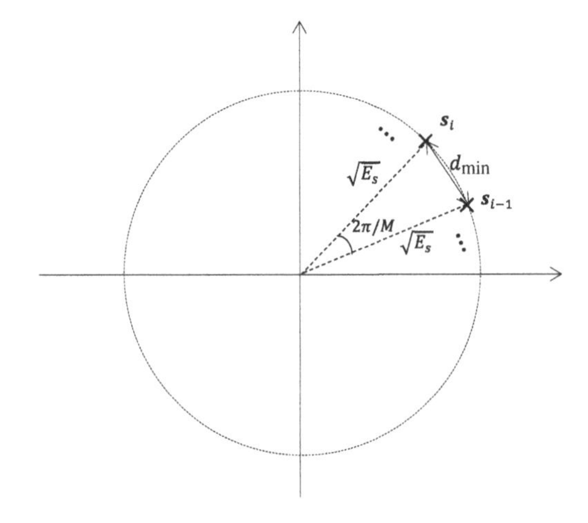

Figure 6.8 Minimum distance of M-PSK signal constellation.

With M-ary signaling, each symbol corresponds to $\log_2 M$ bits, hence $E_s = \log_2(M)E_b$ and $\gamma_s = \log_2(M)\gamma_b$, where γ_b is the signal-to-noise ratio per bit. Therefore, we can also write

$$P_e \leq 2Q\left(\sqrt{2\log_2(M)\gamma_b}\,\sin(\pi/M)\right), \tag{6.67}$$

for the average symbol error probability.

Let us now consider the bit error rate analysis with M-PSK signaling. To proceed further, we must specify the mapping from the sequence of bits to the PSK symbols. As in the case of QPSK, Gray mapping is of particular interest, which ensures that the adjacent symbols differ in only one bit. For instance, a Gray mapping for 8-PSK is given in Fig. 6.10.

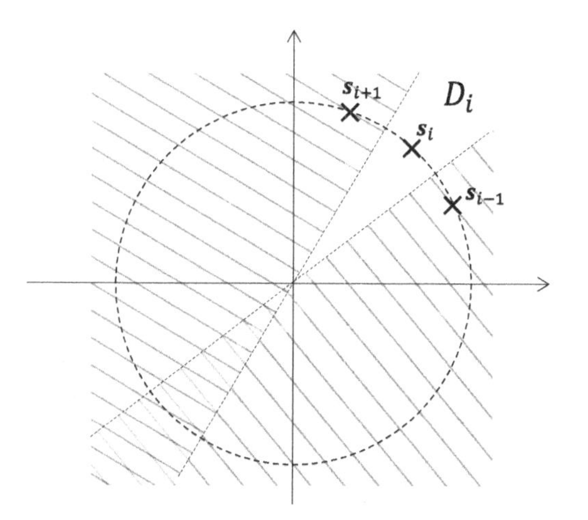

Figure 6.9 Error region for transmission of s_i as the union of two half-spaces.

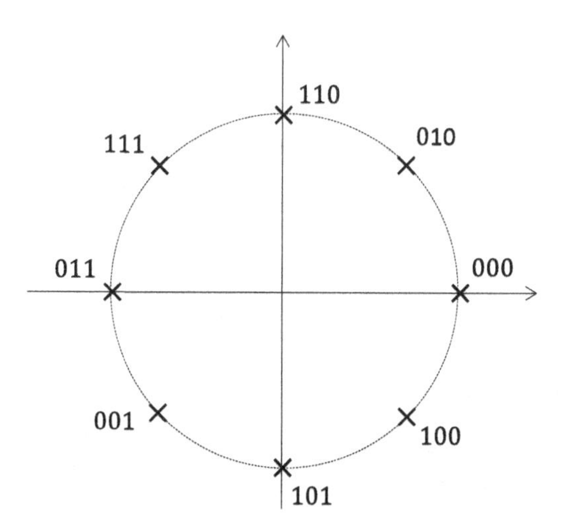

Figure 6.10 Gray mapping for 8-PSK.

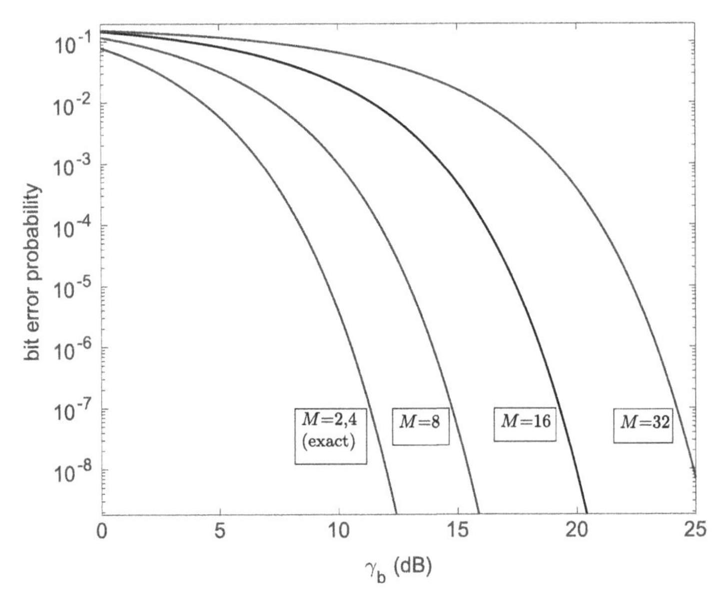

Figure 6.11 Bit error probability of M-PSK signaling with Gray mapping for different constellation sizes. The results for $M = 2$ and 4 are exact, while the others are approximate.

Assume that Gray mapping is employed. Then, most errors (for high signal-to-noise ratios) will be in favor of a neighboring symbol, which differs in only one bit among the $\log_2 M$ transmitted bits; therefore, we deduce that the average bit error probability can be approximated as

$$P_b \approx \frac{2}{\log_2 M} Q\left(\sqrt{2\log_2(M)\gamma_b}\sin(\pi/M)\right). \tag{6.68}$$

Figure 6.11 depicts the average bit error probability for M-PSK signaling for $M = 2, 4, 8, 16$, and 32. As the constellation size increases, the transmission rate increases; however, the error probability also increases. In fact, for large constellation sizes, doubling the constellation size (i.e., each additional bit of information) results in a loss of approximately 6 dB in the error probability performance. This is because, for large M, the minimum squared Euclidean distance is proportional to $\sin^2(\pi/M) \approx \left(\frac{\pi}{M}\right)^2$, hence doubling M reduces it by a factor of 4, which corresponds to $10\log_{10} 4 \approx 6$ dB.

6.2.3 Quadrature Amplitude Modulation

In the previous two subsections, we have considered bandpass PAM and PSK signaling, for which information is embedded in the amplitude and phase of a carrier signal, respectively. As an alternative, we can embed digital information both in the amplitude and the phase, resulting in quadrature amplitude modulation (QAM).

For M-ary QAM signaling, the ith message is represented by the signal

$$s_i(t) = A_i p(t)\cos(2\pi f_c t + \theta_i) \tag{6.69}$$

$$= A_i \cos\theta_i p(t)\cos(2\pi f_c t) - A_i \sin\theta_i p(t)\sin(2\pi f_c t), \tag{6.70}$$

$t \in [0, T_s)$, $i = 1, 2, \ldots, M$, where A_i is the amplitude and θ_i is the phase assigned to the ith message.

It is clear that this is a two-dimensional signaling scheme, and an orthonormal basis is given by

$$\psi_1(t) = \sqrt{\frac{2}{E_p}}p(t)\cos(2\pi f_c t) \text{ and } \psi_2(t) = -\sqrt{\frac{2}{E_p}}p(t)\sin(2\pi f_c t), \tag{6.71}$$

for $t \in [0, T_s)$. As the basis functions are the same, we notice that this is the same signal space as in PSK signaling. Hence, the vector representation of $s_i(t)$ is

$$\mathbf{s}_i = \begin{bmatrix} s_{i1} \\ s_{i2} \end{bmatrix} = \begin{bmatrix} A_i \cos\theta_i \\ A_i \sin\theta_i \end{bmatrix}, \tag{6.72}$$

and its energy is given by

$$E_{si} = \|\mathbf{s}_i\|^2 = A_i^2, \tag{6.73}$$

and the average energy of the signal constellation becomes

$$\bar{E}_s = \frac{1}{M}\sum_{i=1}^{M} A_i^2. \tag{6.74}$$

For M-QAM signaling, selecting square constellations (when $\log_2 M$ is even) and cross constellations (when $\log_2 M$ is odd) is common. 16-ary square and 8-ary cross QAM constellations are depicted in Fig. 6.12.

The techniques developed in Chapter 5 can be used to obtain the optimal receiver structure and conduct a performance analysis. While there is no simple solution for

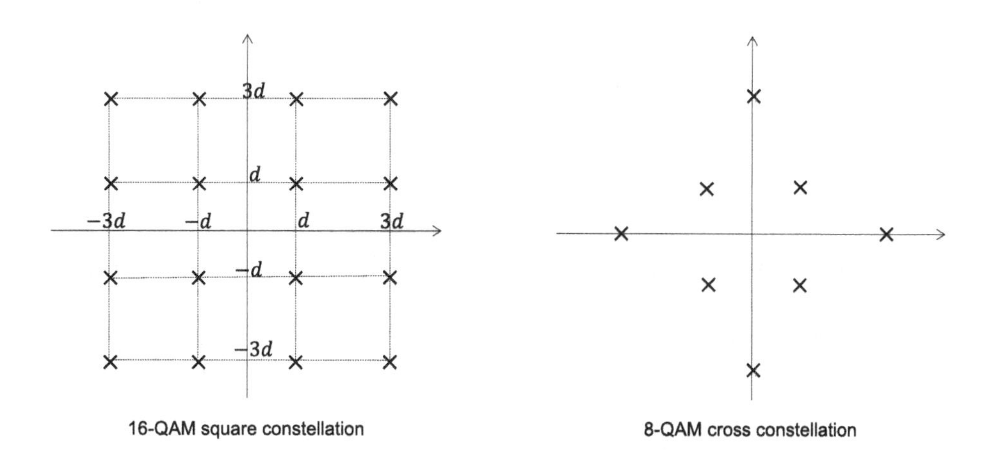

Figure 6.12 16-ary square and 8-ary cross QAM constellations.

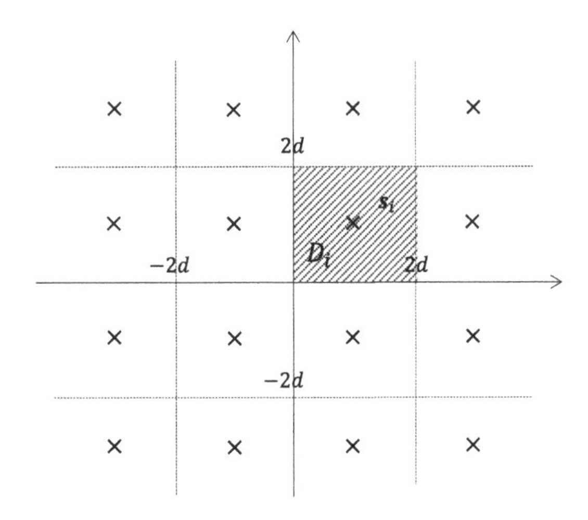

Figure 6.13 Decision regions for 16-ary signaling with a square constellation.

the general case, the case of an M-ary square constellation is straightforward to study, as illustrated in the following.

We depict the decision regions for 16-QAM signaling (with a square constellation) over an AWGN channel in Fig. 6.13. Assuming equally likely signaling, the ML receiver for an AWGN channel first obtains the vector form for the received signal $\boldsymbol{r} = \begin{bmatrix} r_1 \\ r_2 \end{bmatrix}$ (in the same way as given in Fig. 6.4), and employs the rule specified by the decision regions as shown in the figure.

Let us determine the average symbol error probability of the M-QAM signaling scheme with a square constellation (assuming that $\log_2 M$ is even).

Since $\log_2 M$ is even, $M = L^2$ for some L (which is a power of 2). We notice that M-QAM signaling with a square constellation is equivalent to two L-PAM constellations (each with amplitude levels $(2i - 1 - L)d$, $i = 1, 2, \ldots, L$), one utilizing $p(t)\cos(2\pi f_c t)$, the other employing $p(t)\sin(2\pi f_c t)$. These two schemes do not interfere with each other as they are in orthogonal directions (and the AWGN terms affecting them are independent).

Recall that the probability of a correct decision for L-PAM signaling is given by

$$1 - \frac{2(L-1)}{L} Q\left(\sqrt{\frac{2d^2}{N_0}}\right). \tag{6.75}$$

As the error probabilities of the two L-PAM signals in orthogonal directions are independent, we can write the error probability for M-QAM with a square constellation as

$$P_e = 1 - \mathbb{P}(\text{correct decision}) \tag{6.76}$$

$$= 1 - \left(1 - \frac{2(L-1)}{L} Q\left(\sqrt{\frac{2d^2}{N_0}}\right)\right)^2 \tag{6.77}$$

$$= \frac{4(\sqrt{M}-1)}{\sqrt{M}} Q\left(\sqrt{\frac{2d^2}{N_0}}\right) - \frac{4(\sqrt{M}-1)^2}{M} Q^2\left(\sqrt{\frac{2d^2}{N_0}}\right). \quad (6.78)$$

This can be approximated for large signal-to-noise ratios as

$$P_e \approx 4\left(1-\frac{1}{\sqrt{M}}\right) Q\left(\sqrt{\frac{2d^2}{N_0}}\right). \quad (6.79)$$

To express this in terms of the average signal-to-noise ratio $\bar{\gamma}_s = \bar{E}_s/N_0$, we compute the average energy of the constellation as

$$\bar{E}_s = \frac{1}{M}\sum_{i=1}^{M} \|\boldsymbol{s}_i\|^2 \quad (6.80)$$

$$= \frac{4}{M}\sum_{j=1}^{L/2}\sum_{l=1}^{L/2}((2j-1)^2 + (2l-1)^2)d^2 \quad (6.81)$$

$$= \frac{2d^2(M-1)}{3}. \quad (6.82)$$

Hence, we obtain

$$P_e \approx 4\left(1-\frac{1}{\sqrt{M}}\right) Q\left(\sqrt{\frac{3\bar{\gamma}_s}{M-1}}\right). \quad (6.83)$$

6.3 Lowpass-Equivalent Representations of Bandpass Signals

In this section, we provide a convenient way of representing bandpass signals through their lowpass equivalents. We will find this representation instrumental in describing the implementation of practical transmitters and receivers, and in developing a simplified mathematical treatment, as will be clarified in the rest of this chapter and in the following chapters.

We say that a real signal $x(t)$ with Fourier transform $X(f)$ is a bandpass signal if $|X(f)| = 0$ for $|f| \notin (f_1, f_2)$ for some frequencies f_1 and f_2, as illustrated in Fig. 6.14.

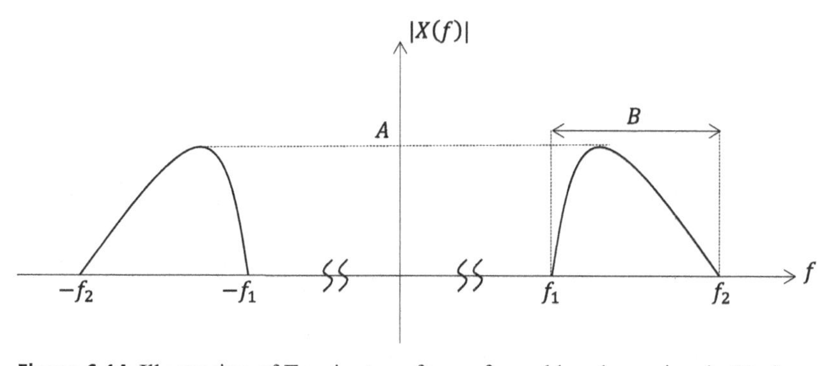

Figure 6.14 Illustration of Fourier transform of a real bandpass signal $x(t)$. As $x(t)$ is real, the Fourier transform $X(f)$ is conjugate symmetric.

The signal bandwidth is $B = f_2 - f_1$, and $x(t)$ is called a *narrowband* signal if $B \ll f_1, f_2$.

We define the pre-envelope of a bandpass signal $x(t)$, denoted $x_+(t)$, as the signal with Fourier transform

$$X_+(f) = 2U(f)X(f) = \begin{cases} 2X(f), & \text{if } f > 0, \\ 0, & \text{if } f \leq 0, \end{cases} \tag{6.84}$$

where $U(f)$ is the unit step function. We do not worry about $f = 0$, as $x(t)$ is assumed to be a bandpass signal. Noting that the inverse Fourier transform of $U(f)$ is given by

$$u(t) = \delta(t) + \frac{j}{\pi t}, \tag{6.85}$$

we can write the pre-envelope signal in the time domain as

$$x_+(t) = x(t) + j\hat{x}(t), \tag{6.86}$$

where $\hat{x}(t)$ is the Hilbert transform of $x(t)$, recall that $\hat{x}(t) = \frac{1}{\pi t} * x(t)$. Recalling that the bandpass signal $x(t)$ is a real function, we observe that its Hilbert transform is also real and that the real part of its pre-envelope signal $x_+(t)$ is $x(t)$ itself.

We define the lowpass equivalent of $x(t)$, denoted $\tilde{x}(t)$, as the signal whose Fourier transform is a shifted version of $X_+(f)$ by $-f_c$, that is,

$$\tilde{X}(f) = X_+(f + f_c). \tag{6.87}$$

The relationship among the Fourier transforms of the original signal, its pre-envelope, and lowpass equivalents is illustrated in Fig. 6.15. It should be clear that with an appropriate selection of the reference frequency f_c, the lowpass-equivalent signal has frequency content near the origin, that is, it is slowly time-varying. This is a useful property for generating bandpass signals, as will be shown shortly.

In the time domain, we have

$$\tilde{x}(t) = x_+(t)e^{-j2\pi f_c t}. \tag{6.88}$$

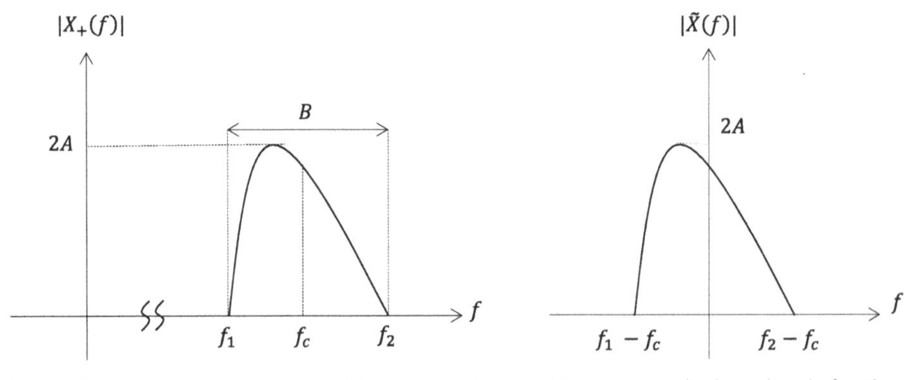

Figure 6.15 Fourier transforms of the pre-envelope and lowpass-equivalent signals for the bandpass signal depicted in Fig. 6.14.

Since $x(t) = Re\{x_+(t)\}$, we obtain

$$x(t) = \mathcal{R}\left\{\tilde{x}(t)e^{j2\pi f_c t}\right\}. \tag{6.89}$$

We refer to the real part of the complex envelope as the in-phase component, and its imaginary part as the quadrature component of the bandpass signal. Namely,

$$\tilde{x}(t) = x_I(t) + jx_Q(t), \tag{6.90}$$

where $x_I(t)$ is the in-phase component and $x_Q(t)$ is the quadrature component. Using these two (complex) signals, we obtain the original bandpass signal as

$$x(t) = x_I(t)\cos(2\pi f_c t) - x_Q(t)\sin(2\pi f_c t). \tag{6.91}$$

Since the in-phase and quadrature components are slowly varying signals, it is clear that the *rapid* time variations of the bandpass signal are due to the pure sinusoidal terms $\cos(2\pi f_c t)$ and $\sin(2\pi f_c t)$. This is a highly convenient way of expressing the bandpass signal $x(t)$ since (being slowly varying signals), $x_I(t)$ and $x_Q(t)$ are easy to generate and process. For instance, we can generate samples of $x_I(t)$ and $x_Q(t)$ (digitally) and then, via a simple (and cheap) digital-to-analog converter, we can obtain the in-phase and quadrature components of the bandpass signal (as a continuous-time signal). We can then use a local oscillator, phase shifter, and mixers to obtain the desired bandpass signal. This process is illustrated in Fig. 6.16.

Given a bandpass signal (e.g., received signal in a bandpass communication system), obtaining the in-phase and quadrature signals is easy. All we need to do is multiply the signal with $\cos(2\pi f_c t)$ and pass it through a lowpass filter to obtain the in-phase component and multiply by $-\sin(2\pi f_c t)$ and lowpass filter to obtain the quadrature component. That is,

$$x(t)\cos(2\pi f_c t) = (x_I(t)\cos(2\pi f_c t) - x_Q(t)\sin(2\pi f_c t))\cos(2\pi f_c t) \tag{6.92}$$

$$= \frac{1}{2}x_I(t) + \frac{1}{2}x_I(t)\cos(4\pi f_c t) - \frac{1}{2}x_Q(t)\sin(4\pi f_c t), \tag{6.93}$$

where the frequency contents of the second and third terms are concentrated around $\pm 2f_c$ (recall that $x_I(t)$ and $x_Q(t)$ are lowpass signals), hence only the

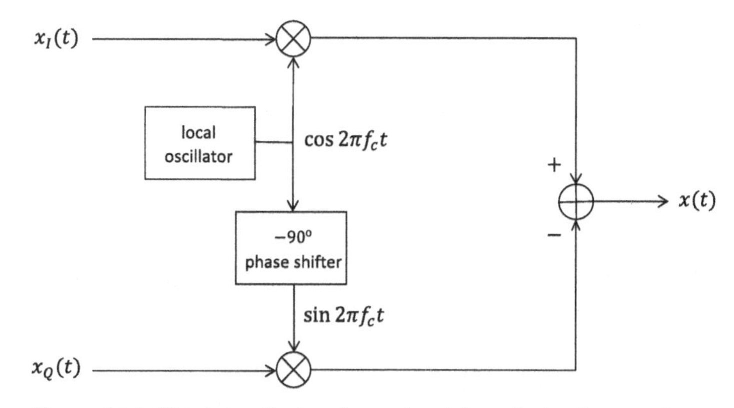

Figure 6.16 Obtaining the bandpass signal from its in-phase and quadrature components.

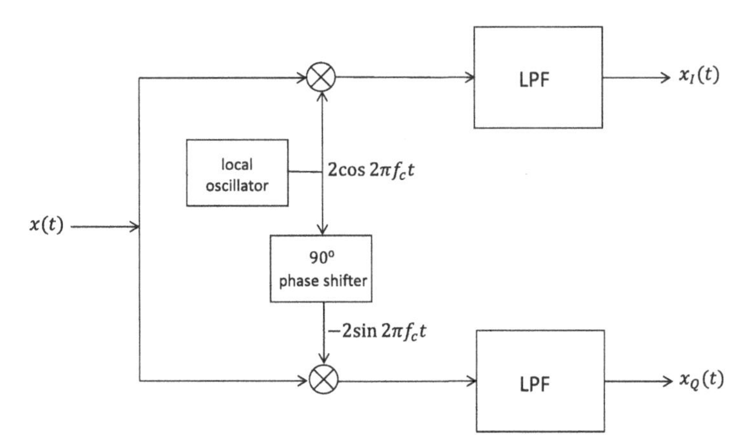

Figure 6.17 Obtaining the in-phase and quadrature components of a bandpass signal.

in-phase component remains after lowpass filtering. Similarly,

$$-x(t)\sin(2\pi f_c t) = -(x_I(t)\cos(2\pi f_c t) - x_Q(t)\sin(2\pi f_c t))\sin(2\pi f_c t) \tag{6.94}$$

$$= -\frac{1}{2}x_I(t)\sin(4\pi f_c t) + \frac{1}{2}x_Q(t) - \frac{1}{2}x_Q(t)\cos(4\pi f_c t), \tag{6.95}$$

where the first and third terms are eliminated by the lowpass filter, resulting in the quadrature component of the signal only.

We illustrate the process of obtaining the in-phase and quadrature components of a bandpass signal in Fig. 6.17.

Example 6.2

Consider a (real) bandpass signal $x(t) = p(t)\cos(2\pi f_c t - \theta)$, where $p(t)$ is a (real) lowpass signal, f_c is the carrier frequency, and θ is a constant phase. Determine the lowpass equivalent of $x(t)$.

Solution

We can write

$$x(t) = \mathcal{R}\left\{p(t)e^{-j\theta}e^{j2\pi f_c t}\right\}, \tag{6.96}$$

hence, we identify the lowpass-equivalent signal as

$$\tilde{x}(t) = p(t)e^{-j\theta}. \tag{6.97}$$

When $\theta = 0$, we have $x(t) = p(t)\cos(2\pi f_c t)$ with a lowpass equivalent of $\tilde{x}(t) = p(t)$, that is, the in-phase component is $p(t)$ and the quadrature component is 0. On the other hand, when $\theta = -\frac{\pi}{2}$, we have $x(t) = -p(t)\sin(2\pi f_c t)$ with a lowpass equivalent of $\tilde{x}(t) = jp(t)$, that is, the quadrature component is $p(t)$ while the in-phase component is 0.

6.4 Bandpass White Gaussian Noise

In the previous section, we have described how deterministic bandpass signals can be represented using their lowpass equivalents. With the objective of studying bandpass communication systems via their lowpass equivalents, in this section, we describe how a similar characterization can be accomplished for random signals. Specifically, we focus on bandpass Gaussian noise and demonstrate how it can be represented using lowpass random processes.

Consider a bandpass communication system for which the transmitted signal $s(t)$ is limited to the frequency band $[f_1,f_2]$. The received signal through an AWGN channel will be of the form $s(t) + n(t)$, where $n(t)$ is a white Gaussian noise with power spectral density $N_0/2$. We envision the presence of a front-end bandpass filter with a passband of $[f_1,f_2]$; then, at the filter output, we will have the transmitted signal corrupted by the filtered version of the white Gaussian noise. Noting that the transmitted signal will remain intact as its spectral content is within the passband of the filter, the received signal at the filter output will be given by

$$r(t) = s(t) + w(t), \tag{6.98}$$

where $w(t)$ is the *filtered* white Gaussian noise process obtained using an ideal bandpass filter with the passband $[f_1,f_2]$. Note also that even if the front-end filter we have assumed is not present (or it does not have such a strict passband behavior), equivalent filtering operations will take place later in the communication chain; hence, for all intents and purposes, we can assume its presence in our study of bandpass communication systems.

Let us focus on bandpass filtering of white Gaussian noise $n(t)$ resulting in $w(t)$. From our discussion of random processes in Chapter 2, since the AWGN is zero mean, the filtered noise is also zero mean, that is, $\mathbb{E}[w(t)] = 0$, and its power spectral density is given by

$$S_w(f) = \begin{cases} \frac{N_0}{2}, & \text{for } |f| \in [f_1,f_2], \\ 0, & \text{else}. \end{cases} \tag{6.99}$$

That is, the power spectral density of the filtered noise is equal to $N_0/2$ within the passband of the bandpass filter, and it is 0 outside. Also, $w(t)$ is a Gaussian random process as the input process is Gaussian, and the bandpass filtering operation is linear. We refer to the random process $w(t)$ as a *bandpass white Gaussian noise*. The filtering operation and the resulting PSD are illustrated in Fig. 6.18.

Similar to the case of deterministic bandpass signals, we can write a (wide-sense stationary) bandpass random process $X(t)$ in terms of its in-phase and quadrature components. That is, with a reference frequency f_c (typically within the band $[f_1,f_2]$), we have

$$X(t) = X_I(t)\cos(2\pi f_c t) - X_Q(t)\sin(2\pi f_c t), \tag{6.100}$$

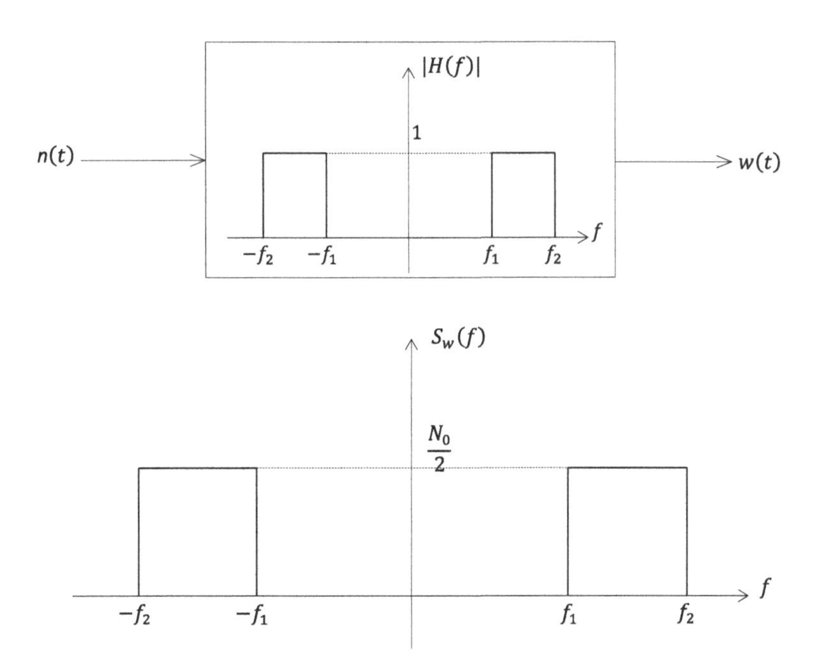

Figure 6.18 Filtering of a zero-mean, wide-sense stationary white Gaussian noise process by a linear time-invariant bandpass filter results in a zero-mean, wide-sense stationary bandpass Gaussian random process.

where $X_I(t)$ is the in-phase noise component and $X_Q(t)$ is the quadrature noise component. These processes are jointly wide-sense stationary, and their power spectral densities are given by

$$S_{X_I}(f) = S_{X_Q}(f) = \begin{cases} S_X(f - f_c) + S_X(f + f_c), & \text{for } -B \leq f \leq B, \\ 0, & \text{else}, \end{cases} \tag{6.101}$$

where $B = \max\{|f_2 - f_c|, |f_1 - f_c|\}$. Clearly, with a typical selection of the reference frequency f_c, the power spectral densities of the in-phase and quadrature components are both concentrated around $f = 0$; hence, these processes are lowpass.

Let us now specialize the above expression to the case of a bandpass white Gaussian noise process with a reference frequency taken in the middle of its frequency band. This will be sufficient for studying bandpass communication systems for our purposes. Consider a bandpass white noise $w(t)$ with power spectral density

$$S_w(f) = \begin{cases} \frac{N_0}{2}, & \text{for } |f| \in [f_c - W/2, f_c + W/2], \\ 0, & \text{else}, \end{cases} \tag{6.102}$$

where W is the bandwidth of the (random) bandpass signal. We write this signal as

$$w(t) = w_I(t)\cos(2\pi f_c t) - w_Q(t)\sin(2\pi f_c t), \tag{6.103}$$

with

$$S_{w_I}(f) = S_{w_Q}(f) = \begin{cases} N_0, & \text{for } -W/2 \le f \le W/2, \\ 0, & \text{else,} \end{cases} \tag{6.104}$$

where $w_I(t)$ and $w_Q(t)$ are the in-phase and quadrature components of the bandpass white Gaussian noise $w(t)$. Both $w_I(t)$ and $w_Q(t)$ are lowpass random processes (with a bandwidth of $W/2$); they are jointly wide-sense stationary and Gaussian. In fact, with the above selection of the reference frequency and due to the symmetry of $S_w(f)$ (around f_c), the random processes $w_I(t)$ and $w_Q(t)$ are independent.

Figure 6.19 depicts the power spectral density of the bandpass white noise as well as the power spectral densities of its in-phase and quadrature components.

To summarize, we now have a lowpass-equivalent representation of the bandpass white Gaussian noise process $w(t)$ as a complex Gaussian random process given by

$$w_I(t) + jw_Q(t), \tag{6.105}$$

where the real and imaginary parts are zero-mean, stationary Gaussian random processes with power spectral densities as given in (6.104), and they are independent of each other.

We can determine the average powers of $w(t)$, $w_I(t)$, and $w_Q(t)$ by integrating the corresponding power spectral densities and obtain

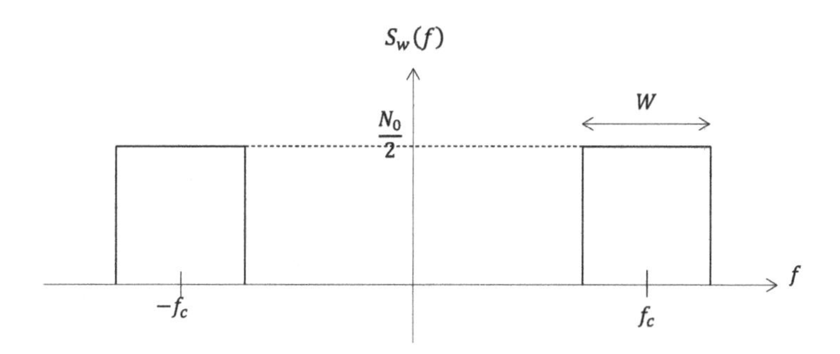

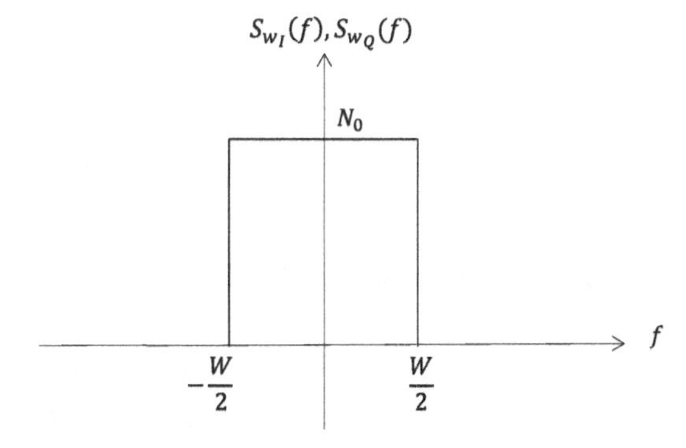

Figure 6.19 Power spectral densities of a bandpass white noise and its in-phase and quadrature components with a reference frequency taken at the middle of its frequency band.

$$P_w = P_{w_I} = P_{w_Q} = N_0 W, \tag{6.106}$$

and the average power of the lowpass-equivalent random signal $w_I(t) + jw_Q(t)$ as $2N_0 W$.

It should also be clear that the bandwidth of the lowpass-equivalent random process is $W/2$, that is, half of the bandwidth of the bandpass random process owing to the selection of the reference frequency f_c at the center of the frequency band of the original bandpass signal.

Example 6.3

Consider a bandpass white Gaussian noise $v(t)$ with power spectral density

$$S_v(f) = \begin{cases} \frac{N_0}{2}, & \text{for } |f| \in [f_c, f_c + W], \\ 0, & \text{else,} \end{cases} \tag{6.107}$$

where W is the bandwidth of the (random) bandpass signal. Determine the power spectral densities of its in-phase and quadrature components with respect to the reference frequency f_c. (Notice that the reference frequency is not selected at the center of the frequency band of $v(t)$.)

Solution

Following the same approach, we can determine the power spectral density of the in-phase and quadrature components of $v(t)$ as

$$S_{v_I}(f) = S_{v_Q}(f) = \begin{cases} S_v(f - f_c) + S_v(f + f_c), & \text{for } -W \leq f \leq W, \\ 0, & \text{else,} \end{cases} \tag{6.108}$$

namely,

$$S_{v_I}(f) = S_{v_Q}(f) = \begin{cases} \frac{N_0}{2}, & \text{for } -W \leq f \leq W, \\ 0, & \text{else.} \end{cases} \tag{6.109}$$

We observe that this is different from the one obtained by selecting the reference frequency at the center of the band (see (6.104)).

Note that a more significant difference is the following: with this reference frequency selection, the in-phase and quadrature components are not uncorrelated (which was the case for white noise when the reference frequency was selected at the center of the frequency band). We omit the details of this result.

6.5 Study of Bandpass Communications Through Lowpass Equivalents

In Sections 6.3 and 6.4, we developed the lowpass-equivalent representations of bandpass signals. To summarize, both deterministic and random signals can be written as a superposition of (lowpass) in-phase and quadrature components multiplied with sinusoidal signals, facilitating their generation and processing.

Therefore, this representation provides a convenient means of studying bandpass communication systems. This section explores this idea further and considers single-carrier modulation schemes using lowpass-equivalent representations.

Consider an M-ary modulation scheme using signals from the space spanned by the two basis functions

$$\psi_1(t) = \sqrt{\frac{2}{E_p}} p(t) \cos(2\pi f_c t) \tag{6.110}$$

and

$$\psi_2(t) = -\sqrt{\frac{2}{E_p}} p(t) \sin(2\pi f_c t), \tag{6.111}$$

for $t \in [0, T_s)$, with $f_c \gg \frac{1}{T_s}$, as in PSK and QAM. This formulation also covers the PAM signaling with the understanding that the second components of the transmitted signal vectors are selected as 0. Let us denote the signal used for transmission of the mth message as

$$s_m(t) = A_{m,r}\psi_1(t) + A_{m,i}\psi_2(t), \quad m = 1, 2, \ldots, M, \tag{6.112}$$

where $A_{m,r}$ and $A_{m,i}$ are constants used to select the specific constellation points. The lowpass-equivalent signals (with reference frequency f_c) are then

$$s_{m,l}(t) = \sqrt{\frac{2}{E_p}} (A_{m,r} + jA_{m,i}) p(t). \tag{6.113}$$

That is, all the M (lowpass-equivalent) signals can be written as the scaled version of a basic pulse shape $p(t)$, that is, this signal set is one-dimensional with the basis function

$$\psi_l(t) = \sqrt{\frac{1}{E_p}} p(t) \tag{6.114}$$

and, correspondingly, the vector form of the (lowpass-equivalent) signals becomes

$$s_{m,l} = \sqrt{2}(A_{m,r} + jA_{m,i}), \quad m = 1, 2, \ldots, M. \tag{6.115}$$

Therefore, we have a complex number representing each constellation point. For PAM signaling, these points are selected on the real line (symmetric around 0 and with uniform separation between consecutive points). In contrast, for PSK, they are chosen on a circle, and for rectangular QAM, they are selected on a rectangular grid in the complex plane.

Assume that we transmit a bandpass signal $s_m(t)$ through an AWGN channel and that a front-end bandpass filter removes the noise components outside the frequency band of the signal. With coherent detection, that is, when the channel phase is estimated and compensated perfectly, the received signal can be written as follows:

$$r(t) = s_m(t) + w(t), \tag{6.116}$$

where $w(t)$ is the bandpass white Gaussian noise (see Section 6.4). Using the reference frequency f_c, the lowpass-equivalent received signal becomes

$$r_l(t) = s_{m,l}(t) + w_l(t), \tag{6.117}$$

where $r_l(t)$ is the lowpass-equivalent received signal, $s_{m,l}(t)$ is the lowpass-equivalent transmitted signal, and $w_l(t)$ is a zero-mean complex white Gaussian noise process, which is the lowpass equivalent of the noise process $w(t)$.

We recall from our discussion in the previous section that a bandpass stationary white Gaussian noise process with power spectral density $N_0/2$ has a lowpass equivalent with independent real and imaginary parts, each with a power spectral density N_0 within the band of interest. Also, the real and imaginary parts are independent stationary Gaussian processes with zero mean. In other words, we can consider $w_l(t)$ as a complex process whose real and imaginary parts are zero-mean independent Gaussian random processes, each with power spectral density N_0. That is,

$$w_l(t) = w_I(t) + jw_Q(t), \tag{6.118}$$

with the autocorrelation functions of the in-phase component $w_I(t)$ and the quadrature component $w_Q(t)$ being $N_0\delta(\tau)$.

We can correlate the received signal with the basis function and write a mathematically equivalent form:

$$r_l = s_{m,l} + w_l, \tag{6.119}$$

where

$$r_l = \langle r_l(t), \psi_l(t) \rangle = \int_0^{T_s} r_l(t)\psi_l^*(t)dt \tag{6.120}$$

and

$$w_l = \langle w(t), \psi_l(t) \rangle = \int_0^{T_s} w(t)\psi_l^*(t)dt. \tag{6.121}$$

That is, r_l is the complex number representing the received signal and $s_{m,l}$ is the complex number representing the transmitted constellation point. We can show that w_l is a complex Gaussian random variable with zero mean, and its real and imaginary parts are independent, each with variance N_0.

Let us illustrate the last point more explicitly by the following example.

Example 6.4

Let $w_l(t)$ be a stationary complex Gaussian random process with zero mean and assume that the power spectral densities of its real and imaginary components are N_0 each. Also, assume that the real and imaginary parts are independent. Determine the variance of the real part of the random variable defined as $w_l = \langle w(t), \psi_l(t)\rangle$, where $\psi_l(t)$ is a normalized complex function, that is, $\int_0^{T_s} |\psi_l(t)|^2 dt = 1$.

Solution

Let us write $w_l(t)$ and $\psi_l(t)$ explicitly as

$$w_l(t) = w_I(t) + jw_Q(t) \text{ and } \psi_l(t) = \psi_I(t) + j\psi_Q(t), \tag{6.122}$$

where $w_I(t), w_Q(t), \psi_I(t)$, and $\psi_Q(t)$ are all real. We can then write the real part of w_l as

$$\mathcal{R}\{w_l\} = \int_0^{T_s} w_I(t)\psi_I(t)dt - \int_0^{T_s} w_Q(t)\psi_Q(t)dt. \tag{6.123}$$

Let us compute the variance of this random variable:

$$\text{Var}(\mathcal{R}\{w_l\}) = \mathbb{E}\left[\left(\int_0^{T_s} w_I(t)\psi_I(t)dt - \int_0^{T_s} w_Q(t)\psi_Q(t)dt\right)\right. \tag{6.124}$$

$$\left.\left(\int_0^{T_s} w_I(\tau)\psi_I(\tau)d\tau - \int_0^{T_s} w_Q(\tau)\psi_Q(\tau)d\tau\right)\right] \tag{6.125}$$

$$= \int_0^{T_s}\int_0^{T_s} \mathbb{E}[w_I(t)w_I(\tau)]\psi_I(t)\psi_I(\tau)dtd\tau \tag{6.126}$$

$$+ \int_0^{T_s}\int_0^{T_s} \mathbb{E}[w_Q(t)w_Q(\tau)]\psi_Q(t)\psi_Q(\tau)dtd\tau \tag{6.127}$$

$$- \int_0^{T_s}\int_0^{T_s} \mathbb{E}[w_I(t)w_Q(\tau)]\psi_I(t)\psi_Q(\tau)dtd\tau \tag{6.128}$$

$$- \int_0^{T_s}\int_0^{T_s} \mathbb{E}[w_Q(t)w_I(\tau)]\psi_Q(t)\psi_I(\tau)dtd\tau. \tag{6.129}$$

As $w_I(t)$ and $w_Q(t)$ are independent zero-mean random processes, the last two terms evaluate to 0. Hence, we can write the variance of the real part of w_l as

$$\text{Var}(\mathcal{R}\{w_l\}) = \int_0^{T_s} \psi_I(\tau)\left(\int_0^{T_s} N_0\delta(t-\tau)\psi_I(t)dt\right)d\tau \tag{6.130}$$

$$+ \int_0^{T_s} \psi_Q(\tau)\left(\int_0^{T_s} N_0\delta(t-\tau)\psi_Q(t)dt\right)d\tau \tag{6.131}$$

$$= N_0 \int_0^{T_s} \psi_I^2(\tau) + \psi_Q^2(\tau) d\tau \tag{6.132}$$

$$= N_0 \int_0^{T_s} |\psi_l(\tau)|^2 d\tau \tag{6.133}$$

$$= N_0. \tag{6.134}$$

In the same fashion, we can show that the imaginary part of w_l has a variance of N_0 and that the real and imaginary parts are independent.

Previously, we developed the receiver structures and computed the average probability of errors for PAM, PSK, and QAM signaling using the bandpass signals. The same development can be done through the lowpass-equivalent signals with the new (complex) noise model. We illustrate this with a specific example.

Example 6.5

Let us consider the QPSK signaling we studied earlier in the chapter. We stated that such signaling can be characterized using the original bandpass signals with the two basis functions

$$\psi_1(t) = \sqrt{\frac{2}{E_p}} p(t) \cos(2\pi f_c t) \text{ and } \psi_2(t) = -\sqrt{\frac{2}{E_p}} p(t) \sin(2\pi f_c t), \tag{6.135}$$

with E_p denoting the energy of the pulse $p(t)$ and the constellation points given by

$$\begin{bmatrix} \sqrt{E_s} \\ 0 \end{bmatrix}, \begin{bmatrix} 0 \\ \sqrt{E_s} \end{bmatrix}, \begin{bmatrix} -\sqrt{E_s} \\ 0 \end{bmatrix}, \begin{bmatrix} 0 \\ -\sqrt{E_s} \end{bmatrix}, \tag{6.136}$$

where $E_s = \frac{A^2 E_p}{2}$ is the energy per symbol (with A being the amplitude of the signals used).

Describe the same signaling scheme using lowpass-equivalent representations.

Solution

The lowpass-equivalent forms of the four signals used are

$$Ap(t), \ jAp(t), \ -Ap(t), \ -jAp(t). \tag{6.137}$$

This is a one-dimensional signal set with a normalized basis function $p(t)/\sqrt{E_p}$. Noting that $A\sqrt{E_p} = \sqrt{2E_s}$, the signal constellation becomes

$$\sqrt{2E_s}, \ j\sqrt{2E_s}, \ -\sqrt{2E_s}, \ -j\sqrt{2E_s}. \tag{6.138}$$

We observe that with the use of the lowpass-equivalent representations, the energy of each signal is $2E_s$, which is twice the energy we computed previously for the actual bandpass signals transmitted. We note that there is no contradiction here as the lowpass-equivalents are only mathematical representations, and the actual energy transmitted needs to be computed from the original bandpass signals. Note also that this does not change the probability of error results, as the variance of the lowpass equivalent noise term is also doubled (compared to the original bandpass representations).

To see the last point more clearly, consider transmission over an AWGN channel. The received signal (in mathematically equivalent form) is given by

$$r_l = s_l + w_l, \tag{6.139}$$

where s_l is the transmitted signal (in the lowpass-equivalent form) and w_l is a circularly symmetric zero-mean complex Gaussian random variable with variance N_0 per dimension.

As an example, let us compute a pairwise error probability $P\left(\sqrt{2E_s} \rightarrow -\sqrt{2E_s}\right)$. This is the same as the probability of the real part of w_l being less than $-\sqrt{2E_s}$, hence we can write

$$P\left(\sqrt{2E_s} \rightarrow -\sqrt{2E_s}\right) = Q\left(\frac{\sqrt{2E_s}}{\sqrt{N_0}}\right) = Q\left(\sqrt{\frac{2E_s}{N_0}}\right). \tag{6.140}$$

Let us verify that this result matches the result with our previous approach for computing the error probability. As the squared Euclidean distance between the signals being considered here (in the original bandpass representation) is $2\sqrt{E_s}$, using the formula developed for the pairwise error probability in Chapter 5, the pairwise error probability under consideration becomes

$$Q\left(\frac{2\sqrt{E_s}}{\sqrt{2N_0}}\right) = Q\left(\sqrt{\frac{2E_s}{N_0}}\right), \tag{6.141}$$

which is the same as the result we obtained using the lowpass-equivalent representation in (6.140).

The model in (6.116) and (6.117) is valid when there is no phase introduced by the channel, that is, when the channel phase is estimated at the receiver and is compensated for perfectly. We know from our discussion at the beginning of this chapter that when $p(t)\cos(2\pi f_c t)$ is transmitted, the received signal will be of the form $p(t)\cos(2\pi f_c t + \phi)$, where ϕ is the phase introduced by the channel (due to residual error in the propagation delay estimate). More precisely, when the bandpass signal

$$s_m(t) = A_{m,r}\sqrt{\frac{2}{E_p}}p(t)\cos(2\pi f_c t) - A_{m,i}\sqrt{\frac{2}{E_p}}p(t)\sin(2\pi f_c t) \tag{6.142}$$

is transmitted, the received signal is given by

$$r(t) = A_{m,r}\sqrt{\frac{2}{E_p}}p(t)\cos(2\pi f_c t + \phi) - A_{m,i}\sqrt{\frac{2}{E_p}}p(t)\sin(2\pi f_c t + \phi) + n(t). \tag{6.143}$$

Since the lowpass equivalent of the transmitted signal $s_m(t)$ is given by

$$s_{m,l}(t) = A_{m,r}\sqrt{\frac{2}{E_p}}p(t) + jA_{m,i}\sqrt{\frac{2}{E_p}}p(t), \tag{6.144}$$

and the lowpass equivalent of $p(t)\cos(2\pi f_c t + \phi)$ is $p(t)e^{j\phi}$, and that of $p(t)\sin(2\pi f_c t + \phi)$ is $-jp(t)e^{j\phi}$, we can write the lowpass equivalent of the received signal (without noise) as $e^{j\phi}s_{m,l}(t)$. Therefore, we can write the lowpass-equivalent received signal as

$$r_l(t) = e^{j\phi}s_{m,l}(t) + w_l(t). \tag{6.145}$$

In terms of the mathematically equivalent form, this is nothing but

$$r_l = e^{j\phi}s_{m,l} + w_l. \tag{6.146}$$

We observe that the effect of the channel phase is to rotate the complex signal by an angle ϕ. In other words, the signal constellation gets rotated by the channel phase ϕ in the complex plane. If the phase term is available at the receiver (i.e., for coherent detection), the lowpass-equivalent received signal is multiplied by $e^{-j\phi}$ to remove the channel phase effect.

We further note that a (zero-mean, circularly symmetric) complex white Gaussian noise process multiplied by the complex number $e^{-j\phi}$ has the same statistics as the original one $w_l(t)$. Hence, there is no change in the study of the noise effects (see Problem 6.15).

6.6 Two Variations of QPSK Signaling

$\frac{\pi}{4}$-QPSK. With standard QPSK, when transmitting a long sequence of symbols, there would be many $\pm 180°$ phase shifts for both in-phase and quadrature components of the transmitted signal. These rapid phase changes are not desirable from an engineering perspective as they cause an abrupt change in the signal (causing large spectral sidelobes).[1] With this motivation, alternatives to standard QPSK signaling are also developed.

In $\frac{\pi}{4}$-QPSK, two QPSK constellations (i.e., the standard one and the rotated constellation) are used to modulate even and odd-indexed symbols, respectively. Namely, the transmitted signals selected for two consecutive symbols are from alternate constellations. As such, $\pm 180°$ phase shifts are avoided, and the phase shifts are limited to $\pm 135°$. Note that the implementation complexity of $\frac{\pi}{4}$-QPSK is the same as that of standard QPSK, and so is the error probability.

[1] We will discuss the spectral occupancy of digitally modulated signals in the next chapter.

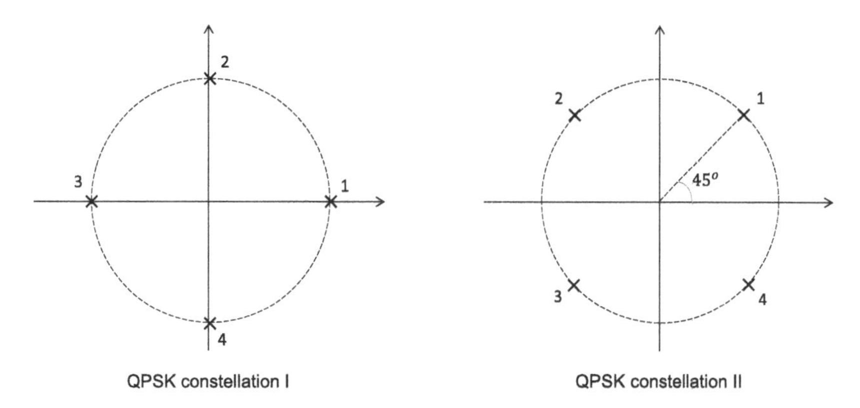

Figure 6.20 Two signal constellations used in $\frac{\pi}{4}$-QPSK signaling.

For example, consider the two signal constellations depicted in Fig. 6.20. Assuming that the transmitted sequence of symbols is

$$1, 2, 1, 3, 2, 4, 2, 3, 1, 4, 1, 2, \tag{6.147}$$

the phases of the transmitted signals with $\frac{\pi}{4}$-QPSK are

$$0, \frac{3\pi}{4}, 0, -\frac{3\pi}{4}, \frac{\pi}{2}, -\frac{\pi}{4}, \frac{\pi}{2}, -\frac{3\pi}{4}, 0, -\frac{\pi}{4}, 0, \frac{3\pi}{4}. \tag{6.148}$$

We observe that the use of two different constellations for alternating symbol periods limits the phase transitions, as intended.

Offset QPSK. We now describe offset QPSK (OQPSK) as another variation of QPSK signaling, eliminating $\pm 180°$ phase transitions. Let us express the symbols transmitted as bits and consider Gray mapping. With OQPSK, the two bits making up each 4-ary symbol are transmitted in the in-phase and quadrature components, which are staggered by half a symbol duration.

Denoting the symbol duration by T_s, half the symbol duration by $T_b = T_s/2$, the shaping pulse by $p(t)$, $t \in [0, 2T_b)$, the even-indexed bits by b_{2n}, and the odd-indexed bits by b_{2n+1}, $n \in \mathbb{Z}$, we can write the lowpass equivalent of the entire transmitted signal with OQPSK as

$$s_l(t) = A \sum_{n=-\infty}^{\infty} (2b_{2n} - 1)p(t - 2nT_b) + jA \sum_{n=-\infty}^{\infty} (2b_{2n+1} - 1)p(t - 2nT_b - T_b), \tag{6.149}$$

where mapping of the form $0 \to -1$ and $1 \to 1$ is employed.

In Fig. 6.21, the in-phase and quadrature components of the transmitted signal with OQPSK corresponding to a rectangular pulse shape are illustrated for transmission of the bit sequence $b_0, b_1, b_2, \ldots$ given by

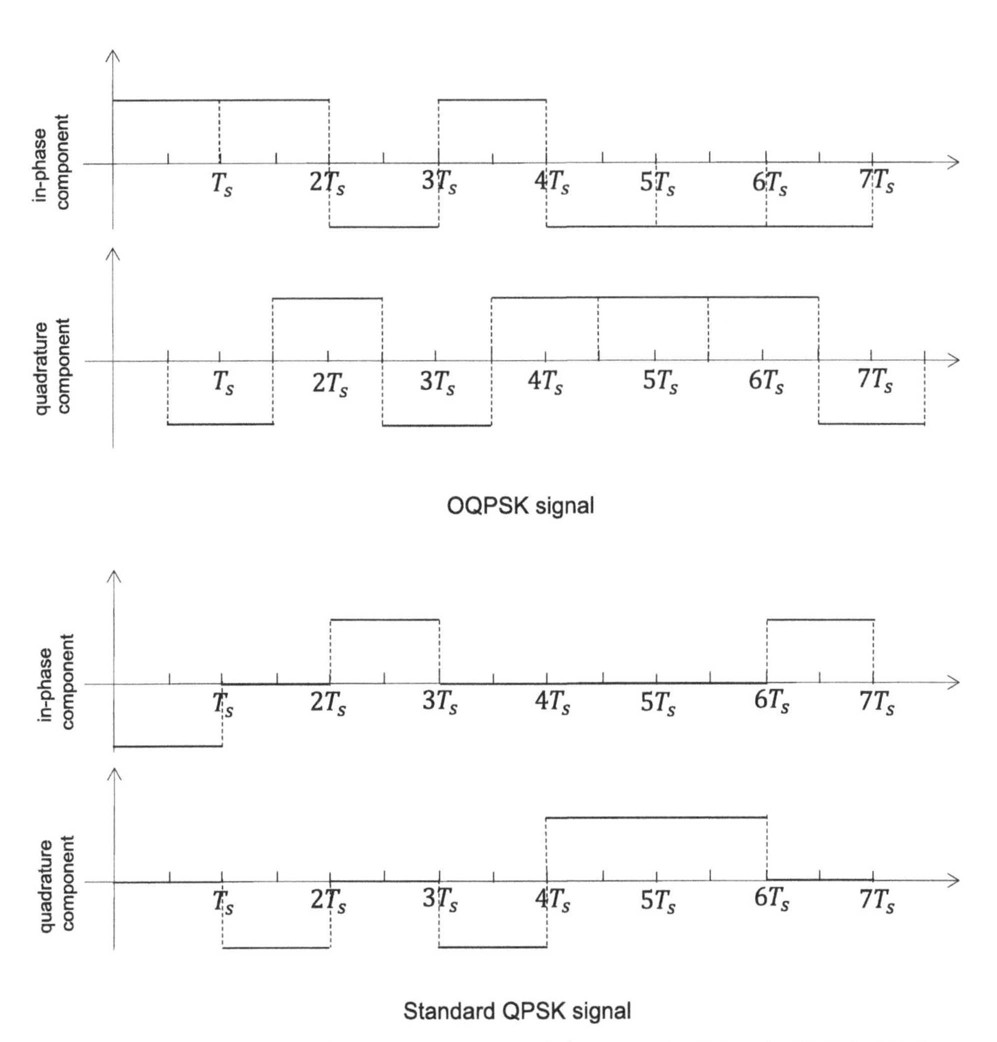

Figure 6.21 In-phase and quadrature components of the transmitted signal with OQPSK for the sequence given. Also depicted are the in-phase and quadrature components for the signal obtained with the standard QPSK.

$$1, 0, 1, 1, 0, 0, 1, 1, 0, 1, 0, 1, 0, 0. \tag{6.150}$$

Also shown in the figure are the in-phase and quadrature components of the transmitted signal with the standard QPSK transmission with $00 \rightarrow 1, 01 \rightarrow j, 10 \rightarrow -1$, and $11 \rightarrow -j$.

We observe that while there are $\pm 180°$ phase shifts with the standard QPSK transmission, with OQPSK, the phase transitions are limited to $\pm 90°$; however, they occur more frequently (twice as frequently as standard QPSK). Note also that, with perfect synchronization and compensation of the channel phase, the in-phase and quadrature components of the transmitted signal do not interfere, and they can be detected separately without bringing any additional implementation complexity.

6.7 Coherent versus Non-coherent and Differential Detection

As explained earlier in the chapter, the effect of propagation delay on bandpass communications is a channel phase shift. At this point, we know how to characterize this in terms of the actual bandpass signals and their lowpass equivalents. For example, in the absence of noise, when the transmitted signal is of the form $Ap(t)\cos(2\pi f_c t)$ (with lowpass equivalent $Ap(t)$), the received signal will be $Ap(t)\cos(2\pi f_c t + \phi)$ (whose lowpass equivalent is $Ae^{j\phi}p(t)$).

Based on the knowledge of the channel phase at the receiver, we distinguish among three different receiver structures: (1) coherent receiver, (2) non-coherent receiver, and (3) differentially coherent receiver.

Coherent Detection. When the channel phase is available and is compensated for at the receiver, the resulting receiver is referred to as *coherent*. This is applicable when the channel phase is constant or slowly changing, so that it can be tracked accurately at the receiver. There are highly effective techniques to accomplish this. In this chapter, we have focused on coherent detection up to now.

Non-coherent Detection. In this case, the channel phase is not available at the receiver, and hence, it is not employed in the detection process. Using the lowpass-equivalent notation, the relevant channel model is given by

$$r_l(t) = e^{j\phi} s_{m,l}(t) + w_l(t), \tag{6.151}$$

where ϕ is a random variable, usually taken as uniform on $[0, 2\pi)$, which corresponds to having no information about the phase introduced by the channel. Since the channel phase (i.e., phase rotation of the transmitted signal) is unknown, no information can be carried in the signal phase in this case. Thus, we cannot employ the usual PAM, PSK, or QAM signaling schemes. On the other hand, we can embed digital information on signal amplitudes.

Let us give an example employing on–off signaling.

Example 6.6

Consider a binary modulation scheme which uses $s_0(t) = 0$ (to transmit the bit 0) and $s_1(t) = Ap(t)\cos(2\pi f_c t)$ (to transmit the bit 1), $t \in [0, T_b)$, where A is an amplitude level, T_b is the bit period (with $T_b \gg 1/f_c$), $p(t)$ is a lowpass shaping pulse, and f_c is the carrier frequency. This is nothing but an on–off signaling scheme. Assume that the transmission takes place over an AWGN channel, and the transmitted bits are equally likely. Also, assume that the bandwidth of $p(t)$ is W with $W \ll f_c$ and that non-coherent detection is employed.

Determine the optimal receiver structure that minimizes the average bit error probability.

Solution

In terms of the lowpass equivalents, we have $s_{0,l}(t) = 0$ and $s_{1,l}(t) = Ap(t)$. Using the basis function $\psi_l(t) = \frac{p(t)}{\sqrt{E_p}}$ with E_p denoting the energy of the pulse $p(t)$, and noting that the average energy per bit is $E_b = \frac{1}{4}A^2E_p$, we can write the two signals in terms of their mathematically equivalent forms:

$$s_{0,l} = 0 \text{ and } s_{1,l} = 2\sqrt{E_b}. \tag{6.152}$$

Then, the received signal is given by

$$r_l = e^{j\phi}s_{i,l} + w_l, \tag{6.153}$$

when $s_i(t)$ is transmitted. Here, ϕ is a uniform random variable on $[0, 2\pi)$ (as non-coherent detection is employed), and w_l is a circularly symmetric complex Gaussian noise with zero mean and variance N_0 per dimension.

To determine the optimal receiver structure, we employ the ML decision rule (as the bits are equally likely), that is, we compare for conditional PDFs $f(r_l|0 \text{ is sent})$ and $f(r_l|1 \text{ is sent})$, and pick the bit corresponding to the larger one as our decision. We can write these conditional PDFs as

$$f(r_l|0 \text{ is sent}) = \frac{1}{2\pi N_0}\exp\left\{-\frac{|r_l|^2}{2N_0}\right\} \tag{6.154}$$

and

$$f(r_l|1 \text{ is sent}) = \int_0^{2\pi} \frac{1}{2\pi N_0}\exp\left\{-\frac{|r_l - 2\sqrt{E_b}e^{-j\theta}|^2}{2N_0}\right\}\frac{1}{2\pi}d\theta \tag{6.155}$$

$$= \frac{1}{2\pi N_0}\exp\left\{-\frac{|r_l|^2}{2N_0}\right\}e^{-2E_b/N_0}\int_0^{2\pi}\frac{1}{2\pi}\exp\left\{\frac{2\sqrt{E_b}\mathcal{R}(r_l e^{j\theta})}{N_0}\right\}d\theta. \tag{6.156}$$

Simplifying, we obtain

$$\frac{f(r_l|1 \text{ is sent})}{f(r_l|0 \text{ is sent})} = e^{-2E_b/N_0}\frac{1}{2\pi}\int_0^{2\pi}\exp\left\{\frac{2\sqrt{E_b}|r_l|\cos(\theta + \angle r_l)}{N_0}\right\}d\theta \tag{6.157}$$

$$= e^{-2E_b/N_0}\frac{1}{2\pi}\int_0^{2\pi}\exp\left\{\frac{2\sqrt{E_b}|r_l|\cos(\theta)}{N_0}\right\}d\theta \tag{6.158}$$

$$= e^{-2E_b/N_0}I_0\left\{\frac{2\sqrt{E_b}}{N_0}|r_l|\right\}, \tag{6.159}$$

where $I_0(x)$ is the modified Bessel function of the first kind of order 0, defined as

$$I_0(x) = \frac{1}{2\pi}\int_0^{2\pi} e^{x\cos\theta}d\theta. \tag{6.160}$$

To summarize, the ML decision rule (for a non-coherent receiver) becomes

$$I_0 \left\{ \frac{2\sqrt{E_b}}{N_0} |r_l| \right\} \underset{0}{\overset{1}{\gtrless}} e^{2E_b/N_0}. \tag{6.161}$$

Noting that $I_0(x)$ is an increasing function, we notice that this rule is nothing but a comparison of the magnitude of the received signal with a threshold (which can be written using the inverse of the modified Bessel function of the first kind of order zero).

Using standard approaches, we can also compute the average bit error probability. However, the resulting expressions are not easy to manipulate, and the results cannot be written in closed form.

Let us close our discussion on non-coherent detection by noting that since there is no information on the phase introduced by the channel and we are using less information than the coherent case, the resulting performance is inferior to that of a coherent receiver. We also note that non-coherent detection is not typical for a single-carrier communication system. This is because, in most cases, we have (at least some) information on the channel phase, and we can utilize this information for signal detection at the receiver.

Differential Detection. For the case of differential detection, we assume that the channel phase is still unknown but slowly changing so that it can be taken as a constant over two consecutive symbol transmissions. As the channel phase is identical over consecutive symbol intervals, we can embed information on the phase differences between successive symbols and detect the phase changes among them at the receiver. Such a receiver is called a differential receiver (or differentially coherent receiver). We discuss this approach in some detail in the next section.

6.8 Differential Phase-Shift Keying

In differential phase-shift keying (DPSK), information is carried on the phase changes of the sinusoidal signal in consecutive time intervals. For instance, considering M-ary signaling, the phase is changed by $2(i-1)\pi/M$ for the ith symbol, $i = 1, 2, \ldots, M$. Let θ_k denote the phase used in the kth time interval, that is, the transmitted signal is given by

$$x(t) = Ap(t)\cos(2\pi f_c t + \theta_k), \quad \text{for } t \in [(k-1)T_s, kT_s). \tag{6.162}$$

Assuming that the next symbol being transmitted is i, the signal transmitted in the next time interval is given by

$$x(t) = Ap(t)\cos(2\pi f_c t + \theta_{k+1}), \quad \text{for } t \in [kT_s, (k+1)T_s), \tag{6.163}$$

where $\theta_{k+1} = \theta_k + \phi_i$, with $\phi_i = \frac{2\pi}{M}(i-1)$. Using the lowpass-equivalent representations, we have

$$x_l(t) = Ap(t)e^{j\theta_k}, \text{ for } t \in [(k-1)T_s, kT_s) \tag{6.164}$$

and

$$x_l(t) = Ap(t)e^{j\theta_{k+1}}, \text{ for } t \in [(kT_s, (k+1)T_s). \tag{6.165}$$

Using the basis function $\frac{p(t)}{\sqrt{E_p}}$ and noting that $E_s = \frac{A^2 E_p}{2}$, with E_p denoting the energy of the pulse $p(t)$ in one symbol interval, the mathematically equivalent transmitted signals (in consecutive time intervals) are given by the complex numbers

$$\sqrt{2E_s}e^{j\theta_k} \text{ and } \sqrt{2E_s}e^{j\theta_{k+1}}. \tag{6.166}$$

The received signals for the two consecutive symbol periods are given by

$$r_{l,k} = \sqrt{2E_s}e^{j(\theta_k+\phi)} + w_{l,k} \tag{6.167}$$

and

$$r_{l,k+1} = \sqrt{2E_s}e^{j(\theta_{k+1}+\phi)} + w_{l,k+1}, \tag{6.168}$$

where ϕ is the unknown channel phase.

At the receiver, we multiply the received signal in the $(k+1)$th time interval with the conjugate of the one for the kth interval to cancel out the unknown channel phase and recover the transmitted symbol. Namely, we compute

$$\begin{aligned} r_{l,k+1}r_{l,k}^* = {} & 2E_s e^{j(\theta_{k+1}-\theta_k)} + \sqrt{2E_s}\left(e^{j(\theta_{k+1}+\phi)}w_{l,k}^* + e^{-j(\theta_k+\phi)}w_{l,k+1}\right) \\ & + w_{l,k+1}w_{l,k}^*, \end{aligned} \tag{6.169}$$

and make a decision based on the angle of the complex number $r_{l,k+1}r_{l,k}^*$.

An exact analysis of the resulting error probability is difficult to carry out (due to the product of two complex Gaussian random variables). However, a simple approximation is easy to obtain. For the practical scenario of high signal-to-noise ratios, the product of two independent noise terms can be neglected as the noise terms will be small. Also, scaling by $\sqrt{2E_s}$, we can approximate the decision variable as

$$\approx \sqrt{2E_s}e^{j\phi_i} + e^{j(\theta_{k+1}+\phi)}w_{l,k}^* + e^{-j(\theta_k+\phi)}w_{l,k+1}. \tag{6.170}$$

Noting that for a circularly symmetric complex Gaussian random variable, complex conjugation or multiplication by a complex exponential does not change its statistics, the effective noise is the sum of two independent complex Gaussian random variables each with variance N_0 per dimension. In other words, the approximate decision variable is given by

$$\sqrt{2E_s}e^{j\phi_i} + w', \tag{6.171}$$

with w' being a complex Gaussian random variable with variance $2N_0$ per dimension. This is the same as the received signal for coherent detection of PSK signals, except that the noise variance is doubled. Therefore, we can assert that the error probability with differential detection is approximately 3 dB worse than that of coherent PSK (which was studied in detail in Section 6.2.2).

This approximation is reasonably accurate for higher-order constellations (when $M > 4$); however, it is overly pessimistic for the binary case. An exact analysis for the binary case shows that the error probability with differential detection is

$$P_b = \frac{1}{2}e^{-\gamma_b}, \tag{6.172}$$

which is, for high signal-to-noise ratios, less than 1 dB worse than the coherent detection case for BPSK (for which $P_b = Q(\sqrt{2\gamma_b})$).

As the loss with BPSK is small, and the complexity of carrier phase estimation is eliminated, binary DPSK is a good choice in practice compared to coherently detected BPSK. For higher-order PSK schemes, the loss is closer to 3 dB.

6.9 Carrier Phase Synchronization

As discussed earlier in this chapter, a communication channel will introduce a phase shift for bandpass communications due to the (residual) propagation delay. For phase-coherent demodulation, the introduced channel phase must be estimated and compensated for at the receiver. In this section, we briefly describe how this can be accomplished by first considering the simplified case of an unmodulated carrier transmission and then discussing the more interesting case of modulated signals.

Tracking the Phase of an Unmodulated Carrier. Let us first consider the received signal when an unmodulated carrier signal is transmitted over an AWGN channel. That is,

$$r(t) = A_c \cos(2\pi f_c t + \phi) + n(t), \tag{6.173}$$

where A_c is the carrier amplitude and f_c is the carrier frequency, and ϕ is the unknown channel phase, which is to be estimated. Since the noise term is white Gaussian, the maximum likelihood estimator minimizes the squared Euclidean distance between the received signal and the "clean" carrier signal, that is, we need to select ϕ that minimizes

$$\int_{T_o} (r(t) - A_c \cos(2\pi f_c t + \phi))^2 dt, \tag{6.174}$$

where T_o is the observation window. Equivalently, we maximize

$$\int_{T_o} r(t)\cos(2\pi f_c t + \phi)dt = \left(\int_{T_o} r(t)\cos(2\pi f_c t)dt\right)\cos(\phi) - \left(\int_{T_o} r(t)\sin(2\pi f_c t)dt\right)\sin(\phi). \tag{6.175}$$

Differentiating with respect to ϕ and setting the result to zero, we obtain

$$\hat{\phi} = -\tan^{-1}\left\{\frac{\int_{T_o} r(t)\sin(2\pi f_c t)dt}{\int_{T_o} r(t)\cos(2\pi f_c t)dt}\right\} \tag{6.176}$$

as the ML estimate of the channel phase.

We can also estimate the channel phase using a phase-locked loop (PLL), depicted in Fig. 6.22. Selecting the voltage controlled oscillator output as $-2\sin(2\pi f_c t + \hat{\phi}(t))$ with $\hat{\phi}(t) = K\int_{-\infty}^{t} v(\tau)d\tau$, and ignoring the noise component and taking $A_c = 1$, the error signal becomes

$$e(t) = -2\cos(2\pi f_c t + \phi)\sin(2\pi f_c t + \hat{\phi}) \tag{6.177}$$

$$= \sin(\phi - \hat{\phi}) - \sin(4\pi f_c t + \phi + \hat{\phi}). \tag{6.178}$$

With a lowpass loop filter, the double frequency term in the error signal is removed, giving rise to the simplified model depicted in Fig. 6.23.

Let us give an intuitive explanation of how the PLL works. Assume that the actual channel phase ϕ is larger than the estimated one $\hat{\phi}$. If the phase difference is not too high, this would imply that $\sin(\phi - \hat{\phi})$ is larger than zero; hence, the integrator input will be larger than zero as well, thereby increasing the value of the estimated phase, and making it closer to the actual one. If the actual phase is smaller than the estimated one, assuming that the difference is not too high, $\sin(\phi - \hat{\phi})$ will be negative, hence making the integrator input negative, reducing the value of the estimated phase (i.e., making it closer to the actual channel phase). Therefore, in

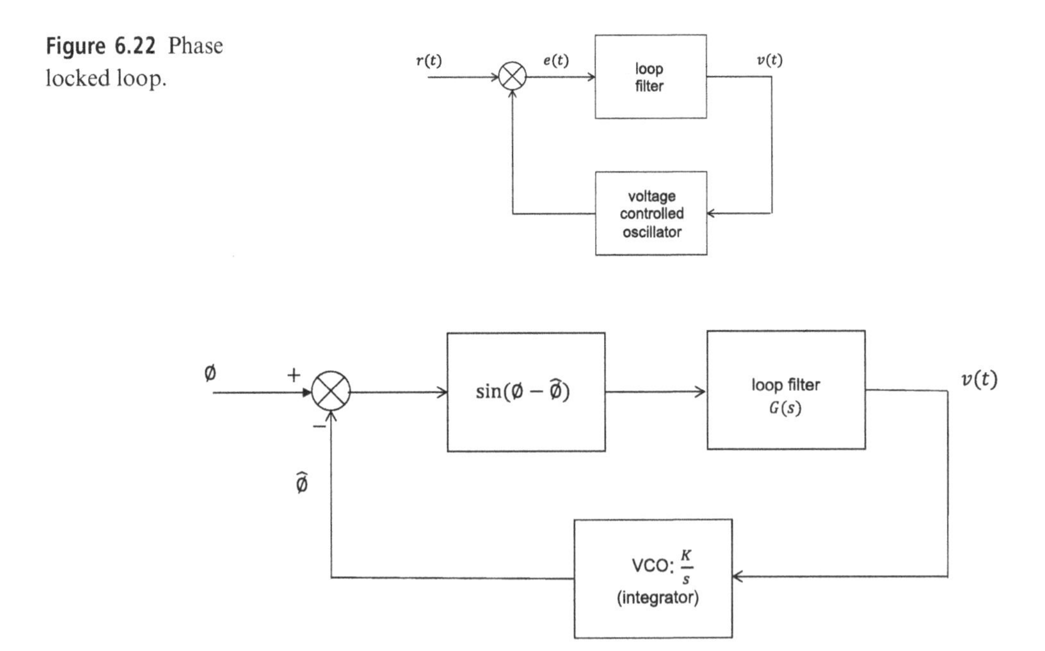

Figure 6.22 Phase locked loop.

Figure 6.23 Modeling of a PLL.

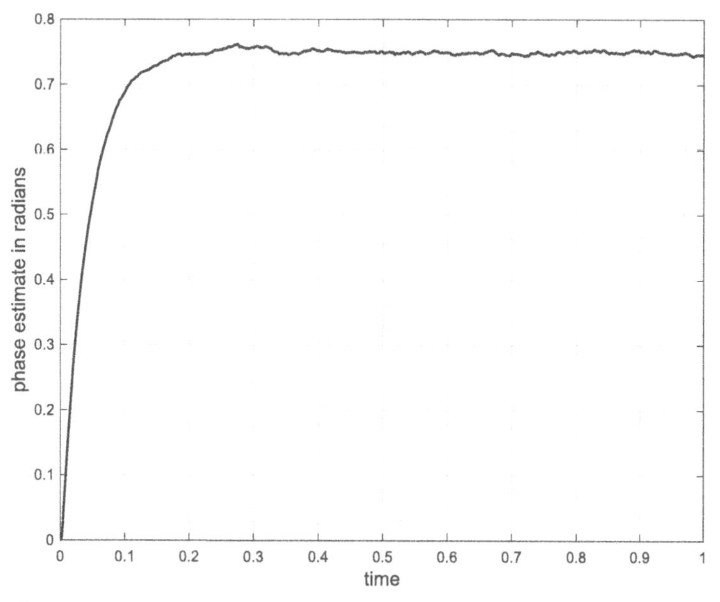

Figure 6.24 Sample output of a PLL tracking the phase of an unmodulated carrier with $\phi = 0.75$ radians in white Gaussian noise.

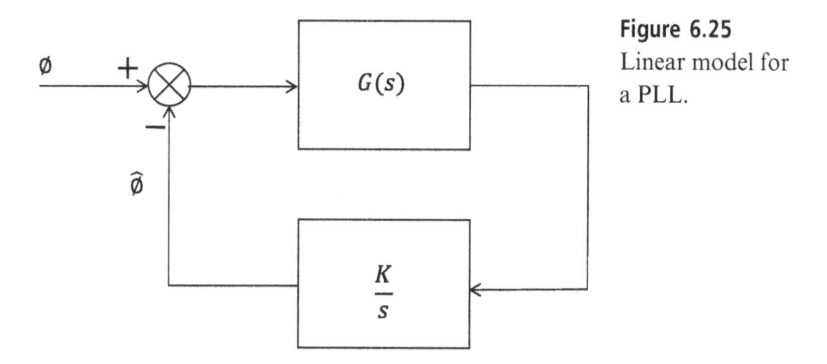

Figure 6.25 Linear model for a PLL.

either case, the phase estimate will become closer and closer to the actual channel phase as time passes.

A sample output of a PLL used to estimate the phase of an unmodulated carrier (with $\phi = 0.75$ radians) in the presence of AWGN is shown in Fig. 6.24. The estimated phase starts with an initial guess of $\hat{\phi} = 0$, then catches up with the correct channel phase, remaining at lock despite the channel noise.

When the PLL is in lock, at steady state, $\hat{\phi} \approx \phi$, hence we have $\sin(\phi - \hat{\phi}) \approx \phi - \hat{\phi}$, and we obtain a linear approximation to the PLL as shown in Fig. 6.25.

The overall transfer function (from ϕ to $\hat{\phi}$) can easily be computed to be

$$H(s) = \frac{KG(s)/s}{1 + KG(s)/s}. \tag{6.179}$$

With a sufficiently high loop gain K and a proper selection of the loop filter, we can conclude that $H(s) \approx 1$, hence $\hat{\phi} \approx \phi$ as desired.

We also note that, in practice, the channel phase will vary (slowly) with time, and a PLL can be used to track these variations effectively.

Tracking the Phase of a Modulated Signal. We now focus on estimating the carrier phase for a modulated signal. Assume that the received signal is of the form

$$r(t) = A(t)\cos(2\pi f_c t + \phi) + n(t), \tag{6.180}$$

where $A(t)$ is a zero-mean random process and $n(t)$ is the AWGN term. This model is suitable for both analog transmission with double-sideband suppressed carrier AM and for digital transmission through pulse amplitude modulation. We are interested in estimating the unknown phase ϕ.

Since the random signal $A(t)$ is a zero-mean process, it cannot be used as input to a PLL; hence, we need to take a different approach. One approach is to use a *squaring loop*, that is, square the received signal, bandpass filter the resulting signal using a narrowband filter tuned to $2f_c$, then feed it to a PLL. A simple calculation shows that this approach makes it possible to estimate 2ϕ. As a drawback, since we are estimating twice the channel phase, there would be an ambiguity of π in the phase estimate.

Another (non-decision-directed) approach is to use a *Costas loop*. The idea is to multiply the received signal by both cosine and sine signals, and, from these two products, to obtain a reference signal that can be used to derive a phase-locked loop.

In the case of digital modulation, it is also possible to develop decision-directed methods for carrier phase recovery. In this case, some transmitted bits/symbols can be employed as pilots (i.e., reference signals known a priori), and an estimate of the channel phase can be obtained, similar to the ML estimation of the unmodulated carrier phase. Once an estimate is available, it can be used for demodulation, and the decisions on the bits/symbols can be utilized to track the channel phase following the same approach. We do not go into the details of decision-directed carrier phase recovery methods.

6.10 Chapter Summary

In this chapter, we studied the basics of bandpass communications using single-carrier modulation techniques, including the common examples of PSK and QAM signaling. We described the effects of communication channels on bandpass signals and gave a way of representing bandpass signals using lowpass equivalents with respect to a reference frequency. The lowpass-equivalent representations offer a convenient way to study bandpass communication systems, as will be further exploited in subsequent chapters. In addition, we presented variations of QPSK, which find applications in practical communication systems. We distinguished between coherent, non-coherent, and differentially coherent reception of PSK signals and studied differential PSK in some detail. Finally, we concluded the chapter with a brief coverage of carrier phase synchronization methods.

PROBLEMS

6.1 We are interested in designing a bandpass PAM scheme using a pulse with a single sideband. Recall our discussion in Chapter 3 (specifically, in Section 3.1 on the generation of upper-sideband and lower-sideband SSB AM signals), and determine the pulses that can be employed for PAM with

(a) the upper-sideband signal,

(b) the lower-sideband signal.

Express the pulses to be used in terms of a basic pulse $p(t)$, its Hilbert transform $\hat{p}(t)$, and the carrier frequency f_c.

6.2 Compare the 16-PSK and 16-QAM (with a square constellation) in terms of their error rate performance. Which one is more power efficient? By how much (expressed in dB)?

6.3 Consider binary PAM with transmitted signal of the form

$$\pm Ap(t)\cos(2\pi f_c t),$$

where A is a constant, $p(t)$ is a basic pulse of duration T_b, and f_c is the carrier frequency. Assume that the transmitted bits are equally likely.

This scheme is employed over an AWGN channel where the channel phase introduced is ϕ. Assume that the receiver estimates the phase with some error and obtains $\hat{\phi} = \phi + \Delta\phi$. The receiver treats the estimated phase $\hat{\phi}$ as correct and implements the ML receiver for demodulation.

Determine the average bit error probability in terms of E_b/N_0 and $\Delta\phi$. What happens if $\Delta\phi = \pi/2$? How about when $\Delta\phi = 0$?

6.4 This problem is similar to Problem 6.3, except that we will employ a single sideband pulse for binary PAM. Consider a binary PAM scheme with transmitted signal of the form

$$\pm A\left(p(t)\cos(2\pi f_c t) + \hat{p}(t)\sin(2\pi f_c t)\right),$$

where A is a constant, $p(t)$ is a basic pulse of duration T_b, f_c is the carrier frequency, and $\hat{p}(t)$ is the Hilbert transform of $p(t)$. As this uses a single-sideband signal for transmission, it is more advantageous in terms of bandwidth under ideal conditions.

This scheme is employed to transmit equally likely bits over an AWGN channel where the channel phase introduced is ϕ. Assume that the receiver estimates the phase with some error and obtains $\hat{\phi} = \phi + \Delta\phi$. The receiver treats the estimated phase $\hat{\phi}$ as correct and implements the ML detection.

Determine the average bit error probability in terms of the given parameters. (Note that the result will be in a more complicated form compared to the solution of the previous problem, and it will involve the correlation of $p(t)$ and $\hat{p}(t)$ over the bit duration.) Specialize your answer to the case with $\Delta\phi = 0$.

6.5 Show that if $x_l(t)$ is the lowpass equivalent of the bandpass signal $x(t)$, then its real and imaginary parts (i.e., the in-phase and quadrature components) can be written as

$$x_I(t) = x(t)\cos(2\pi f_c t) + \hat{x}(t)\sin(2\pi f_c t),$$
$$x_Q(t) = \hat{x}(t)\cos(2\pi f_c t) - x(t)\sin(2\pi f_c t),$$

where $\hat{x}(t)$ is the Hilbert transform of $x(t)$.

6.6 Find the lowpass equivalent of the following bandpass signals. Assume that the reference frequency is f_c. (Assume that $g(t), g_1(t), g_2(t)$ are lowpass signals.)
(a) $g(t)\cos(2\pi f_c t)$.
(b) $g(t)\sin(2\pi f_c t)$.
(c) $Ag_1(t)\cos(2\pi f_c t) + Bg_2(t)\sin(2\pi f_c t)$.

6.7 Determine the lowpass equivalent of the following signals with respect to the reference frequency f_c:
(a) $\cos(2\pi(f_c + \Delta f)t + \theta)$;
(b) $\sin(2\pi(f_c + \Delta f)t + \theta)$.

6.8 Consider a bandpass white Gaussian noise $v(t)$ with power spectral density

$$S_v(f) = \begin{cases} \frac{N_0}{2}, & \text{for } |f| \in [f_c - W_1, f_c + W_2], \\ 0, & \text{else.} \end{cases}$$

Clearly, the bandwidth of the random process is $W_1 + W_2$. Determine the power spectral densities of the in-phase and quadrature components with respect to the reference frequency f_c. Also, determine their average power contents.

6.9 Four equally likely signals are transmitted over an additive noise channel using the waveforms

$$s_1(t) = 0,$$
$$s_2(t) = -2A\cos(2\pi f_0 t),$$
$$s_3(t) = 2A\sin(2\pi f_0 t),$$
$$s_4(t) = 2A\cos(2\pi f_0 t),$$

for $t \in [0, T_s)$, T_s being the symbol period. The noise is white Gaussian with power spectral density $N_0/2$, and $f_0 = N/T_s$ where $N \gg 1$ is an integer. Assume that the receiver employs coherent detection and there is no synchronization error, that is, the received signal is given by

$$r(t) = s_i(t) + n(t),$$

where $s_i(t)$ is the transmitted signal and $n(t)$ is the noise process.
(a) Find an orthonormal set of basis functions for this signal set and plot the signal constellation.
(b) Describe the optimal receiver structure (give a complete block diagram, and be specific).
(c) Determine the exact (conditional) probability of error given that $s_1(t)$ is transmitted. Express your answer in terms of $\bar{E}_b/N_0$, where $\bar{E}_b$ is the average energy per bit.

6.10 Solve Problem 6.9 using the lowpass-equivalent representations of the transmitted signals and the additive Gaussian noise (using the reference frequency as f_0).

6.11 Consider a ternary (i.e., $M = 3$) communication system employing

$$s_1(t) = g(t)\cos(2\pi f_c t),$$
$$s_2(t) = g(t)\sin(2\pi f_c t),$$
$$s_3(t) = -g(t)\cos(2\pi f_c t) - g(t)\sin(2\pi f_c t),$$

for $t \in [0, T_s)$, where T_s is the symbol period, to transmit the symbols 1, 2, and 3, respectively, over an AWGN channel.

(a) Find the vector representation of the transmitted signals (using an orthonormal basis) and plot the signal constellation.

(b) Determine the optimal receiver structure. Simplify as much as possible.

6.12 Four equally likely symbols are transmitted using QPSK modulation. Assume that the channel phase is estimated with an error of $\Delta\phi$, and the estimated phase is used as if it is the correct one at the receiver to demodulate the symbols. Determine a tight upper bound on the symbol probability in terms of E_b/N_0 and $\Delta\phi$, and compare it with the case with no error in the channel phase estimation.

6.13 Consider the 8-QAM cross constellation shown in Fig. 6.12. The constellation points are selected on two different circles. The points on the inner circle of radius r_i are picked with phases $\frac{\pi}{4}, \frac{3\pi}{4}, \frac{5\pi}{4}, \frac{7\pi}{4}$, while the four points on the outer circle with radius r_o are picked with phases $0, \frac{\pi}{2}, \pi, \frac{3\pi}{2}$.

Determine the optimal ratio r_o/r_i to maximize the normalized minimum distance of the constellation, that is, to minimize the average error probability at high signal-to-noise ratios over an AWGN channel.

6.14 Consider a 4-ary signaling scheme using the signals

$$p(t)\cos(2\pi f_c t + \theta_k),$$

where θ_k is selected as $0, \frac{\pi}{4}, \pi, \frac{7\pi}{4}$ to represent the four symbols. Assume that the symbols are equally likely.

Determine the union bound on the average error probability and compare it with that of the standard QPSK modulation. Express your answers in the signal-to-noise ratio E_s/N_0, and comment on your result.

6.15 Let X be a circularly symmetric complex Gaussian random variable with zero mean and unit variance, and θ be an arbitrary (deterministic) phase. Show that the random variable $Y = e^{-j\theta}X$ has the same distribution as the random variable X.

6.16 Four equally likely symbols are transmitted using QPSK over an AWGN channel. Assuming that the mapping of bits to symbols follows natural binary coding, that is, 00, 01, 10, and 11 are mapped to the phases $0, \pi/2, \pi$, and $3\pi/2$, respectively, determine the average bit error probability in terms of the signal-to-noise ratio per bit $\gamma_b = E_b/N_0$. Compare the result with the case of Gray mapping.

6.17 Determine the phases of the transmitted signals with $\frac{\pi}{4}$-QPSK corresponding to the sequence of symbols

$$3, 1, 2, 2, 3, 3, 2, 2, 4, 1, 2, 4, 1, 2,$$

assuming that the two QPSK constellations given in Fig. 6.20 are used.

6.18 Plot the in-phase and quadrature components of a transmitted signal with OQPSK modulation corresponding to the transmission of the bit sequence

$$1, 1, 0, 0, 0, 1, 0, 1, 1, 0, 0$$

(a) using a rectangular shaping function,
(b) using a half-sinusoid shaping pulse given by

$$p(t) = \sin\left(\frac{\pi t}{2T_b}\right), \quad \text{for } t \in [0, 2T_b),$$

where T_b is the bit duration.

6.19 Determine the sequence of phases transmitted with 4-ary differential PSK corresponding to the sequence of symbols

$$1, 4, 1, 3, 1, 4, 2, 2, 4, 1, 3.$$

6.20 Consider a DPSK scheme with $M = 4$. Assume that the sequence of phases transmitted is

$$\pi, \pi, 0, \frac{\pi}{2}, \frac{3\pi}{2}, \frac{3\pi}{2}, \pi, 0, \frac{\pi}{2}, \pi, 0, \pi,$$

with an initial phase of 0. Determine the sequence of symbols being transmitted.

6.21 Consider the on–off signaling scheme studied in Example 6.6. Compute the conditional error probability given that bit 0 is transmitted.

6.22 Consider the on–off signaling scheme of Example 6.6. Assume that the prior probabilities of the transmitted bits are p_0 and $p_1 = 1 - p_0$, respectively. Determine the optimal decision rule to minimize the bit error probability.

6.23 Assume that for a particular channel, the lowpass-equivalent received signal is given by

$$r_l(t) = e^{j\phi} s_l(t) + z(t),$$

where $s_l(t)$ is the lowpass-equivalent transmitted signal, ϕ is the phase that the channel introduces, and $z(t)$ is a zero-mean complex Gaussian random process. The real and imaginary parts of $z(t)$ are independent, and each component has a power spectral density of N_0.

Assume that the channel phase is estimated as $\hat{\theta}$; however, there is an ambiguity of π. Namely, if the channel phase is θ for a particular transmission, we cannot differentiate between θ and $\theta + \pi$ at the receiver. (Assume that we have the correct phase with probability 1/2; it is changed by π with the same probability.)

(a) Without making any computations, do you think BPSK is a proper modulation scheme over this channel? How about orthogonal signaling? How about on–off signaling?

(b) Assume that binary orthogonal signaling with $s_1(t) = \sin(2\pi f_c t)$ and $s_2(t) = \cos(2\pi f_c t)$ is used over this channel. Describe the optimal ML detector at the receiver.

6.24 In this problem, the objective is to determine the error probability of differentially detected BPSK signaling with equally likely bits. To accomplish this, we note that this problem is equivalent to the non-coherent detection of orthogonal signals over two symbol periods. That is, one can consider the lowpass-equivalent form of the transmitted signal vectors as $[\sqrt{2E_b}\ \ \sqrt{2E_b}]$ and $[\sqrt{2E_b}\ \ -\sqrt{2E_b}]$. Therefore, with a change of basis, equivalently, we have the transmitted signals as $[2\sqrt{E_b}\ \ 0]$ and $[0\ \ 2\sqrt{E_b}]$. Assuming that the first signal is transmitted, the received signal becomes

$$[r_1\ \ r_2] = \left[2e^{j\phi}\sqrt{E_b} + w_1\ \ \ w_2\right],$$

where ϕ is the channel phase and w_1 and w_2 are independent zero-mean complex Gaussian random variables with variance $2N_0$ each. The error probability with non-coherent detection is then given by the probability of the event that $|e^{j\phi}2\sqrt{E_b} + w_1| < |w_2|$.

(a) Determine the probability density functions of $|2e^{j\phi}\sqrt{E_b} + w_1|$ and $|w_2|$, and show that the first random variable is Ricean while the second one is Rayleigh distributed.

(b) Show that the error probability of differentially detected BPSK is given by $P_b = \frac{1}{2}e^{-E_b/N_0}$ by computing the probability

$$\mathbb{P}\left(|e^{j\phi}\sqrt{2E_b} + w_1| < |w_2|\right).$$

6.25 Assume that a phase-locked loop is used to estimate the carrier phase ϕ in demodulation of a binary PSK signal received in white Gaussian noise.

(a) Determine the effect of a phase error $\phi - \hat{\phi}$ on the probability of error, that is, what is the probability of error in terms of E_b/N_o and the phase error $\phi - \hat{\phi}$?

(b) What is the loss in SNR if the phase error $\phi - \hat{\phi} = 45°$ (expressed in dB)?

(c) Assume that the phase error $\phi - \hat{\phi}$ (in degrees) is a discrete random variable with probability density function $f_X(x) = 0.8\delta(x) + 0.1\delta(x - 30) + 0.1\delta(x + 30)$. Determine the average error rate as a function of the signal-to-noise ratio E_b/N_o. Compare your result with the case of no phase error (at high SNRs). How much is the loss (in dB)? (Note that this distribution for the phase error is not realistic. Nevertheless, this PDF is selected to make the calculations easier.)

COMPUTER PROBLEMS

6.26 Using the mathematically equivalent model for a single-carrier bandpass communication system over an AWGN channel (i.e., employing suitable two-dimensional signal constellations along with the equivalent noise vector), simulate the symbol error rate performance of phase-shift keying with $M = 2, 4, 8$, and 16. Also, simulate the bit error rate performance of these modulation schemes with Gray mapping and compare them with the theoretical bit error probabilities (which are available in an exact form for $M = 2$ and 4, and as an upper bound for $M = 8$ and 16).

6.27 Simulate the error probabilities of QAM with square constellations with $M = 16$ and 64 over an AWGN channel. Compare the simulation results with the analytical ones.

6.28 Consider the on–off signaling scheme studied in Example 6.6 where non-coherent reception was employed. Using the lowpass-equivalent representations, write a MATLAB code to simulate the error probability of the absolute value detector derived in the example. Compare the simulation results with the case of coherent reception and estimate the performance difference between coherent and non-coherent detection for on–off signaling. You may use the theoretical results for the coherent reception case for simplicity.

6.29 Simulate the differentially detected binary PSK modulation performance over an AWGN channel and compare your simulation results with theoretical expectations.

6.30 Simulate the symbol error rate performance of differentially detected M-ary PSK modulation with $M = 8$ and 16 over an AWGN channel. How much degradation do you observe with respect to coherent detection at high signal-to-noise ratios?

6.31 In this problem, the objective is to study carrier phase synchronization for a single-carrier bandpass communication system. We simulate a continuous-time system in discrete time with a sufficiently high sampling rate. Below is a MATLAB code that implements a phase-locked loop to track the phase of a sinusoidal signal.

```
fc = 1000;   % Frequency of the sinusoid
delta_t = 2*10^-4;   % Time step used in the simulation (greater than the
Nyquist rate)
sigma = .05;   % Standard deviation of the additive Gaussian noise for each
sample
T = 1;   % Duration of the signal
t = 0:delta_t:T;   % time vector
theta = 0.75;   % Phase of the sinusoid (to be tracked)
s = cos(2*pi*fc*t + theta);   % Noiseless sinusoidal signal
r = s + sigma*randn(1,length(s));   % Observation in white Gaussian noise
% PLL implementation follows. Note that an FIR loop filter is used.
N_ord = 20;   % Order of the filter
cut_off = 0.1;   % Normalized cut-off frequency of the filter
```

```
b = fir1(N_ord,cut_off);     % Filter coefficients are determined using the fir1
function
K = 50;    % gain of the integrator (for the VCO)
int_of_v = 0;    % Integrator output initialized
e = zeros(1,N_ord);    % Error signal (input to the loop filter) is initialized
theta_prime = [];    % Initialization of the estimated sinusoidal phase values
for i=1:length(t)
    theta_prime = [theta_prime K*int_of_v];    % Estimate of the phase at the
current step
    vco_out = -sin(2*pi*fc*t(i)+theta_prime(i));    % output of the VCO
    e = [e r(i)*vco_out];    % Error signal is extended with the next sample
    v_of_t = 0;    % loop filter output for the current time step is initialized
    for j=1:N_ord+1
        v_of_t = v_of_t + b(j)*e(i+N_ord-j+1);
    end
    % Loop filter output sample is obtained at this point
    int_of_v = int_of_v + delta_t*v_of_t;    % Integrator output is obtained
end
% Plotting the PLL output
plot(t,theta_prime,'b','LineWidth',1.5);
```

(a) Run the code with different phase values of the sinusoidal signal and observe the convergence of the estimated phase to the correct one. Also, play with the parameters of the PLL (e.g., the gain of the VCO, filter cut-off frequency, filter order, etc.) and noise variance to determine their effects on the phase estimates. Comment on your results.

(b) Modify the code in part (a) to study a scenario with a time-varying phase. That is, assume that the phase is a slowly varying function of time (instead of being a constant) of your choosing, and run the modified code to see if these variations in the channel phase can be tracked. Try different sets of parameters and comment on your results.

(c) We now consider a BPSK-modulated signal instead of a pure sinusoid over a channel with an unknown phase. Write a Matlab code to simulate a "squaring loop" to track the unknown (but constant) channel phase. You can use the given PLL code as a reference and make the necessary modifications for this scenario. Run several experiments, and comment on your results. Are you always able to lock onto the correct phase in this case? Why or why not?

7 Spectrum of Digitally Modulated Signals

In the previous two chapters, we have studied the basics of digital modulation techniques, including transmission using single-carrier bandpass signals. While our study was detailed, we have only considered transmission over ideal additive white Gaussian noise channels. We assumed that the transmitted signal is received perfectly with no distortion or channel effects other than deterministic channel attenuation and additive white Gaussian noise. In other words, ignoring the additive Gaussian noise at the receiver, the channel is a simple ideal all-pass filter. A more realistic assumption for a practical communication channel is a bandlimited linear time-invariant filter, necessitating the design of communication systems with this constraint in mind.

In this chapter, our first goal is to study digital transmission over bandlimited channels and the design of transmission pulses that guarantee no intersymbol interference. As a by-product, we will also determine the spectral efficiencies of different modulation schemes. Stated differently, we will be able to compute the minimum channel bandwidth necessary (in Hertz) for a given transmission rate (in bits/second) with different modulation techniques in such a way that there is no intersymbol interference (ISI). In addition, we will consider the design of practical pulses for communications over bandlimited channels. With the motivation of transmitting digital information over a given frequency band, it is also crucial to compute the power spectral density of digitally transmitted signals, which can be accomplished by building on the basic concepts in random processes.

The chapter is organized as follows. We distinguish between baseband and bandpass communications and consider signal transmission in a given frequency band in Section 7.1. We argue that if a time-limited pulse is used for transmission (as done in the previous chapters), a bandlimited channel will distort the transmitted signal and cause intersymbol interference. We show that it is possible to avoid ISI as long as the transmission rate is not above the Nyquist rate, and we address the pulse design for no intersymbol interference. We also introduce the raised cosine pulse as a widely used solution to avoid ISI. In addition to the ideal bandlimited channels, we also touch upon the design of transmit and receive pulses when the channel is not ideal. We build upon these results and study the spectral efficiency of different modulation schemes for both baseband and bandpass communications in Section 7.2. We then compute the power spectral density of the transmitted signals in Section 7.3. We accomplish this by showing that, with linear modulation, the transmitted signal can be modeled as cyclostationary random processes and by extending the concept of power spectral density to such random signals. We explore how the selection of transmit pulses or the introduction of correlation among transmitted symbols can shape the spectral content of transmitted signals. We conclude the chapter in Section 7.4.

7.1 Transmission Over Bandlimited Channels

In a typical communication system, whether baseband or bandpass, we are allocated a frequency band of operation over which the transmission must occur. For instance, in a baseband communication scheme in which the allocated frequency band for communication is $[0, B]$ Hz, the transmitted signals must be selected such that their spectral content is limited to this band. Similarly, in a bandpass communication system, we might be allocated channel frequencies between f_1 and f_2, with a channel bandwidth of $f_2 - f_1$ Hz, in which case the transmitted signals must have a spectral content within the frequencies f_1 and f_2.

The restriction on the frequencies of operation may be due to physical constraints, for example, for a given medium, transmission of signals using certain frequencies may not be feasible. For instance, transmission and reception of radio waves at very low frequencies would require huge antennas, making the use of such frequencies highly impractical. Or, the restriction on the frequency of operation may be due to the requirements of a regulatory body or international standards. For instance, Wi-Fi systems are restricted to operate at a frequency band of around 2.4, 5, or 6 GHz. Regardless of the reason, we need to pay attention to the communication signal design and ensure that the transmitted signals are confined to the desired frequency band of operation.

We can model an ideal baseband communication channel as a linear time-invariant filter with a constant frequency response (let us take it as unity) and linear phase within $[-B, B]$, where B is the channel bandwidth. That is,

$$H(f) = e^{-j2\pi f_o t} \quad \text{for } |f| \leq B, \tag{7.1}$$

and 0 otherwise. Similarly, for an ideal bandpass communication channel with a passband between the frequencies f_1 and f_2,

$$H(f) = e^{-j2\pi f_o t} \quad \text{for } f_1 \leq |f| \leq f_2, \tag{7.2}$$

and 0 outside this band. The magnitude responses of ideal baseband and ideal bandpass channels are illustrated in Fig. 7.1.

The only effect of an ideal bandlimited channel on a transmitted signal whose spectral content is within the channel passband is a constant delay, i.e., there is no signal distortion apart from the usual AWGN. In other words, as long as the signals are selected appropriately, the receiver structures developed in the previous two chapters apply. On the other hand, if the channel is not ideal (i.e., $|H(f)|$ is not

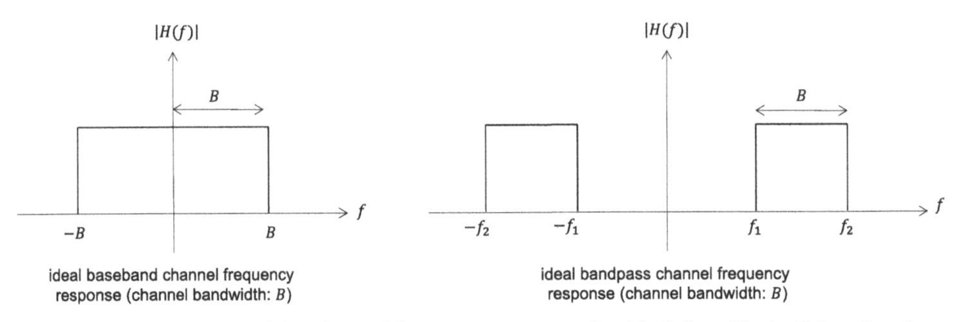

Figure 7.1 Magnitudes of the channel frequency response for ideal (bandlimited) baseband and bandpass channels of bandwidth B.

constant or $H(f)$ is not linear phase), or if the spectral occupancy of the transmitted signal is not within the passband of the bandlimited channel, the transmitted signal will experience distortion. That is, the received signal (excluding the channel noise) will be the convolution of the transmitted signal and the impulse response of the channel (which is modeled as an LTI system), and there will be ISI.

Noting that the presence of ISI may be detrimental to the communication system performance, in this section, we will describe how the communication pulses can be selected to make sure that there is no ISI. As a by-product, we will also be able to quantify the spectral efficiency of different modulation schemes; namely, we will be able to determine the amount of channel bandwidth necessary for a given transmission rate, as will be described in Section 7.2.

7.1.1 Pulse Design for No Intersymbol Interference

In this section, we consider baseband and single-carrier bandpass communication systems and explore the design of communication pulses.

Let us first consider a baseband communication system employing a transmit pulse $p_T(t)$. Let us denote the symbol period as T. Unlike the previous two chapters, we do not constrain the pulse $p_T(t)$ to be limited to the symbol duration. This is because a time-limited pulse would result in a transmitted signal whose spectral content is unlimited, and it may not be suitable for transmission over a strictly bandlimited channel.

Let us denote the transmitted signal as

$$v(t) = \sum_{n=-\infty}^{\infty} a_n p_T(t - nT), \tag{7.3}$$

where $\{a_n\}$ is the sequence of transmitted amplitude levels. In other words, we are considering a baseband PAM scheme.

Assume that transmission takes place over an ideal bandlimited channel and that the transmit pulse $p_T(t)$ has a Fourier transform limited to the passband of the channel. Ignoring the channel noise and assuming that a timing recovery algorithm is used at the receiver to compensate for the channel delay perfectly, the received signal will be the same as the transmitted signal. Let us assume that a receiver filter

$p_R(t)$ is employed, and the filter output is sampled periodically with the sample at time mT corresponding to the mth symbol. Namely, in the absence of noise, the received signal is given by

$$r(t) = \sum_{n=-\infty}^{\infty} a_n g(t - nT), \tag{7.4}$$

where $g(t) = p_T(t) * p_R(t)$ is the combined transmit and receive pulse. While we take the pulse used as the receive filter to be arbitrary, it should be clear that the selection $p_R(t) = p_T(-t)$ corresponds to a matched filter.[1] The receive filter output sample at time mT, denoted by r_m, is given by

$$r_m = r(mT) = \sum_{n=-\infty}^{\infty} a_n g_{m-n}, \tag{7.5}$$

where $g_k = g(kT)$. That is, the filter output sample (excluding the noise) based on which we will decide on the symbol transmitted at the mth time interval is given by

$$r_m = a_m g_0 + \sum_{n \neq m} a_n g_{m-n}. \tag{7.6}$$

Clearly, the first term is the signal corresponding to the desired symbol, while the second term is the ISI caused by the other symbols.

In the rest of this section, our objective is to design transmit and receive pulses in such a way that there is no ISI. Recall from Chapter 6 that we can study bandpass transmission schemes using the lowpass equivalent representations with the carrier frequency f_c taken as reference. Hence, adopting the lowpass representations, the equivalent transmitted signal is given by

$$v_l(t) = \sum_{n=-\infty}^{\infty} a_n p_T(t - nT), \tag{7.7}$$

where a_n is a complex number representing the symbol transmitted during the nth time interval. The real part of a_n determines the signal transmitted in the in-phase component, while its imaginary part specifies the transmitted signal in the quadrature component. Noticing that this is precisely the same expression as in the baseband case, except that the coefficient a_n is complex as opposed to a real number, we argue that the same expression is obtained at the receive filter output, that is, the decision variable (excluding noise) to be used for the symbol transmitted in the mth time interval is also in the form of (7.6). The only difference here is that the coefficients, in this case, assume complex values.

In short, the model in (7.6) is general, and it can be used for both baseband and bandpass communication systems, covering the baseband PAM, bandpass PAM, PSK, and QAM signaling equally well. For the PAM case, the sequence denoting the transmitted signals is real-valued; however, for the case of PSK or QAM, it is

[1] Note that we do not take $p_R(t) = p_T(T - t)$ since we take the receive filter sample at time nT for the nth symbol as opposed to the time instance $(n + 1)T$.

complex. Note also that for the case of bandpass communications, we assume that, along with timing recovery, carrier phase recovery is also performed, the channel phase is estimated and compensated for, that is, coherent detection is adopted.

To make sure that there is no ISI, we should select the combined transmit and receive pulse in such a way that

$$g_k = g(kT) = 0 \quad \text{for } k \neq 0, \tag{7.8}$$

and, without loss of generality, we can take $g_0 = g(0) = 1$.

Nyquist Criterion for Zero ISI. A necessary and sufficient condition for $g(t)$ to satisfy $g(0) = 1$ and $g(kT) = 0$ for $k \neq 0$ is that

$$\sum_{l=-\infty}^{\infty} G\left(f + \frac{l}{T}\right) = T, \tag{7.9}$$

where $G(f)$ denotes the Fourier transform of $g(t)$.

Let us prove this result. We recall the inverse Fourier transform relationship as

$$g(t) = \int_{-\infty}^{\infty} G(f)e^{j2\pi ft}\, df. \tag{7.10}$$

Evaluating this expression at $t = kT$, we obtain

$$g(kT) = \int_{-\infty}^{\infty} G(f)e^{j2\pi fkT}\, df \tag{7.11}$$

$$= \sum_{n=-\infty}^{\infty} \int_{(2n-1)/2T}^{(2n+1)/2T} G(f)e^{j2\pi fkT}\, df, \tag{7.12}$$

which follows by splitting the integration over the entire real line into integrals over non-overlapping intervals of length $1/T$. With a change of variable $f' = f - \frac{n}{T}$, and noting that $e^{j2\pi nk} = 1$ (and renaming f' as f), we can write

$$g(kT) = \int_{-1/2T}^{1/2T} \left[\sum_{n=-\infty}^{\infty} G\left(f + \frac{n}{T}\right)\right] e^{j2\pi fkT}\, df. \tag{7.13}$$

We observe that the function

$$G'(f) = \sum_{n=-\infty}^{\infty} G\left(f + \frac{n}{T}\right) \tag{7.14}$$

is periodic with period $1/T$. Its Fourier series expansion can be written as

$$G'(f) = \sum_{m=-\infty}^{\infty} g'_m e^{j2\pi fmT}, \tag{7.15}$$

with

$$g'_m = T \int_{-1/2T}^{1/2T} G'(f) e^{-j2\pi f m T} df. \tag{7.16}$$

Comparing this with (7.13), we identify that

$$g'_m = Tg(-mT). \tag{7.17}$$

Therefore, the condition for no ISI, that is,

$$g(mT) = \begin{cases} 1, & \text{for } m = 0, \\ 0, & \text{else} \end{cases} \tag{7.18}$$

is satisfied if and only if

$$g'_m = \begin{cases} T, & \text{for } m = 0, \\ 0, & \text{else,} \end{cases} \tag{7.19}$$

which is equivalent to

$$G'(f) = \sum_{l=-\infty}^{\infty} G\left(f + \frac{l}{T}\right) = T, \tag{7.20}$$

hence, we obtain the statement of the Nyquist criterion for no ISI.

Signal Design for No ISI. Let us focus on transmission over an ideal bandlimited channel with a bandwidth of B Hz. We will employ the Nyquist criterion just proved to design combined transmit and receive pulses, which guarantees no ISI. Our development is valid for both baseband and bandpass transmission schemes. For the latter case, the pulses will determine the lowpass-equivalent signals from which the actual bandpass signals can be generated. Also note that if the reference frequency is selected as the middle of the frequency band, a bandwidth of B for the lowpass-equivalent signal corresponds to a bandwidth of $2B$ for the bandpass signal.

We will consider three cases:

(1) very small symbol period, that is, $T < \frac{1}{2B}$;
(2) symbol period matched to the Nyquist rate, that is, $T = \frac{1}{2B}$;
(3) a larger symbol period, that is, $T > \frac{1}{2B}$.

Case 1: Transmission Above the Nyquist Rate $\left(T < \frac{1}{2B}\right)$. In this case, the transmission rate is too high, making it impossible to satisfy the Nyquist criterion for no ISI. This point is illustrated in Fig. 7.2. It is clear from the figure that if we insist on making the symbol period smaller than the Nyquist rate ($1/2B$), then we have to deal with the resulting ISI, potentially increasing the complexity of the communication system.

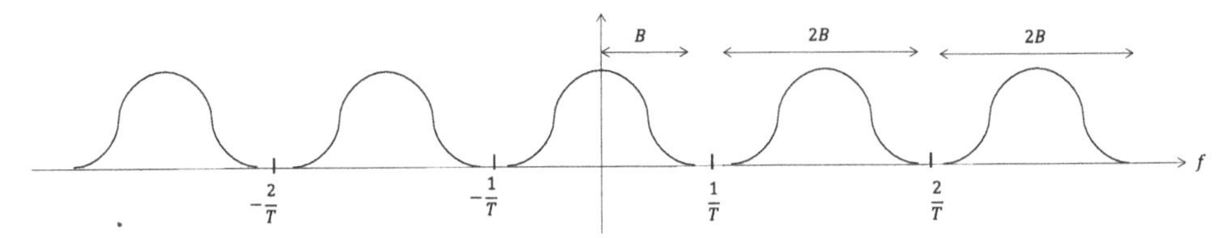

Figure 7.2 If the transmission rate is too high, that is, $T < 1/2B$, it is impossible to satisfy the Nyquist criterion for no ISI as it is impossible to make the sum of the replicas of $G(f)$ add to a constant.

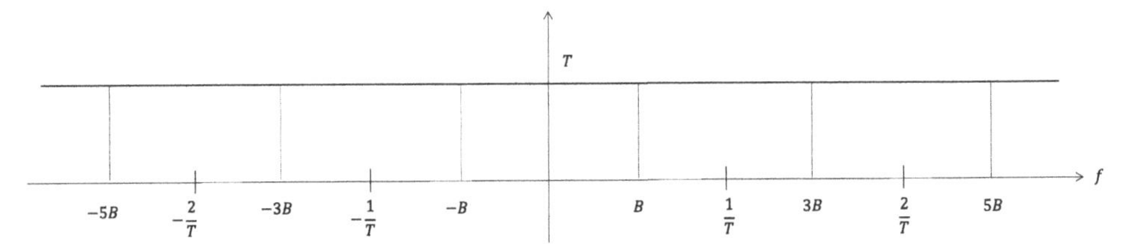

Figure 7.3 When $T = 1/2B$, there is a unique selection for the transmission pulse to make sure there is no ISI, that is, a pulse with a rectangular Fourier transform extending from $-B$ to B with amplitude T should be used.

Case 2: Transmission at the Nyquist Rate $\left(T = \frac{1}{2B}\right)$. There is a unique solution that results in no ISI when transmitting at the Nyquist rate (see Fig. 7.3). Namely, with

$$G(f) = \begin{cases} T, & \text{if } |f| \leq B, \\ 0, & \text{else,} \end{cases} \tag{7.21}$$

or, in the time domain

$$g(t) = \text{sinc}(t/T), \tag{7.22}$$

we have a transmit/receive pulse, which results in no ISI. Combining this with the previous case, we conclude that the sinc pulse is optimal in terms of bandwidth efficiency if no ISI is desired. When transmitting at a rate of $1/T$ symbols per second, a channel bandwidth of $B = 1/2T$ Hz is required. Or, for every symbol per second increase in the transmission rate, we need an additional bandwidth of $1/2$ Hz.[2]

We illustrate a sinc transmit/receive pulse being used in digital modulation in Fig. 7.4. We observe that when the sampling instances are picked at integer multiples of the symbol period T, there is only a contribution due to the symbol being transmitted within the corresponding time interval, with no interference from the other symbols.

While the sinc pulse is optimal in terms of bandwidth efficiency, it is not very practical. This is because the tails of the sinc function decay very slowly, only with $\sim 1/|t|$. Even if there is a slight misalignment in the sampling instances, there will be

[2] Indeed, we had previously stated this result when studying the minimum bandwidth requirements of pulse-code modulation without giving proof. We now know how it follows.

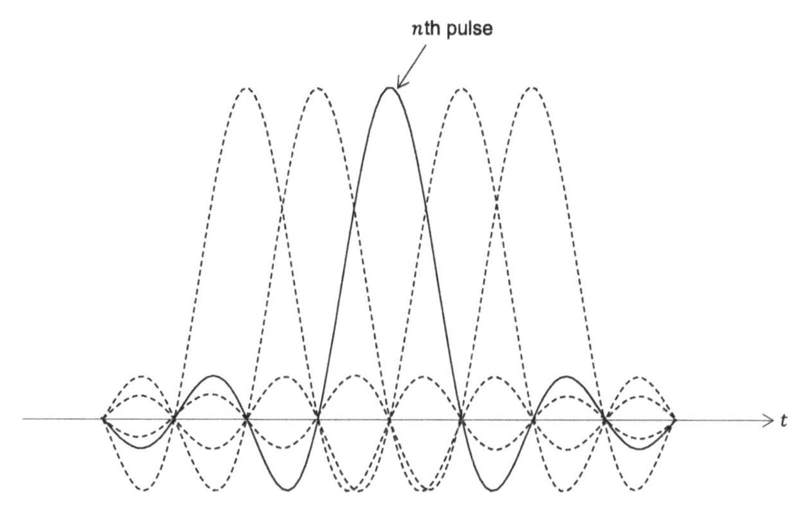

Figure 7.4 Illustration of a sinc pulse used for digital modulation. Along with the nth pulse, two previous and two following pulses are also shown. It is observed that there is no interference at the peak of the nth pulse from the other symbols.

significant (in fact, non-convergent) interference from the other transmitted symbols. Therefore, the use of a sinc pulse is not desirable. Instead, as explained below, we need to reduce the transmission rate and design other pulses that satisfy the Nyquist criterion for no ISI with faster-decaying tails.

Case 3: Transmission Below the Nyquist Rate $\left(T > \frac{1}{2B}\right)$. In this case, there are infinitely many solutions to the problem of pulse selection for no ISI, and it is possible to select well-behaving pulses with rapidly decaying tails among them, making this scenario much more practical.

In what follows, we describe the particular class of raised cosine pulses for transmission with no ISI over bandlimited channels.

An appealing and widely used transmit/receive pulse for no ISI is the raised cosine pulse with the Fourier transform

$$G_{rc}(f) = \begin{cases} T, & \text{if } |f| \leq \frac{1-\alpha}{2T}, \\ \frac{T}{2}\left[1 + \cos\left(\frac{\pi T}{\alpha}\left(|f| - \frac{1-\alpha}{2T}\right)\right)\right], & \text{if } \frac{1-\alpha}{2T} \leq |f| \leq \frac{1+\alpha}{2T}, \\ 0, & \text{if } |f| \geq \frac{1+\alpha}{2T}, \end{cases} \tag{7.23}$$

where $\alpha \in [0, 1]$ is called the roll-off factor. It is straightforward to verify that this selection satisfies the Nyquist criterion; hence, the pulse does not result in ISI. The raised cosine pulse in the Fourier domain is illustrated in Fig. 7.5. The case with $\alpha = 0$ corresponds to the sinc pulse. We also observe that the bandwidth of the raised cosine pulse is $\frac{1+\alpha}{2T}$, that is, $(1+\alpha)$ times the minimum required bandwidth for no ISI of $1/2T$. When the roll-off factor $\alpha = 1$, the required bandwidth is doubled compared to the smallest possible one; hence, the excess bandwidth is 100%.

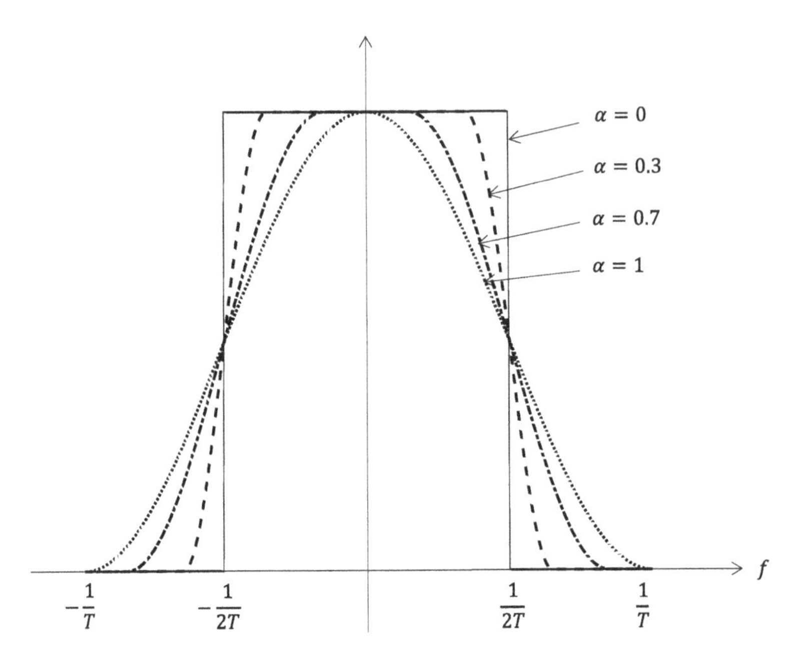

Figure 7.5 Fourier transform of the raised cosine pulse for four different roll-off factors. The Fourier transform of the pulse becomes smoother with an increasing roll-off factor; hence, it decays faster in the time domain.

The raised cosine pulse in the time domain is given as

$$g_{rc}(t) = \text{sinc}(t/T)\frac{\cos(\pi\alpha t/T)}{1 - 4\alpha^2 t^2/T^2}, \tag{7.24}$$

which shows that the tails of the pulse decay with $1/|t|^3$ (provided that $\alpha > 0$). This important observation demonstrates that the raised cosine pulse is practical: the ISI due to timing errors is guaranteed to be finite, contrary to the case with a sinc pulse, for which the resulting ISI could be unbounded.

We illustrate the raised cosine pulse in the time domain for several values of the roll-off factor in Fig. 7.6. We observe that the pulse values at the integer multiples of the symbol period T (except for $t = 0$) are zero; hence, with the correct sampling instances, there is no ISI. We also observe that with increasing values of α, the tails of the pulse decay faster, which is highly preferable in terms of limiting the ISI when there is a timing error.

Finally, note that the raised cosine pulse is for the combination of the transmit and receive pulses. A practical solution for determining the transmit and receive pulses individually would be to pick each as a square-root raised cosine pulse, that is,

$$G_T(f) = G_R^*(f) = \sqrt{|G_{rc}(f)|}e^{-j2\pi f t_0}. \tag{7.25}$$

Of course, other choices for $G_T(f)$ and $G_R(f)$ are also possible to obtain an overall combined raised cosine pulse.

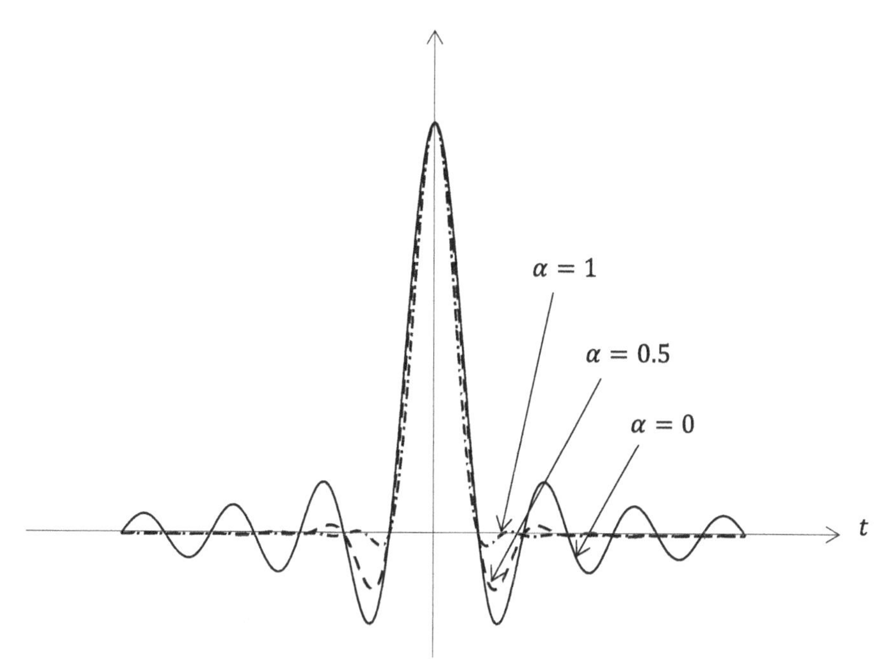

Figure 7.6 Raised cosine pulse in the time domain for three different roll-off factors. The tails of the pulse decay faster with increasing roll-off factor, making it more practical (while the required bandwidth is increased).

7.1.2 Transmission Over Non-ideal Channels

So far, we have considered communication over an ideal bandlimited channel and described how the transmit and receive pulses can be designed to ensure no intersymbol interference. We now touch upon the case of non-ideal channels. More precisely, the channel is still modeled as a linear time-invariant filter; however, it has a non-constant magnitude response within its passband, a non-linear phase, or both.

Let us denote the impulse response of the (non-ideal) bandlimited channel by $h'(t)$ and its frequency response by $H'(f)$. A representative magnitude plot of the frequency response is depicted in Fig. 7.7. With a transmit pulse of $g_T(t)$ and a receive pulse of $g_R(t)$ (with Fourier transforms $G_T(f)$ and $G_R(f)$, respectively), the received signal can be written as

$$r(t) = \sum_{n=-\infty}^{\infty} a_n g(t - nT), \tag{7.26}$$

where a_n is the sequence of transmitted symbols and

$$g(t) = g_T(t) * h'(t) * g_R(t) \tag{7.27}$$

is the effective pulse at the receiver side. Note that this formulation covers the baseband PAM and bandpass PAM, PSK, and QAM by employing the

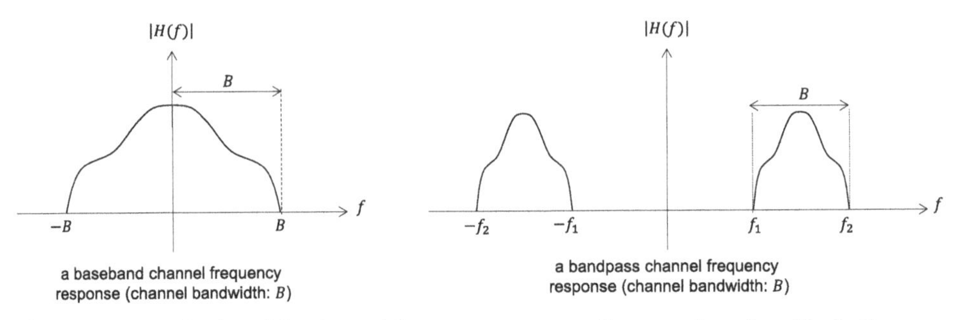

Figure 7.7 Magnitudes of the channel frequency responses for exemplary (bandlimited) non-ideal baseband and bandpass channels of bandwidth B.

lowpass-equivalent signals and the equivalent lowpass channel impulse response. The Fourier transform of the effective pulse is given by

$$G(f) = G_T(f)H'(f)G_R(f). \tag{7.28}$$

We can employ the result of the previous subsection to design the transmit and receive pulses for no ISI as long as $T \geq 1/2B$. If $T < 1/2B$, there is no way of eliminating the ISI at the receiver output. For instance, assuming that the symbol period is sufficiently small (i.e., $T \geq 1/2B$), we can pick the transmit and receive pulses so that the overall (effective) pulse at the receiver is a raised cosine function, that is, we can select $G_T(f)$ and $G_R(f)$ such that

$$G(f) = G_T(f)H'(f)G_R(f) = G_{rc}(f). \tag{7.29}$$

It is possible to perform pre-compensation of the channel at the transmitter, split the channel compensation between the transmitter and the receiver, or compensate for the non-ideal channel at the receiver. For example, by targeting an overall raised cosine pulse, we can select

$$G_T(f) = \frac{\sqrt{|G_{rc}(f)|}e^{-j2\pi f t_0}}{H'(f)}, \quad G_R(f) = \sqrt{|G_{rc}(f)|}e^{j2\pi f t_0} \tag{7.30}$$

for precompensation at the transmitter. Or, we can select

$$G_T(f) = \sqrt{|G_{rc}(f)|}e^{-j2\pi f t_0}, \quad G_R(f) = \frac{\sqrt{|G_{rc}(f)|}e^{j2\pi f t_0}}{H'(f)} \tag{7.31}$$

for channel compensation at the receiver.

We close this section by noting that intersymbol interference is present in many practical communication schemes and must be dealt with. It may not always be possible to select the transmit and receive pulses to eliminate the ISI. In such cases, channel equalization can be used to alleviate the effects of ISI for reliable communication. For example, a linear equalizer, that is, an LTI filter designed using a suitable criterion, can be employed at the receiver. Non-linear equalization techniques (e.g., decision feedback equalization) can also be used. Coverage of channel equalization techniques is beyond our scope.

7.2 Spectral Efficiency of PAM, PSK, and QAM

The spectral efficiency of a digital modulation scheme is defined as the ratio of the transmission rate to the required bandwidth, that is,

$$\text{spectral efficiency} = \frac{\text{transmission rate}}{\text{required bandwidth}}, \tag{7.32}$$

with a unit of bits/sec per Hz, or bits/s/Hz. This is a measure of how efficiently the channel bandwidth resources are being utilized as it explicitly shows the transmission rate in bits/second per unit channel bandwidth in Hertz.

Using the main result of the previous section, we are in a position to determine the spectral efficiency of different transmission schemes.

Spectral Efficiency of PAM. Let us first consider baseband PAM transmission with a symbol period of T. With M-ary signaling, each symbol carries $\log_2 M$ bits, hence the transmission rate is $\frac{\log_2 M}{T}$ bits/sec. From the previous section, we know that the optimal pulse shape in terms of bandwidth efficiency is a sinc function with a bandwidth of $1/2T$. Therefore, the spectral efficiency becomes

$$\frac{\log_2 M/T}{1/2T} = 2\log_2 M \ \text{ bits/s/Hz}. \tag{7.33}$$

Next, let us consider bandpass PAM. The minimum bandwidth of the lowpass-equivalent pulse for no ISI is $1/2T$. Assuming that double-sideband signaling is employed (i.e., with the pulse $p_T(t)\cos(2\pi f_c t)$), the channel bandwidth required becomes $1/T$. However, we can employ single-sideband signaling since the information is embedded only in the amplitude. With the single-sideband PAM signal given by

$$p_T(t)\cos(2\pi f_c t) \pm \hat{p}_T(t)\sin(2\pi f_c t), \tag{7.34}$$

where $\hat{p}_T(t)$ is the Hilbert transform of $p_T(t)$, a transmission bandwidth of only $1/2T$ is required. Therefore, the spectral efficiency of bandpass PAM becomes the same as that of baseband PAM.

To summarize, the spectral efficiency for M-PAM signaling is simply

$$r_{\text{M-PAM}} = 2\log_2 M \ \text{ bits/s/Hz}. \tag{7.35}$$

We observe that by increasing M, the spectral efficiency is increased, that is, the spectral resources are utilized more effectively. On the other hand, recalling our results in Chapter 5, this comes at the cost of increased signal-to-noise ratio requirements: the scheme becomes less power efficient. With an increased constellation size, a higher signal-to-noise ratio is needed to obtain the same error probability. In other words, there is a trade-off between power and bandwidth requirements.

Spectral Efficiency of PSK and QAM. As BPSK is the same as binary PAM, it is already covered above, and the spectral efficiency is 2 bits/s/Hz. Hence, we focus on the case with $M > 2$.

With M-ary PSK or QAM ($M > 2$), the minimum bandwidth required for no ISI is $1/T$ Hz (twice the bandwidth of the lowpass-equivalent waveform). Therefore, the spectral efficiency becomes

$$r_{M\text{-PSK}} = r_{M\text{-QAM}} = \frac{\log_2 M/T}{1/T}, \tag{7.36}$$

or simply

$$r_{M\text{-PSK}} = r_{M\text{-QAM}} = \log_2 M \ \text{ bits/s/Hz}. \tag{7.37}$$

We note that, as in M-PAM signaling, the spectral efficiency increases as the constellation size M increases while the power efficiency reduces (as observed from the error probability results for M-PSK and M-QAM in Chapter 6).

Spectral Efficiency with Raised Cosine Pulses. The above spectral efficiency results assume that a bandwidth optimal sinc pulse is used for transmission. If, instead, a more practical pulse is employed, we would obtain lower spectral efficiencies. For instance, with a raised cosine pulse with roll-off factor α, all the required bandwidth expressions above should be increased by a factor of $(1+\alpha)$, and hence, the spectral efficiencies should be reduced by the same factor $(1+\alpha)$.

Example 7.1

Determine the maximum transmission rate possible for a channel bandwidth of $B = 40$ MHz for the following modulation schemes: BPSK, 8-PAM, 8-PSK, and 256-QAM.

Solution

With BPSK: the spectral efficiency is $r = 2$ bits/s/Hz, hence a transmission rate of up to $rB = 80$ Mbps can be supported with the given bandwidth.

With 8-PAM: the spectral efficiency is $r = 2\log_2(8) = 6$ bits/s/Hz, hence a transmission rate of up to $rB = 240$ Mbps can be supported.

With 8-PSK: the spectral efficiency is $r = \log_2(8) = 3$ bits/s/Hz, hence a transmission rate of up to $rB = 120$ Mbps can be supported.

With 256-QAM: $r = \log_2(256) = 8$ bits/s/Hz, hence a transmission rate of up to $rB = 320$ Mbps can be supported.

Example 7.2

Consider the previous problem but assume that a raised cosine pulse with a roll-off factor of 0.5 is employed. Determine the resulting transmission rates.

Solution

With the raised cosine pulse, the spectral efficiencies will reduce by a factor of $1+\alpha = 1.5$ since the bandwidth of the transmit pulse is $(1+\alpha)/2T$ as opposed to that of the ideal Nyquist pulse of bandwidth $1/2T$. Therefore, given that the transmission bandwidth is fixed at 40 MHz, all the transmission rates will need to be divided by 1.5. That is, the supported rates become (1) for BPSK: 53.3 Mbps, (2) for 8-PAM: 160 Mbps, (3) for 8-PSK: 80 Mbps, and (4) for 256-QAM: 213.3 Mbps.

7.3 Power Spectral Density of Transmitted Signals

In this section, we compute the average power content of digitally modulated signals as a function of frequency, that is, their power spectral density. Through this calculation, we can determine whether transmission occurs in the intended frequency band or how much of the signal power spills out and interferes with other transmissions taking place in different frequency bands.

Digitally modulated waveforms are random processes since the data symbols transmitted are random variables. In what follows, we will demonstrate that a linearly modulated signal is a cyclostationary random process (assuming that the symbol sequence is wide-sense stationary). Recall from Chapter 2 that, for a cyclostationary random process, the mean function is periodic, and so is the autocorrelation function, and averaging the autocorrelation function over its period gives the average autocorrelation function (denoted by $\bar{R}(\tau)$).

The power spectral density of a cyclostationary random process is determined by taking the Fourier transform of $\bar{R}(\tau)$. The interpretation of this quantity is the same as that of the power spectral density of wide-sense stationary random processes. For instance, by integrating it over a given frequency band, we can compute the power contained within the band. Therefore, we can determine some very important quantities, including the amount of interference caused in other frequency bands.

7.3.1 Power Spectral Efficiency of Linearly Modulated Signals

Let us start with a linearly modulated signal given by

$$V(t) = \sum_{n=-\infty}^{\infty} a_n g_T(t - nT), \tag{7.38}$$

where $\{a_n\}$ is the sequence of transmitted symbols, T is the symbol duration, and $g_T(t)$ is the transmit pulse. $V(t)$ is a random process as $\{a_n\}$ is a random sequence. $V(t)$ is a lowpass signal representing the actual transmitted signal for baseband transmission or the lowpass equivalent of the transmitted signal for single-carrier bandpass transmission. Therefore, PAM, PSK, and QAM can be studied simultaneously with this signal model. For PAM signaling, $\{a_n\}$ are real, however, they are complex for M-PSK and M-QAM transmission (for $M > 2$).

Let us assume that the transmitted symbol sequence is WSS. That is, its mean sequence $\mathbb{E}[a_n]$ is a constant, and its autocorrelation sequence $\mathbb{E}[a_{n_1} a_{n_2}^*]$ is only a function of the time difference $n_1 - n_2$. Let us denote the mean by $m_a = \mathbb{E}[a_n]$, and the autocorrelation sequence by $R_a[m] = \mathbb{E}[a_{n+m}a_n^*]$. This assumption is quite general and covers all the important cases of interest to us. For instance, we typically model the transmitted symbols as i.i.d., which is certainly WSS with $R_a[m] = \sigma_a^2\delta[m]$, with σ_a^2 being the symbol power and $\delta[m]$ denoting a discrete impulse function.

We determine the mean function of the random process $V(t)$ as

$$\mathbb{E}[V(t)] = \mathbb{E}\left[\sum_{n=-\infty}^{\infty} a_n g_T(t - nT)\right] \tag{7.39}$$

$$= \sum_{n=-\infty}^{\infty} \mathbb{E}[a_n] g_T(t - nT) \tag{7.40}$$

$$= m_a \sum_{n=-\infty}^{\infty} g_T(t - nT), \tag{7.41}$$

where the last line follows since the transmitted sequence has a constant mean due to our assumption of having a wide-sense stationary transmitted sequence $\{a_n\}$. Clearly, $\mathbb{E}[V(t)]$ is a periodic function with period T.

Let us now compute the autocorrelation function of the process $V(t)$. We can write

$$R_V(t + \tau, t) = \mathbb{E}[V(t + \tau)V^*(t)] \tag{7.42}$$

$$= \mathbb{E}\left[\sum_{m=-\infty}^{\infty} a_m g_T(t + \tau - mT) \sum_{n=-\infty}^{\infty} a_n^* g_T(t - nT)\right] \tag{7.43}$$

$$= \sum_{m=-\infty}^{\infty}\sum_{n=-\infty}^{\infty} \mathbb{E}[a_m a_n^*] g_T(t + \tau - mT) g_T(t - nT) \tag{7.44}$$

$$= \sum_{m=-\infty}^{\infty}\sum_{n=-\infty}^{\infty} R_a[m - n] g_T(t + \tau - mT) g_T(t - nT). \tag{7.45}$$

Defining $m' = m - n$ and renaming m' again as m, we obtain

$$R_V(t + \tau, t) = \sum_{m=-\infty}^{\infty} R_a[m] \sum_{n=-\infty}^{\infty} g_T(t + \tau - (m + n)T) g_T(t - nT). \tag{7.46}$$

It is easily verified that

$$R_V(t + \tau, t) = R_V(t + T + \tau, t + T), \tag{7.47}$$

namely, the autocorrelation function of $V(t)$ is periodic in the t variable with period T. As the mean and autocorrelation functions are periodic, we conclude that $V(t)$ is a cyclostationary random process.

Next, we compute the average autocorrelation function of $V(t)$, that is, $\bar{R}_V(\tau)$:

$$\bar{R}_V(\tau) = \frac{1}{T}\int_{-T/2}^{T/2} R_V(t + \tau, t)dt \tag{7.48}$$

$$= \sum_{m=-\infty}^{\infty} R_a[m] \sum_{n=-\infty}^{\infty} \frac{1}{T} \int_{-T/2}^{T/2} g_T(t+\tau-(m+n)T) g_T(t-nT) dt \quad (7.49)$$

$$= \frac{1}{T} \sum_{m=-\infty}^{\infty} R_a[m] R_g(\tau - mT), \quad (7.50)$$

where

$$R_g(\tau) = \int_{-\infty}^{\infty} g_T(t) g_T(t+\tau) dt. \quad (7.51)$$

To determine the power spectral density of the transmitted linearly modulated signal $V(t)$ denoted by $S_V(f)$, we need to compute the Fourier transform of $\bar{R}_V(\tau)$. Hence, we obtain

$$S_V(f) = \int_{-\infty}^{\infty} \frac{1}{T} \sum_{m=-\infty}^{\infty} R_a[m] R_g(\tau - mT) e^{-j2\pi f \tau} d\tau \quad (7.52)$$

$$= \frac{1}{T} \sum_{m=-\infty}^{\infty} R_a[m] \int_{-\infty}^{\infty} R_g(\tau - mT) e^{-j2\pi f \tau} d\tau. \quad (7.53)$$

Since $R_g(\tau) = g_T(\tau) * g_T(-\tau)$, the Fourier transform of $R_g(\tau)$ is simply $|G_T(f)|^2$. Using the time-shifting property of the Fourier transform, we can write

$$S_V(f) = \frac{1}{T} \left(\sum_{m=-\infty}^{\infty} R_a[m] e^{-j2\pi f m T} \right) |G_T(f)|^2 \quad (7.54)$$

$$= \frac{1}{T} S_a(f) |G_T(f)|^2, \quad (7.55)$$

where $S_a(f)$ is defined as

$$S_a(f) = \sum_{m=-\infty}^{\infty} R_a[m] e^{-j2\pi f m T}. \quad (7.56)$$

We observe that the power spectral density of the transmitted signal is a function of the autocorrelation sequence of the transmitted symbols and the Fourier transform of the transmit pulse. If the transmit pulse is smoother, then its Fourier transform is more compact, and hence, the power spectral density of the emitted signal is more compact; more of the emitted power is within the desired frequency band, and there is less interference to other communication systems in different adjacent bands.

Special Case (Transmission of a Zero-Mean Uncorrelated Sequence). Let us consider the special case of a zero-mean uncorrelated transmitted sequence. Since this is a proper model for the transmission of equally likely and uncorrelated PAM, PSK, and QAM symbols, it is of high practical significance.

We have $R_a[m] = \sigma_a^2 \delta[m]$, hence $S_a(f) = \sigma_a^2$ and

$$S_V(f) = \frac{\sigma_a^2}{T}|G_T(f)|^2. \tag{7.57}$$

We observe that the power spectral density of the transmitted signal is only a function of the transmit pulse (up to a scaling factor). If the transmit pulse is smoother, the PSD is more compact.

Example 7.3

Consider the transmission of a zero-mean uncorrelated sequence of symbols with $\sigma_a^2 = 1$. Determine and plot the power spectral density of the transmitted signal corresponding to the three different transmit pulses:

(1) $g_{T,1}(t) = \frac{1}{\sqrt{T}}\Pi\left(\frac{t}{T}\right)$;

(2) $g_{T,2}(t) = \sqrt{\frac{3}{T}}\Lambda\left(\frac{2t}{T}\right)$;

(3) $g_{T,3}(t) = \sqrt{\frac{2}{3T}}\,(1 + \cos(2\pi t/T))$ if $t \in [-T/2, T/2]$, and 0 otherwise.

Here, T is the symbol period and the pulses are normalized so that the basic pulse energies are all unity.

Solution

We can compute the Fourier transforms of these three pulses in a straightforward manner as

$$G_{T,1}(f) = \sqrt{T}\mathrm{sinc}(fT), \tag{7.58}$$

$$G_{T,2}(f) = \sqrt{\frac{3T}{4}}\mathrm{sinc}^2\left(\frac{fT}{2}\right), \tag{7.59}$$

$$G_{T,3}(f) = \sqrt{\frac{2T}{3}}\frac{\mathrm{sinc}(fT)}{1 - f^2T^2}. \tag{7.60}$$

Therefore, the power spectral densities of modulated signals with the uncorrelated, zero-mean and unit-variance symbols (e.g., with binary PAM symbols using $+1$ and -1) become

$$S_{V,1}(f) = \mathrm{sinc}^2(fT), \tag{7.61}$$

$$S_{V,2}(f) = \frac{3}{4}\mathrm{sinc}^4\left(\frac{fT}{2}\right), \tag{7.62}$$

$$S_{V,3}(f) = \frac{2}{3}\frac{\mathrm{sinc}^2(fT)}{(1 - f^2T^2)^2}. \tag{7.63}$$

We observe that the PSD of the scheme, which employs the first (rectangular) transmit pulse, decays with $\sim 1/f^2$ for large frequencies. This is a slow decay, meaning that this communication system may cause high interference in adjacent bands. The decay of the PSD for the second pulse is $\sim 1/f^4$, and for the third one is $\sim 1/f^6$, which is much more favorable.

The three power spectral densities are depicted in Fig. 7.8. For instance, if the communication band allocated is $2/T$, then we can consider the PSD in the interval $(-2/T, 2/T)$ as being due to the transmission within the desired band of operation, while the PSD outside this range is due to the signal components spilling out of the allocated band (potentially causing interference on other communication systems).

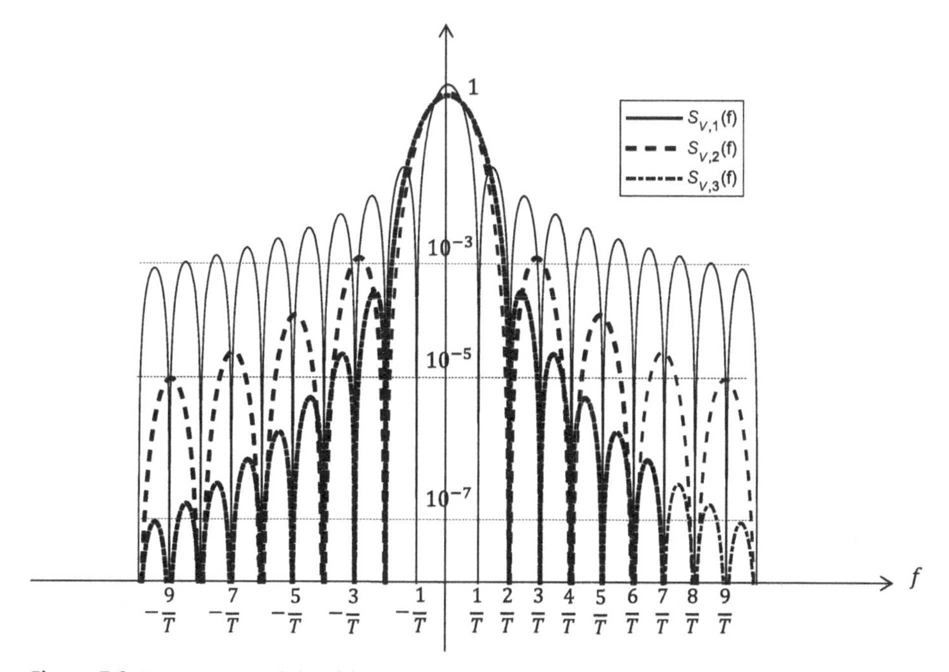

Figure 7.8 Power spectral densities of the transmitted signals with the three pulses given in the example. The smoother pulses (in the time domain) result in much more compact PSDs, that is, the interference caused in the adjacent bands is small.

One can compute the fraction of average power within the desired band versus total power by integrating the power spectral density over the respective frequency bands. This ratio is around 96% for the first pulse, 99.7% for the second one, and 99.95% for the third one. In other words, with a rectangular pulse, almost 4% of the total transmit power will spill out of the frequency band of operation, and hence, it will cause significant interference in other frequency bands. With the second and third pulses, the amount of interference to other communication systems operating in different frequency bands is much less.

Example 7.4

As another illustration, let us consider the transmission of a sequence of symbols with a non-zero mean. Assume that equally likely and uncorrelated binary symbols $a_n = 0$ or $a_n = 1$ are transmitted using PAM. Determine the power spectral density of the transmitted signal.

Solution

We readily determine

$$R_a[0] = \mathbb{E}[a_n^2] = \frac{1}{2}, \tag{7.64}$$

and for $m \neq 0$

$$\begin{aligned} R_a[m] &= \mathbb{E}[a_{n+m}a_n] && (7.65)\\ &= \mathbb{E}[a_{n+m}]E[a_n] && (7.66)\\ &= \frac{1}{4}. && (7.67) \end{aligned}$$

In other words,

$$R_a[m] = \frac{1}{4} + \frac{1}{4}\delta[m]. \tag{7.68}$$

From which, we obtain

$$\begin{aligned} S_a(f) &= \sum_{m=-\infty}^{\infty} \left(\frac{1}{4} + \frac{1}{4}\delta[m]\right) e^{-j2\pi fmT} && (7.69)\\ &= \frac{1}{4} + \frac{1}{4} \sum_{m=-\infty}^{\infty} e^{-j2\pi fmT}. && (7.70) \end{aligned}$$

Interpreting the infinite sum of the complex exponentials as a Fourier series expansion of a periodic signal (of f) with period $1/T$, we can rewrite it as an impulse train and obtain

$$S_a(f) = \frac{1}{4} + \frac{1}{4T} \sum_{m=-\infty}^{\infty} \delta\left(f - \frac{m}{T}\right). \tag{7.71}$$

Finally, the power spectral density of the transmitted signal is determined as

$$\begin{aligned} S_V(f) &= \frac{1}{T} S_a(f)|G_T(f)|^2 && (7.72)\\ &= \frac{1}{T}\left(\frac{1}{4} + \frac{1}{4T}\sum_{m=-\infty}^{\infty} \delta\left(f - \frac{m}{T}\right)\right)|G_T(f)|^2 && (7.73)\\ &= \frac{1}{4T}|G_T(f)|^2 + \frac{1}{4T^2}\sum_{m=-\infty}^{\infty} \left|G_T\left(\frac{m}{T}\right)\right|^2 \delta\left(f - \frac{m}{T}\right). && (7.74) \end{aligned}$$

Since the transmitted symbols are not of zero mean, the power spectral density of the transmitted signal contains impulses at multiples of $1/T$.

Example 7.5

Consider transmission of uncorrelated and equally likely binary symbols with $a_n = 0$ and $a_n = 1$ using PAM with the basic transmission pulse

$$g_T(t) = \frac{1}{\sqrt{T}} \Pi\left(\frac{t}{T}\right). \tag{7.75}$$

Determine the power spectral density of the transmitted signal.

Solution

Using the result of the previous example and the Fourier transform of the transmit pulse, the power spectral density of the transmitted signal is readily computed as

$$S_V(f) = \frac{1}{4T} T\mathrm{sinc}^2(fT) + \frac{1}{4T^2} \sum_{m=-\infty}^{\infty} T\mathrm{sinc}^2\left(\frac{m}{T}T\right) \delta\left(f - \frac{m}{T}\right) \tag{7.76}$$

$$= \frac{\mathrm{sinc}^2(fT)}{4} + \frac{1}{4T}\delta(f). \tag{7.77}$$

It is clear that due to spectral nulls of the employed transmit pulse at multiples of $1/T$, only the impulse at $f = 0$ survives. The resulting power spectral density is depicted in Fig. 7.9.

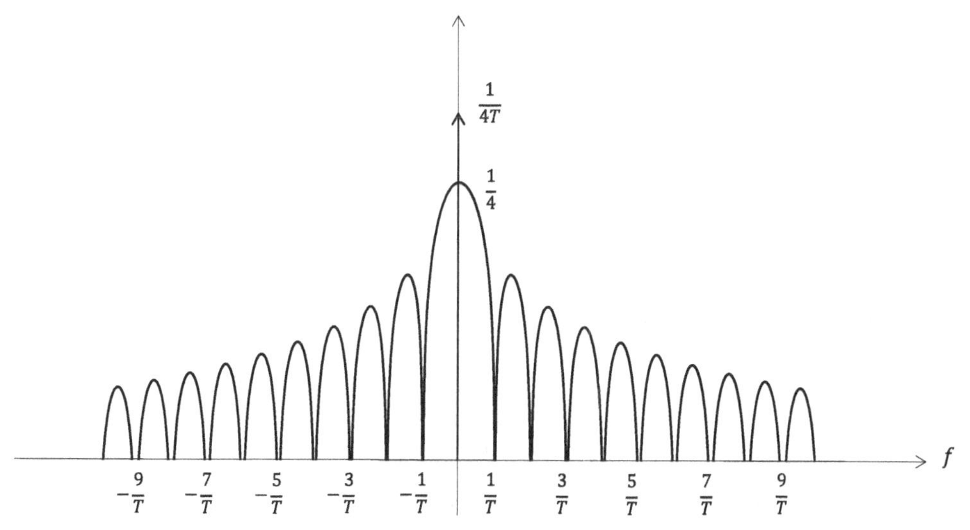

Figure 7.9 Power spectral density of binary on–off signaling with uncorrelated symbols using a rectangular transmit pulse.

The following two examples illustrate how a transmitted symbol sequence can be used to shape the power spectral density of a digitally modulated signal.

Example 7.6

Assume that the elements of the sequence $\{b_n\}_{n=-\infty}^{\infty}$ are independent binary random variables taking values of ± 1 with equal probability. This data is precoded as

$$a_n = b_n + \alpha b_{n-2}, \tag{7.78}$$

and the resulting sequence is used to modulate a pulse $g_T(t)$, that is, the resulting transmitted signal is given by

$$V(t) = \sum_{n=-\infty}^{\infty} a_n g_T(t - nT). \tag{7.79}$$

It is desired to have a null in the power spectral density of the transmitted signal at frequencies $f = \pm\frac{1}{4T}$. Determine the value of α to accomplish this goal.

Solution
We will determine the desired value of α by making sure that $S_a(f)$ is zero for $f = \pm 1/4T$. The autocorrelation function of the transmitted sequence $\{a_n\}$ is given by

$$R_a[m] = \mathbb{E}[a_{n+m}a_n] \tag{7.80}$$
$$= \mathbb{E}[(b_{n+m} + \alpha b_{n+m-2})(b_n + \alpha b_{n-2})] \tag{7.81}$$
$$= (1+\alpha^2)\delta[m] + \alpha\delta[m-2] + \alpha\delta[m+2], \tag{7.82}$$

where we have used $R_b[m] = \mathbb{E}[b_{n+m}b_n] = \delta[m]$. Therefore, we obtain

$$S_a(f) = \sum_{m=-\infty}^{\infty} R_a[m]e^{-j2\pi fmT} \tag{7.83}$$
$$= 1 + \alpha^2 + \alpha e^{-j4\pi fT} + \alpha e^{j4\pi fT} \tag{7.84}$$
$$= 1 + \alpha^2 + 2\alpha\cos(4\pi fT). \tag{7.85}$$

With $S_a(1/4T) = S_a(-1/4T) = 0$, we obtain

$$1 + \alpha^2 - 2\alpha = 0, \tag{7.86}$$

resulting in $\alpha = 1$.

Example 7.7

Assume that the sequence $\{b_n\}$ is WSS with autocorrelation function

$$R_b(m) = \delta(m) + \frac{1}{2}\delta(m-1) + \frac{1}{2}\delta(m+1). \tag{7.87}$$

Also assume that precoding of the form $a_n = b_n - b_{n-1}$ is employed, and the resulting $\{a_n\}$ sequence is transmitted using PAM with transmit pulse $g(t)$. Determine the power spectral density of the transmitted signal (in terms of the symbol period T and the Fourier transform of the transmit pulse $G(f)$).

Solution
We first need to compute the autocorrelation function of $\{a_n\}$, that is,

$$R_a[m] = \mathbb{E}[a_{n+m}a_n] \tag{7.88}$$
$$= \mathbb{E}[(b_{n+m} - b_{n+m-1})(b_n - b_{n-1})] \tag{7.89}$$
$$= 2R_b[m] - R_b[m-1] - R_b[m+1] \tag{7.90}$$
$$= \delta[m] - \frac{1}{2}\delta[m-2] - \frac{1}{2}\delta[m+2]. \tag{7.91}$$

We then obtain

$$S_a(f) = \sum_{m=-\infty}^{\infty} R_a[m]e^{-j2\pi fmT} \tag{7.92}$$
$$= 1 - \frac{1}{2}e^{-j4\pi fT} - \frac{1}{2}e^{j4\pi fT} \tag{7.93}$$

$$= 1 - \cos(4\pi fT) \tag{7.94}$$
$$= 2\sin^2(2\pi fT). \tag{7.95}$$

Therefore, the power spectral density of the transmitted signal is given by

$$S_V(f) = \frac{2\sin^2(2\pi fT)}{T}|G(f)|^2. \tag{7.96}$$

Through this example, we observe that we can shape the power spectral density of the transmitted signal using precoding.

7.3.2 Power Spectral Density of Bandpass Signals

So far, we have focused on the calculation of the power spectral density of transmitted signals via linear modulation. Our computation was for both baseband and bandpass transmission schemes. For the latter, what we computed was the power spectral density of the lowpass-equivalent signal. Let us now explicitly determine the power spectral density of the transmitted bandpass signal.

The transmitted signal with bandpass PAM, PSK, or QAM is of the form

$$U(t) = Re\{V(t)e^{j2\pi f_c t}\}, \tag{7.97}$$

where

$$V(t) = \sum_{n=-\infty}^{\infty} a_n g_T(t - nT) \tag{7.98}$$

is the lowpass-equivalent transmitted signal and f_c is the carrier frequency. In general, the $\{a_n\}$ are complex symbols for the case of PSK and QAM, while they are real for bandpass PAM.

The power spectral density of the transmitted bandpass signal $U(t)$, denoted by $S_U(f)$, is given as

$$S_U(f) = \frac{1}{4}(S_V(f - f_c) + S_V(f + f_c)). \tag{7.99}$$

Therefore, the development of the previous subsection is readily applicable, that is, we can first calculate the power spectral density of the lowpass-equivalent signal, and then simply use the above formula to obtain the PSD of the actual transmitted bandpass signal.

Let us give a simple illustration.

Example 7.8

Consider the transmission of uncorrelated and equally likely 16-QAM symbols $\{a_n\}$, with $\mathbb{E}[|a_n|^2] = 2\sigma_a^2$. Assuming that the shaping pulse used is rectangular, $g_T(t) = \frac{1}{\sqrt{T}}\Pi\left(\frac{t}{T}\right)$, and that the carrier frequency is f_c, determine the power spectral density of the transmitted signal.

Solution

The power spectral density of the lowpass-equivalent transmitted signal is given by (see Example 7.3.1)

$$S_V(f) = \sigma_a^2 \mathrm{sinc}^2(fT). \tag{7.100}$$

Therefore, the power spectral density of the actual transmitted (bandpass) signal becomes

$$S_U(f) = \frac{\sigma_a^2}{4}\mathrm{sinc}^2((f - f_c)T) + \frac{\sigma_a^2}{4}\mathrm{sinc}^2((f + f_c)T). \tag{7.101}$$

$S_U(f)$ is depicted in Fig. 7.10. The figure shows that the transmitted signal occupies a frequency band around f_c, as expected.

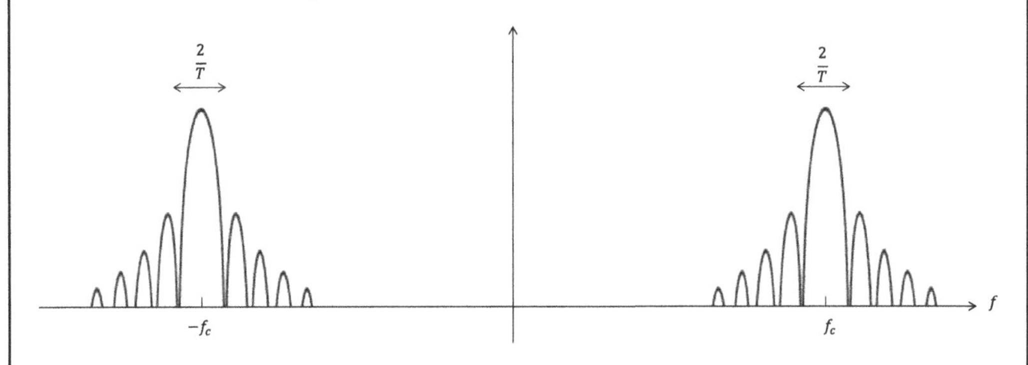

Figure 7.10 Power spectral density of 16-QAM signaling with a rectangular shaping pulse.

7.4 Chapter Summary

In this chapter, we focused on the spectral occupancy of digitally modulated signals. Through this development, we can determine how a channel bandwidth can be shared among different digital communication systems, how much of the total transmitted power is within the allocated communication band, and what fraction spills out of the band, causing interference to other communication systems operating in adjacent frequency bands. We also described a way of transmit and receive pulse design to avoid intersymbol interference in a digital communication system. As a by-product, we identified the minimum channel bandwidth required for sending digitally modulated pulses with no intersymbol interference and determined the spectral efficiencies of PAM, PSK, and QAM.

PROBLEMS

7.1 A bandlimited AWGN channel has frequency response

$$C(f) = \frac{1}{|1 + \exp(-j\pi f/W)|}$$

for $|f| < W$, and 0 otherwise (the channel phase is zero). Consider the use of a transmit pulse with frequency response $G_T(f)$ and a receive pulse with frequency response $G_R(f)$ to transmit binary PAM signals over this channel. We consider three alternatives to remove the effects of the non-ideal channel (to make sure that there is no intersymbol interference):

Case 1: Precompensate the effect of the channel at the transmitter, that is, $G_T(f) = 1/C(f)$ and $G_R(f) = 1$.

Case 2: Split the compensation of the channel between the transmitter and the receiver by using $G_T(f) = G_R(f) = 1/\sqrt{C(f)}$.

Case 3: Compensate the effect of the channel at the receiver, that is, $G_T(f) = 1$ and $G_R(f) = 1/C(f)$.

Assume there is no timing error. Since the ISI is completely removed, we use the receive filter outputs directly for demodulation (using a simple threshold rule).

Compare the above three alternatives in terms of the average error rates. Rank them from best to worst and indicate the difference in their performance (expressed in dB).

7.2 Consider transmission over a (baseband) channel with impulse response $c(t)$ using a transmit pulse $g_T(t)$ and a receive pulse $g_R(t)$. Define $x(t) = g_T(t) * c(t) * g_R(t)$. Denote the transmitted symbols by the sequence $\{a_m\}$, $m \in \mathbb{Z}$. We sample the receive filter output $y(t)$ periodically and obtain $y_m = y(mT)$, where T is the symbol duration and there is no timing error. The channel bandwidth is W.

We decide to use partial response signaling and obtain

$$y_m = a_m + a_{m-1} + \text{noise}$$

(instead of obliterating the intersymbol interference).

(a) Assuming that $2WT = 1$, find the Fourier transform of $x(t)$, denoted by $X(f)$, to satisfy the given partial response signaling condition.

(b) Determine the corresponding time-domain pulse $x(t)$. How do the tails of $x(t)$ decay? Would the resulting ISI be finite if there is a mistiming in sampling the filter output at the receiver?

7.3 Assume that the transmit and receive pulses used over a communication channel are selected such that there is intersymbol interference and the received signal is given by

$$y_m = \frac{1}{3}a_{m-1} + a_m + \frac{1}{3}a_{m+1} + w_m, \quad m = \cdots, -2, -1, 0, 1, 2, \cdots,$$

where the $\{a_m\}$ take on $+1$ or -1 with $1/2$ probability each, and they are independent. The noise samples $\{w_m\}$ are independent and identically distributed Gaussian random variables with zero mean and variance σ^2.

(a) As a first option, we employ symbol-by-symbol detection using the equivalent received signal sequence $\{y_m\}$. In other words, we employ a threshold rule on the corresponding channel output y_m to decide on the bit a_m. Compute the resulting probability of error.

(b) As a second option, we use a three-tap linear filter and form

$$z_m = -\frac{3}{7}y_{m+1} + \frac{9}{7}y_m - \frac{3}{7}y_{m-1},$$

and decide on the mth transmitted bit using a threshold rule applied to z_m.

Compute the resulting error probability. Does the use of the filter help to reduce the error probability?

7.4 Assume that the transmit and receive pulses used over a communication channel are such that an equivalent channel with intersymbol interference is observed. That is, the channel input–output relationship is given by

$$y_m = \frac{3}{10}a_{m-1} + a_m + \frac{3}{10}a_{m+1} + w_m, \quad m = \cdots, -2, -1, 0, 1, 2, \cdots,$$

where the $\{a_m\}$ are the symbols transmitted and the $\{w_m\}$ represent noise.

We decide to use a linear filter at the receiver and obtain

$$z_m = c_{-2}y_{m+2} + c_{-1}y_{m+1} + c_0 y_m + c_1 y_{m-1} + c_2 y_{m-2},$$

where the c_n are the filter tap coefficients.

Determine the set of equations to be solved (expressed in matrix form) to eliminate the intersymbol interference from the four nearest symbols, namely, from $a_{m-2}, a_{m-1}, a_{m+1}$, and a_{m+2} (in determining a_m).

7.5 Determine the maximum possible transmission rates (in Mbps) over a channel with a bandwidth of 16 MHz with the following modulation schemes:

(a) QAM with $M = 256$;
(b) PSK with $M = 32$;
(c) bandpass PAM with $M = 64$ (using a single-sideband pulse);
(d) bandpass PAM with $M = 64$ (using a double-sideband pulse).

7.6 Solve Problem 7.5 assuming that a raised cosine pulse with 50% excess bandwidth is used as the shaping pulse.

7.7 Assume that we have a bandwidth of 200 kHz available for transmission over a (bandpass) channel. Determine the maximum transmission rate possible with the following modulation schemes: 16-PSK, 256-QAM, assuming that (for the baseband-equivalent transmitted signal) a sinc pulse (in the time domain) with a bandwidth equal to the channel bandwidth is used.

7.8 Solve Problem 7.7 assuming that a raised cosine pulse with a 25% excess bandwidth is used.

7.9 Show that the raised cosine pulse satisfies the Nyquist criterion for no intersymbol interference.

7.10 The Fourier transform of the raised cosine pulse is given in (7.23). Compute the inverse Fourier transform to show that the raised cosine pulse in the time domain is as given in (7.24).

7.11 Assume that a transmitted signal is given by

$$V(t) = A\sum_{n=-\infty}^{\infty} a_n g(t - nT),$$

where A is a constant, $g(t)$ is a shaping pulse, and T is the symbol period. Also, $a_n = b_n - b_{n-3}$, where the $\{b_n\}$ are i.i.d. symbols with $\mathbb{P}(b_n = 1) = p$ and $\mathbb{P}(b_n = -1) = 1 - p$.

Compute the power spectral density of $V(t)$.

7.12 A baseband signal is given by

$$V(t) = \sum_n a_n g(t - nT),$$

where $g(t)$ is a shaping pulse, and T is the symbol period. Suppose that $a_n = b_n - b_{n-2}$, where $b_n \in \{-1, 0, 1\}$. Also assume that the $\{b_n\}$ are uncorrelated, and $\mathbb{P}(b_n = -1) = \mathbb{P}(b_n = 0) = \mathbb{P}(b_n = 1) = 1/3$.

(a) Determine the autocorrelation sequence of $\{a_n\}$.

(b) Determine the power spectral density of $V(t)$.

7.13 A baseband PAM signal is given by

$$V(t) = \sum_n a_n g(t - nT),$$

where $g(t)$ is a shaping pulse and T is the symbol period. Suppose that $a_n = b_n - b_{n-2}$, where b_n is a sequence of uncorrelated binary $+1$- or -1-valued random variables with $\mathbb{P}(b_n = -1) = \mathbb{P}(b_n = 1) = 1/2$.

(a) Determine the autocorrelation sequence of $\{a_n\}$.

(b) Determine the power spectral density of $V(t)$.

(c) Repeat part (b) if the possible values of b_n are 0 or 1 (again with equal probabilities).

7.14 Prove the relationship between the power spectral density of a bandpass signal and that of its lowpass equivalent given in (7.99).

7.15 Assume that

$$V(t) = A \sum_{n=-\infty}^{\infty} a_n g(t - nT)$$

is the lowpass equivalent of the modulated signal for a particular digital communication scheme, where $g(t)$ is a shaping pulse and T is the symbol period. The transmitted symbols are

$$a_n = \frac{1}{2}(b_n - b_{n-1}),$$

where the b_n are independent with $\mathbb{P}(b_n = -1) = \mathbb{P}(b_n = +1) = 1/2$.

(a) Determine the power spectral density of the transmitted bandpass signal assuming that the reference frequency is f_c.

(b) In this part, assume that $a_n = 2(b_n \oplus b_{n-1}) - 1$, where $\{b_n\}$ are independent with $\mathbb{P}(b_n = 0) = \mathbb{P}(b_n = 1) = 1/2$ and $\oplus$ denotes modulo 2 addition. What is the power spectral density of the modulated signal? Which one (the signal in part (a) or that in part (b)) is more appropriate to shape the power spectral density of the transmitted signal?

7.16 Assume that

$$V(t) = A \sum_{n=-\infty}^{\infty} a_n g(t - nT)$$

is the lowpass equivalent of a transmitted signal, where A is a constant, $g(t)$ is a shaping pulse, and T is the symbol period. The transmitted sequence is given by

$$a_n = -b_{n-1} + 2b_n - b_{n+1},$$

where the $\{b_n\}$ are i.i.d. with $\mathbb{P}(b_n = 1) = \mathbb{P}(b_n = -1) = 1/2$.

(a) Find the power spectral density of $V(t)$.

(b) Find the power spectral density of the bandpass signal transmitted, assuming that the reference frequency is f_c.

COMPUTER PROBLEMS

7.17 Generate a realization of a 4-ary baseband PAM signal, including 20–30 consecutive symbol transmissions using a square-root raised cosine pulse with an excess bandwidth of 25%. Take a sampling rate at least 20 times the symbol rate. Do not consider any additive noise.

(a) Pass the signal generated through a filter matched to the transmit pulse, and mark the correct sampling instances to demodulate the transmitted symbols. What do you observe? Are you able to recover the transmitted symbols correctly?

(b) Assume that there is a timing error, and the sampling instance is erroneous by $T_s/10$. What do you observe in this case? Is there an intersymbol interference? How large is the ISI?

(c) Repeat parts (a) and (b) using a rectangular shaping pulse.

(d) Repeat parts (a) and (b) using a square-root raised cosine pulse with an excess bandwidth of 100%.

7.18 Generate a long realization of a binary baseband PAM modulated signal using a sampling period $T_b/10$, where T_b is the bit period. Assume that the bits are equally likely. Using MATLAB's *pwelch* function, estimate the power spectral density of the resulting signals

(a) using a rectangular shaping pulse,

(b) using a half-sinusoid shaping pulse.

Compare the power spectral density estimates with the theoretical expectations.

7.19 Generate a long realization of a binary on–off modulated signal using a sampling period $T_b/10$, where T_b is the bit period. Assume that the bits are equally likely. Using MATLAB's *pwelch* function, estimate the power spectral density of the resulting signals

(a) using a rectangular shaping pulse,

(b) using a smoother pulse of your choice.

Compare the power spectral density estimates with the theoretical expectations.

7.20 Generate a long realization of a binary PAM signal using a raised cosine shaping pulse assuming equally likely bits. Consider at least 11 symbol periods for each pulse and a sampling rate at least ten times the bit rate. Using the long realization you have generated, estimate the power spectral density of the resulting PAM signal for an excess bandwidth of

(a) 10%,
(b) 25%,
(c) 50%,
(d) 100%.

You may use MATLAB's *pwelch* command. Are your results consistent with theoretical expectations?

7.21 Consider a baseband signal given by

$$V(t) = \sum_n a_n g(t - nT),$$

where $g(t)$ is a shaping pulse, T is the symbol period, and $a_n = b_n - b_{n-2}$, where the $\{b_n\}$ are uncorrelated and $\mathbb{P}(b_n = -1) = \mathbb{P}(b_n = 1) = 1/2$.

Generate a long realization of the modulated signal assuming a rectangular pulse and a sampling rate around ten times the symbol rate. Using MATLAB's *pwelch* command, estimate the power spectral density of the resulting modulated signal. Do you observe that inducing correlations among the transmitted symbols can shape the power spectral density? Is the PSD more compact than the case of no correlation?

8 Multicarrier Digital Modulation

In the previous three chapters, we covered the fundamentals of digital modulation systems (specifically, those employing linear modulation), single-carrier bandpass transmission, and the spectral content of digitally modulated signals. We have argued that there are different methods of embedding digital information on different attributes of the transmitted signals. For instance, in pulse amplitude modulation, the amplitude of a basic (baseband or bandpass) pulse determines the transmitted symbol, while the phase of a sinusoidal signal identifies the transmitted symbol in phase-shift keying; both amplitude and phase determine the transmitted symbol in quadrature amplitude modulation.

In this chapter we demonstrate that it is also possible to perform digital modulation using carrier signals with different frequencies, resulting in multicarrier communications. For instance, one can pick multiple sinusoidal signals with different frequencies to convey digital information, resulting in frequency-shift keying (FSK). While this is also a bandpass communication scheme, it is fundamentally different from those studied in Chapter 6, which employ a single carrier frequency. In addition, we can divide the available frequency band into subbands and transmit over these subchannels simultaneously using, for example, quadrature amplitude modulation, through a scheme called orthogonal frequency-division multiplexing (OFDM). Our objective in this chapter is to cover FSK and one of its variants, and OFDM in some detail.

The chapter is organized as follows. We describe the basics of frequency-shift keying in Section 8.1. Specifically, we focus on the case of orthogonal FSK and study both coherent and non-coherent reception over AWGN channels. We then focus on a particular case of binary orthogonal FSK which preserves the phase continuity at the symbol boundaries, namely, minimum-shift keying (MSK) in Section 8.2. We study OFDM in some depth in Section 8.3. Along with a basic description, we highlight how OFDM can be efficiently implemented through fast Fourier transform (FFT) and its inverse. Also, we describe addition of a cyclic prefix as a way to avoid intersymbol interference over dispersive channels and compute the spectral efficiency of OFDM. The chapter is concluded in Section 8.4.

8.1 Frequency-Shift Keying

In Chapter 6, we have introduced bandpass digital communication schemes based on differentiating different symbols using the amplitude and/or phase of a single-frequency sinusoidal signal. Another attribute of a carrier signal which can be used to differentiate among different symbols is its frequency. Frequency-shift keying is based on this principle, namely, the digital information is embedded in the frequency of a sinusoidal carrier signal.

For M-ary FSK signaling, to transmit the ith symbol, we employ the signal

$$s_i(t) = A\cos(2\pi f_i t), \quad t \in [0, T_s), \tag{8.1}$$

where T_s is the symbol period, $f_i \gg 1/T_s$, and $i = 1, 2, \ldots, M$. That is, a sinusoid with frequency f_i is used to represent the ith symbol.

When FSK signals with different frequencies are transmitted through a channel, even though the propagation delays are the same, they will undergo different phase shifts. This is because the amount of phase shift is equal to $2\pi f_i \tau$, with τ denoting the propagation delay. This is different from the case of single-carrier signaling considered in Chapter 6, for which the introduced channel phase shifts corresponding to the transmitted signals for different symbols were identical. In the case of FSK, the channel phase shifts experienced at different frequencies are different.

Neglecting noise and attenuation, the received signal corresponding to transmission of $s_i(t) = A\cos(2\pi f_i t)$ is given by

$$A\cos(2\pi f_i t + \phi_i), \quad \text{for } t \in [0, T_s), \tag{8.2}$$

where ϕ_i denotes the phase shift corresponding to the frequency f_i, $i = 1, 2, \ldots, M$.

We differentiate between two cases: (1) the channel phases are estimated for all the frequencies used and employed in the detection process (i.e., a coherent receiver is adopted); and (2) the channel phases are not estimated, hence, they are not available at the receiver (i.e., a non-coherent receiver is employed). We will study these two cases separately as they require different treatments.

Coherent Receiver. With coherent reception, the channel phases for different frequencies used in FSK modulation are estimated, and they are available at the receiver, as such they can be compensated. That is, the received signal corresponding to each symbol will be the same as the transmitted one:

$$s_i(t) = A\cos(2\pi f_i t), \quad \text{for } t \in [0, T_s) \tag{8.3}$$

(neglecting channel attenuation and noise effects). Therefore, we can directly work with this signal set using the tools developed in Chapter 5 to identify the signal space and develop the optimal receiver structure.

An important special case is orthogonal FSK with coherent detection, for which

$$\langle s_i(t), s_j(t) \rangle = 0, \tag{8.4}$$

for $i \neq j$, $i,j = 1,2,\ldots,M$. It can be shown that this orthogonality can be achieved by selecting the frequency separations as integer multiples of $\frac{1}{2T_s}$. To see this: take $f_i - f_j = \frac{k}{2T_s}$ for $k \in \mathbb{Z}$, and $k \neq 0$, and compute

$$\langle s_i(t), s_j(t)\rangle = \int_0^{T_s} A^2 \cos(2\pi f_i t)\cos(2\pi f_j t)dt \tag{8.5}$$

$$= \frac{A^2}{2}\int_0^{T_s} \cos\left(2\pi(f_i + f_j)t\right) + \cos\left(2\pi(f_i - f_j)t\right)dt \tag{8.6}$$

$$= \frac{A^2}{2}\int_0^{T_s} \cos\left(2\pi(f_i + f_j)t\right) + \cos\left(\frac{\pi k t}{T_s}\right)dt, \tag{8.7}$$

where the first term evaluates to 0 since $f_i + f_j \gg 1/T_s$ (recall our arguments in Chapter 6 when evaluating similar integrals), and the second term can be computed in closed form, resulting in

$$\langle s_i(t), s_j(t)\rangle = \frac{A^2 T_s}{2\pi k}\sin(\pi k t/T_s)\Big|_0^{T_s} \tag{8.8}$$

$$= 0. \tag{8.9}$$

This result establishes that, with coherent detection, we can select the frequencies with separation $1/2T_s$ to obtain orthogonal FSK. Namely, we choose

$$f_1,\ f_2 = f_1 + \Delta f,\ f_3 = f_1 + 2\Delta f, \ldots, f_M = f_1 + (M-1)\Delta f, \tag{8.10}$$

with $\Delta f = 1/2T_s$. Since all the signals used are orthogonal, the signal space is M-dimensional, and we can obtain a basis for it by simply normalizing them. The M orthonormal basis functions for the employed signal set are

$$\psi_i(t) = \sqrt{2/T_s}\cos(2\pi f_i t), \quad i = 1,2,\ldots,M. \tag{8.11}$$

Hence, we obtain the vector representation for the orthogonal FSK signals as

$$s_1 = \begin{bmatrix} A\sqrt{T_s/2} \\ 0 \\ 0 \\ \vdots \\ 0 \end{bmatrix},\quad s_2 = \begin{bmatrix} 0 \\ A\sqrt{T_s/2} \\ 0 \\ \vdots \\ 0 \end{bmatrix},\quad s_3 = \begin{bmatrix} 0 \\ 0 \\ A\sqrt{T_s/2} \\ \vdots \\ 0 \end{bmatrix},\ldots,\quad s_M = \begin{bmatrix} 0 \\ 0 \\ 0 \\ \vdots \\ A\sqrt{T_s/2} \end{bmatrix}. \tag{8.12}$$

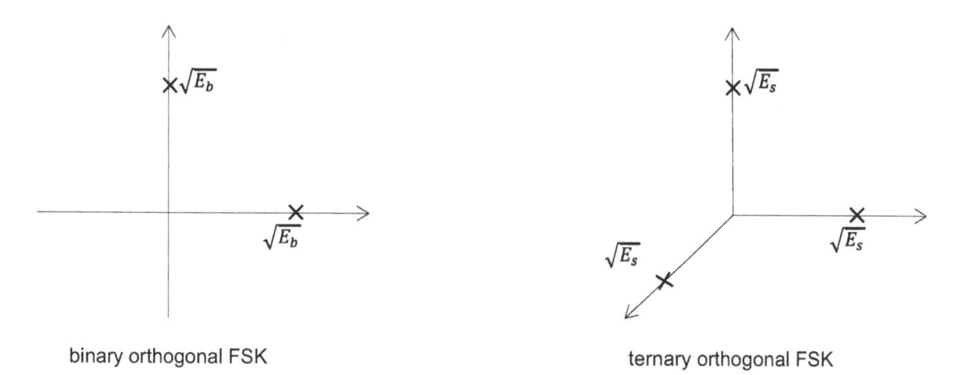

Figure 8.1 Orthogonal FSK signal constellations for $M = 2$ and $M = 3$.

All the signals have equal energy $E_s = A^2T_s/2$, hence we can also write

$$s_1 = \begin{bmatrix} \sqrt{E_s} \\ 0 \\ 0 \\ \vdots \\ 0 \end{bmatrix}, \quad s_2 = \begin{bmatrix} 0 \\ \sqrt{E_s} \\ 0 \\ \vdots \\ 0 \end{bmatrix}, \quad s_3 = \begin{bmatrix} 0 \\ 0 \\ \sqrt{E_s} \\ \vdots \\ 0 \end{bmatrix}, \ldots, \quad s_M = \begin{bmatrix} 0 \\ 0 \\ 0 \\ \vdots \\ \sqrt{E_s} \end{bmatrix}. \tag{8.13}$$

The signal constellations for binary FSK and FSK with $M = 3$ cases are depicted in Fig. 8.1.

We are now in a position to derive the ML decision rule for orthogonal FSK with coherent detection. Assuming an AWGN channel with power spectral density $N_0/2$, the received signal when $\boldsymbol{s}_i$ is transmitted is given by

$$\boldsymbol{r} = \boldsymbol{s}_i + \boldsymbol{n}, \tag{8.14}$$

where $\boldsymbol{n}$ is the $M \times 1$ noise vector whose elements are i.i.d. Gaussian random variables with zero mean and variance $N_0/2$. (Recall our development in Chapter 5 for details.)

We need to minimize the squared Euclidean distance between the received vector and the constellation points. The squared Euclidean distance between $\boldsymbol{r}$ and $\boldsymbol{s}_i$ is given by

$$||\boldsymbol{r} - \boldsymbol{s}_i||^2 = \left(r_i - \sqrt{E_s}\right)^2 + \sum_{k=1,k\neq i}^{M} r_k^2 \tag{8.15}$$

$$= E_s - 2\sqrt{E_s}r_i + \sum_{k=1}^{M} r_k^2. \tag{8.16}$$

Since the first and third terms are independent of $\boldsymbol{s}_i$, minimizing this expression over the candidate symbols is equivalent to minimizing the middle term, or simply maximizing r_i. That is, the ML decision rule becomes

$$\hat{m} = \underset{m=1,2,\ldots,M}{\arg\max}\ r_m. \tag{8.17}$$

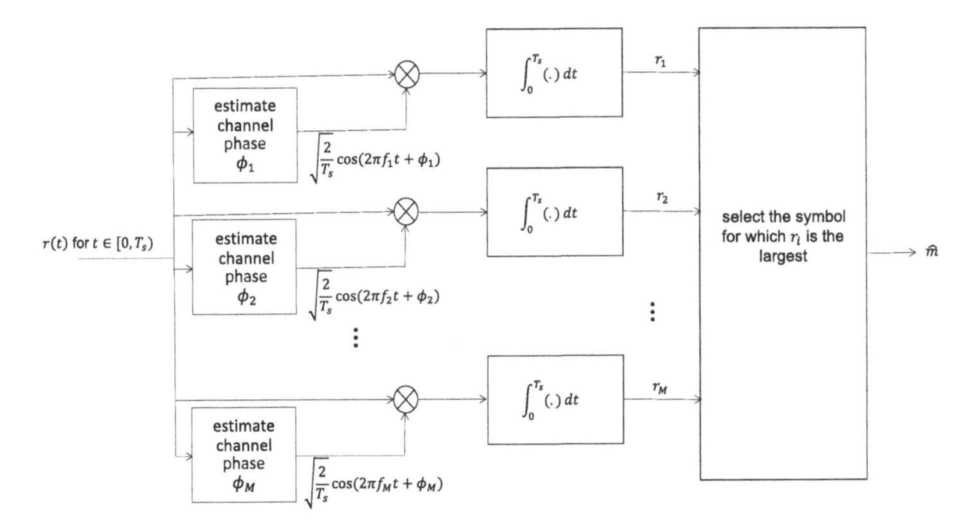

Figure 8.2 ML receiver structure for orthogonal FSK with coherent detection.

In short, the receiver simply picks the symbol with the largest correlator output (or matched filter output) as its decision.

The overall ML receiver structure is depicted in Fig. 8.2. Note that compensating for the channel phase in the received signal (as loosely stated above) is equivalent to utilizing functions with proper phase shifts accounting for the channel phase shift in the correlation-type receiver when producing the decision variables. This is explicitly depicted in the figure.

Let us compute the error probability for orthogonal FSK with coherent detection for the binary case ($M = 2$) with equally likely priors. In this case, we have a two-dimensional signal set with constellation points

$$s_1 = \begin{bmatrix} \sqrt{E_b} \\ 0 \end{bmatrix}, \quad s_2 = \begin{bmatrix} 0 \\ \sqrt{E_b} \end{bmatrix}, \tag{8.18}$$

with E_b denoting the energy per bit. Since there are only two constellation points, the bit error probability computation becomes the same as the pairwise error probability computation carried out in Chapter 5. Hence,

$$P_b = Q\left(\sqrt{\frac{||s_1 - s_2||^2}{2N_0}}\right) = Q\left(\sqrt{\frac{E_b}{N_0}}\right) = Q(\sqrt{\gamma_b}), \tag{8.19}$$

where $\gamma_b = E_b/N_0$ is the signal-to-noise ratio per bit. Comparing this result with that of BPSK from Chapter 6, we observe that binary orthogonal FSK (BFSK) is 3 dB inferior. That is, we need a 3 dB higher signal-to-noise ratio with BFSK to achieve the same error probability as BPSK.

Error probability calculations can be performed for higher-order signaling as well; however, the resulting expressions are more involved, and they are omitted.

Non-coherent Receiver. We now consider the case of non-coherent FSK, that is, the case when the channel phases are not estimated; hence, they are not available on the receiver side. Since the phase estimation step is avoided, this receiver is less complex than the coherent one, and it may be cheaper to implement. However, its performance is worse as less information about the received signal is utilized by the receiver.

For non-coherent FSK, the received signals (excluding channel attenuation and noise) are of the form

$$A\cos(2\pi f_i t + \phi_i), \quad \text{for } t \in [0, T_s), \tag{8.20}$$

$i = 1, 2, \ldots, M$, where ϕ_i is the channel phase introduced at frequency f_i. We model the phases as random variables on their full range, that is, ϕ_i is uniform on $[0, 2\pi)$. Notice that since the frequencies of the transmitted signals are not the same, the corresponding channel phase shifts (ϕ_i) are different for different symbols, and they can be modeled to be independent of each other.

Consider two sinusoids $x_1(t) = \cos(2\pi f_i t + \phi_i)$ and $x_2(t) = \cos(2\pi f_j t + \phi_j)$ (on the interval $[0, T_s)$) with two different frequencies and two different phases. We can show that these two sinusoids are orthogonal if their frequency separation is an integer multiple of $1/T_s$, that is, if $f_j = f_i + k/T_s$ for some non-zero $k \in \mathbb{Z}$. We have

$$\langle x_1(t), x_2(t)\rangle = \int_0^{T_s} \cos(2\pi f_i t + \phi_i)\cos(2\pi f_j t + \phi_j)dt \tag{8.21}$$

$$= \int_0^{T_s} \cos(2\pi(f_i + f_j)t + \phi_i + \phi_j) + \cos(2\pi(f_i - f_j)t + \phi_i - \phi_j)dt \tag{8.22}$$

$$= \int_0^{T_s} \cos(2\pi(f_i + f_j)t + \phi_i + \phi_j) + \cos(2\pi kt/T_s + \phi_i - \phi_j)dt \tag{8.23}$$

$$= 0 + \left.\frac{\sin(2\pi kt/T_s + \phi_i - \phi_j)}{2\pi k/T_s}\right|_0^{T_s} \tag{8.24}$$

$$= 0, \tag{8.25}$$

where the first integral in the third line is 0, since we take the frequencies significantly larger than the inverse of the symbol period (i.e., $f_i + f_j \gg 1/T_s$), and the second one is directly evaluated. This result shows that two sinusoids over a period of T_s seconds, each with an arbitrary phase, are orthogonal as long as their frequency separation is a multiple of $1/T_s$. Note also that this is indeed the minimum separation between the frequencies for the two sinusoids with arbitrary phases to be orthogonal.

Recall from the previous discussion that for the coherent FSK case, since the phases can be aligned, a frequency separation of $1/2T_s$ is sufficient for orthogonality; however, for non-coherent FSK, we need twice this separation to make the signals orthogonal.

An important special case is orthogonal FSK with non-coherent detection. In this case, we can select the M frequencies as

$$f_1,\ f_1+\frac{1}{T_s},\ f_1+\frac{2}{T_s},\ \ldots,\ f_1+\frac{M-1}{T_s}. \tag{8.26}$$

The received signal through an AWGN channel when $s_i(t) = A\cos(2\pi f_i t + \phi_i)$ is transmitted is given by

$$r(t) = A\cos(2\pi f_i t + \phi_i) + n(t) \quad \text{for } t \in [0, T_s), \tag{8.27}$$

which can also be written as

$$r(t) = A\cos(\phi_i)\cos(2\pi f_i t) - A\sin(\phi_i)\sin(2\pi f_i t) + n(t) \quad \text{for } t \in [0, T_s). \tag{8.28}$$

As the phase information is not available, to determine the decision variable for the ith symbol, the receiver correlates $r(t)$ with $\cos(2\pi f_i t)$ and $-\sin(2\pi f_i t)$, respectively, to collect all the energy of the signal. Consider

$$r(t)\cos(2\pi f_i t) = A\cos(\phi_i)\cos^2(2\pi f_i t) - A\sin(\phi_i)\sin(2\pi f_i t)\cos(2\pi f_i t) + \text{noise} \tag{8.29}$$

$$= \frac{A}{2}\cos(\phi_i) + \frac{A}{2}\cos(\phi_i)\cos(4\pi f_i t) - \sin(\phi_i)\sin(4\pi f_i t) + \text{noise} \tag{8.30}$$

$$= \frac{A}{2}\cos(\phi_i) + \text{high-frequency terms} + \text{noise}. \tag{8.31}$$

Similarly,

$$-r(t)\sin(2\pi f_i t) = \frac{A}{2}\sin(\phi_i) + \text{high-frequency terms} + \text{noise}. \tag{8.32}$$

By integrating the above two products over the symbol period (or simply lowpass filtering), the high-frequency terms are eliminated and (ignoring noise effects) we can recover

$$\frac{A}{2}T_s\cos(\phi_i) \ \text{ and } \ \frac{A}{2}T_s\sin(\phi_i), \tag{8.33}$$

respectively. Therefore, by squaring these terms and adding them up, we obtain $\frac{A^2}{2}T_s^2$, regardless of the phase term. Following the same steps with the sinusoids with incorrect (but orthogonal) frequencies will result in noise only. Therefore, the outputs of these correlators can be employed to decide on the transmitted symbol. The decision rule will simply pick the branch with the largest decision variable.

The non-coherent receiver structure for orthogonal FSK is depicted in Fig. 8.3. While we only provided the intuition behind this receiver structure, we note that it is indeed the ML solution.

It is difficult but possible to conduct performance analysis and determine the average error probability of orthogonal FSK with non-coherent detection. For the special case of $M = 2$ (i.e., for binary orthogonal FSK), the resulting error probability is given by

$$P_b = e^{-\gamma_b/2}, \tag{8.34}$$

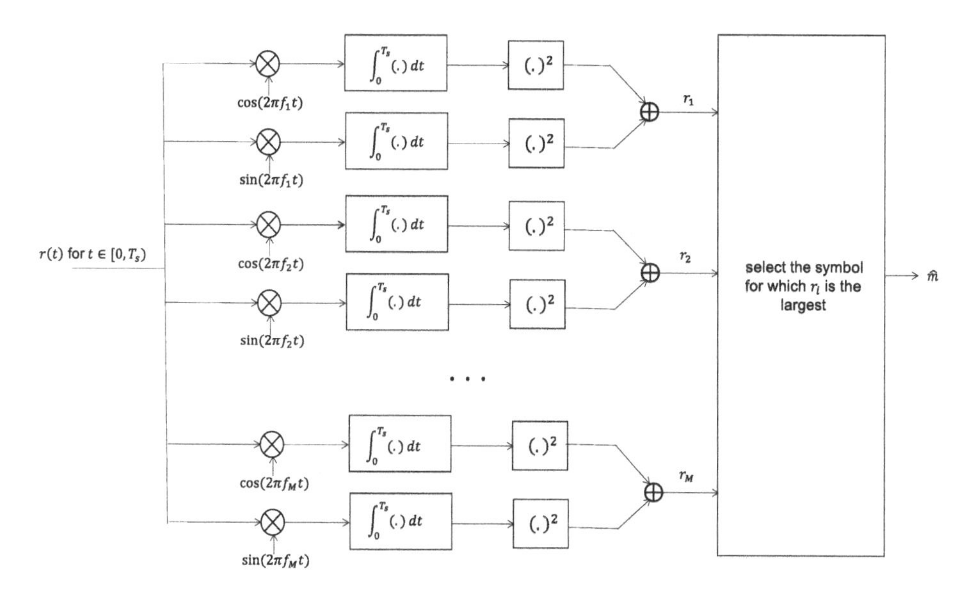

Figure 8.3 ML receiver structure for orthogonal FSK with non-coherent detection.

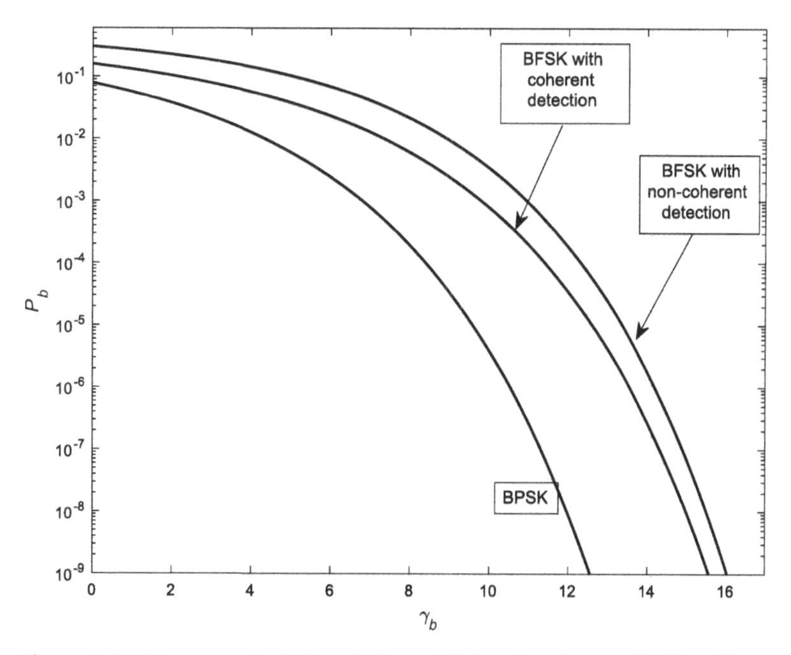

Figure 8.4 Bit error probability of binary PSK and orthogonal FSK with coherent and non-coherent detection.

with $\gamma_b = E_b/N_0$ denoting the signal-to-noise ratio per bit. Comparing this result with the error probability of orthogonal FSK with coherent detection (whose error probability is $Q(\sqrt{\gamma_b})$, we see that non-coherent detection is only slightly inferior (by about 0.8 dB for high signal-to-noise ratios). We depict the bit error probability of orthogonal FSK with coherent and non-coherent detection (along with BPSK signaling) in Fig. 8.4.

Spectral Efficiency of Orthogonal FSK. We now determine the spectral efficiency of orthogonal FSK with coherent and non-coherent detection at the receiver. Since the minimum frequency separation for orthogonal FSK with coherent detection is $1/2T$, where T is the symbol duration, the transmission bandwidth with M-ary signaling becomes $M/2T$. As the transmission rate is $\frac{\log_2 M}{T}$ bits/sec, the spectral efficiency is computed as

$$r_{\text{M-FSK, coherent}} = \frac{\log_2 M/T}{M/2T} \tag{8.35}$$

$$= \frac{2\log_2 M}{M} \text{ bits/s/Hz.} \tag{8.36}$$

For orthogonal FSK with non-coherent detection, the minimum frequency separation is $1/T$ as opposed to $1/2T$; hence, the required bandwidth is doubled, and the spectral efficiency is halved compared to that of coherently detected FSK. That is,

$$r_{\text{M-FSK, non-coherent}} = \frac{\log_2 M}{M} \text{ bits/s/Hz.} \tag{8.37}$$

We observe that with orthogonal FSK signaling, the spectral efficiency reduces with increasing constellation size. On the other hand, power efficiency increases as the minimum signal-to-noise ratio required for a given error probability reduces.[1] We observe that there is a trade-off between the power and bandwidth requirements of the system: with increasing M, the bandwidth efficiency reduces, however, the power efficiency increases, and vice versa.

Example 8.1

Determine the maximum transmission rate possible for a channel bandwidth of $B = 40$ MHz for

(1) 8-ary orthogonal FSK with coherent detection,
(2) 16-ary orthogonal FSK with non-coherent detection.

Solution

(1) With 8-ary FSK with non-coherent detection: the spectral efficiency is $r = 2\log_2(8)/8 = 3/4$ bits/s/Hz, hence a transmission rate of up to $rB = 30$ Mbps can be supported.
(2) With 16-ary FSK with non-coherent detection: $r = \log_2(16)/16 = 1/4$ bits/s/Hz, hence a transmission rate of up to $rB = 10$ Mbps can be supported.

[1] We did not cover the error probability with M-FSK except for the binary case.

8.2 Minimum–Shift Keying

In Section 8.1, we studied orthogonal frequency-shift keying. A related modulation scheme is called *minimum-shift keying*, which is a binary orthogonal FSK scheme in which phase continuity at the symbol boundaries is preserved. This section describes MSK in some detail.

Let us denote the transmitted sequence of bits by $\{b_n\}$, and define $a_n = 2b_n - 1$, that is, $b_n = 0$ is represented by $a_n = -1$, and $b_n = 1$ is represented by $a_n = +1$. Let us also denote the bit period by T_b. Then, with MSK, the transmitted signal for the nth bit is given by

$$s(t) = A\cos\left(2\pi f_c t + \pi a_n \frac{t - nT_b}{2T_b} + \theta_n\right), \quad \text{for } t \in [nT_b, (n+1)T_b). \tag{8.38}$$

Here, θ_n is the phase of the transmitted signal at the beginning of the symbol period corresponding to the nth bit. At the end of the symbol period, the phase becomes $\theta_n + \frac{\pi}{2}$ or $\theta_n - \frac{\pi}{2}$, depending on the value of the transmitted bit. Also, the phase transition throughout the bit duration is linear. See Fig. 8.5 for the phase of the MSK signal as a function of time for a given bit sequence.

By a simple rearrangement, the transmitted signal can also be written as

$$s(t) = A\cos\left(2\pi\left(f_c + a_n \frac{1}{4T_b}\right)t + \theta_n - \frac{\pi n}{2} a_n\right), \quad \text{for } t \in [nT_b, (n+1)T_b). \tag{8.39}$$

This expression clearly shows that MSK is nothing but an orthogonal FSK scheme that uses the two frequencies: $f_c + \frac{1}{4T_b}$ and $f_c - \frac{1}{4T_b}$. The difference from standard FSK is the phase continuity constraint. As the two frequencies are separated by $1/2T_b$, which is the minimum frequency separation needed to make the sinusoids orthogonal, this modulation scheme is called minimum-shift keying.

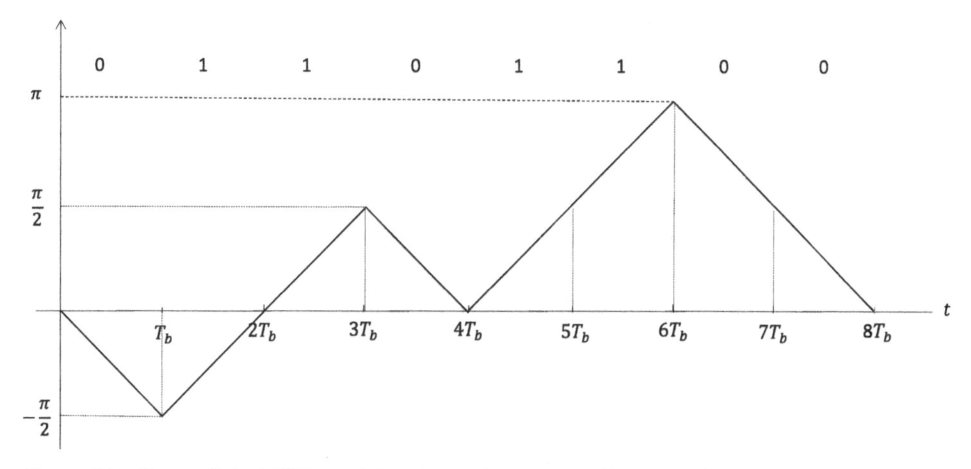

Figure 8.5 Phase of the MSK modulated signal corresponding to a bit sequence b_n of 0, 1, 1, 0, 1, 1, 0, 0. As clearly depicted, the phase transitions in each bit interval are linear.

Note that minimum-shift keying is a special case of a more general (non-linear) modulation scheme called *continuous-phase frequency-shift keying (CPFSK)*, which is a special case of *continuous-phase modulation (CPM)*. These schemes are also used in practice; however, they are beyond our scope.

As is clear from the expression of $s(t)$, the transmitted signal with MSK has a constant envelope. Also, it should be evident that MSK is a modulation scheme with memory as the previously transmitted bits are also important to generate the transmitted signal for the nth symbol period. Since the signaling scheme has memory, the demodulation process does not look straightforward, unlike the modulation schemes we have covered thus far.[2] On the other hand, we can develop an alternate representation of the transmitted signal and show that the detection of MSK signals can be greatly simplified.

Consider the MSK signal $s(t)$ given earlier:

$$s(t) = A\cos\left(2\pi f_c t + \frac{a_n \pi}{2T_b}t + \theta_n - \frac{\pi n}{2}a_n\right), \tag{8.40}$$

for $t \in [nT_b, (n+1)T_b)$. Using a trigonometric identity, we can write

$$s(t) = A\cos(2\pi f_c t)\cos\left(\frac{a_n\pi}{2T_b}t + \alpha_n\right) - A\sin(2\pi f_c t)\sin\left(\frac{a_n\pi}{2T_b}t + \alpha_n\right), \tag{8.41}$$

where $\alpha_n = \theta_n - \frac{\pi n}{2}a_n$, for $t \in [nT_b, (n+1)T_b)$.

With the reference frequency taken as f_c, the in-phase component of $s(t)$ is given by

$$s_I(t) = A\cos\left(\frac{a_n\pi}{2T_b}t + \alpha_n\right) \tag{8.42}$$

$$= A\cos\left(\frac{a_n\pi}{2T_b}t\right)\cos(\alpha_n) - A\sin\left(\frac{a_n\pi}{2T_b}t\right)\sin(\alpha_n). \tag{8.43}$$

Similarly, its quadrature component is given by

$$s_Q(t) = A\sin\left(\frac{a_n\pi}{2T_b}t + \alpha_n\right) \tag{8.44}$$

$$= A\sin\left(\frac{a_n\pi}{2T_b}t\right)\cos(\alpha_n) + A\cos\left(\frac{a_n\pi}{2T_b}t\right)\sin(\alpha_n). \tag{8.45}$$

Let us examine the α_n term more closely. We can write

$$\alpha_n = \theta_n - \frac{\pi n}{2}a_n \tag{8.46}$$

$$= \theta_{n-1} + a_{n-1}\frac{\pi}{2} - \frac{\pi n}{2}a_n \tag{8.47}$$

$$= \theta_{n-1} - (n-1)a_{n-1}\frac{\pi}{2} + n\frac{\pi}{2}(a_{n-1} - a_n) \tag{8.48}$$

$$= \alpha_{n-1} + n\frac{\pi}{2}(a_{n-1} - a_n). \tag{8.49}$$

[2] Sequence detection algorithms can be used to demodulate the modulation schemes with memory (CPM and CPFSK); however, coverage of these algorithms is beyond our scope.

Recall that $a_n \in \{-1, +1\}$, hence $a_{n-1} - a_n$ is an even number. Also, taking $\alpha_0 = 0$, we see that α_n is either 0 or π, and that α_n can only change for odd values of n. As $\alpha_n = 0$ or π, $\sin(\alpha_n) = 0$. Also, defining $d_{I,n} = \cos(\alpha_n)$ which takes on -1 or $+1$, we obtain

$$s_I(t) = A\cos(\alpha_n)\cos\left(\frac{a_n\pi}{2T_b}t\right) \tag{8.50}$$

$$= A\cos(\alpha_n)\cos\left(\frac{\pi t}{2T_b}\right) \tag{8.51}$$

$$= A\cos(\alpha_n)\sin\left(\frac{\pi(t+T_b)}{2T_b}\right), \tag{8.52}$$

for $t \in [nT_b, (n+1)T_b)$. Note that the second step follows since cosine is an even function and $a_n = \pm 1$, and the third step is due to a simple relationship between sine and cosine. Similarly, we obtain

$$x_Q(t) = A\cos(\alpha_n)\sin\left(\frac{a_n\pi}{2T_b}t\right) \tag{8.53}$$

$$= Aa_n\cos(\alpha_n)\sin\left(\frac{\pi t}{2T_b}\right), \tag{8.54}$$

for $t \in [nT_b, (n+1)T_b)$, which follows since $a_n = \pm 1$.

Let us define a pulse $p(t)$ as a half-sinusoid on the interval $[0, 2T_b)$, that is,

$$p(t) = \sin\left(\frac{\pi t}{2T_b}\right) \text{ for } t \in [0, 2T_b), \tag{8.55}$$

and 0 otherwise. Then,

$$s_I(t) = A\cos(\alpha_n)p(t+T_b) \text{ for } t \in [nT_b, (n+1)T_b) \tag{8.56}$$

and

$$s_Q(t) = Aa_n\cos(\alpha_n)p(t) \text{ for } t \in [nT_b, (n+1)T_b). \tag{8.57}$$

Recall that $\alpha_n = 0$ or π, hence $\cos(\alpha_n) = \pm 1$ and $a_n\cos(\alpha_n) = \pm 1$. Let us define these two binary quantities as

$$c_{I,n} = \cos(\alpha_n) \text{ and } c_{Q,n} = a_n\cos(\alpha_n) = a_n c_{I,n}. \tag{8.58}$$

We have shown above that α_n can only change for odd values of n, in other words, $c_{I,n}$ can change values only for odd n values, and remains the same for even n. In other words, for even n, $c_{I,n}$ remains identical to $c_{I,n-1}$, while $c_{Q,n}$ may or may not be the same as $c_{Q,n-1}$.

We can also argue that $c_{Q,n}$ can only change values for even n. This is because, due to the α_n recursion written in (8.49), α_n can change from π to 0 (or vice versa) only if a_n changes. This is equivalent to $c_{I,n}$ changing. On the other hand, $c_{Q,n} = a_n c_{I,n}$, which means that $c_{Q,n}$ remains the same for odd n. In other words, for odd n, $c_{Q,n}$ is the same as $c_{Q,n-1}$, while $c_{I,n}$ may or may not be the same as $c_{I,n-1}$.

To sum up, we obtain the in-phase and quadrature components of the transmitted signal as

$$s_I(t) = Ac_{I,n}p(t + T_b) \text{ and } s_Q(t) = Ac_{Q,n}p(t), \tag{8.59}$$

for $t \in [nT_b, (n+1)T_b)$. We note that the bits $c_{I,n}$ and $c_{Q,n}$ (can) change every $2T_b$ seconds, and their transitions are staggered by T_b with respect to each other.

For ease of notation, let us denote the transmitted bit sequence (in ± 1 form) as d_k and assume that the even-indexed bits are transmitted through the in-phase component while the odd-indexed bits are transmitted through the quadrature component. Then, the overall MSK-modulated signal can be written as

$$\begin{aligned} s(t) = & \left\{ A \sum_{n=-\infty}^{\infty} d_{2k}p(t - 2kT_b + T_b) \right\} \cos(2\pi f_c t) \\ & - \left\{ A \sum_{k=-\infty}^{\infty} d_{2k+1}p(t - 2kT_b) \right\} \sin(2\pi f_c t). \end{aligned} \tag{8.60}$$

We notice that this is nothing but an offset QPSK signal (with a half-sinusoid shaping pulse, $p(t) = \sin(\pi t/2T_b)$ for $t \in [0, T_s)$), with symbol period $T_s = 2T_b$, where the in-phase and quadrature components are staggered by T_b.

An exemplary MSK signal along with the in-phase and quadrature components are plotted in Fig. 8.6. While it is unrealistic, the carrier frequency is selected to be $f_c = 1/T_b$ for ease of illustration. The transmitted d_k bit sequence is selected as $\{1, -1, 1, -1, -1, 1, 1, 1, -1, -1, -1\}$. Hence, the bits corresponding to the in-phase component are $\{-1, -1, 1, 1, -1\}$ and the quadrature component are $\{1, 1, -1, 1, -1, -1\}$. We identify $c_{I,n}$ bits for the given signal duration from $-5T_b$ to $5T_b$ as $\{-1, -1, -1, -1, 1, 1, 1, 1, -1, -1\}$, and the $c_{Q,n}$ bits as $\{1, 1, 1, -1, -1, 1, 1, -1, -1, -1\}$. Given the relationship established earlier, $c_{Q,n} = a_n c_{I,n}$ and have $a_n = c_{I,n}c_{Q,n}$, and determine the corresponding transmitted bit sequence b_n as $\{0, 0, 0, 1, 0, 1, 1, 0, 1, 1\}$. One can match and verify from Fig. 8.6 that transmission of a 0 bit is accomplished using the frequency $f_c - \frac{1}{4T_b} = \frac{3}{4T_b}$, while transmission of a 1 bit is via the frequency $f_c + \frac{1}{4T_b} = \frac{5}{4T_b}$, as expected from the original form of the MSK signal.

We also note that the in-phase and quadrature components being transmitted are not constant envelop signals, however, their sum is.

Let us also comment on the MSK receiver structure for an AWGN channel and the corresponding bit error probability performance. We know from our development in the previous chapters that for transmission over an AWGN channel, the in-phase and quadrature components do not interfere. Therefore, it should be clear from the alternate representation given in (8.60) that, using the in-phase component of the received signal, we can detect the even-indexed bits; and, with the quadrature components, we can detect the odd-indexed bits, straightforwardly (using the standard detector for phase-shift keying).

With the transmitted bits being given by ± 1, we also conclude that the error probability performance is the same as that of BPSK. In other words,

$$P_b = Q\left(\sqrt{2\gamma_b}\right), \tag{8.61}$$

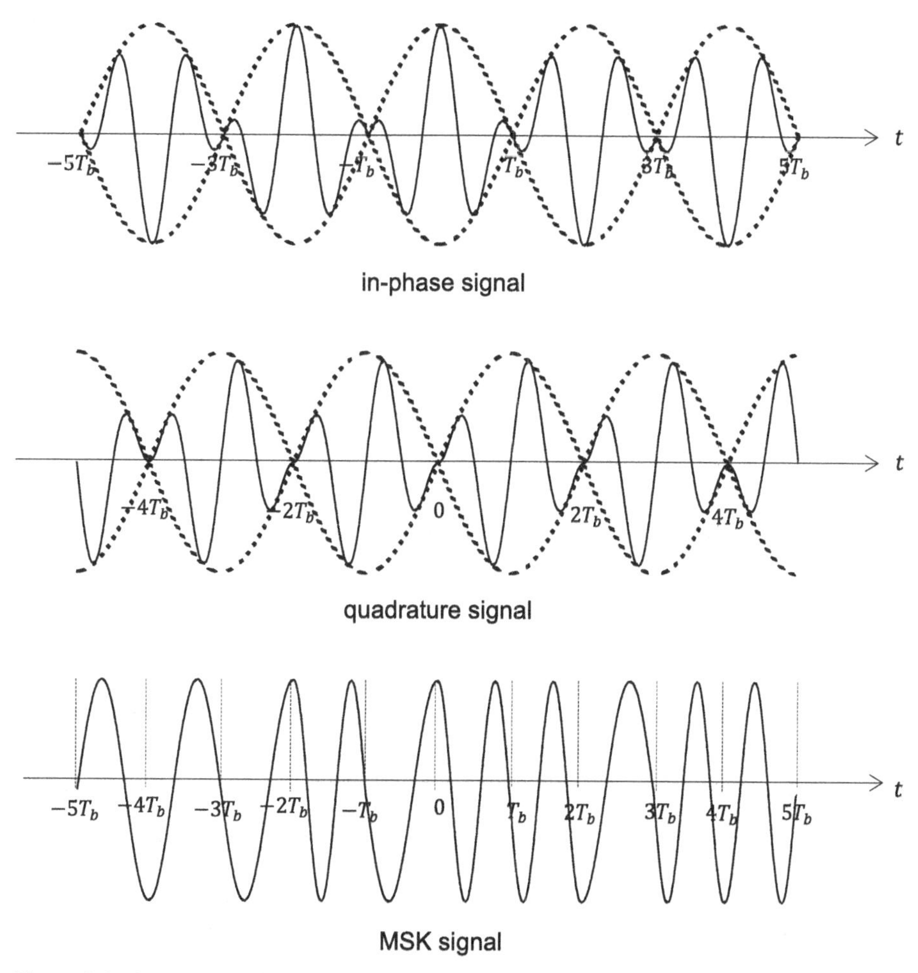

Figure 8.6 Illustration of an MSK signal. Top: in-phase component, middle: quadrature component, bottom: overall MSK signal. For clarity of depiction, f_c is taken as $1/T_b$, that is, the two frequencies used are $3/4T_b$ and $5/4T_b$.

where $\gamma_b = E_b/N_0$ denotes the signal-to-noise ratio per bit, with E_b being the energy per bit.

In the previous section, we studied the error performance of orthogonal FSK and noted that it is 3 dB inferior to that of BPSK. Since MSK is nothing but an orthogonal FSK scheme with phase continuity, the above error rate result may look counterintuitive. This is explained as follows: MSK contains memory, and the optimal detector described above effectively exploits this memory. If, as a suboptimal approach, the decisions are only based on the received signal for each bit interval separately (still with coherent detection), the same bit error probability as the orthogonal FSK would be obtained, which is 3 dB worse.

Finally, let us also comment on the spectral characteristics of MSK signaling. Assuming that the transmitted bits are i.i.d. with equal probabilities, the power spectral density of an MSK-modulated signal is given by

$$S_{U,MSK}(f) = \frac{4A^2 T_b}{\pi^2}\left(\left(\frac{\cos(2\pi T_b(f-f_c))}{1-16T_b^2(f-f_c)^2}\right)^2 + \left(\frac{\cos(2\pi T_b(f+f_c))}{1-16T_b^2(f+f_c)^2}\right)^2\right) \quad (8.62)$$

(see Problem 8.8). We observe that the tails of the power spectral density decay with the fourth power of frequency. This is a much more rapid decay compared to BPSK or QPSK (with a rectangular shaping function); hence, the resulting out-of-band spillage is limited, and the interference caused to adjacent frequency bands is small. This is due to the fact that there are no discontinuities in the transmitted signal.

Owing to its desirable properties, MSK signaling (more precisely, a variation of it called Gaussian MSK in which the instantaneous frequency transitions are smoother and the power spectral density is more compact) is employed in practical applications, including the Global System for Mobile (GSM).

8.3 Orthogonal Frequency–Division Multiplexing

Another transmission scheme with a wide range of applications from asymmetric digital subscriber lines (ADSL) to 4G and 5G wireless systems is orthogonal frequency-division multiplexing. The main idea in OFDM is to divide the overall frequency band of operation into many subbands and employ PAM, PSK, or QAM for each subchannel. The subcarriers are selected to be orthogonal to each other – as will be explained shortly – hence the name, orthogonal frequency-division multiplexing.

OFDM can be implemented efficiently using FFT at the receiver side and inverse FFT at the transmitter side. Furthermore, when properly designed, OFDM provides a simple way of transmitting over non-ideal (frequency-selective) channels, namely, it is possible to eliminate the effects of the channel's frequency selectivity (and avoid intersymbol interference altogether). We will discuss both of these points in the rest of this section.

8.3.1 Basic Description

Consider transmission over a bandpass channel with a bandwidth of W whose frequency response is as depicted in Fig. 8.7. We focus on the description for a bandpass channel (as employed in wireless communications), however, similar ideas hold for baseband transmission as well (e.g., as in asymmetric digital subscriber lines). A channel with such a frequency response affects the different frequency components of a transmitted signal differently, hence it is called *frequency-selective*. If single-carrier transmission is used over a frequency-selective channel, there will be intersymbol interference, which may complicate the receiver design and implementation. With proper design, OFDM eliminates intersymbol interference and provides a simple way of signaling over frequency-selective channels.

Let us proceed with a basic description of OFDM signaling. The overall frequency band is divided into N subbands, each with a bandwidth of Δf, that is,

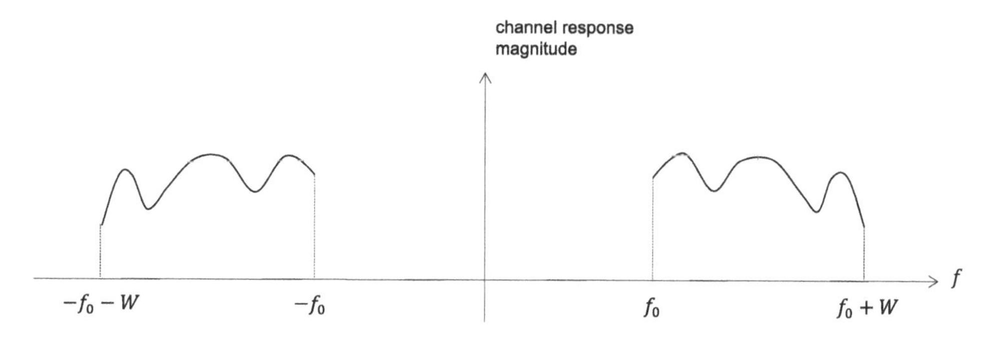

Figure 8.7 Magnitude response of an exemplary frequency selective channel.

$$N = \frac{W}{\Delta f} \tag{8.63}$$

is the number of subcarriers. If Δf is taken sufficiently small (by using a large number of subcarriers), the channel becomes almost ideal (with a flat frequency response) within each subband, and this makes the signaling particularly simple.

Let us denote the kth subcarrier frequency by f_k, with

$$f_k = f_0 + k\Delta f, \tag{8.64}$$

for $k = 0, 1, 2, \ldots, N - 1$. We select the OFDM signal duration (also called the OFDM word duration) as $T = 1/\Delta f$. All the subcarriers become orthogonal with this selection. Namely, as shown in Section 8.1,

$$\langle \cos(2\pi f_i t + \phi_i), \cos(2\pi f_j t + \phi_j) \rangle = \int_0^T \cos(2\pi f_i t + \phi_i) \cos(2\pi f_j t + \phi_j) dt = 0, \tag{8.65}$$

with $i, j \in \{0, 1, \ldots, N - 1\}$ and $i \neq j$, for arbitrary subcarrier phases ϕ_i, ϕ_j.

We employ PAM, PSK, or QAM for transmission over each subchannel. In other words, we transmit N different symbols from the selected constellation in parallel. As an example, if QAM is being used, and (the complex representation of) the transmitted symbol is given by

$$X_k = A_k e^{j\theta_k} = A_k \cos(\theta_k) + jA_k \sin(\theta_k) \tag{8.66}$$

for the kth subcarrier, the corresponding time-domain signal is given by

$$A_k \cos(\theta_k) \cos(2\pi f_k t) - A_k \sin(\theta_k) \sin(2\pi f_k t), \quad \text{for } t \in [0, T). \tag{8.67}$$

We also note that the channel effect at different subcarriers will be different. For instance, if the channel frequency response is denoted by $C(f)$, then the complex coefficient $C_k = C(f_k)$ will determine the effective channel attenuation and phase rotation for the kth subchannel. An illustration of the channel magnitude across different subchannels is depicted in Fig. 8.8.

We further note that since the channel gains (hence, the corresponding signal-to-noise ratios) for different subchannels are different, the modulation schemes can be

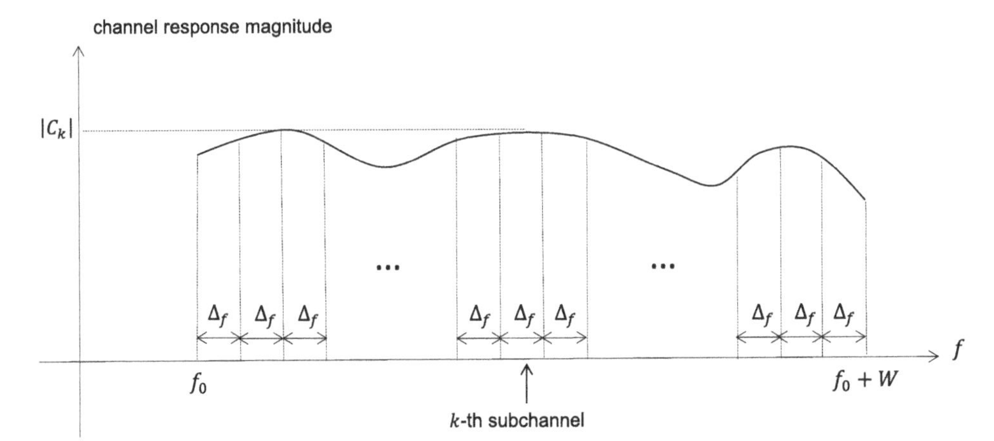

Figure 8.8 Illustration of a typical magnitude response of a frequency selective channel and division of the entire frequency band into subchannels (which are almost non-selective). The channel gain (in the frequency domain) for the kth subchannel is denoted by $|C_k|$.

tailored for them differently. For instance, for a given OFDM word, a high-order QAM can be used over a frequency band experiencing a high signal-to-noise ratio, while for a band with a lower signal-to-noise ratio, smaller constellations, such as QPSK, could be more suitable.

8.3.2 OFDM Implementation via FFT and IFFT

Consider an OFDM signal with number of subcarriers N. The continuous-time transmitted signal is given by a superposition of the signals transmitted over all the subchannels, that is,

$$x(t) = \sum_{n=0}^{N-1} A_n \cos\left(2\pi\left(f_0 + \frac{n}{T}\right)t + \theta_n\right), \quad \text{for } t \in [0, T), \tag{8.68}$$

where T is the OFDM word duration and $X_n = A_n \exp(j\theta_n)$ is the two-dimensional (complex) symbol carried on the nth subcarrier. The subcarriers employed are f_0, $f_0 + \frac{1}{T}, \ldots, f_0 + \frac{N-1}{T}$. We can (roughly) say that the frequency band occupied by this signal is from $f_0 - \frac{1}{2T}$ to $f_0 + \frac{N-1}{T} + \frac{1}{2T}$, and the OFDM word has a bandwidth of $W = \frac{N}{T}$.

Let us denote the lowpass-equivalent transmitted signal with respect to the reference frequency selected at the middle of the frequency band over which transmission takes place, that is, $f_c = f_0 + \frac{W}{2}$, by $x_l'(t)$. The signal $x_l'(t)$ can be written as

$$x_l'(t) = \left(\sum_{n=0}^{N-1} X_n e^{j2\pi nt/T}\right) e^{-j2\pi Wt/2}. \tag{8.69}$$

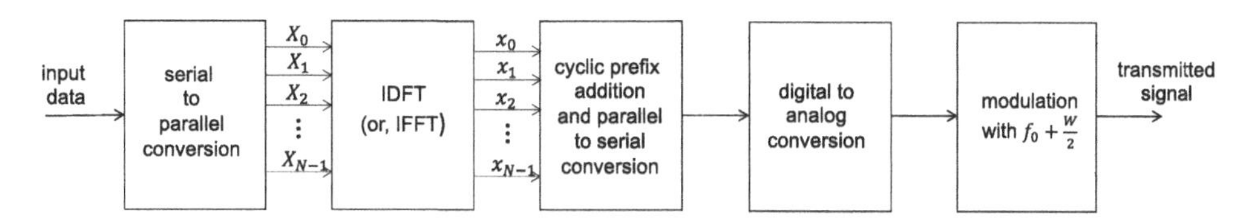

Figure 8.9 Block diagram of an OFDM transmitter.

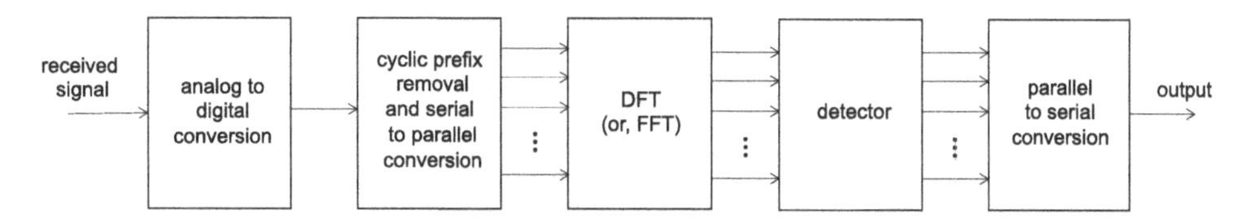

Figure 8.10 Block diagram of an OFDM receiver.

Consider the samples of the (related) complex signal

$$x_l(t) = \sum_{n=0}^{N-1} X_n e^{j2\pi nt/T}, \tag{8.70}$$

taken T/N seconds apart. Due to the sampling theorem, these samples are sufficient to reconstruct the actual continuous-time signal $x_l(t)$ as it has a bandwidth of $W/2$ and the sampling rate corresponds to the Nyquist rate. We can write

$$x_l(t)\Big|_{t=kT/N} = \sum_{n=0}^{N-1} X_n e^{j2\pi nk/N}, \tag{8.71}$$

which is nothing but a scaled version of the N-point inverse discrete Fourier transform (IDFT) of the length-N discrete-time sequence $\{X_n\}$. Let us denote these time-domain samples by $x_k = x_l(kT/N)$, $k = 0, 1, 2, \ldots, N-1$.

The above argument shows that the lowpass-equivalent version of the transmitted OFDM signal can be obtained by taking an inverse discrete Fourier transform of the transmitted (complex) symbol sequence and then converting the digital signal into an analog form. The transmitted OFDM signal can then be obtained by a simple up-conversion to the desired frequency band. Since it is very easy to obtain the time samples of the lowpass-equivalent transmitted signal through simple IDFT or inverse FFT (IFFT) operations, the resulting transmitter (and the corresponding receiver) are very simple to implement, as depicted in Figs. 8.9 and 8.10. The block diagrams also depict the cyclic prefix addition and removal processes, which are described in the next subsection.

8.3.3 Cyclic Prefix Addition

Consider the transmission of an OFDM-modulated signal over a channel for which the frequency response is not a constant over the transmission bandwidth (i.e., we have a frequency-selective channel). This is typical for many communication

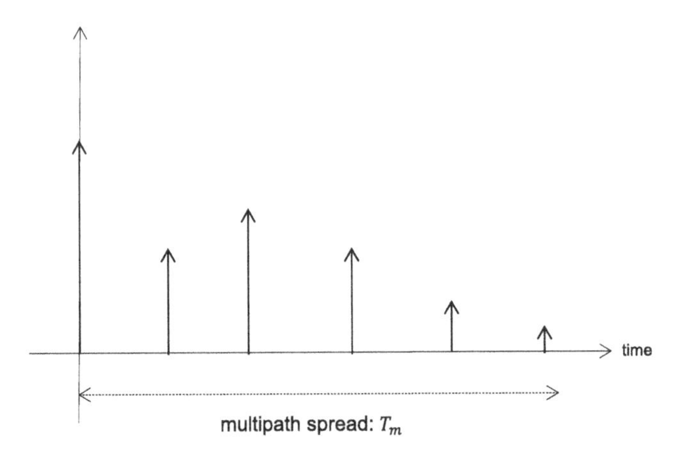

Figure 8.11 Illustration of the magnitude of the impulse response of a typical multipath channel. The impulse response extends for the duration of the multipath spread of the channel (denoted by T_m). We emphasize that this is an illustration of only the magnitude of the impulse response; in fact, the channel tap gains are modeled as complex (assuming that the baseband-equivalent model is adopted).

scenarios, including transmission over wideband wireless channels. OFDM is well motivated for exactly these types of channels. This is because, effectively, OFDM-based transmission divides the frequency band into smaller subbands by employing many subcarriers, and the resulting subchannels become almost flat over the corresponding subbands. This enables an easier receiver structure when used together with a cyclic prefix.

In the time domain, we have a channel that is modeled as a linear time-invariant filter whose impulse response extends for a certain duration, say T_m seconds. This is illustrated in Fig. 8.11. Here, we have nothing but a multipath channel for which the signal is received through several (resolvable) paths. Indeed, the path delays can be taken as multiples of $1/T$, which is the Nyquist rate (considering the bandwidth of the overall OFDM signal). Note that the channel gains of different arrivals (in the time domain) are taken as complex numbers, as we will be working with the lowpass-equivalent signals.

Given that the channel impulse response does not extend beyond T_m seconds, we can employ a cyclic prefix of length T_m to eliminate the intersymbol interference. We show this explicitly in the following. Concerning the samples of the OFDM word in the time domain (i.e., the IDFT outputs), we add a cyclic prefix by appending the m samples at the end of the IDFT block to the beginning of the OFDM word, where m is selected to be no less than $\frac{T_m}{T}N$ to cover the multipath spread of the channel. In other words, we transmit

$$\underbrace{x_{N-m}, x_{N-m+1}, \ldots, x_{N-1}}_{\text{cyclic prefix of length } m}, \underbrace{x_0, x_1, x_2, \ldots, x_{N-1}}_{\text{IDFT output}}. \tag{8.72}$$

A typical OFDM signal with a cyclic prefix is depicted in Fig. 8.12.

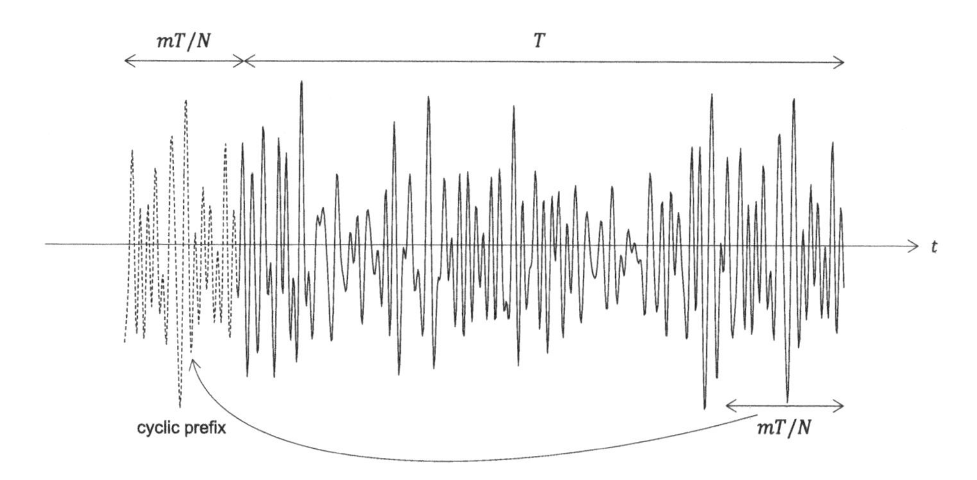

Figure 8.12 A typical OFDM signal in the time domain (with cyclic prefix).

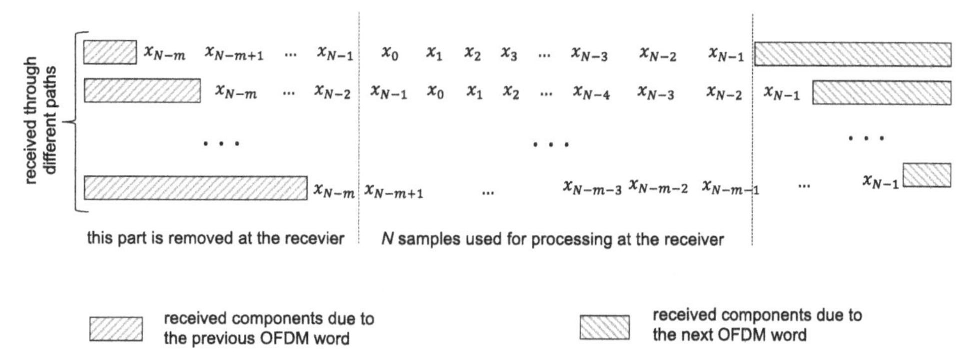

Figure 8.13 Illustration of the OFDM signal transmission through a multipath channel (with different path delays). The cyclic prefix is removed, and the DFT/FFT window is selected in such a way that there is no contribution from the previous or next OFDM words, that is, there is no intersymbol interference.

With the use of the cyclic prefix, the channel input–output relationship is governed through a circular convolution, and the input–output relationship in the frequency domain is particularly simple as the intersymbol interference is completely removed. For an illustration of how the ISI is eliminated, see Fig. 8.13.

Let us write the corresponding input–output relationship and establish the channel model with OFDM. Assume that the cyclic prefix duration (which is m in terms of the number of samples taken with period T/N) is taken to be longer than the duration of the channel impulse response (i.e., the multipath spread), and let us denote the channel tap coefficients by $h[k]$, for $k = 0, 1, \ldots, m-1$. We can write the received signal in the time domain as

$$y_n = \sum_{k=0}^{m-1} h[k]x[n-k]_N + \text{noise}, \tag{8.73}$$

where $x[l]_N$ is defined as x of $(l \mod N)$. This is nothing but

$$y_n = h[n] \circledast x[n] + \text{noise}, \tag{8.74}$$

where $\circledast$ denotes circular convolution. Taking the discrete Fourier transform (DFT), we can write the channel input–output relationship (in the frequency domain) as

$$Y_k = C_k X_k + \text{noise}, \quad k = 0, 1, \ldots, N-1, \tag{8.75}$$

where C_k is the channel frequency response at the kth subcarrier (obtained by the DFT of the time-domain channel coefficients $h[k]$). Note also that for an AWGN channel, since the time-domain noise is a white Gaussian process, the corresponding noise in the frequency domain also becomes white Gaussian. The received signal corresponding to the signal transmitted on the kth subchannel does not see any interference from the other transmitted symbols in the same OFDM word or from other OFDM words; hence, detecting the signals transmitted through different subchannels one by one is optimal.

Let us also determine the spectral efficiency of an OFDM transmission scheme. Assume that an M-ary complex signal constellation is used for each subcarrier (e.g., 16-QAM). With N subchannels, the total number of transmitted bits with one OFDM word is $\log_2(M)N$. With a cyclic prefix length of m samples, these bits are transmitted in $\left(1 + \frac{m}{N}\right) T$ seconds. Therefore, the transmission rate becomes

$$\frac{\log_2(M)N}{\left(1 + \frac{m}{N}\right) T}, \tag{8.76}$$

in bits/second. The amount of bandwidth needed for this transmission is N/T Hz, and therefore, the spectral efficiency becomes

$$\frac{N}{N+m} \log_2 M \quad \text{bits/s/Hz}. \tag{8.77}$$

We observe that adding a cyclic prefix reduces the spectral efficiency by a factor of $\frac{N}{N+m}$; otherwise, the spectral efficiency is the same as that of the modulation scheme employed on each subcarrier.

Example 8.2

Let us design an OFDM transmission scheme that employs 16-QAM on each subchannel for communicating over a bandpass channel in the frequency band 100–120 MHz. Assume that the multipath spread of the channel is 5 μs, and a cyclic prefix is to be used to remove the interference among OFDM words. Complete the design for two different cases:

(1) number of subcarriers $N = 1024$,
(2) number of subcarriers $N = 4096$.

Solution

(1) With the given channel bandwidth of 20 MHz, for the first scheme, the subcarrier separation is $\Delta f = \frac{20\text{MHz}}{1024} \approx 19.53$ kHz, and the OFDM word duration excluding

the cyclic prefix is 51.2 μs. Sampling of the OFDM word at the Nyquist rate corresponds to a sampling period of $T/N = 0.05$ μs. Since a cyclic prefix of length at least the multipath spread of the channel is needed, the minimum required CP length becomes 100 samples. Let us take the CP length as 128 samples, which corresponds to 6.4 μs. Therefore, the OFDM word duration including the CP becomes 57.6 μs. With $N = 1024$ subcarriers, employing 16-QAM, each OFDM word carries $1024 \times 4 = 4096$ bits. The overall transmission rate becomes 71.1 Mbps, or the spectral efficiency becomes $71.1/20 = 3.56$ bits/s/Hz.

Note that we could have used the derived expression in (8.77) for the spectral efficiency to directly obtain $\frac{N}{N+m} \log_2 M = \frac{1024}{1024+128} \log_2(16) = 3.56$ bits/s/Hz.

(2) For the second scheme, the number of subcarriers is $N = 4096$, which corresponds to $\Delta f = 4.88$ kHz, and an OFDM word duration (excluding the CP) of 204.8 μs. We can select the CP length as 128 samples, which corresponds to $128 \times 204.8\mu\text{s}/4096 = 6.4\mu\text{s}$ (as a value greater than the multipath spread of the channel). The OFDM word length including the CP becomes 211.2 μs. With 16-QAM, the transmission rate becomes $4096 \times 4/211.2 = 77.58$ Mbps, and the corresponding spectral efficiency is 3.88 bits/s/Hz.

8.3.4 Further Remarks and Challenges with OFDM

We close this section by making several brief comments about OFDM transmission. So far, we have assumed that the channel exhibits multipath, but it is time-invariant. In many practical (wireless communication) scenarios, the time-invariance assumption may not hold due to the movement of the transmitters and receivers, as well as the changes in the environment. A detailed discussion of time-varying channels is beyond our scope; however, we make a brief comment that such time variations will destroy the orthogonality of the subcarriers in OFDM transmission and will cause interference among adjacent subcarriers. Such interference is referred to as *intercarrier interference (ICI)*. Therefore, when designing an OFDM system, the potential time variations should be taken into account, that is, the subcarrier separation should be selected sufficiently further apart (reducing the OFDM word length in the time domain) so that the channel remains almost constant over the OFDM word, and the orthogonality of the subcarriers is preserved. Otherwise, the resulting ICI needs to be mitigated, increasing the receiver complexity.

We note that time and frequency synchronization is critical in OFDM systems. Uncompensated frequency offset may cause significant ICI, deteriorating the system performance. If there are timing offsets or timing errors, there will be ICI and ISI at the receiver output, again deteriorating the system performance. A detailed discussion of these effects and suitable synchronization algorithms is beyond our scope.

Finally, we note that since OFDM signals are obtained by adding a large number of sinusoids with different frequencies, the resulting transmitted signal in the time domain may have a large *peak-to-average power ratio (PAPR)*. Transmission of a signal with a large PAPR may saturate the power amplifiers at the transmitter, causing intermodulation distortion. Looking at the problem differently, it becomes costly to implement power amplifiers with large linear regions. Therefore, PAPR reduction techniques may be needed, which are also beyond our scope.

8.4 Chapter Summary

This is the fourth and final chapter dealing with digital modulation techniques. Specifically, we went beyond single carrier transmission and described communication schemes employing carrier signals with different frequencies. We first dealt with frequency-shift keying, with special emphasis on orthogonal FSK with both coherent and non-coherent detection. We then described minimum-shift keying, which is a method of frequency modulation with phase continuity at the symbol boundaries. MSK is a modulation scheme with memory, and its power spectral density is compact, which is highly desirable in practical communication systems. Finally, we studied multicarrier trasmission through orthogonal frequency-division multiplexing, which finds widespread application both in wireline and wireless communications. We described the basic transmission scheme along with the addition of a cyclic prefix and the resulting channel input–output relationship. We also highlighted that other practical issues should be taken into account for OFDM signal design.

PROBLEMS

8.1 Determine the maximum transmission rate possible over a channel bandwidth of 100 MHz using orthogonal FSK with coherent detection with a constellation size of
(a) $M = 2$,
(b) $M = 16$,
(c) $M = 64$.

8.2 Determine the maximum transmission rate possible over a channel bandwidth of 100 MHz using orthogonal FSK with non-coherent detection with a constellation size of
(a) $M = 2$,
(b) $M = 16$,
(c) $M = 64$.

8.3 Which of the below modulation schemes will support a transmission rate of at least 10 Mbps over a bandwidth of 60 MHz?
(a) Binary orthogonal FSK with coherent detection.
(b) Binary orthogonal FSK with non-coherent detection.
(c) 8-ary orthogonal FSK with coherent detection.
(d) 8-ary orthogonal FSK with non-coherent detection.
(e) 32-ary orthogonal FSK with coherent detection.
(f) 32-ary orthogonal FSK with non-coherent detection.

8.4 Consider binary orthogonal FSK with equally likely bits over an AWGN channel with power spectral density $N_0/2$. Show that the average error probability with non-coherent detection is given by

$$P_b = e^{-\gamma_b/2},$$

where $\gamma_b = E_b/N_0$, with E_b being the energy per bit.

8.5 Consider binary FSK with a frequency separation of Δf and bit duration of T. Assume that coherent detection is employed. What is the minimum value of the frequency separation Δf in terms of T so that the average error probability is the lowest? Compare the error probability corresponding to this FSK scheme with that of orthogonal FSK with coherent detection.

8.6 Consider M-ary orthogonal FSK with coherent demodulation with an average symbol energy E_s. Assume that it is used over an AWGN channel with power spectral density $N_0/2$.

(a) Determine an integral expression for the average symbol error probability with equally likely signaling. Do not evaluate it.

(b) Compute the union bound on the symbol error probability in terms of E_s/N_0.

8.7 Plot the phase of the signal transmitted for an MSK-modulated waveform corresponding to the sequence of bits

$$-1, 1, 1, 1, -1, -1, -1, 1.$$

8.8 Show that the power spectral density of MSK with independent and equally likely bits is given by

$$S_{U,MSK}(f) = \frac{4A^2 T_b}{\pi^2}\left(\left(\frac{\cos(2\pi T_b(f - f_c))}{1 - 16T_b^2(f - f_c)^2}\right)^2 + \left(\frac{\cos(2\pi T_b(f + f_c))}{1 - 16T_b^2(f + f_c)^2}\right)^2\right).$$

8.9 Consider an OFDM transmission scheme that employs 256-QAM for each subband for communicating over a bandpass channel in the frequency band 50–80 MHz. Assume that the multipath spread of the channel is 10 μs, and a cyclic prefix is used to remove the interference among OFDM words. Complete the design assuming that

(a) the number of subcarriers $N = 1024$,

(b) the number of subcarriers $N = 4096$.

What are the resulting transmission rates and spectral efficiencies?

8.10 Consider a multipath channel (an LTI system) with impulse response

$$c(t) = \sum_{k=0}^{L-1} h_k \delta(t - \tau_k),$$

where L is the number of arrivals, h_k and τ_k are the gains and delays of these arrivals.

Consider a specific channel with $L = 4$, gains [1 0.8 − 0.7 0.5], and corresponding delays (in ms) [0 10 25 35]. We are given the 12–20 kHz frequency band to communicate over this channel.

(a) Estimate the ISI length assuming that single-carrier signaling is used in the entire band. Note that there is no unique answer as this will depend on the selected symbol duration (which has to be consistent with the available bandwidth).

(b) Design an OFDM system to communicate over this channel that will make sure that there is no ISI among different OFDM words. Identify the number of subcarriers, the cyclic prefix length, as well as any other parameters needed to complete the design.

(c) What is the transmission rate (in bits/sec) for the system you have designed in part (b) assuming that QPSK is used for each subchannel? How would your answer change if 16-QAM is used?

8.11 We communicate over a frequency-selective channel with impulse response

$$c(t) = 0.5 + 0.2\delta(t - 0.001) - 0.3\delta(t - 0.0025)$$

(where t is taken in seconds). We employ an OFDM scheme using the frequency band 10–20 kHz with $N = 2048$ subcarriers. The first 1024 subcarriers use 8-PSK while the subcarriers from 1025 to 2048 employ 16-PSK. The average symbol energies are the same for all the subcarriers.

Rank order the symbols transmitted on the 100th, 600th, 1100th, and 1600th subcarriers in terms of the symbol error probabilities (assuming that the transmission takes place over an AWGN channel).

8.12 We communicate over a frequency-selective channel with the (magnitude of the) frequency response as shown in Fig. 8.14. Note that $c(t)$ is real, hence its Fourier transform is conjugate symmetric, but only the positive frequency portion of the magnitude spectrum is shown. Assume that the transmit filter and receive filter frequency responses in the passband of the channel are $|G_T(f)| = |G_R(f)| = 1$ (for 100–104 MHz). Also, assume that the noise is a WSS white Gaussian process with power spectral density $S_n(f) = \frac{N_0}{2} = 10^{-14}$ (W/Hz) (at the input to the receive filter).

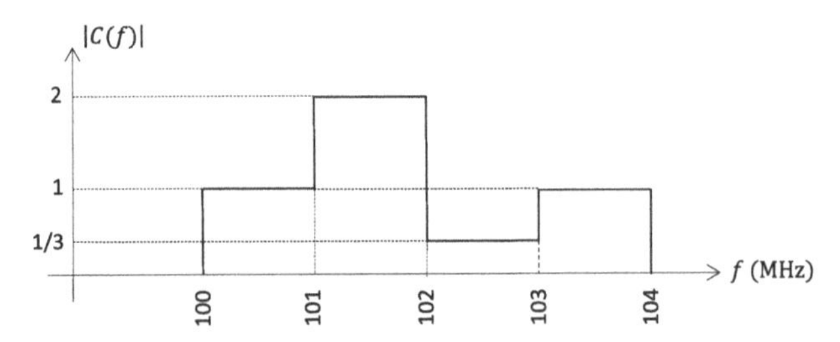

Figure 8.14 Magnitude response of the channel in Problem 8.12.

(a) We decide to use OFDM transmission over this 4 MHz bandwidth channel with $N = 1024$ subcarriers. What is the OFDM word duration excluding any cyclic prefix?

(b) Consider the OFDM scheme in part (a). Assume that a cyclic prefix of sufficient length is used to remove any ISI. We have four different sources S_1, S_2, S_3, S_4 to transmit over the channel. We allocate the subcarriers 1–256 to S_1, 257–512 to S_2, 513–768 to S_3, and 769–1024 to S_4. We use QPSK on each subchannel for all the sources. Also, we use the same power levels.

Rank the error rate performances for the different source transmissions (at the receiver) from best to worst (i.e., which source would be transmitted most reliably, which one is second best, etc.). Also indicate the performance differences in terms of SNR, that is, to obtain the same error rate (e.g., for S_1) how much more SNR would be needed (in dB) compared to S_2, for example.

(c) Consider part (b). Assume that for transmission of S_1, we employ QPSK; for S_2, we employ 8-PSK; and for S_3 and S_4, we employ BPSK. Also, assume that the cyclic prefix length used is 100 μs. Compute the transmission rates of the four sources in bits/sec.

COMPUTER PROBLEMS

8.13 Simulate binary (orthogonal) FSK with coherent detection over an AWGN channel by working directly with the transmitted signals (not the equivalent simplified models). Namely, pick the bit duration and the frequencies to be used, generate the transmitted signals as a function of time, including the channel effect (Gaussian noise), and implement a correlation-type receiver. Assume that the channel phase is known, that is, coherent detection is performed, and estimate the error rates.

8.14 Repeat the previous problem for the case of non-coherent detection. In this case, make sure to include the effects of the channel propagation delay (i.e., the introduction of the channel phase effects), simulate the envelope detector, and estimate the resulting error rates. Compare the results with those of coherent detection.

8.15 Simulate the orthogonal FSK with $M = 8$ using the mathematically equivalent signal representations and plot the resulting symbol error probabilities for the following cases:

(a) with coherent detection when the channel phase is perfectly known;

(b) with coherent detection when the channel phase is imperfectly available with a phase error of ϕ for three different values: $\phi = \frac{\pi}{60}$, $\frac{\pi}{20}$, and $\frac{\pi}{10}$;

(c) with non-coherent detection (assuming a uniform channel phase over $[0, 2\pi)$).

8.16 Consider a multipath channel (an LTI system) with impulse response

$$c(t) = \sum_{k=0}^{L-1} h_k \delta(t - \tau_k),$$

where L is the number of channel taps, h_k and τ_k are the gains and delays of these taps.

Consider a specific channel with $L = 4$, gains [1 0.4 − 0.7 0.5], and corresponding delays [0 2.356 3.214 9.645] (given in ms). Therefore, the channel multipath duration is about 10 ms. We are given the 10–20 kHz frequency band to communicate over this channel. Notice that if we use single-carrier signaling, there will be severe intersymbol interference.

(a) Design an OFDM system to communicate over this channel (with a cyclic prefix) that will make sure that there is no ISI among OFDM words. Decide on the number of subcarriers, length of cyclic prefix, and so on, and indicate the effective transmission rate (assuming that QPSK signaling is employed).

(b) Plot the magnitude spectrum of the channel over the frequency band given above.

(c) Assuming QPSK transmission, generate three examples of OFDM signals (corresponding to a specific sequence of symbols being transmitted).

(d) For the exemplary signals, estimate the peak-to-average power ratio (PAPR).

(e) Conduct simulations (using 100,000 or more bits) to estimate the error rates as a function of the noise level (assuming that transmission takes place over an additive white Gaussian noise channel with power spectral density $N_0/2$).

(f) Examine where most of the errors occur in the previous part and relate this to the channel magnitude spectrum.

9 Channel Coding

In the previous four chapters, we have studied different transmission techniques over additive white Gaussian noise channels. Along with different transmitter and receiver structures, we have computed the corresponding bit/symbol error probabilities and determined the amount of channel bandwidth necessary for transmission. In our coverage, we have considered the transmission of different bits (or symbols) independently of each other. On the other hand, we recall from the discussion in Chapter 1 that we can add *controlled redundancy* to protect the transmitted messages against channel impairments. This is accomplished through channel coding.

The idea in channel coding is to add redundancy into the transmitted bit or symbol sequence in such a way that even if there are (raw) errors for some bits/symbols, it is possible to correct them using the relationships among the different bits/symbols making up the codewords by exploiting the code constraints. We will cover two basic channel-coding approaches: block coding and convolutional coding.

The chapter is organized as follows. We first explain why channel coding may be helpful by providing two simple examples in Section 9.1. We then focus on the fundamental limits of reliable transmission over AWGN channels as well as a different model called the *binary symmetric channel* in Section 9.2. Through the computation of these ultimate limits, we will be able to assess the performance of a digital transmission scheme and motivate the design and use of channel codes. For instance, we will see that uncoded BPSK transmission is about 8 dB away from theoretical limits (taking a probability of error of 10^{-5} as reliable transmission), and hence, there is a great need to employ powerful channel codes to improve the system performance. With this motivation, we then turn our attention to the study of channel-coding techniques. Section 9.3 is devoted to linear block codes, where we introduce the basics and study encoding and decoding algorithms as well as their performance over basic channel models. We then study convolutional codes by presenting different representations, encoding and decoding algorithms, and performance analysis in Section 9.4. We consider both hard-decision decoding and soft-decision decoding approaches and provide a method of approximate performance analysis. We conclude the chapter in Section 9.5.

9.1 Why Does Channel Coding Help?

In this section, we give two specific examples of channel codes to demonstrate that channel coding is helpful (when employed properly) in reducing transmission error probabilities, hence improving signal fidelity.

Example 9.1

Consider transmission of four equally likely messages over an AWGN channel using BPSK modulation. In the first case, there is no channel coding, hence the four symbols (denoted by bits as "00," "01," "10," and "11" are transmitted via two uses of the channel. Using the mathematically equivalent model, the mapping

$$\begin{aligned} 0\ 0 &\longrightarrow -1 \quad -1 \\ 0\ 1 &\longrightarrow -1 \quad +1 \\ 1\ 0 &\longrightarrow +1 \quad -1 \\ 1\ 1 &\longrightarrow +1 \quad +1 \end{aligned} \tag{9.1}$$

describes the transmitted symbols. For the coded case, assume that each of the four messages is mapped to three bits, and the coded bits are transmitted with BPSK. We select the mapping in such a way that makes the distances among the selected sequences as large as possible. Specifically, assume that the mapping

$$\begin{aligned} 0\ 0 &\longrightarrow 0\ 0\ 0 \longrightarrow -1 \quad -1 \quad -1 \\ 0\ 1 &\longrightarrow 0\ 1\ 1 \longrightarrow -1 \quad +1 \quad +1 \\ 1\ 0 &\longrightarrow 1\ 1\ 0 \longrightarrow +1 \quad +1 \quad -1 \\ 1\ 1 &\longrightarrow 1\ 0\ 1 \longrightarrow +1 \quad -1 \quad +1 \end{aligned} \tag{9.2}$$

identifies how the two bits are represented by 3-bit codewords and the corresponding transmitted sequences. Notice that while there are eight possible binary triplets, we carefully select four of them to transmit the four messages, introducing redundancy.

Let us now compare the uncoded and coded cases in terms of their error performance over an AWGN channel. To do this, we can think of the uncoded and coded cases as two and three-dimensional signal constellations, respectively, with four constellation points each, and employ the tools developed in Chapter 5. That is, we can determine the average energy of each signaling scheme along with its minimum distance, and compare the normalized squared Euclidean distances for a performance assessment.

The average energy for the uncoded case is $E_{\text{av, uncoded}} = 2$, and the minimum squared Euclidean distance is $d^2_{\text{min, uncoded}} = 4$. While for the coded case, we have $E_{\text{av, coded}} = 3$ and $d^2_{\text{min, coded}} = 8$. Normalized minimum squared distances are then

$$\frac{d^2_{\text{min, uncoded}}}{E_{\text{av, uncoded}}} = 2, \qquad \frac{d^2_{\text{min, coded}}}{E_{\text{av, coded}}} = \frac{8}{3}. \tag{9.3}$$

Therefore, we conclude that the coded scheme has a better performance compared to the uncoded case, and the amount of improvement at high signal-to-noise ratios is around $10\log_{10}\left(\frac{8/3}{2}\right) \approx 1.25$ dB.

The above example illustrates that even a simple scheme of adding controlled redundancy, when carefully done, may be helpful to improve transmission fidelity. We also note that this improvement in the error probability comes at the cost of increasing bandwidth. That is to say, since the coded transmission requires three channel uses as opposed to two, if we are to keep the overall signaling rate the same (in bits/second), we need to use narrower transmission pulses for the coded case, hence increasing the overall transmission bandwidth. The bandwidth expansion factor with the coding scheme of the above example is $3/2$.

The next example is another illustration of the benefits of channel coding using a class of codes that will be covered later in this chapter, called Hamming codes.

Example 9.2

Consider a Hamming code with parameters 31 and 26, that is, 26 information bits are mapped to 31 coded bits. The minimum distance of the code is 3, and it can correct exactly 1-bit error among 31 coded bit transmissions (as will be shown later in the chapter). A message of length 26 bits is transmitted

(1) using the uncoded transmission in 26 channel uses,
(2) using the Hamming code in 31 channel uses.

The transmission is over an AWGN channel and the modulation scheme is BPSK. Assume that the noise variance in the equivalent mathematical representation of the channel is $N_0/2$. Also, assume that the Hamming code is decoded as follows: first, a binary decision is made on each bit based on the matched filter output, and then the noisy received binary sequence is decoded.

Compare the probability of error for the uncoded and coded cases.

Solution

Consider first a single-bit transmission over the AWGN channel. Denoting the BPSK signals as $\sqrt{E_s}$ and $-\sqrt{E_s}$, the error probability for the bit transmission becomes $Q(\sqrt{2E_s/N_0})$ (recall the expression in (6.35)). Therefore, we have an equivalent binary-input, binary-output channel for which the cross-over probability (0 being received as 1, and 1 being received as 0) equals $\epsilon = Q(\sqrt{2E_s/N_0})$. This is called a binary symmetric channel (BSC). See Fig. 9.1 for a depiction.

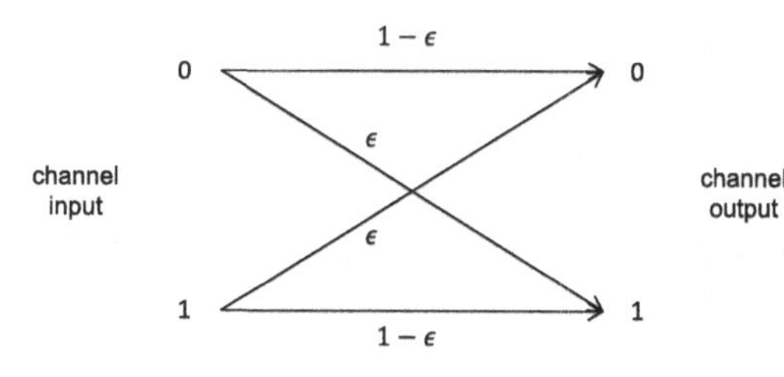

Figure 9.1 Binary symmetric channel with cross-over probability ϵ.

For the uncoded transmission of the 26 bit-long message, denoting energy per bit as E_b, we can write the message error probability as

$$P_{\text{e, uncoded}} = 1 - \left(1 - Q\left(\sqrt{2E_b/N_0}\right)\right)^{26}, \tag{9.4}$$

as even when one bit is in error, the overall 26 bit-long message will be decoded in error.

For the Hamming coded case, since there are 31 coded bits corresponding to 26 bits, the energy per transmitted symbol is $E_s = \frac{26}{31}E_b$. As stated at the beginning of the example, this code can correct exactly one bit in error among 31 coded bits, hence if there is no error in the transmission of the coded bits, or if there is exactly one bit in error, then the message will be received correctly. That is, the message error probability with coding becomes

$$P_{\text{e, coded}} = 1 - \Big(\mathbb{P}(\text{all correct transmissons}) + \mathbb{P}(\text{exactly one incorrect transmission})\Big) \tag{9.5}$$

$$\begin{aligned} &= 1 - \Big(\mathbb{P}(\text{all correct transmissons}) \\ &\qquad + 31\,\mathbb{P}(\text{any fixed } k\text{th transmission being incorrect}) \\ &\qquad \times \mathbb{P}(\text{remaining 30 being correct})\Big) \end{aligned} \tag{9.6}$$

$$= 1 - \left(1 - Q\left(\sqrt{\frac{52E_b}{31N_0}}\right)\right)^{31} - 31Q\left(\sqrt{\frac{52E_b}{31N_0}}\right)\left(1 - Q\left(\sqrt{\frac{52E_b}{31N_0}}\right)\right)^{30}. \tag{9.7}$$

Denoting the signal-to-noise ratio per bit as $\gamma_b = E_b/N_0$, and assuming that $\gamma_b \gg 1$ (high signal-to-noise ratios), the Q-function terms will be small and their higher powers can be omitted for an accurate approximation of the error probability. That is, we can write

$$P_{\text{e, uncoded}} \approx 26Q\left(\sqrt{2\gamma_b}\right) \tag{9.8}$$

and

$$P_{\text{e, coded}} = 1 - \left(1 - Q\left(\sqrt{\frac{52}{31}\gamma_b}\right)\right)^{31} - 31Q\left(\sqrt{\frac{52}{31}\gamma_b}\right)\left(1 - Q\left(\sqrt{\frac{52}{31}\gamma_b}\right)\right)^{30} \tag{9.9}$$

$$\begin{aligned} &\approx 31Q\left(\sqrt{\frac{52}{31}\gamma_b}\right) - \binom{31}{2}Q^2\left(\sqrt{\frac{52}{31}\gamma_b}\right) - 31Q\left(\sqrt{\frac{52}{31}\gamma_b}\right) \\ &\quad + 31 \times 30Q^2\left(\sqrt{\frac{52}{31}\gamma_b}\right) \end{aligned} \tag{9.10}$$

$$\approx 475Q^2\left(\sqrt{\frac{52}{31}\gamma_b}\right). \tag{9.11}$$

Considering that the Q-function terms decay exponentially in γ_b, it is clear that the coded scheme is significantly more advantageous compared to the uncoded one in terms of message reliability. For instance, for $\gamma_b = 7$ dB, we evaluate the uncoded and coded message error probabilities as

$$P_{\text{e, uncoded}} \approx 0.02 \text{ and } P_{\text{e, coded}} \approx 0.0017. \tag{9.12}$$

Similarly, for $\gamma_b = 10$ dB, the error probabilities with uncoded and coded cases become

$$P_{\text{e, uncoded}} \approx 10^{-4} \text{ and } P_{\text{e, coded}} \approx 2.1 \times 10^{-7}. \tag{9.13}$$

As in the previous example, there is a trade-off: the advantage in the error probability is obtained with an increase in the bandwidth requirements. For this example, the bandwidth expansion factor is $\frac{31}{26} \approx 1.19$.

Examples 9.1 and 9.2 clearly illustrate that channel coding can improve the reliability of transmitted messages. In the next section, we will discuss the ultimate limits of reliable transmission over a noisy channel, and in the rest of the chapter, we will study two important classes of channel codes, namely, linear block codes and convolutional codes.

9.2 Ultimate Limits of Transmission

Claude E. Shannon, in his celebrated 1948 paper, proved a channel-coding theorem that states that it is possible to obtain an arbitrarily small error probability when transmitting over a noisy channel and determined the highest rate of reliable communication. We now state this fundamental result.

Given a channel, a rate R is achievable (transmittable) if there exists a sequence (R, ϵ_n), with $\epsilon_n \to 0$ as $n \to \infty$, such that for every n, there exist:

(1) A set of messages $\mathcal{M}(n) \stackrel{\text{def}}{=} \{1, 2, 3, \ldots, M(n)\}$ such that $|\mathcal{M}(n)| = M(n) \geq 2^{Rn}$.
(2) A channel encoder

$$\mathcal{E}_n : \mathcal{M}(n) \to \mathcal{M}^n \tag{9.14}$$

and a decoder $\mathcal{D}_n : \mathcal{M}'^n \to \mathcal{M}(n)$, with average error probability

$$P_e \stackrel{\text{def}}{=} \frac{1}{|\mathcal{M}(n)|} \sum_{c \in \mathcal{M}(n)} \mathbb{P}\Big(\mathcal{D}_n(r) \neq c \;\Big|\; c \text{ is transmitted}\Big) \leq \epsilon_n, \tag{9.15}$$

with r denoting the channel output.

Channel-Coding Theorem. For noisy channels, there exists a quantity called the channel capacity (denoted by C) for which all rates R less than C are achievable, that is, reliable transmission with an arbitrarily small error probability at rates $R < C$ is possible. Furthermore, if the transmission rate is larger than the channel capacity ($R > C$), the error probability is bounded away from 0; hence, no reliable transmission is possible.

The channel capacity C is a strict limit of reliable transmission. Regardless of the complexity of the coding scheme, it is not possible to exceed this limit and achieve reliable communication at higher transmission rates than the channel capacity.

The proof of the channel-coding theorem or computation of the channel capacity for different channel models is beyond our scope. We only give the capacity of a binary symmetric channel (introduced in the previous section) and that of an AWGN channel in our coverage.

9.2.1 Capacity of Binary Symmetric Channels

The capacity of a BSC with cross-over probability ϵ is given by

$$C_{\text{BSC}} = 1 - H_b(\epsilon), \tag{9.16}$$

in bits/channel use, where $H_b(\epsilon) = -\epsilon \log \epsilon - (1 - \epsilon) \log (1 - \epsilon)$ is the binary entropy function. The capacity is plotted as a function of the cross-over probability ϵ in Fig. 9.2.

We observe that the channel capacity is 0 for $\epsilon = 1/2$ since in this case the (binary) output sequence becomes independent of the channel input, and it is impossible to distinguish any messages regardless of the coding rate. For $\epsilon = 0$ or for $\epsilon = 1$, the capacity is 1 bits/channel use as there is no channel error (the $\epsilon = 1$ case is simply a renaming of the output bits). As another example, for $\epsilon = 0.1$, that is, when 10% of the bits are randomly received in error (with i.i.d. channel errors among different uses of the BSC), the capacity evaluates to $C \approx 0.53$ bits/channel use. In other words, for $\epsilon = 0.1$, there is a channel code of rate smaller than 0.53 (it could be arbitrarily close) that results in an arbitrarily low probability of error. And, for any transmission rate above 0.53 bits/channel use, the error probability cannot be made near zero, that is, reliable transmission is not possible.

Let us give an intuitive argument on the channel capacity for a BSC being $C = 1 - H_b(\epsilon)$. Consider transmission of n bits over a BSC with cross-over probability ϵ in n channel uses. For large n, for a specific input sequence of length n,

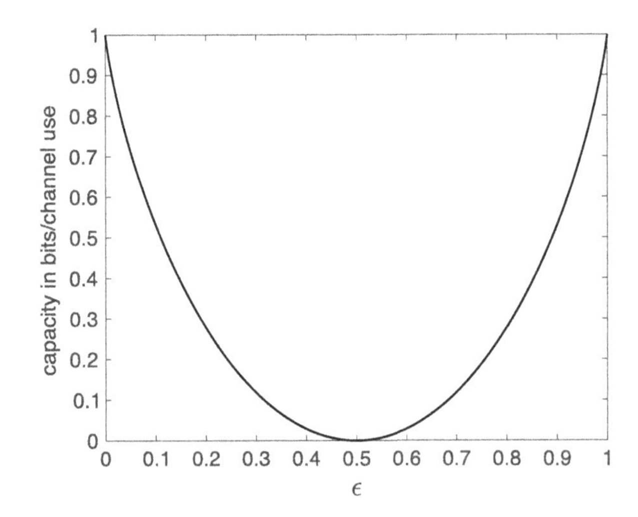

Figure 9.2 Capacity of a binary symmetric channel as a function of the cross-over probability ϵ.

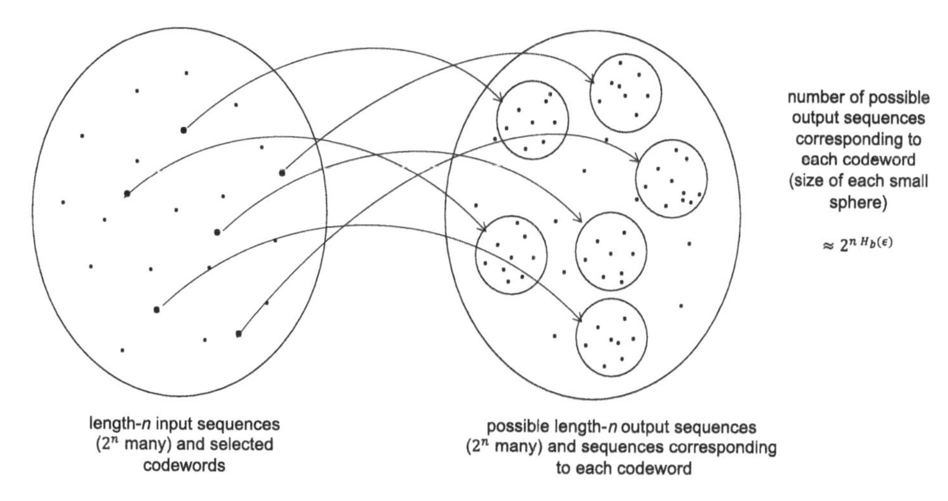

Figure 9.3 Illustration of the BSC capacity derivation.

there will be around $n\epsilon$ bits in error. Assuming that $n\epsilon$ is an integer, the number of error patterns is approximately $\binom{n}{n\epsilon}$. Using Stirling's approximation, which states that $\log(n!) \approx n \log n - n \log e$, we have

$$\binom{n}{n\epsilon} \approx 2^{nH_b(\epsilon)}. \tag{9.17}$$

That is, there are roughly $2^{nH_b(\epsilon)}$ possible sequences that would be observed for transmission of a length-n input sequence. This means that when one of these n-tuples is received, the channel code should allow us to decode the original transmitted sequence with no ambiguity. This is depicted in Fig. 9.3.

Since there are a total of 2^n binary n-tuples (as possible output sequences), and each selected length-n codeword rules out about $2^{nH_b(\epsilon)}$ output sequences, we can select

$$\frac{2^n}{2^{nH_b(\epsilon)}} \tag{9.18}$$

transmitted sequences (representing as many messages), which can be distinguished at the receiver. Taking the logarithm of this quantity and dividing by n, we can express the number of bits per channel use that can be reliably transmitted, resulting in $1 - H_b(\epsilon)$, which is the BSC capacity.

9.2.2 Capacity of a Discrete-Time AWGN Channel

A discrete-time AWGN channel is described by the input–output relationship

$$Y = X + N, \tag{9.19}$$

where X is the channel input, N is a zero-mean Gaussian random variable with variance P_N, and Y is the channel output. The noise terms are independent of each

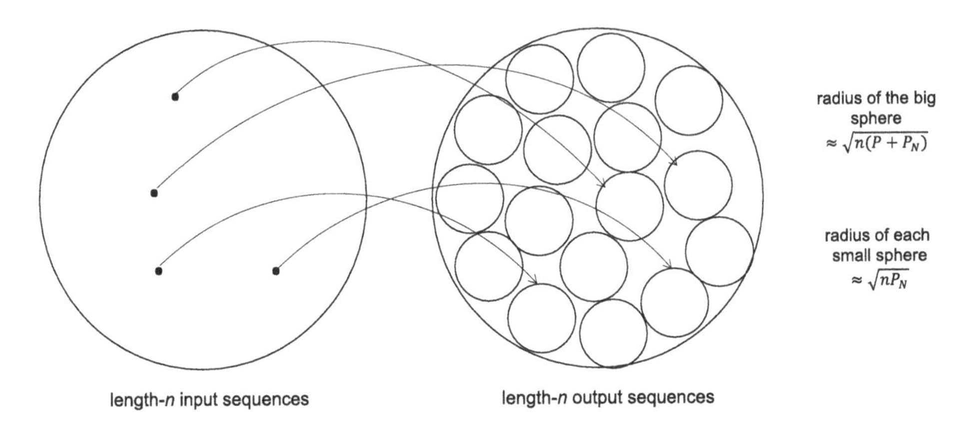

Figure 9.4 Intuitive explanation of the AWGN channel capacity.

other for different uses of the channel. This is nothing but the simple mathematical model for a one-dimensional modulation scheme over an AWGN channel. We also assume that the average input power is limited to P, that is, $\mathbb{E}[X^2] \leq P$. The capacity of this channel is given by

$$C_{\text{discrete time, AWGN}} = \frac{1}{2}\log\left(1 + \frac{P}{P_N}\right) \tag{9.20}$$

in bits/channel use. We observe that the channel capacity increases logarithmically with the ratio P/P_N, which can be considered as the signal-to-noise ratio.

We can give a *sphere-packing* interpretation of the capacity of an AWGN channel (in a similar fashion to that of a BSC). With n uses of the channel, the total input power transmitted is nP and the total noise power is nP_N. The average power of the length-n output sequence is $n(P + P_N)$. Therefore, the n-dimensional received signals can be considered to be contained in an n-dimensional sphere of radius $\sim \sqrt{n(P + P_N)}$. Each received signal conditioned on the transmitted message will be within a sphere of radius $\sim \sqrt{nP_N}$ as they will be corrupted by Gaussian noise. This is illustrated in Fig. 9.4.

Since the small spheres should not overlap for reliable transmission, the number of distinct messages we can transmit (while ensuring reliable transmission) becomes the ratio of the volume of the large sphere to that of the small one, that is,

$$\left(\frac{\sqrt{n(P + P_N)}}{\sqrt{nP_N}}\right)^n. \tag{9.21}$$

Taking the logarithm of this expression and dividing the result by n, we determine the number of bits per channel use that can be transmitted. Namely, we obtain

$$\frac{1}{n}\log\left(\frac{\sqrt{n(P + P_N)}}{\sqrt{nP_N}}\right)^n = \frac{1}{2}\log\left(1 + \frac{P}{P_N}\right), \tag{9.22}$$

which is nothing but the capacity expression for the discrete-time AWGN channel. Note that this is only an intuitive explanation, not a formal proof (which can be found in more advanced texts on information theory).

9.2.3 Capacity of Waveform Gaussian Channel

Let us now consider the more familiar *waveform Gaussian channel* model, namely,

$$Y(t) = X(t) + n(t), \tag{9.23}$$

where $X(t)$ is the channel input (modeled as a random process) with average power constraint P and bandwidth constraint B, $n(t)$ is a zero-mean WSS AWGN process with power spectral density $S_n(f) = N_0/2$. The capacity of the waveform Gaussian channel is given by

$$C_{\text{waveform, AWGN}} = B \log\left(1 + \frac{P}{N_0 B}\right) \tag{9.24}$$

bits/second.

The average AWGN noise power within the frequency band $[-B, B]$ is $N_0 B$, hence the ratio $\frac{P}{N_0 B}$ can be considered as the signal-to-noise ratio. We observe that the channel capacity increases logarithmically with this ratio.

As a simple example, if $B = 3$ kHz, a signal-to-noise ratio of $\frac{P}{N_0 B} = 39$ dB results in a channel capacity of $C = 38.8$ kbps.

Let us give an intuitive explanation of the channel capacity of the waveform Gaussian channel. To do this, we need to determine the number of bits that can be transmitted in T units of time using a bandwidth of B Hz. Given the power constraint, the total energy of the input signal is bounded by PT units. Since the noise power within the bandwidth of B Hz is $\frac{N_0}{2} 2B = N_0 B$, the total energy due to the noise becomes $N_0 BT$ units.

We refer to a result from information theory without proof, which states that there are approximately $2BT$ orthonormal basis functions for functions that are approximately time-limited to T and approximately bandlimited to B. Since we are considering transmission for T units of time using a bandwidth of B units, using this result, we can equivalently consider $2BT$ parallel Gaussian channels each with an input power

$$\frac{PT}{2BT} = \frac{P}{2B}, \tag{9.25}$$

and the AWGN power

$$\frac{N_0 BT}{2BT} = \frac{N_0}{2}. \tag{9.26}$$

Using the result of Section 9.2.2, the capacity of each of these channels is given by

$$\frac{1}{2} \log\left(1 + \frac{P/2B}{N_0/2}\right) \tag{9.27}$$

in bits/channel use. Recalling that each channel use corresponds to T seconds, and there are $2BT$ such parallel channels, the entire transmission rate corresponds to

$$\frac{1}{T}2BT\frac{1}{2}\log\left(1+\frac{P/2B}{N_0/2}\right)=B\log\left(1+\frac{P}{N_0B}\right) \tag{9.28}$$

in bits/seconds, which is the result given in (9.24).

We observe that for a fixed channel bandwidth and noise power spectral density level, the capacity grows logarithmically as a function of the input power P. On the other hand, for a fixed transmit power and noise PSD, the capacity approaches an upper bound as the channel bandwidth increases. That is,

$$\lim_{B\to\infty} B\log\left(1+\frac{P}{N_0B}\right)=\frac{P}{N_0}\log(e)\approx 1.44\frac{P}{N_0}\text{ bits/second.} \tag{9.29}$$

Let us now consider the spectral efficiency r. Recall from the previous chapter that the spectral efficiency is defined as

$$r=\frac{R}{B}\text{ bits/s/Hz,} \tag{9.30}$$

where R is the transmission rate in bits/second. Since, for reliable communications, the transmission rate needs to be less than the channel capacity, (i.e., $R<C$), we obtain

$$r<\frac{C}{B}=\log\left(1+\frac{P}{N_0B}\right). \tag{9.31}$$

Since the energy per bit is defined as the ratio of total energy to the number of bits transmitted, we obtain $E_b=\frac{PT}{RT}$ with T denoting the transmission duration. That is, $E_b=\frac{P}{R}$, from which we obtain

$$r<\log\left(1+r\frac{E_b}{N_0}\right) \tag{9.32}$$

or

$$\frac{E_b}{N_0}>\frac{2^r-1}{r} \tag{9.33}$$

as a condition for reliable communication. In other words, for a given spectral efficiency r, there is a lower limit on the required signal-to-noise ratio E_b/N_0. If E_b/N_0 is below this limit, it is impossible to have reliable communication.

The asymptotic case of $r\to 0$ deserves special attention. As $r\to 0$, we obtain $E_b/N_0>\ln 2$, or

$$E_b/N_0\Big|_{dB}>-1.6\text{ dB.} \tag{9.34}$$

That is, it is impossible to transmit reliably at any spectral efficiency below -1.6 dB for E_b/N_0.

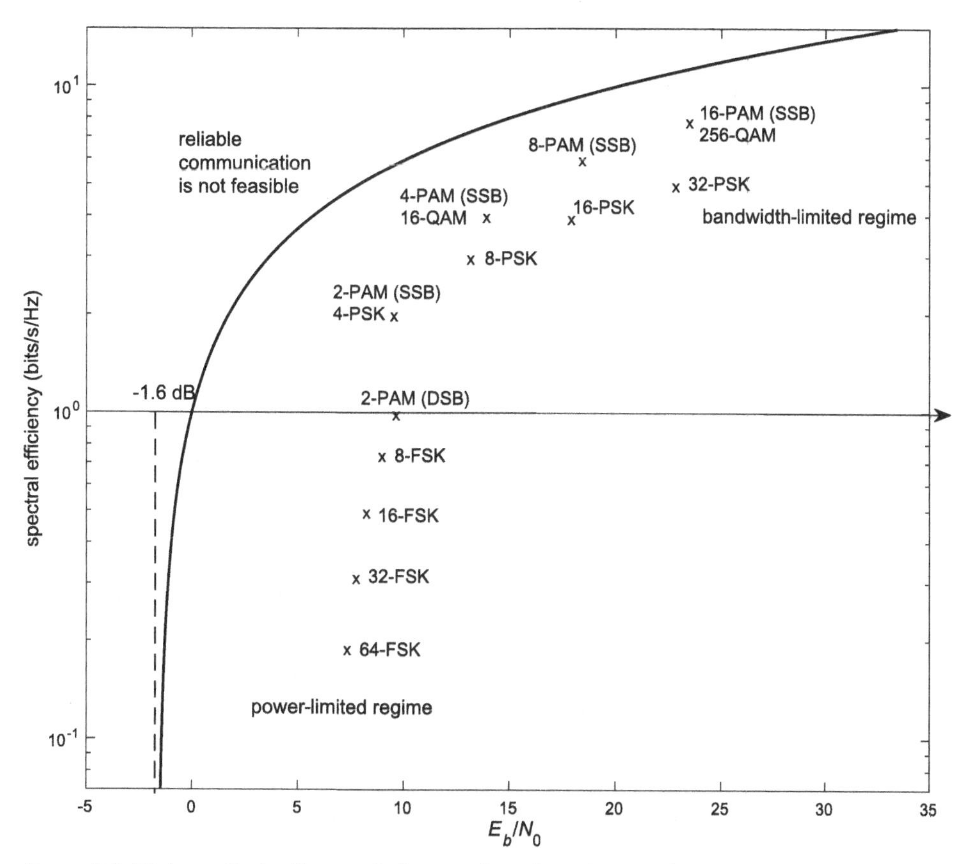

Figure 9.5 Ultimate limit of transmission as a function of E_b/N_0 for an AWGN channel. Also depicted are the E_b/N_0 values for various modulation schemes for an error probability of 10^{-5} (along with their spectral efficiencies).

We illustrate the channel capacity bound as a function of E_b/N_0 in Fig. 9.5. There is a highest spectral efficiency achievable for any given E_b/N_0, or expressed differently, for any value of the spectral efficiency, there is a minimum required E_b/N_0 for reliable transmission. For instance, for $r = 1$ bits/s/Hz, $E_b/N_0 > 0$ dB is required. It is also depicted in the figure that reliable communication is not feasible for $E_b/N_0 < -1.6$ dB.

The channel capacity result in (9.24) also shows that there is a trade-off between bandwidth and power. When there is limited channel bandwidth for transmission (e.g., in wireless communications), the bandwidth is a premium; hence, higher spectral efficiencies would be required, necessitating larger transmit power levels. When channel bandwidth is abundant, spectral efficiency can be reduced while meeting the data rate requirements; hence, the transmit power levels can be reduced. The bandwidth-limited and power-limited regimes are illustrated in Fig. 9.5.

Also depicted in Fig. 9.5 is the spectral efficiency versus required power trade-off for different modulation schemes. The required E_b/N_0 values for an error probability of 10^{-5} are marked for several modulation schemes.[1] Clearly, for bandwidth efficiency, one can employ M-PSK or M-QAM with a large M, while for power efficiency, one can use M-FSK. It is also clear that the performance of these modulation schemes is far from information-theoretic limits. For instance, for $r = 2$ bits/s/Hz, an E_b/N_0 of 10 dB is required with uncoded BPSK for an average error probability of 10^{-5}. On the other hand, the information-theoretic limit dictates that 1.8 dB should be sufficient.

To close the gap between the information-theoretic limits and the practical digital modulation performance with no coding, one should employ channel-coding techniques. In the rest of this chapter, our goal is to introduce the two fundamental classes of channel codes, namely, linear block codes and convolutional codes, and study them in some detail.

Note that there have been tremendous developments in channel coding over the last few decades. For instance, turbo codes, which are linear block codes obtained by concatenating convolutional codes in parallel or in series via an interleaver, were invented in the 1990s and studied in depth in subsequent years. Shortly after, low-density parity-check (LDPC) codes were reinvented (after first being discovered in the early 1960s by Robert G. Gallager in his doctoral thesis) and widely studied in the last two decades. Both of these channel-coding schemes are very powerful; specifically, they can attain a very close performance to the information-theoretical limits (when long block lengths are used). More recently, Erdal Arıkan invented polar codes, which are the first class of codes that provably achieve the Shannon limit for channels with certain symmetries. In short, we now have practical means of implementing codes with a performance close to the ultimate theoretical limits, at least for certain communication channels. Having said that, coverage of turbo, LDPC, or polar codes is beyond the scope of this text. Instead, we focus only on the basics of linear block codes and convolutional codes.

9.3 Linear Block Codes

We define an (n, k) block code as a collection of $M = 2^k$ distinct binary sequences of length n (called *codewords*). Here, k can be thought of as the length of the (binary) message sequence. There are 2^k possible message sequences; hence, there are as many codewords. While generalizations are possible, we will exclusively consider coding over the binary field, that is, both the messages and codewords will be sequences of 0s and 1s.

With an (n, k) block code, we map k information (or message) bits to n coded bits. We denote the codewords as

$$c_1, c_2, \ldots, c_M, \tag{9.35}$$

[1] Somewhat arbitrarily, we consider a probability of error of 10^{-5} as reliable communication.

where each codeword is a binary n-tuple. The code rate is $R_c = k/n$, and hence the bandwidth expansion factor is n/k.

Linear Block Codes. If the set of codewords $\boldsymbol{c}_1, \boldsymbol{c}_2, \ldots, \boldsymbol{c}_M$ form a subspace of all n-tuples, then the code is called a linear block code. (Note that all operations are in the binary field.) An equivalent definition is as follows: if the componentwise modulo-2 sum of any two codewords is also a codeword, then the code is a linear block code. As a simple consequence, we notice that the all-zero n-tuple is always a codeword for a linear block code. This is because adding a codeword with itself (which is also a codeword) results in the all-zero sequence.

Let us give a simple example.

Example 9.3

An even-parity code is obtained by adding a parity bit to the sequence of message bits (of length k) to make the total number of 1s even. The code length is $n = k + 1$. One can check that this is a linear block code by verifying that the sum of any two codewords is also a codeword.

As a specific example, for $k = 5$, if the message sequence is given by

$$0, 1, 1, 0, 1, \tag{9.36}$$

we append a parity bit of 1 to make the total number of 1s an even number, resulting in the coded sequence

$$0, 1, 1, 0, 1, 1, \tag{9.37}$$

where the last bit denotes the parity bit.

Note that with an even-parity code, we can detect single-bit errors in the transmission of the codeword. This is because, with a single-bit error, the number of 1s in the received sequence will be an odd number; hence, we will readily identify that the received sequence is not a codeword.

The even-parity code introduced in Example 9.3 can detect single-bit errors; however, it cannot identify the position of the error, hence it cannot be used for error correction. In the next example, we develop a code that can correct exactly one bit in error in the transmission of a codeword through a noisy binary-input, binary-output channel.

Example 9.4

Consider a coding scheme in which the message bits are written in a two-dimensional array format, and an even-parity code is applied to all the rows and columns of the array including the parity bits. As an instance, consider mapping of 25 message bits to 36 coded bits as illustrated in Fig. 9.6.

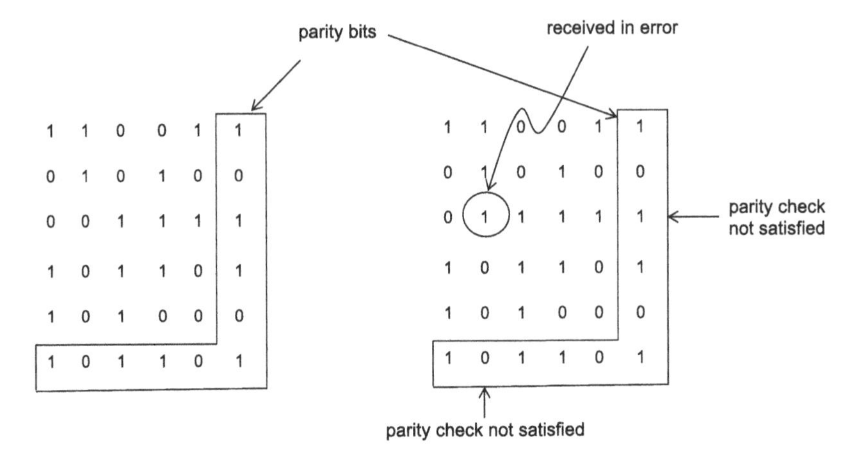

Figure 9.6 A (36, 25) linear block code obtained by applying an even parity-check code on the rows and columns of a matrix representing the message bits. The codeword corresponding to the 25 message bits is given on the left-hand side. The received sequence of bits at the output of a noisy communication channel is given on the right-hand side. The bit (marked by a circle) is received in error; hence, the corresponding row and column parities are not satisfied.

It can easily be verified that this is a linear block code. Also, the code can correct exactly one bit in error (for the 36-bit codeword transmission) as the row index and the column index of the single-bit error can be identified by the even-parity codes applied to the rows and columns separately.

Let us give another example.

Example 9.5

A $(5, 2)$ code with codewords

$$\{00000,\ 10100,\ 01111,\ 11011\} \tag{9.38}$$

is a linear block code, which can easily be verified. For encoding, different mappings from the pairs of bits comprising the message sequences to the codewords can be used, for example,

$$\begin{aligned} 00 &\longrightarrow 00000 \\ 01 &\longrightarrow 01111 \\ 10 &\longrightarrow 10100 \\ 11 &\longrightarrow 11011 \end{aligned} \tag{9.39}$$

is one such mapping.

9.3.1 Generator and Parity-Check Matrices

Recall that a linear block code is a k-dimensional subspace of all binary n-tuples. Therefore, we can describe this subspace using k basis vectors (codeword sequences of length n bits). The generator matrix of an (n, k) linear block code, denoted by $\boldsymbol{G}$, is a $k \times n$ matrix of 0s and 1s, where the rows form a basis for the k-dimensional code subspace. Note that the generator matrix of a linear block code is not unique as different basis vectors can be used to specify the same subspace.

Let us now present a way of encoding message sequences using the generator matrix of a linear block code. Even though many different mappings from the message sequences to the codewords are possible, we will select a particular one and describe the encoding process accordingly. Towards this goal, define the length-k row vectors

$$\begin{aligned} \boldsymbol{e}_1 &= [1, 0, 0, \ldots, 0], \\ \boldsymbol{e}_2 &= [0, 1, 0, \ldots, 0], \\ \boldsymbol{e}_3 &= [0, 0, 1, \ldots, 0], \\ &\vdots \\ \boldsymbol{e}_k &= [0, 0, 0, \ldots, 1]. \end{aligned} \tag{9.40}$$

Denote the rows of the generator matrix $\boldsymbol{G}$ as $1 \times n$ vectors $\boldsymbol{g}_i$, that is,

$$\boldsymbol{G} = \begin{bmatrix} \boldsymbol{g}_1 \\ \boldsymbol{g}_2 \\ \vdots \\ \boldsymbol{g}_k \end{bmatrix}. \tag{9.41}$$

To fix the mapping from the message bits to the codewords, let us pick the codeword corresponding to $\boldsymbol{e}_i$ as $\boldsymbol{g}_i$, for $i = 1, 2, \ldots, k$. To complete the description of the encoding process, consider an arbitrary message vector $\boldsymbol{x}$ as

$$\boldsymbol{x} = [x_1, x_2, \ldots, x_k], \tag{9.42}$$

with $x_i \in \{0, 1\}$. We can write

$$\boldsymbol{x} = \sum_{i=1}^{k} x_i \boldsymbol{e}_i. \tag{9.43}$$

We assign to the message vector $\boldsymbol{x}$ the codeword

$$\boldsymbol{c} = \sum_{i=1}^{k} x_i \boldsymbol{g}_i, \tag{9.44}$$

obtained by the same linear combination. In other words, we can write the encoding process as a matrix multiplication (over the binary field) as

$$\boldsymbol{c} = \boldsymbol{x}\boldsymbol{G}. \tag{9.45}$$

We re-emphasize that this is not the only possible encoding process. A linear block code is specified by its codewords only, and there are many possible mappings from the message sequences to the codewords. The above construction is only one such example. Having said that, whenever needed, we will refer to this encoding procedure in the rest of the section unless otherwise specified.

Example 9.6

Consider the simple code described in Example 9.3. We can select a generator matrix for the code as

$$\boldsymbol{G} = \begin{bmatrix} 1 & 0 & 1 & 0 & 0 \\ 0 & 1 & 1 & 1 & 1 \end{bmatrix}. \tag{9.46}$$

Using the above procedure for encoding, for instance, for $\boldsymbol{x} = [1 \;\; 1]$, the codeword is obtained as

$$\boldsymbol{c} = [1 \;\; 1] \begin{bmatrix} 1 & 0 & 1 & 0 & 0 \\ 0 & 1 & 1 & 1 & 1 \end{bmatrix} = [1\ 1\ 0\ 1\ 1]. \tag{9.47}$$

Systematic Codes. A linear block code is called systematic if $n - k$ parity bits are appended to the k information bits to form the n-bit codewords. With the encoding procedure described above, this is equivalent to saying that the generator matrix is of the form

$$\boldsymbol{G} = [\boldsymbol{I}_k \;\; \boldsymbol{P}], \tag{9.48}$$

where $\boldsymbol{I}_k$ is a $k \times k$ identity matrix and the matrix $\boldsymbol{P}$ is $k \times (n - k)$. Hence, the codeword (which is a $1 \times n$ vector) to the corresponding $1 \times k$ message vector $\boldsymbol{x}$ is given by

$$\boldsymbol{c} = \boldsymbol{x}\boldsymbol{G} = [\boldsymbol{x} \;\; \boldsymbol{x}\boldsymbol{G}]. \tag{9.49}$$

Clearly, the first k elements of the codeword are the message bits, and the $1 \times (n-k)$ vector $\boldsymbol{x}\boldsymbol{G}$ is composed of the parity-check bits. We can explicitly write

$$c_i = x_i, \quad \text{for } i = 1, 2, \ldots, k, \tag{9.50}$$

$$c_i = \sum_{j=1}^{k} P_{ji} x_j, \quad \text{for } i = k+1, k+2, \ldots, n, \tag{9.51}$$

where P_{ji} is the jith element of $\boldsymbol{P}$.

Example 9.7

The (4, 2) linear block code described by $c_1 = x_1$, $c_2 = x_2$, $c_3 = x_1 + x_2$, and $c_4 = x_1$ is a systematic linear block code with generator matrix

$$\boldsymbol{G} = \begin{bmatrix} 1 & 0 & 1 & 1 \\ 0 & 1 & 1 & 0 \end{bmatrix}. \tag{9.52}$$

Parity-Check Matrix. The parity-check matrix of a linear block code, denoted by $\boldsymbol{H}$, is an $(n-k) \times n$ matrix of 0s and 1s whose rows form a basis for the null space of $\boldsymbol{G}$. The rows of $\boldsymbol{H}$ are orthogonal to the rows of $\boldsymbol{G}$ (and, hence, to any codeword). That is, for any codeword $\boldsymbol{c}$, we have

$$\boldsymbol{c}\boldsymbol{H}^T = \boldsymbol{0}, \tag{9.53}$$

where the $\boldsymbol{0}$ vector on the right-hand side is $1 \times (n-k)$. We can also write

$$\boldsymbol{G}\boldsymbol{H}^T = \boldsymbol{0}. \tag{9.54}$$

For systematic linear block codes with generator matrix $\boldsymbol{G} = [\boldsymbol{I}_k \ \boldsymbol{P}]$, the parity-check matrix becomes

$$\boldsymbol{H} = \left[\boldsymbol{P}^T \ \boldsymbol{I}_{n-k}\right]. \tag{9.55}$$

To see this, we note that

$$\begin{aligned} \boldsymbol{G}\boldsymbol{H}^T &= [\boldsymbol{I}_k \ \boldsymbol{P}]\left[\boldsymbol{P}^T \ \boldsymbol{I}_{n-k}\right]^T & (9.56)\\ &= [\boldsymbol{I}_k \ \boldsymbol{P}]\begin{bmatrix} \boldsymbol{P} \\ \boldsymbol{I}_{n-k} \end{bmatrix} & (9.57)\\ &= \boldsymbol{P} + \boldsymbol{P} & (9.58)\\ &= \boldsymbol{0}. & (9.59) \end{aligned}$$

In other words, the rows of the $\boldsymbol{H}$ matrix so formed are orthogonal to the rows of the generator matrix, hence to all the codewords of the linear block code. Noting also that the rows of the $\boldsymbol{H}$ matrix are linearly independent, we can say that they form a basis for the null space of the code, hence the $\boldsymbol{H}$ matrix is the parity-check matrix of the linear block code with $\boldsymbol{G} = [\boldsymbol{I}_k \ \boldsymbol{P}]$.

Example 9.8

For the code described in Example 9.7, we can simply write the parity-check matrix as

$$\boldsymbol{H} = \begin{bmatrix} 1 & 1 & 1 & 0 \\ 1 & 0 & 0 & 1 \end{bmatrix}. \tag{9.60}$$

9.3.2 Minimum Distance of a Code

Our objective in this section is to define the minimum distance of a code, specifically, a linear block code. The minimum distance is a measure of how powerful the channel code is, and it can be used for a performance assessment as we will demonstrate later in the chapter.

We define the Hamming distance between any two codewords (or, more generally, any two binary vectors of the same length) $\boldsymbol{c}_i$ and $\boldsymbol{c}_j$, denoted by $d_H(\boldsymbol{c}_i, \boldsymbol{c}_j)$, as the number of positions at which the two codewords differ. For instance, for

$$\boldsymbol{c}_i = [1\,0\,1\,1\,0\,1\,1] \tag{9.61}$$

and

$$c_j = [0\ 1\ 0\ 0\ 1\ 0\ 1], \tag{9.62}$$

we obtain $d_H(c_i, c_j) = 6$.

The Hamming weight of a codeword c_i (or, more generally, a binary vector), denoted by $w_H(c_i)$, is the number of 1s in the codeword. For instance, for $c_i = [1\ 0\ 1\ 1\ 0\ 1\ 1]$, we have $w_H(c_i) = 5$.

We define the minimum distance of a block code (not necessarily linear) as the minimum Hamming distance between any two codewords, namely,

$$d_{\min} = \min_{c_i, c_j \in \mathcal{C}, i \neq j} d_H(c_i, c_j), \tag{9.63}$$

where $\mathcal{C}$ denotes the set of codewords. Similarly, we define the minimum weight of a code as the minimum weight of its non-zero codewords, that is,

$$w_{\min} = \min_{c_i \in \mathcal{C}, c_i \neq \mathbf{0}} w_H(c_i). \tag{9.64}$$

For a linear block code, the minimum distance of the code is the same as its minimum weight. One can prove this statement as follows. Consider a codeword with minimum weight $w_{\min}$. Since the Hamming distance between this codeword and the all-zero codeword has to be larger than or equal to $d_{\min}$, we can write $w_{\min} \geq d_{\min}$. Conversely, take two codewords c_i and c_j with a Hamming distance of $d_{\min}$. Since their componentwise binary sum $c_i + c_j$ is another codeword of the linear block code, and since its weight is equal to the Hamming distance of c_i and c_j, that is, $d_{\min}$, we conclude that $w_{\min} \leq d_{\min}$. Combining the two results, we establish that $w_{\min} = d_{\min}$.

Another interesting result is the following. For a linear block code, $d_{\min}$ is the smallest number of columns of $\boldsymbol{H}$ that add to $\mathbf{0}$. The proof of this result follows by observing that there is a one-to-one relationship between the columns of the parity-check matrix that add to zero and codewords of the linear block code. We leave the details of the proof of this statement as an exercise.

Example 9.9

Consider the linear block code in the previous example with generator and parity-check matrices

$$\boldsymbol{G} = \begin{bmatrix} 1 & 0 & 1 & 1 \\ 0 & 1 & 1 & 0 \end{bmatrix} \quad \text{and} \quad \boldsymbol{H} = \begin{bmatrix} 1 & 1 & 1 & 0 \\ 1 & 0 & 0 & 1 \end{bmatrix}, \tag{9.65}$$

respectively. We see that the second and third columns of $\boldsymbol{H}$ add to zero, hence $d_{\min} = 2$. We can also verify this fact by listing all the non-zero codewords (linear combinations of the rows of $\boldsymbol{G}$) as

$$\begin{aligned} &[1\ 0\ 1\ 1], \\ &[0\ 1\ 1\ 0], \\ &[1\ 1\ 0\ 1], \end{aligned} \tag{9.66}$$

and noting that the minimum weight among these three non-zero codewords is 2.

9.3.3 Hamming Codes

Hamming codes are a class of codes with

$$n = 2^m - 1 \text{ and } k = 2^m - m - 1, \tag{9.67}$$

where $m \geq 3$ is an integer. Namely, Hamming codes with code parameters

$$(n, k) = (7, 4),\ (15, 11),\ (31, 26),\ (63, 57), \ldots \tag{9.68}$$

exist. The rate of the code is

$$R_c = \frac{k}{n} = \frac{2^m - m - 1}{2^m - 1}. \tag{9.69}$$

One way to define a Hamming code is by taking the parity-check matrix of the code as an $(n-k) \times n$ matrix of all $2^m - 1$ non-zero binary m-tuples. The order in which the columns are listed is not important, as it simply renames the coordinates of the codewords. The following example gives explicitly the parity-check matrix of a $(7, 4)$ Hamming code.

Example 9.10

Picking $m = 3$, hence $n = 7$ and $k = 4$, the parity-check matrix

$$\boldsymbol{H} = \begin{bmatrix} 0 & 1 & 1 & 1 & 1 & 0 & 0 \\ 1 & 0 & 1 & 1 & 0 & 1 & 0 \\ 1 & 1 & 0 & 1 & 0 & 0 & 1 \end{bmatrix} \tag{9.70}$$

defines a Hamming code (since the columns are all the non-zero binary triplets). The parity-check matrix is given in a systematic form; thus, we can readily obtain a generator matrix as

$$\boldsymbol{G} = \begin{bmatrix} 1 & 0 & 0 & 0 & 0 & 1 & 1 \\ 0 & 1 & 0 & 0 & 1 & 0 & 1 \\ 0 & 0 & 1 & 0 & 1 & 1 & 0 \\ 0 & 0 & 0 & 1 & 1 & 1 & 1 \end{bmatrix}. \tag{9.71}$$

The minimum distance of a Hamming code (regardless of the code dimensions) is $d_{\min} = 3$. To see this, we notice that there is no all-zero column of $\boldsymbol{H}$, that is, $d_{\min}$ cannot be 1. Also, there are no two identical columns; hence, no two columns add to zero, and $d_{\min} > 2$. On the other hand, there are many combinations of three columns of $\boldsymbol{H}$ that add to zero. For instance, take the m-tuple with the first two elements 1 and the rest as 0, and add this to two columns with a single 1 in the first and second positions, respectively. Therefore, three columns of the parity-check matrix add to zero, and we conclude that $d_{\min} = 3$.

9.3.4 Decoding of Linear Block Codes

In this section, we explore the decoding of linear block codes over noisy channels, specifically over binary symmetric channels. For transmission over an AWGN channel, this means that the receiver first makes a raw decision on the coded bits based on the matched filter output sample, and then these *hard* decisions are fed to a channel decoder, in other words, hard-decision decoding is used.

Decoding Rule for Transmission Over a BSC. Consider transmission of a codeword over a BSC with cross-over probability ϵ, and without loss of generality, assume that $\epsilon < 1/2$. Assume also that the channel errors are independent of each other. Denoting the received binary sequence as

$$\boldsymbol{r} = [r_1\, r_2\, r_3 \ldots r_n], \tag{9.72}$$

the channel decoder's function is to determine the codeword that minimizes the probability of error. Assuming that the codewords are equally likely to be transmitted, this is the usual maximum likelihood decoding rule (similar to ML detection in the context of digital modulation). Namely, the decoder will declare

$$\hat{\boldsymbol{c}}_{\text{opt}} = \operatorname*{argmax}_{\boldsymbol{c}_i \in C} \mathbb{P}(\boldsymbol{r}|\boldsymbol{c}_i) \tag{9.73}$$

as the decoded codeword. Denoting the ith codeword as $\boldsymbol{c}_i = [c_{i1}\ c_{i2}\ \ldots\ c_{in}]$, c_{ij} representing bits, we can write

$$\hat{\boldsymbol{c}}_{\text{opt}} = \operatorname*{argmax}_{\boldsymbol{c}_i \in C} \prod_{j=1}^{n} \mathbb{P}(r_j|c_{ij}) \tag{9.74}$$

$$= \operatorname*{argmax}_{\boldsymbol{c}_i \in C} \epsilon^{d_H(\boldsymbol{c}_i,\boldsymbol{r})}(1-\epsilon)^{n-d_H(\boldsymbol{c}_i,\boldsymbol{r})} \tag{9.75}$$

$$= \operatorname*{argmax}_{\boldsymbol{c}_i \in C} \left(\frac{\epsilon}{1-\epsilon}\right)^{d_H(\boldsymbol{c}_i,\boldsymbol{r})} \tag{9.76}$$

$$= \operatorname*{argmin}_{\boldsymbol{c}_i \in C} d_H(\boldsymbol{c}_i,\boldsymbol{r}). \tag{9.77}$$

The third line follows by ignoring a factor common to all the codewords, and the last line follows since $\frac{\epsilon}{1-\epsilon} < 1$. In other words, the decoder will need to pick the codeword that is closest to the received binary vector in the Hamming distance sense.

Note that we can also obtain a decoding rule for transmission over an AWGN channel (using the matched filter outputs directly), namely, *soft-decision decoding*; however, we will not explore this approach in the context of linear block codes. In the next section, when studying convolutional codes, we will consider both hard-decision decoding and soft-decision decoding.

While the minimization of the Hamming distance between the received vector and all the codewords is easy to state as a decoding rule, it is not easy to implement as there are a very large number of codewords even for relatively small codes. For instance, for $k = 100$, there are 2^{100} codewords, which is an astronomical figure.

The decoding process is simplified for the case of linear block codes, as discussed below.

Syndrome Table Decoding. Consider communication over a BSC using an (n, k) linear block code. Denote the transmitted codeword by $\boldsymbol{c}$ and the received binary sequence by $\boldsymbol{r}$. We define the error pattern as the difference of $\boldsymbol{r}$ and $\boldsymbol{c}$, denoted as

$$\boldsymbol{e} = \boldsymbol{r} + \boldsymbol{c}, \tag{9.78}$$

since the operations are in the binary field. The error pattern contains "1" in the positions with channel errors, and "0" in the positions with no errors.

We define the syndrome of a received sequence $\boldsymbol{r}$ as the length-$(n-k)$ vector

$$\boldsymbol{s} = \boldsymbol{r}\boldsymbol{H}^T. \tag{9.79}$$

Since we have $\boldsymbol{r} = \boldsymbol{c} + \boldsymbol{e}$, and $\boldsymbol{c}\boldsymbol{H}^T = \boldsymbol{0}$ for any codeword $\boldsymbol{c}$, we can write

$$\boldsymbol{s} = \boldsymbol{e}\boldsymbol{H}^T, \tag{9.80}$$

namely, the syndrome depends only on the error pattern, not on the specific codeword being transmitted.

From our discussion on the parity-check matrix of linear block codes, we know that if the syndrome $\boldsymbol{s} = \boldsymbol{0}$, then the received sequence is a codeword. If it is a non-zero vector, then $\boldsymbol{r}$ is not a codeword, and the presence of errors during transmission is detected. Based on the expression in (9.77), the objective of the channel decoder is to find the codeword closest to the received sequence in the Hamming distance sense. This is equivalent to determining the most likely error pattern with the syndrome $\boldsymbol{s}$. With the BSC cross-over probability less than $1/2$, the most likely error pattern is the one with the lowest Hamming weight.

Since $\boldsymbol{s}$ is $1 \times (n-k)$, there are a total of 2^{n-k} possible syndromes. To determine the most likely error pattern, we use a syndrome table, listing all the possible syndromes along with the lowest weight error patterns to which they correspond. Note that this table is not necessarily unique. In any case, one can select any lowest weight error pattern for each syndrome to complete the table, and the decoding result will be optimal, that is, the result will be a codeword with the lowest Hamming distance from the received sequence.

The decoding process works as follows: once the syndrome corresponding to the received sequence is computed, the corresponding error pattern is identified in the table, and it is added to the received sequence to obtain the most likely codeword. The following example illustrates the syndrome table decoding process for a Hamming code.

Example 9.11

Construct the syndrome table for the $(7,4)$ Hamming code whose generator and parity check matrices were given in Example 9.10.

Solution

Since $n - k = 3$ for this example, there are $2^3 = 8$ different syndromes, and the error patterns they correspond to are the all-zero error pattern, and seven weight-1 error patterns. The syndrome table is given in Table 9.1.

Table 9.1 Syndrome table for the (7, 4) Hamming code.

syndrome	error pattern
0 0 0	0 0 0 0 0 0 0
0 0 1	0 0 0 0 0 0 1
0 1 0	0 0 0 0 0 1 0
0 1 1	1 0 0 0 0 0 0
1 0 0	0 0 0 0 1 0 0
1 0 1	0 1 0 0 0 0 0
1 1 0	0 0 1 0 0 0 0
1 1 1	0 0 0 1 0 0 0

As an example, assume that the received sequence is $\boldsymbol{r} = [1\ 0\ 1\ 0\ 1\ 0\ 0]$. Then the syndrome is

$$\boldsymbol{s} = \boldsymbol{r}\boldsymbol{H}^T = [0\ 0\ 1]. \tag{9.81}$$

We read the (most likely) error pattern from the syndrome table as $\boldsymbol{e} = [0\ 0\ 0\ 0\ 0\ 0\ 1]$, and obtain the maximum likelihood decoder output as $\boldsymbol{r} + \boldsymbol{e} = [1\ 0\ 1\ 0\ 1\ 0\ 1]$.

In Example 9.11, we notice that the number of weight-1 or fewer error patterns is the same as the number of different syndromes; hence, there is no ambiguity in determining the specific error patterns to be corrected. This is not always the case, as illustrated by the next example.

Example 9.12

Determine a syndrome table for the (4, 2) linear block code with parity check matrix

$$\boldsymbol{H} = \begin{bmatrix} 1 & 1 & 1 & 0 \\ 1 & 0 & 0 & 1 \end{bmatrix}. \tag{9.82}$$

Solution

Since $n - k = 2$, there are four different syndromes. The all-zero syndrome corresponds to the all-zero error pattern (i.e., no errors during transmission), and the three remaining syndromes will be used to correct three different weight-1 error patterns. Since there are a total of four error patterns with weight 1, there will be a weight-1 error pattern that will not be listed in the syndrome table, that is, it will not be correctable. In fact, the error patterns [0 1 0 0] and [0 0 1 0] have the same syndrome, therefore we will need to arbitrarily pick one of them to be in the syndrome table, and leave the other one out. Either way, we will have maximum likelihood decoding.

Selecting the error pattern [0 1 0 0] to be correctable, the syndrome table is given in Table 9.2.

Table 9.2 Syndrome table for the (4, 2) linear block code.

syndrome	error pattern
0 0	0 0 0 0
0 1	0 0 0 1
1 0	0 1 0 0
1 1	1 0 0 0

As an example, if the received sequence of bits is [1 1 1 1], then the corresponding syndrome is $\boldsymbol{s} = [1\ 1\ 1\ 1]\boldsymbol{H}^T = [1\ 0]$. Reading from the table the most likely error pattern as $\boldsymbol{e} = [0\ 1\ 0\ 0]$, we obtain the decoder output as $\boldsymbol{r} + \boldsymbol{e} = [1\ 0\ 1\ 1]$.

9.3.5 Performance of Linear Block Codes

In this section, we cover the performance evaluation of linear block codes. We first discuss the error correction and detection capabilities of linear block codes, then define their weight distribution and utilize it for error probability analysis over a binary symmetric channel.

Error Correction and Detection Capabilities. We now relate the minimum distance of a code to its error detection and correction capabilities. Most of the discussion that follows applies to general block codes, not just linear block codes.

Our first result is as follows. Assume that a channel code is used for error detection (as opposed to forward error correction). Applications of such usage include transmission over the internet with TCP-IP packets in which channel coding is employed by adding some parity bits to determine if the received packet contains any errors. Since the transmitted vector is certainly a codeword, if the received sequence is not a codeword, then we readily identify that there are errors in the transmission process. With this observation, we immediately conclude that a block code with minimum distance $d_{\min}$ can certainly detect error patterns with $d_{\min} - 1$ or fewer 1s. This is because the transmitted codeword cannot be transformed to another codeword with $d_{\min} - 1$ or fewer errors (otherwise, the minimum distance cannot be $d_{\min}$). Note that this is true for any block code, not necessarily linear.

When the code is used for error detection, many more error patterns can be detected in addition to those with weight $d_{\min} - 1$ or less. The only undetectable error patterns are the non-zero error patterns that transform the transmitted codeword to another codeword. There are exactly $2^k - 1$ such patterns, which constitute the non-detectable error patterns. The total number of detectable error patterns is $2^n - 1 - (2^k - 1) = 2^n - 2^k$. Note that this result is also valid for any block code.

We now consider a block code used for forward error correction. That is, given a received binary vector, the objective of the decoder is to determine the most likely codeword and correct the errors introduced by the channel. We already noted that this is done by minimizing the Hamming distance between the received vector and the codewords. The main result is the following: if a block code of minimum distance d_{min} is used for random error correction, then all the error patterns with t or fewer 1s with

$$2t + 1 \leq d_{\text{min}} \leq 2t + 2, \tag{9.83}$$

that is,

$$t = \left\lfloor \frac{d_{\text{min}} - 1}{2} \right\rfloor, \tag{9.84}$$

can be corrected. For this result, too, linearity is not needed; it applies to general block codes.

To explain this result, we notice that when a channel code is used for error correction, each codeword is associated with a decision region (i.e., the set of $\boldsymbol{r}$ vectors that are decoded as that codeword). This decision region certainly contains a Hamming sphere of radius t centered at the codeword (i.e., the set of n-tuples that are within a Hamming distance of t or less from the codeword). For a code of minimum distance d_{min}, with $t = \left\lfloor \frac{d_{\text{min}}-1}{2} \right\rfloor$, the elements of the Hamming spheres associated with different codewords cannot have any n-tuples as common elements. See Fig. 9.7 for an illustration. In other words, error patterns with t or fewer 1s are correctable.

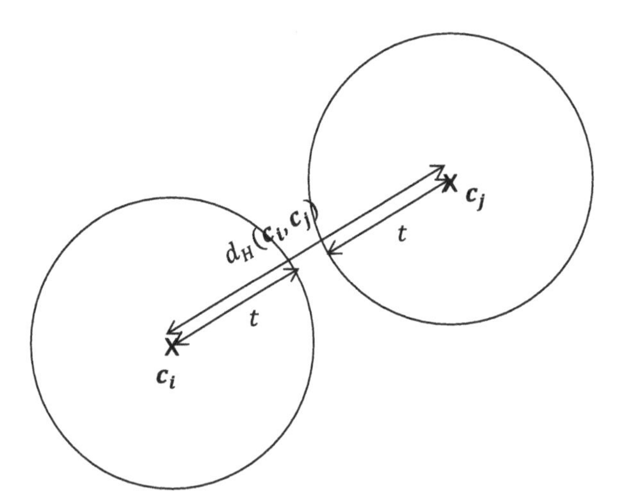

Figure 9.7 Illustration of the Hamming spheres of radius t for two different codewords. The Hamming spheres do not overlap, and they are certainly within the decision regions of the corresponding codewords as long as t is smaller than or equal to $\left\lfloor \frac{d_{\text{min}}-1}{2} \right\rfloor$.

Note also that there is at least one error pattern with $t+1$ 1s that cannot be corrected. This can be seen by carefully studying two codewords separated by the minimum distance of the code and constructing an error pattern with weight $t+1$, which will cause the received vector to be closer to the incorrect codeword.

One other result, applicable to linear block codes, is the following: since there are $(2^{n-k})-1$ non-zero syndromes, each corresponding to one correctable error pattern, we can correct exactly that many error patterns.

Weight Distribution of a Linear Block Code. Consider a linear block code used for transmission over a noisy channel. Pairwise Hamming distances among all its codewords are important to determine the error probability performance of the code. In fact, we can enumerate the distances of all the codewords from the reference codeword and use this list for a reasonably tight performance assessment for coded transmission. In particular, for linear block codes, this enumeration is identical for any reference codeword since the sum of two codewords is another codeword. Therefore, we can simply define what is called the *weight distribution* of a linear block code.

Let A_i be the number of codewords with weight i for an (n,k) linear block code. The list of these counts $A_0, A_1, A_2, \ldots, A_n$ form the weight distribution of the code. It is convenient to represent these weights using a polynomial, called the weight enumerator polynomial of the code, as

$$A(z) = A_0 + A_1 z + A_2 z^2 + \cdots + A_n z^n, \tag{9.85}$$

where z is a dummy variable. There is one codeword with 0 weight, and no non-zero codewords with weights less than d_{min}, therefore, the weight enumerator polynomial is of the form

$$A(z) = 1 + A_{d_{\text{min}}} z^{d_{\text{min}}} + A_{d_{\text{min}}+1} z^{d_{\text{min}}+1} + \cdots + A_n z^n. \tag{9.86}$$

It is clear that the weight enumerator contains more information than the minimum distance of a linear block code, hence it can be used for a more accurate performance assessment compared to the one with the minimum distance alone.

Example 9.13

Consider the $(7,4)$ Hamming code in Example 9.10. We can list all of its codewords by forming all 16 linear combinations of the rows of the generator matrix as follows:

$$\begin{array}{cc}
0000000 & 1001100 \\
1000011 & 0101010 \\
0100101 & 0011001 \\
0010110 & 1101001 \\
0001111 & 1011010
\end{array}$$

$$\begin{array}{cc} 1100110 & 0111100 \\ 1010101 & 1110000 \\ 0110011 & 1111111 \end{array}$$

By computing the Hamming weights of these 16 codewords, we can obtain the weight enumerator as

$$A(z) = 1 + 7z^3 + 7z^4 + z^7. \tag{9.87}$$

In other words, for any codeword of a $(7,4)$ Hamming code, there are seven codewords that are at a Hamming distance of 3, seven codewords that are at a Hamming distance of 4, and one codeword that is at a Hamming distance of 7.

Performance Over a BSC. We now use the weight enumerator of a linear block code to assess the error detection and error correction probabilities over a BSC with crossover probability ϵ.

Assume that a linear block code is used for error detection, and denote the probability of the presence of undetected errors by $P_{\mathrm{u,e}}$. By going over all the undetectable error patterns (i.e., all the non-zero codewords), and adding their probabilities, we obtain

$$P_{\mathrm{u,e}} = \sum_{i=d_{\min}}^{n} A_i \epsilon^i (1-\epsilon)^{n-i}. \tag{9.88}$$

In other words, we can obtain an exact expression for the undetected error probability using the weight enumerator of the code. If the weight distribution is not available, we can obtain an upper bound that uses only the minimum distance of the code by using the fact that

$$A_i \le \binom{n}{i}, \tag{9.89}$$

as there cannot be more than $\binom{n}{i}$ codewords of weight i. That is,

$$P_{\mathrm{u,e}} \le \sum_{i=d_{\min}}^{n} \binom{n}{i} \epsilon^i (1-\epsilon)^{n-i}. \tag{9.90}$$

The right-hand side can also be written in an alternate form as

$$P_{\mathrm{u,e}} \le 1 - \sum_{i=0}^{d_{\min}-1} \binom{n}{i} \epsilon^i (1-\epsilon)^{n-i}. \tag{9.91}$$

Finally, let us consider a linear block code used for forward error correction over a BSC. Denoting the number of correctable error patterns with weight i by α_i, we can write the probability of codeword error $P_{\mathrm{c,e}}$ as

$$P_{\mathrm{c,e}} = 1 - \sum_{i=0}^{n} \alpha_i \epsilon^i (1-\epsilon)^{n-i}. \tag{9.92}$$

Note that $\alpha_0 = 1$ (corresponding to the case of no channel errors). For instance, for the $(7, 4)$ Hamming code, since only the weight-1 error patterns can be corrected, we have $\alpha_1 = 7$ and $\alpha_i = 0$ for $i = 2, 3, \ldots, 7$. Hence, the probability of codeword error over a BSC becomes

$$P_{\mathrm{c,e}} = 1 - (1 - \epsilon)^7 - 7\epsilon(1 - \epsilon)^6. \tag{9.93}$$

If the weight distribution of the linear block code is not available, we can obtain an upper bound on the error probability using only the minimum distance of the code $d_{\min}$ $\left(\text{and, correspondingly, the error correction capability } t = \left\lfloor \frac{d_{\min}-1}{2} \right\rfloor\right)$. Since all the error patterns with weights t or less can be corrected, and since there are exactly $\binom{n}{i}$ error patterns with weight i, we can write

$$P_{\mathrm{c,e}} \leq 1 - \sum_{i=0}^{t} \binom{n}{i} \epsilon^i (1 - \epsilon)^{n-i}. \tag{9.94}$$

Note that this bound is exact for the case of a Hamming code, but not in general, since for most codes there are additional error patterns with weight more than t that can also be corrected.

We close this section by noting that while for some classes of linear block codes, the minimum distances and weight distributions are known, in general, these are very difficult to compute. In other words, even an approximate performance analysis of many interesting classes of codes is very challenging.

9.4 Convolutional Codes

As we studied in Section 9.3, for an (n, k) linear block code, we take k message bits and map them to n information bits, that is, a fixed-length block of message bits is mapped to a fixed-length codeword. In convolutional coding, however, redundancy is added with the help of a finite-state machine. The information bits are input to a circuit that generates the coded bits based on the current and previously transmitted bits using logical operations.

A general rate-k/n convolutional code with constraint length L is depicted in Fig. 9.8. Here, k information bits are input at each step, and these bits, along with the state of the convolutional code (identified by the previously transmitted bits), determine the n coded bits. This is a finite-state machine with $2^{k(L-1)}$ states. The encoders are finite impulse response (FIR) filters in the binary field. There is no prespecified length of the input sequence, hence the encoding process can continue as long as desired.

Let us give a simple example. Figure 9.9 illustrates a constraint length $L = 3$ convolutional code with $k = 1$ and $n = 2$. At each step, one information bit is input to the encoder, and two coded bits are produced at the output. The coded bits are determined by the current information bit and the state of the encoder (which is determined by the previous two information bits). The number of encoder states is 4.

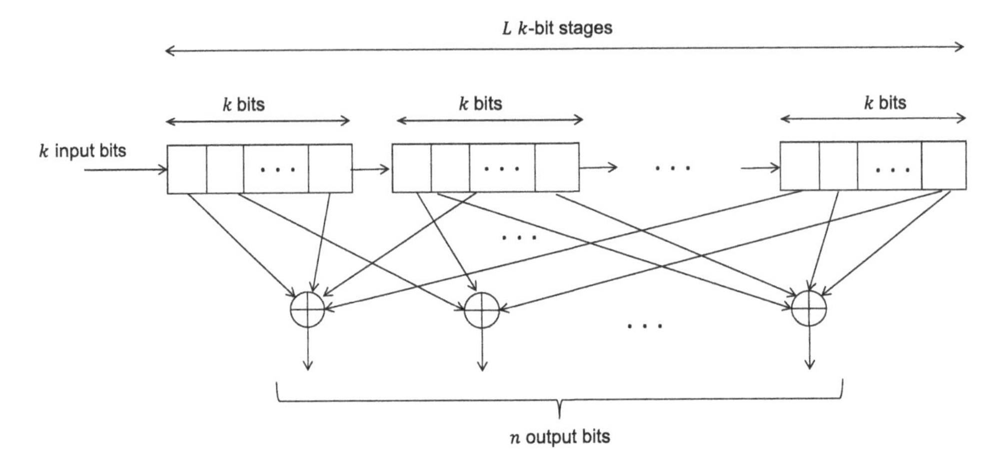

Figure 9.8 Block diagram of a convolutional code.

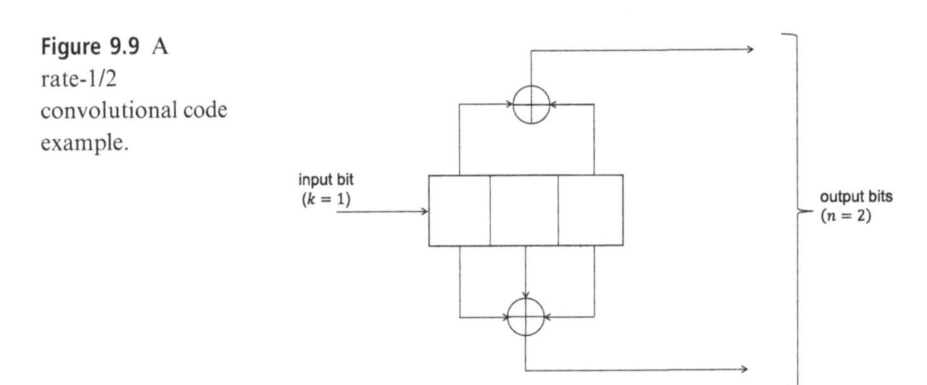

Figure 9.9 A rate-1/2 convolutional code example.

For this example, if the input sequence is

$$1\,0\,1\,1\,0\,1\ldots,$$

then the coded bits are obtained as

$$1\,1,\ 0\,1,\ 0\,0, 1\,0,\ 1\,0,\ 0\,0, \ldots.$$

For simplicity of exposition, in what follows, we will consider convolutional codes with $k = 1$.

9.4.1 Different Representations of Convolutional Codes

Impulse Response and Generator Sequences. Since the encoders are simply FIR filters, they can also be described by their impulse responses, namely, by the output sequence to a single 1 followed by 0s applied at the input. With a constraint length

L, the impulse responses are of length L as well. For instance, for the rate-1/2 convolutional code example given in Fig. 9.9, since there are two output sequences, there are two generators, given by

$$g_1 = [1\ 0\ 1] \tag{9.95}$$

and

$$g_2 = [1\ 1\ 1]. \tag{9.96}$$

Convolution of the binary input sequence with these impulse responses with operations in the binary field gives the coded binary sequence.

We also note that it is customary to use octal notation to specify the generators. For instance, we refer to the above code as the $(5, 7)_{\text{octal}}$ convolutional code.

We call a convolutional code *systematic* if the input bit stream is reproduced at the output as part of the codeword stream. The following example illustrates a systematic convolutional code.

Example 9.14

A rate-1/3 systematic convolutional code example is depicted in Fig. 9.10.

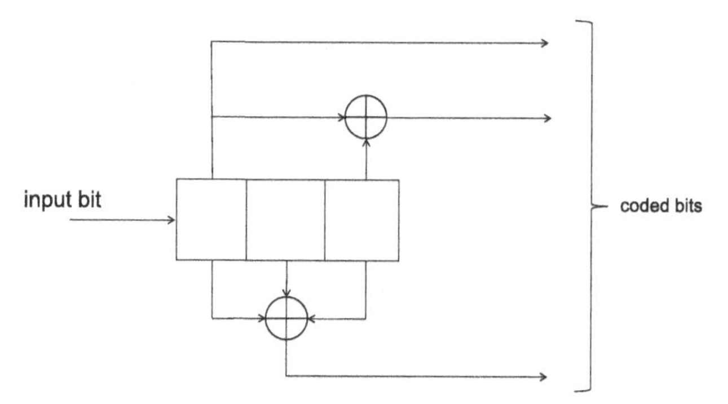

Figure 9.10 A rate-1/3 systematic convolutional code example.

The generators are

$$g_1 = [1\ 0\ 0], \quad g_2 = [1\ 0\ 1], \quad g_3 = [1\ 1\ 1]. \tag{9.97}$$

In octal notation, we refer to this code as the $(4,\ 5,\ 7)_{\text{octal}}$ convolutional code.

State Diagrams. Since convolutional codes are finite-state machines, they can be represented by a state transition diagram as well. A rate-k/n convolutional code with constraint length L is described by a state transition diagram with $2^{k(L-1)}$ states and 2^k transitions out of each state corresponding to 2^k possible input sequences at each step.

Let us give a specific example, focusing on the case with $k = 1$.

Example 9.15

The state diagram for the $(4, 5, 7)_{\text{octal}}$ convolutional code is depicted in Fig. 9.11.

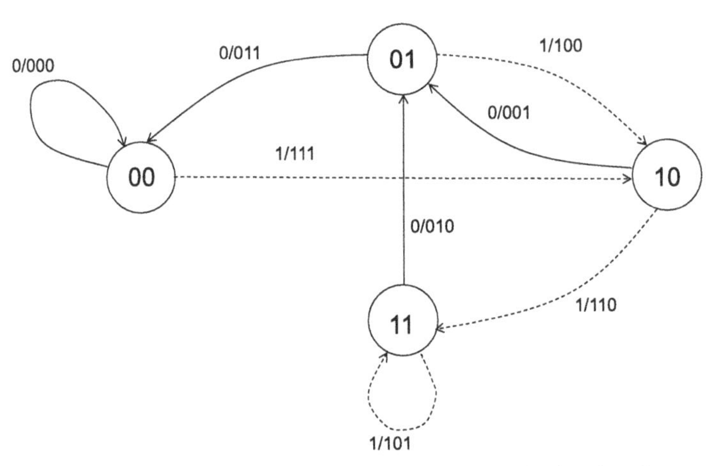

Figure 9.11 The state diagram of the rate-$1/3$ convolutional code example. Dashed lines depict the transitions corresponding to an input bit of 1 while solid lines show transitions corresponding to an input bit of 0.

Note that the states are defined as the contents of the first and second shift registers. Knowing the current state and the current input, we can determine the n output bits and the next state; hence, we can readily determine the encoder output corresponding to the input sequence. For instance, assuming that the initial state of the encoder is the all-zero state, if the input sequence is

$$1, 0, 1, 1, 0, 1, 0, 0, \ldots,$$

we obtain the encoded sequence as

$$111, 001, 100, 110, 010, 100, 001, 011, \ldots$$

by simply traversing through the state diagram with the given input sequence.

Trellis Diagram. Another way of representing convolutional codes is through a trellis diagram, which is nothing but a way of showing the transitions between different states as time evolves.

Example 9.16

The trellis diagram for the $(4, 5, 7)_{\text{octal}}$ convolutional code is depicted in Fig. 9.12.

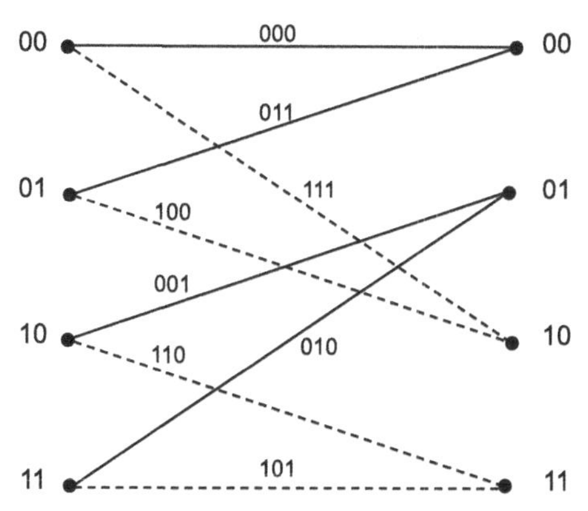

Figure 9.12 The trellis diagram of the rate-1/3 convolutional code example. The 3-bit labels are the corresponding coded bits. Dashed lines correspond to an input bit of 1 while solid lines depict transitions corresponding to an input bit of 0.

The dashed lines correspond to an input bit of 1 while the solid lines depict transitions corresponding to an input bit of 0 in the trellis diagram.

The codewords are *paths* through the trellis diagram. Therefore, we can easily write down the codeword sequence corresponding to a given input sequence by simply tracing the corresponding paths at each step. We will find this observation highly useful when describing the optimal decoding algorithm for convolutional codes.

9.4.2 Catastrophic Convolutional Codes

Certain convolutional codes (called catastrophic convolutional codes) must be avoided as their performance over practical (noisy) channels is poor. We call a convolutional code *catastrophic* if two different message sequences that differ in infinitely many positions are mapped to codewords that differ in only a finite number of positions. For such codes, even some small number of errors (in finite number positions) in the transmitted codeword due to channel noise may result in infinitely many errors.

We can identify catastrophic convolutional codes by examining their state diagrams. Specifically, if the state diagram contains a loop (a path starting and ending in the same state) in which a non-zero input sequence is mapped to an all-zero output, then the code is catastrophic. Let us give an example.

Example 9.17

Consider the rate-$1/2$ convolutional code with generators $g_1 = [1\ 1\ 0]$ and $g_2 = [0\ 1\ 1]$, that is, the $(6,3)_{\text{octal}}$ code. The state diagram is given in Fig. 9.13. We observe that the state diagram contains a loop at state "11" with an input bit 1 (non-zero input) resulting in an all-zero output sequence, hence the code is catastrophic.

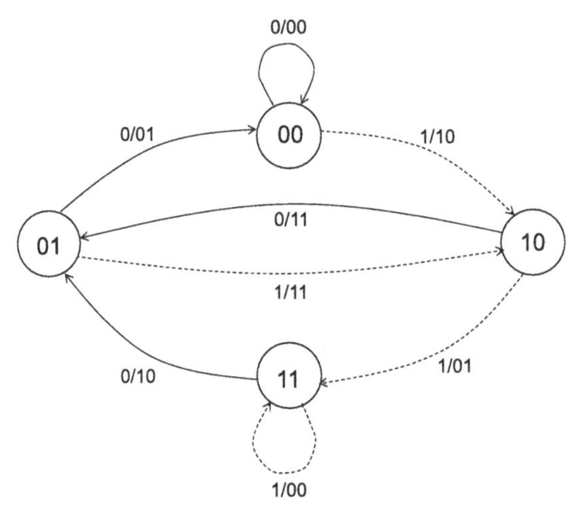

Figure 9.13 The state diagram of the $(6,3)_{\text{octal}}$ convolutional code.

As an illustration, consider the input sequence

$$1, 1, 1, 1, 1, 1, 1, 1, \ldots.$$

The corresponding codeword is given by

$$10, 01, 00, 00, 00, 00, 00, 00, \ldots,$$

which differs from the all-zero codeword (corresponding to the all-zero input sequence) in only two bits: the first and the fourth. Assuming that the all-one message is sent, and the channel noise causes errors in only the first and fourth bits and in no other bit locations, the received sequence will be the all-zero sequence, which is also a codeword (corresponding to the all-zero message sequence). Hence, the decoder will output the all-zero message sequence as its result, and all the bits will be received in error!

9.4.3 Transfer Function of a Convolutional Code

The transfer function of a convolutional code gives information about the paths through the trellis diagram that start with the all-zero state and merge with it for the first time. This information is useful in conducting a performance analysis of the convolutional code over an AWGN or binary symmetric channel (similar to the weight distribution being useful for an error probability analysis of linear block codes).

Let us define the transfer function of a convolutional code as the polynomial

$$T(X,Y,Z) = \sum_{i,j,k} a_{i,j,k} X^i Y^j Z^k, \tag{9.98}$$

where X, Y, Z are auxiliary variables, and $a_{i,j,k}$ is the number of paths (starting from the all-zero state and merging with it for the first time) with input weight i, codeword weight j, and length k. If the length of the path is unimportant, we can set $Z = 1$ and perform the summation over the index k, and define

$$T(X,Y) = \sum_{i,j} a_{i,j} X^i Y^j, \tag{9.99}$$

where

$$a_{i,j} = \sum_{k} a_{i,j,k} \tag{9.100}$$

is the number of paths (starting from the all-zero state and merging with it for the first time) with input weight i and codeword weight j, as the transfer function which does not contain the path-length information.

We can systematically determine the transfer function of a convolutional code by using a labeled flow graph obtained from the state diagram and computing the input–output transfer function using standard techniques. This is best illustrated via an example.

Example 9.18

Compute the transfer function of the $(4,5,7)_{\text{octal}}$ convolutional code.

Solution

To do this we split the all-zero state into two distinct states: one starting all-zero state (denoted by 00) and one terminating all-zero state (denoted by 00′), and then obtain the labeled signal flow graph shown in Fig. 9.14.

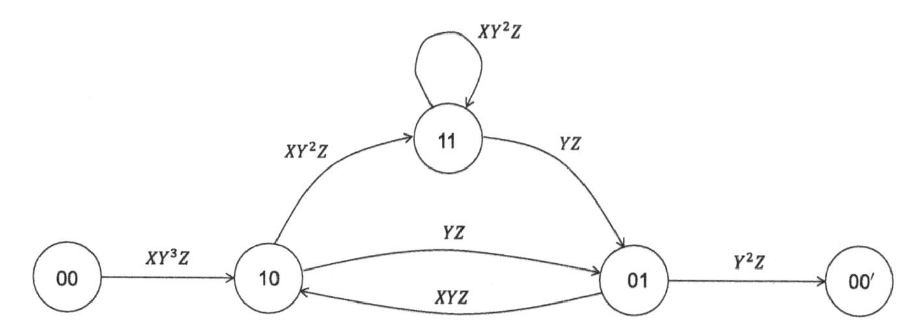

Figure 9.14 Signal flow graph for the convolutional code in the example.

The labels on each path are simply $X^i Y^j Z$, where i is the corresponding input bit weight (0 or 1) and j is the weight of the coded bit sequence for each transition. Z has power 1 since each transition contributes to the path length by one. To determine $T(X, Y, Z)$, we compute the transfer function from the starting all-zero state to the ending one (i.e., from 00 to 00′). Let W_{00}, W_{01}, W_{10}, W_{11}, and $W_{00'}$ denote the node variables. We can write the node equations

$$W_{10} = W_{00}XY^3Z + W_{01}XYZ, \tag{9.101}$$
$$W_{11} = W_{11}XY^2Z + W_{10}XY^2Z, \tag{9.102}$$
$$W_{01} = W_{10}YZ + W_{11}YZ, \tag{9.103}$$
$$W_{00'} = W_{01}Y^2Z. \tag{9.104}$$

Solving for $W_{00'}/W_{00}$, we obtain the transfer function as

$$T(X, Y, Z) = \frac{XY^6Z^3}{1 - XY^2Z - XY^2Z^2}. \tag{9.105}$$

Using long division, we can write the transfer function in polynomial form as

$$T(X, Y, Z) = XY^6Z^3 + XY^6Z^3(XY^2Z + XY^2Z^2) + \dots. \tag{9.106}$$

We observe that there is one path with input weight 1, output weight 6, and length 3. Indeed, this path is obtained by the input "100" resulting in "111, 001, 011." This is the path with the lowest output weight; hence, the minimum distance of the code (also called the free distance) is 6.

Setting $Z = 1$, we obtain

$$T(X, Y) = \frac{XY^6}{1 - 2XY^2} = XY^6 + 2X^2Y^8 + 4X^3Y^{10} + \dots. \tag{9.107}$$

That is, there is a single path with output weight 6 and input weight 1, two paths with output weight 8 and input weight 2, four paths with output weight 10 and input weight 3, and so on.

Note that in the above example, we have computed the transfer function of the code by solving for it explicitly. It is also possible to write it directly from the signal flow graph using Mason's formula, which is omitted. Note also that the transfer function of a convolutional code serves the same purpose as the weight-enumerating function for a linear block code. In other words, we can infer some characteristics of the codewords which allow us to conduct a performance analysis through tight bit error rate bounds, as will be described later in this section.

9.4.4 Maximum Likelihood Decoding of Convolutional Codes

In this section, we consider the maximum likelihood decoding of convolutional codes over AWGN channels. Our main conclusion will be the following: for convolutional codes, optimal decoding (which minimizes the codeword error probability assuming equally likely codewords) can be efficiently implemented through a procedure called the *Viterbi algorithm.*

Consider a rate-k/n convolutional code with constraint length L being used for transmission over an AWGN channel. Assume that the coded bits are modulated using binary PAM (or BPSK), and let us consider the mathematically equivalent channel model. Namely, the bit 0 is transmitted as $-\sqrt{E_s}$, and the bit 1 is transmitted as $+\sqrt{E_s}$, where E_s is the energy per coded bit. The received signal is simply the transmitted signal plus a zero-mean Gaussian random variable with variance $N_0/2$. The noise terms are i.i.d. for different transmissions.

While in principle transmission can take place indefinitely, in practice we will have a finite number of bits to transmit (e.g., a communication packet). Let us assume that mn coded bits are transmitted (for some m). Let us also assume that the initial state of the convolutional code is the all-zero state and that the final state is forced to be the all-zero state as well. That is, there is *trellis termination*, for which the information bit sequence is appended with $k(L-1)$ "0" bits so that the final state is driven to the all-zero state. In other words, $k(m-L+1)$ message bits are encoded into kn coded bits, resulting in an actual transmission rate of

$$\frac{k(m-L+1)}{mn}. \tag{9.108}$$

Clearly, for large m this value will be very close to the convolutional code rate k/n.

Let us establish the notation. We denote by $\underline{\boldsymbol{c}}$ a codeword composed of $\boldsymbol{c}_1, \boldsymbol{c}_2, \ldots, \boldsymbol{c}_m$, which are length-$n$ binary vectors denoting the parts of the codeword corresponding to the trellis sections $1, 2, \ldots, m$, respectively.[2] That is,

$$\underline{\boldsymbol{c}} = [\boldsymbol{c}_1,\ \boldsymbol{c}_2,\ \ldots, \boldsymbol{c}_m]. \tag{9.109}$$

Let $c_{i,j}$ denote the jth coded bit in the ith trellis section, that is,

$$\boldsymbol{c}_i = [c_{i,1},\ c_{i,2},\ \ldots,\ c_{i,n}]. \tag{9.110}$$

Let $c'_{i,j}$ be the modulated version of $c_{i,j}$, that is,

$$c'_{i,j} = \sqrt{E_s}(2c_{i,j} - 1). \tag{9.111}$$

Similarly, $\boldsymbol{c}'_i$ and $\underline{\boldsymbol{c}}'$ are the modulated versions of $\boldsymbol{c}_i$ and $\underline{\boldsymbol{c}}$, respectively.

The received signal for the coded bit $c_{i,j}$ through an AWGN channel is given by

$$y_{i,j} = c'_{i,j} + n_{i,j}, \tag{9.112}$$

where $n_{i,j} \sim \mathcal{N}(0, N_0/2)$. Similarly, $\boldsymbol{y}_i$ and $\underline{\boldsymbol{y}}$ denote the received signal vectors corresponding to the transmission of $\boldsymbol{c}_i$ and $\underline{\boldsymbol{c}}$, respectively.

Finally, let us also define the binary versions of the received signals as well. We denote the binary version of $y_{i,j}$ by $\hat{c}_{i,j}$, that is, $\hat{c}_{i,j} = 1$ if $y_{i,j} \geq 0$, and 0 otherwise. Similarly, the binary versions of $\boldsymbol{y}_i$ and $\underline{\boldsymbol{y}}$ are denoted by $\hat{\boldsymbol{c}}_i$ and $\hat{\underline{\boldsymbol{c}}}$, respectively. The binary forms of the received signals can be thought of as outputs of a binary

[2] We use the underlined boldface notation for codewords to reflect that they are written as a concatenation of the sequence of n coded bits in each trellis section.

symmetric channel (with a certain cross-over probability related to the AWGN noise variance; recall our discussion on this point earlier in this chapter).

Hard-Decision Decoding (HDD). In this case, the decoding process is based on the binary version of the received signal, that is, $\hat{\underline{c}}$. Since this is nothing but transmission over a binary symmetric channel (with a cross-over probability less than 1/2), recall from Section 9.3 that the ML decoding rule is to find the codeword $\underline{c}$ that is closest to $\hat{\underline{c}}$ in the Hamming distance sense. Namely,

$$\underline{c}_{\text{opt}} = \underset{\underline{c}}{\operatorname{argmin}}\, d_H(\underline{c}, \hat{\underline{c}}) \tag{9.113}$$

$$= \underset{\underline{c}}{\operatorname{argmin}} \sum_{i=1}^{m} d_H(c_i, \hat{c}_i), \tag{9.114}$$

which follows since the Hamming distance between a codeword and the received sequence is simply the sum of all the Hamming distances between the received sequence in each trellis section and the corresponding part of the codeword.

Soft-Decision Decoding (SDD). In this case, the decisions are based on the AWGN channel outputs directly, i.e., on $\underline{y}$. We can borrow ideas from multi-dimensional modulation over AWGN channels studied in Chapter 5 to identify the ML decision rule. As the noise terms are i.i.d. zero-mean Gaussian random variables, we notice that the convolutionally coded transmission can be thought of as nothing but a very high-dimensional modulation scheme, and we simply need to minimize the (squared) Euclidean distance between the received vector and (the modulated versions of) the codewords. That is, the decoder output is

$$\underline{c}_{\text{opt}} = \underset{\underline{c}}{\operatorname{argmin}} \left\| \underline{y} - \underline{c}' \right\|^2 \tag{9.115}$$

$$= \underset{\underline{c}}{\operatorname{argmin}} \sum_{i=1}^{m} \left\| y_i - c_i' \right\|^2. \tag{9.116}$$

Or, equivalently, we can maximize the correlation metric:

$$\underline{c}_{\text{opt}} = \underset{\underline{c}}{\operatorname{argmax}}\, \underline{y} \cdot \underline{c}' \tag{9.117}$$

$$= \underset{\underline{c}}{\operatorname{argmax}} \sum_{i=1}^{m} y_i \cdot c_i', \tag{9.118}$$

where "·" denotes the dot product, that is,

$$y_i \cdot c_i' = \sum_{j=1}^{n} y_{i,j} c_{i,j}'. \tag{9.119}$$

Here we again utilized the fact that the squared Euclidean distance between the (modulated version of) the codeword and the received signal is the same as the sum of the squared Euclidean distances of the parts of the modulated version of the codeword and the corresponding received signal in each trellis section.

The Viterbi Algorithm. We notice that for both hard-decision decoding (or communication over a binary symmetric channel) for which the decision metric is the Hamming distance and soft-decision decoding for which the decision metric is either the squared Euclidean distance or the correlation metric, the metrics to be minimized or maximized are additive across the trellis stages. With this observation, optimal decoding can be carried out very efficiently using a procedure called the Viterbi algorithm.

Recall that the codewords of a convolutional code are all the paths through the trellis diagram starting from the all-zero state and ending in the all-zero state after m state transitions, since the trellis is terminated by appending zero bits at the end of the information sequence. Therefore, the job of the optimal decoder is to find the closest path to the received sequence in the Hamming distance sense (for HDD) or the closest path in the squared Euclidean distance sense (or the path with the highest correlation metric) for SDD. The Viterbi algorithm systematically performs this optimization.

The main idea in the Viterbi algorithm is the following: if two (or more) paths are merging at a particular state while going through the trellis, then there is no need to consider further the one(s) with the "worse" (partial) metric. This is because there is no way the "losing" path(s) can ever become better later on, since the metric to be optimized is additive. In other words, we can keep only one surviving path at each state for each time instant along with its metric. Since we terminate the trellis, when we reach the end of the trellis, there will be only one surviving path, and it will be declared as the decoder output. For the case of Hamming distance minimization, occasionally there will be ties when comparing different paths. In such cases, we can keep one of them arbitrarily. We could be eliminating a path that has the same distance as our final decoder output, but this only means that we are finding one of the codewords at the lowest Hamming distance with the received (binary) sequence. With SDD over an AWGN channel, the probability of encountering such a scenario is zero.

Let us illustrate the Viterbi algorithm via examples.

Example 9.19

Consider the rate-$1/3$ $(4, 5, 7)_{\text{octal}}$ convolutional code. Assume that the information sequence $1, 0, 1, 1, 0, 0$ is transmitted, where the last two bits are used for trellis termination. The transmitted codeword is then

$$111,\ 001,\ 100,\ 110,\ 010,\ 011. \tag{9.120}$$

Assume also that the received sequence (through a BSC with a cross-over probability less than $1/2$) is given by

$$\hat{c} = [110,\ 011,\ 100,\ 110,\ 011,\ 011]. \tag{9.121}$$

Note that $\hat{c}$ contains three errors with respect to the transmitted codeword (in the 3rd, 5th, and 15th bits).

The steps of the Viterbi algorithm with HDD are shown in Fig. 9.15.

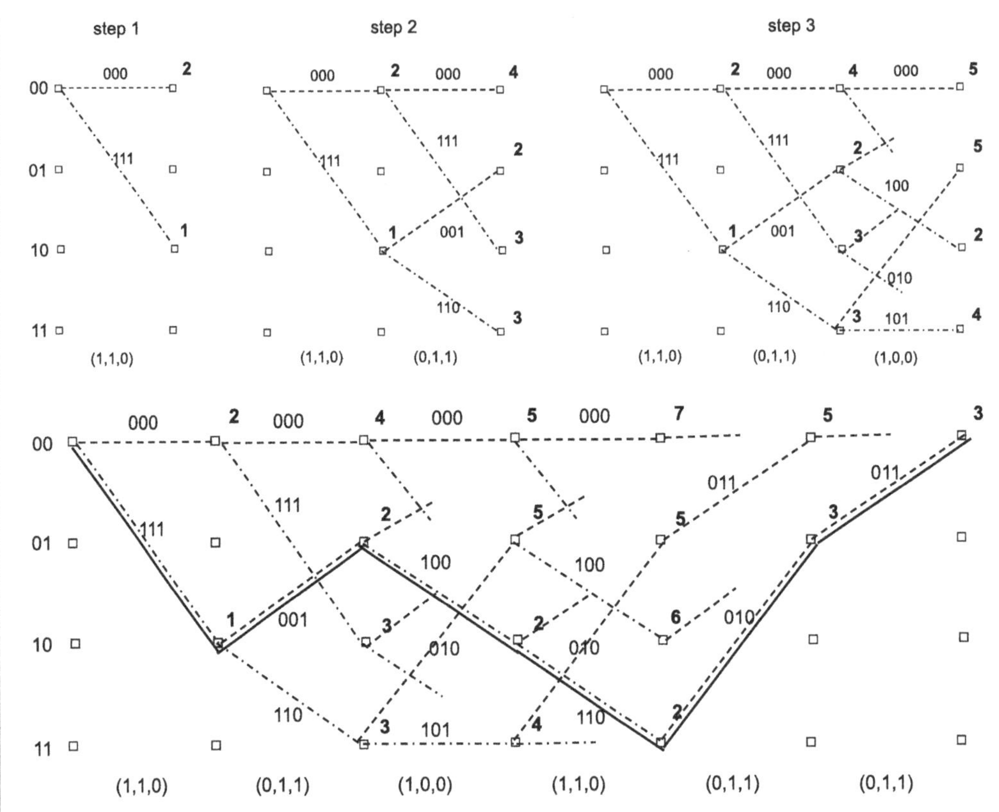

Figure 9.15 Example of hard-decision decoding via the Viterbi algorithm. The first three steps and the complete decoding procedure are shown. The "dashed" transitions correspond to the input bit 0 while the "dash-dotted" transitions are for 1. The decoding result (trace-back) after all the steps are completed is shown using solid lines. The last two bits are used for trellis termination; hence, only the transitions corresponding to the input bit 0 exist. The decoder output is 1,0,1,1,0,0, that is, the correct information sequence is obtained in this example.

In the illustration, the transitions shown in dashed lines correspond to the input bit of 0, and those shown in dash-dotted lines correspond to the input bit of 1. The received sequence as well as the partial Hamming metrics at each step are shown. The decoder result is depicted using solid lines, corresponding to the input sequence of 1, 0, 1, 1 plus the last two trellis termination bits of 0, 0.

Example 9.20

For the same convolutional code as in Example 9.19, let $E_s = 1$ and assume that the received signal through an AWGN channel is given by

$$y = [(0.5, 0.7, -0.6), (-0.5, 0.6, 0.4), (0.2, -0.4, -0.6), (0.4, 0.3, -0.2),$$
$$(-0.5, 0.4, 0.3), (-0.3, 0.2, 0.8)]. \tag{9.122}$$

The Viterbi algorithm steps for decoding using the correlation metric are shown in Fig. 9.16. The decode output is 1, 0, 1, 1.

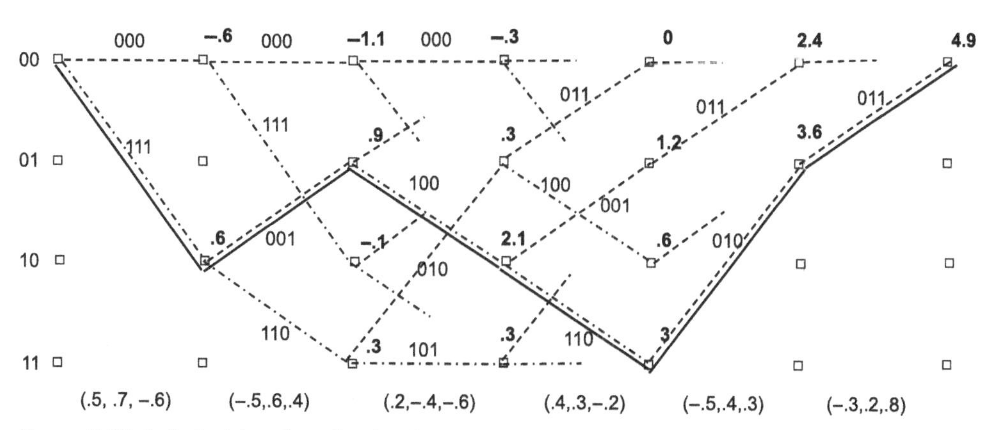

Figure 9.16 Soft-decision decoding implemented using the Viterbi algorithm with the correlation metric. The "dashed" transitions correspond to the input bit 0 while the "dash-dotted" transitions are for 1. The decoder result is shown using solid lines.

Practical Issues. We close this section by highlighting some practical issues about the decoding of convolutional codes. The first point is the following: to reduce the memory requirements and the decoding delay, "path memory truncation" can be used. With a very high probability, the surviving paths at each step "merge" with a certain delay; hence, a decision can be made on the transmitted bits without reaching the end of the trellis with almost no degradation in performance. Typically, a decoding delay of five times the constraint length gives very good results.

As another simplification, SDD can be implemented with "quantization." Indeed, decoding with the quantized versions of the matched filter outputs using three to four bits for each sample is sufficient to obtain almost the same performance as that of the decoder with infinite precision. One bit (or two-level quantization) simply results in HDD.

9.4.5 Performance Analysis of Convolutional Codes

While an exact performance analysis of convolutional codes is not possible, it is feasible to apply the union bound to obtain upper bounds on the bit error probability. We present these results without going through their derivations.

Antipodal Signaling Over AWGN. Assuming that convolutionally coded bits are transmitted using antipodal signaling over an AWGN channel, the bit error probability P_b is bounded as

$$P_b \leq \frac{1}{k} \frac{\partial T(X, Y)}{\partial X} \bigg|_{X=1, Y=\exp(-\gamma_b R_c)}, \tag{9.123}$$

where $R_c = k/n$ is the code rate and $\gamma_b = \frac{E_b}{N_0}$ is the signal-to-noise ratio per information bit. If $k = 1$, that is, for a rate $R_c = 1/n$ code, we obtain

$$P_b \leq \left.\frac{\partial T(X,Y)}{\partial X}\right|_{X=1,\,Y=\exp(-\gamma_b R_c)}. \tag{9.124}$$

For large signal-to-noise ratios ($\gamma_b \gg 1$),

$$P_b \approx \frac{1}{k} b_{d_{\text{free}}} e^{-\gamma_b R_c d_{\text{free}}}, \tag{9.125}$$

where $b_{d_{\text{free}}}$ is the number of paths with the free distance (with the all-zero codeword).

Example 9.21

For the $(4, 5, 7)_{\text{octal}}$ convolutional code,

$$T(X,Y) = \frac{XY^6}{1 - 2XY^2} \tag{9.126}$$

was obtained earlier. Therefore, we can write

$$\left.\frac{\partial T(X,Y)}{\partial X}\right|_{X=1} = \frac{Y^6}{(1-2Y^2)^2}. \tag{9.127}$$

Hence, we obtain

$$P_b \leq \frac{e^{-2\gamma_b}}{(1 - 2e^{-2\gamma_b/3})^2}. \tag{9.128}$$

Or,

$$P_b \approx e^{-2\gamma_b} \tag{9.129}$$

for $\gamma_b \gg 1$.

Compared to the uncoded transmission for which $P_{\text{b, uncoded}} = Q(\sqrt{2\gamma_b})$, which is $\approx e^{-\gamma_b}$ at high signal-to-noise ratios, the above convolutionally coded scheme is better by about 3 dB.

Performance Over a BSC. For a convolutional code used over a BSC with cross-over probability p (with $p < 1/2$), the bit error probability is upper bounded by

$$P_b \leq \frac{1}{k} \left.\frac{\partial T(X,Y)}{\partial X}\right|_{X=1,\,Y=\sqrt{4p(1-p)}}. \tag{9.130}$$

For small p values, we can use the approximation

$$P_b \approx \frac{1}{k} b_{d_{\text{free}}} (4p(1-p))^{d_{\text{free}}/2}. \tag{9.131}$$

Note that this result also covers the case of hard-decision decoding of convolutional codes when used over an AWGN channel. In this case, we simply need to substitute the cross-over probability with the raw error probability for the coded bits, that is, $p = Q(\sqrt{2\gamma_s})$, with $\gamma_s = \gamma_b R_c$ denoting the signal-to-noise ratio per coded bit.

Example 9.22

For the convolutional code in Example 9.21, the bit error rate for transmission over a BSC with cross-over probability p is upper bounded as

$$P_b \leq \left. \frac{Y^6}{(1-2Y^2)^2} \right|_{Y=\sqrt{4p(1-p)}} = \frac{64p^3(1-p)^3}{(1-8p(1-p))^2}. \tag{9.132}$$

For $p \ll 1$, we obtain $P_b \approx 64p^3$. For instance, for $p = 0.01$, $P_b \approx 0.000064$ is obtained for the coded case.

Performance Difference Between SDD and HDD. We can use the approximate analysis above to get an idea of the performance difference between soft-decision decoding and hard-decision decoding of convolutional codes with antipodal signaling over an AWGN channel. For HDD, the raw error probability is

$$p = Q(\sqrt{2\gamma_b R_c}) \approx e^{-\gamma_b R_c}. \tag{9.133}$$

Hence, for $\gamma_b \gg 1$, we can write

$$\sqrt{4p(1-p)} \approx 2\exp\left(-\frac{\gamma_b R_c}{2}\right), \tag{9.134}$$

and we obtain

$$P_b \leq \left. \frac{1}{k} \frac{\partial T(X,Y)}{\partial X} \right|_{X=1,\, Y=\exp(-\gamma_b R_c/2+\ln 2)}. \tag{9.135}$$

Comparing this with the upper bound on the bit error probability for SDD, we observe that there is a performance loss by a factor of

$$\frac{\gamma_b R_c}{\gamma_b R_c/2 - \ln 2} \approx 2 \tag{9.136}$$

in terms of the required signal-to-noise ratio. Therefore, the hard-decision decoding is around 3 dB worse than soft-decision decoding in performance. Note that actual simulation results corroborate this approximate analysis and show that the performance difference is around 2–3 dB.

9.5 Chapter Summary

In this chapter, we studied channel-coding techniques in some depth. We first reviewed the ultimate limits of communications over noisy channels and stated the role of channel coding in approaching these limits. We then covered two major classes of channel codes: linear block codes and convolutional codes. We described specific linear block codes as well as optimal decoding approaches for transmission over a binary symmetric channel as well as an AWGN channel. We also went through some structural properties of linear block codes, allowing for a performance assessment for both error detection and error correction. In our coverage of convolutional codes, we went through different representations, including state and trellis diagrams, described the optimal decoding algorithms for both hard-decision decoding and soft-decision decoding, and developed a way of performance analysis in terms of bit error probabilities over both binary symmetric and AWGN channels.

PROBLEMS

9.1 What is the capacity of a binary symmetric channel with a cross-over probability of 0.01 in bits/channel use? How about its capacity if the cross-over probability is 0.1?

9.2 Compute the capacity of a discrete-time Gaussian channel with an input power constraint of $\mathbb{E}[X^2] \leq 5$ units and noise variances of
(a) $N = 0.1$ units,
(b) $N = 0.5$ units,
(c) $N = 5$ units.

9.3 Consider a waveform Gaussian channel with bandwidth $B = 100$ kHz and an average input power of $P = 2$ W. The noise power spectral density is $S_n(f) = N_0/2$. Determine the channel capacity for
(a) $N_0 = 10^{-6}$ W/Hz,
(b) $N_0 = 2 \times 10^{-5}$ W/Hz,
(c) $N_0 = 5 \times 10^{-4}$ W/Hz.

9.4 Consider an additive white Gaussian noise channel where the input, noise, and output are complex. Namely, we have

$$Y = X + Z,$$

where X is the complex channel input, Z is a circularly symmetric complex Gaussian random variable with zero mean. Assume that the real and imaginary parts of the noise term are independent, and their variances are $N/2$ each. The input power constraint is $\mathbb{E}[|X|^2] \leq P$. Clearly, there are two parallel channels (one for the real part and the other for the imaginary part), and they can be considered separately as they do not interfere. Note also that such a channel model is relevant for bandpass transmission when the lowpass-equivalent representations are used.

(a) Assume that the average power of the real part of the input is P_r and that of the imaginary part of the input is P_i. Argue that the transmission rate is given by the sum of the capacities of the two parallel channels with the individual power constraints, that is,

$$\frac{1}{2}\log\left(1 + \frac{P_r}{N/2}\right) + \frac{1}{2}\log\left(1 + \frac{P_i}{N/2}\right).$$

(b) Considering the overall power constraint of $P_r + P_i \leq P$, show that the transmission rate is maximized for $P_r = P_i = P/2$. What is the resulting channel capacity?

9.5 Consider the $(7, 3)$ binary linear block code in systematic form whose parity-check equations are given by

$$c_4 = x_1 + x_2 + x_3, \quad c_5 = x_1 + x_2, \quad c_6 = x_2 + x_3, \quad c_7 = x_1 + x_3,$$

where x_1, x_2, x_3 are the message digits and c_4, c_5, c_6, c_7 are the parity-check digits.

(a) Find a generator matrix for the code.

(b) Find a parity-check matrix for the code.

(c) What is the minimum distance of this code?

(d) Assume that this code is used over a binary symmetric channel. If it is used for error detection only, how many bit errors can it detect? If it is used for error correction, how many bit errors can it correct?

9.6 Consider an $(8, 5)$ (binary) linear block code. Denote the message bits by $x_1, \ldots, x_5$ and the coded bits by $c_1, c_2, \ldots, c_8$. Assume that the coded bits satisfy the following parity-check equations:

$$\begin{aligned} c_1 + c_4 + c_5 + c_6 &= 0, \\ c_2 + c_3 + c_5 + c_7 &= 0, \\ c_1 + c_2 + c_3 + c_4 + c_8 &= 0. \end{aligned}$$

(a) Determine the parity-check and generator matrices of the code.

(b) What is the codeword corresponding to the message $(x_1, x_2, x_3, x_4, x_5) = (1, 1, 0, 1, 0)$?

(c) We obtain a new code by adding one more parity bit such that all the codewords are of even weight. The new code is a $(9, 5)$ linear block code. What is its parity-check matrix? What is its minimum distance?

9.7 Consider the $(9, 4)$ (binary) linear block code in systematic form whose parity-check equations are given by

$$\begin{aligned} c_0 &= x_1 + x_3, \\ c_1 &= x_0 + x_3, \\ c_2 &= x_0 + x_1 + x_3, \\ c_3 &= x_1 + x_2 + x_3, \\ c_4 &= x_0 + x_1 + x_2 + x_3, \end{aligned}$$

where x_0, x_1, x_2, x_3 are the message digits and c_0, c_1, c_2, c_3, c_4 are the parity-check digits.

(a) Find a generator matrix for the code.

(b) Find a parity-check matrix for the code.

(c) Determine the minimum distance of the code.

9.8 Prove that for a linear block code, $d_{\min}$ is the smallest number of columns of the parity check matrix $\boldsymbol{H}$ that add to $\mathbf{0}$.

9.9 Prove the Singleton bound which states that $d_{\min} \leq n - k + 1$ for an (n, k) linear block code.

9.10 Consider a linear block code with generator matrix

$$\mathbf{G} = \begin{bmatrix} 1 & 0 & 0 & 1 & 0 & 0 & 1 \\ 0 & 1 & 0 & 0 & 1 & 1 & 0 \\ 0 & 0 & 1 & 1 & 0 & 1 & 1 \end{bmatrix}.$$

(a) How many codewords does the code have? List all of them.

(b) What is the minimum distance of this code? What is its weight enumerator?

(c) This code is used over a binary symmetric channel with a cross-over probability of 0.01. Assuming that the code is used for error detection only and that the errors in different channel uses are independent of each other, determine the probability that there will be undetected errors at the decoder output.

9.11 Determine the rate, generator matrix, and minimum distance of the smallest linear block code (i.e., the one with the lowest number of codewords) which contains the five codewords (of length $n = 10$) given below:

$$\begin{aligned} &1\,0\,1\,0\,1\,0\,1\,0\,1\,0, \\ &0\,1\,0\,1\,0\,1\,0\,1\,0\,1, \\ &0\,0\,1\,1\,1\,1\,1\,1\,1\,1, \\ &1\,1\,1\,1\,1\,1\,1\,1\,1\,1, \\ &0\,1\,1\,0\,1\,0\,1\,0\,1\,0. \end{aligned}$$

9.12 Determine the weight enumerator of the $(n, n-2)$ linear block code with parity-check matrix

$$H = \begin{bmatrix} 1 & 0 & 1 & 0 & \cdots & 1 & 0 \\ 0 & 1 & 0 & 1 & \cdots & 0 & 1 \end{bmatrix}$$

where n is even.

9.13 Determine the weight distribution of the $(7, 4)$ Hamming code and using this weight distribution, compute the probability that there will be undetected errors for transmission over a binary symmetric channel with cross-over probability of 0.05 (assuming that the code is used for error detection only).

9.14 For binary (n, k) linear block codes, the Griesmer bound states that

$$n \geq \sum_{i=0}^{k-1} \left\lceil \frac{d_{\min}}{2^i} \right\rceil,$$

where $\lceil a \rceil$ denotes the ceiling of a, that is, the smallest integer equal to or greater than a.

(a) What does this bound imply for the minimum distance of a $(15, 7)$ code?

(b) Prove that this bound implies the Singleton bound (which states that $d_{\min} \leq n - k + 1$).

(c) Prove the Griesmer bound.

9.15 Consider an (n, k) systematic linear block code. Assume that the last message coordinate of the code is removed, resulting in an $(n-1, k-1)$ linear block code. The resulting code is referred to as a shortened code.

(a) How can we obtain the generator matrix of the shortened code in terms of the generator matrix of the original one?

(b) How can we obtain the parity-check matrix of the shortened code from the parity-check matrix of the original one?

(c) Show that the minimum distance of the shortened code cannot be smaller than the minimum distance of the original one.

9.16 For a binary (n, k) Hamming code with $n = 2^m - 1$ and $k = 2^m - m - 1$ (where $m \geq 3$ is an integer), the parity-check matrix contains all the non-zero m-tuples as its columns. Determine whether the all-one sequence of length n is a codeword or not.

9.17 Consider the rate-1/2 convolutional code with generators $g_1 = [1\ \ 1\ \ 0]$ and $g_2 = [1\ \ 1\ \ 1]$.

(a) Show the state diagram of the code.

(b) Show the trellis diagram of the code.

(c) Determine the codeword corresponding to the information sequence

$$1, 1, 0, 0, 1, 0, 1, 1, 0, 1.$$

(Assume that the trellis of the code is terminated by appending two 0 bits at the end of the message sequence.)

9.18 Determine the weight-enumerating function and the free distance of the rate-1/2 convolutional code with generators $g_1 = [1\ \ 0\ \ 1]$ and $g_2 = [1\ \ 1\ \ 1]$. Also determine the union bound on the bit error rate over an AWGN channel with BPSK modulation as a function of the signal-to-noise ratio $\gamma_b = E_b/N_0$.

9.19 Consider the rate-1/2 convolutional code described by the generators $g_1 = [1\ \ 0\ \ 1]$ and $g_2 = [1\ \ 1\ \ 1]$.

(a) Show the state diagram of the code. Is this code catastrophic? Why or why not?

(b) Assume that this code is used over an AWGN channel with BPSK (where the bit 1 is represented by "+1" and the bit 0 is represented by "−1"). The output of the matched filter is given by

$$\underline{r} = [(0.4, 0.1), (-0.5, -0.2), (0.5, -0.3), (-0.9, 1.2), (0.6, -0.5), (0.5, 0.1)].$$

Assume that the initial state is "00" and that the last two information bits are both "0" (they are used to terminate the trellis). What is the result of the soft-decision decoding with the Viterbi algorithm?

Note: Use the "correlation metric" instead of the squared Euclidean distance in the implementation of the Viterbi algorithm.

(c) Consider the same scenario in part (b), but assume that hard-decision decoding is employed. What is the decoding result?

9.20 Consider the rate-1/2 convolutional code described by the generators $g_1 = [1\ \ 0\ \ 1]$ and $g_2 = [1\ \ 1\ \ 1]$ as in the previous problem. However, assume that in this case, the second bits of all even-indexed coded bit pairs are punctured to obtain a different code. For instance, if the codeword is 11,01,00,10,10,00 ..., with puncturing, we transmit 11,0,00,1,10,0 ....

(a) What is the rate of this new code?
(b) If the received signal (under the same setup as in part (b) of Problem 9.19) is

$$\underline{r} = [(-0.3, 0.2), (0.3), (-0.2, 0.3), (0.5), (0.7, -0.5), (0.3)],$$

what is the result of optimal soft-decision decoding?

COMPUTER PROBLEMS

9.21 This problem deals with the performance of a Hamming code over a binary symmetric channel and over an AWGN channel (with BPSK modulation). We will only work with the equivalent mathematical channel models (i.e., we will suppress the actual waveforms being transmitted, matched filtering operations, etc.).

(a) Provide the generator matrices of the $(7, 4)$ and $(15, 11)$ Hamming codes. It is not necessary, but systematic forms of the generator and parity-check matrices can be used for simplicity.
(b) Using MATLAB, generate a large number of information bits randomly (say about 1 million), split the sequence into blocks of length k (4 or 11 for this problem), and using the generator matrices in part (a), encode each message block into a codeword of length n (7 or 15 for the two codes above). Simulate transmission of each codeword over a binary symmetric channel and report the average error rates obtained as a function of the cross-over probability p.
(c) Repeat part (b) assuming that BPSK modulation is used over an AWGN channel. Report the average error rates as a function of the signal-to-noise ratio per information bit $\gamma_b = E_b/N_0$. Also include the uncoded error rate results for comparison. The decoder can be implemented in a brute-force manner by minimizing the Euclidean distance or equivalently by maximizing the correlation metric.

9.22 This problem studies the performance of convolutional codes over an AWGN channel. We focus on a simple convolutional code of rate $1/2$ with only two states, namely, the $(2, 3)_{\text{octal}}$ code (i.e., the generators are $[1 \;\; 0]$ and $[1 \;\; 1]$). Assume that BPSK modulation is used.

(a) Provide the encoder block diagram and compute the transfer function. What is the free distance of the code?
(b) Determine the union bound on the average bit error rate with hard-decision decoding. Also, calculate the upper bound with soft-decision decoding.

(c) Using MATLAB, implement the Viterbi decoding algorithm and simulate the code performance with hard-decision decoding. Plot your results as a function of $\gamma_b = E_b/N_0$, and compare them with the union bound on the bit error rate and with the error probability of uncoded transmission.

(d) Repeat the previous part, assuming that soft-decision decoding is employed.

9.23 Repeat Problem 9.22 with the rate-1/2 four-state convolutional code with generators [1 0 1] and [1 1 1].

10 Topics in Communication System Design

In this short chapter, our objective is to consider some relevant issues in communication system design. Specifically, we study the effects of transmission losses in a communication system and discuss ways of addressing the related challenges. In addition, we review multiple-access techniques that allow different users to efficiently share a communication medium.

We provide a brief exposition to link budget analysis and introduce some important terminology in communication system design in Section 10.1. We consider the use of non-ideal amplifiers to combat the effects of transmission losses and quantify the loss in the signal-to-noise ratio at the amplifier output in Section 10.2. Section 10.3 is devoted to the use of analog and regenerative repeaters for transmission over long distances. Section 10.4 deals with time-division, frequency-division, and code-division multiple-access techniques. The chapter is concluded in Section 10.5.

10.1 Link Budget Analysis

Link budget analysis accounts for the gains and losses in communication systems to determine how far the transmitted signals will reach and what the resulting throughputs will be. This accounting is part of a communication system design and is applicable for transmission over different media, for example, for line-of-sight satellite communications, transmission over coaxial cables, transmission over optical fiber, and so on. We do not give detailed coverage of link budget analysis; instead, we only introduce some important and commonly used terminology.

We can write the following simple relationship between the transmit and receive power levels:

$$\text{received power (dBm)} = \text{transmit power (dBm)} + \text{gains (dB)} - \text{losses (dB)}. \tag{10.1}$$

Here, the power unit dBm refers to the power with respect to a 1 mW level in dB units. For instance, a power level of 20 mW is equal to $10\log_{10}(20\text{ mW}/1\text{ mW}) = 13$ dBm. Similarly, we can use the power units of dBW, which represents the power level with respect to 1 W reference. Gains in (10.1) can be due to the use of directional antennas, while losses can be due to many different sources such as free-space path loss, environmental conditions, transmitter losses, receiver losses, and so on. Specifically, line-of-sight communication is governed by the free-space path loss, which is inversely proportional to the square of the distance between the transmitter and the receiver.

The lowest power level at which the receiver can detect a radio frequency signal and demodulate data is called *receiver sensitivity*. The amount that the received power exceeds the receiver sensitivity (when both are expressed in dB units) is called the *link margin*. For instance, if the receiver sensitivity is −70 dBm, and the received power is −60 dBm, the link margin is 10 dB.

There are also losses due to multipath and time variations, modeled as channel fading (not covered in this book), which cause significant (random) fluctuations in the received power levels. When there is fading, it is impossible to guarantee a link availability with certainty; however, by using a *fade margin*, we can make the communication link available with high probability. For exercises on computing the probability of link availability and on calculating the fade margin for wireless channels modeled as Rayleigh fading, see Problems 10.1 and 10.2.

10.2 Transmission Losses and Noise

As stated in Section 10.1, in a typical communication system, the transmitted signals undergo various losses, and the received power levels become too low for processing by the communication circuitry. Therefore, we need to amplify the received signal before processing. In this section, we study the effects of practical amplifiers on the signal quality measured in terms of the signal-to-noise ratio.

Let the input to an amplifier be white noise with power spectral density $S_n(f) = \frac{N_0}{2}$, as shown in Fig. 10.1. We define the noise-equivalent bandwidth of a lowpass filter as the bandwidth of an ideal lowpass filter with the same output noise power (when the input to the amplifiers is white noise). We assume that the amplifier has a noise-equivalent bandwidth of B_{neq} and a power gain of G.

We refer to an amplifier as ideal if there is no intrinsic noise added to the output signal. Thus, if the amplifier is ideal, the noise power at its output becomes

$$P_{n,o} = GN_0B_{neq}. \tag{10.2}$$

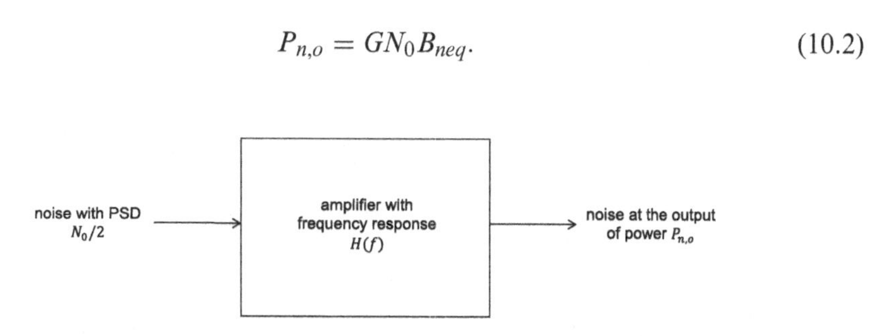

Figure 10.1 While noise input to an amplifier.

In practice, however, the output of the amplifier will be corrupted by additional *intrinsic* noise of the amplifier. That is, the noise power at the output is given by

$$P_{n,o} = GN_0 B_{neq} + P_{n,i}, \tag{10.3}$$

where $P_{n.i}$ denotes the power of the intrinsic noise introduced by the amplifier. From our discussion of thermal (electronic) noise in Chapter 5, we have $N_0 = kT$ (recall (5.3)), where k is the Boltzman constant and T is the temperature in Kelvin. Therefore, we can also write

$$\begin{aligned} P_{n,o} &= GkTB_{neq} + P_{n,i} & (10.4)\\ &= GkB_{neq}\left(T + \frac{P_{n,i}}{GkB_{neq}}\right) & (10.5)\\ &= GkB_{neq}(T + T_e), & (10.6)\end{aligned}$$

where

$$T_e = \frac{P_{n_i}}{GkB_{neq}} \tag{10.7}$$

is defined as the *effective noise temperature* of the amplifier. As $T_e > 0$ for a practical amplifier, the average output noise power is larger than that of an ideal amplifier with the same power gain and noise-equivalent bandwidth.

Let us now examine what happens to the signal-to-noise ratio. Assume that a signal with power $P_{s,i}$ corrupted by white noise is input to a non-ideal amplifier as shown in Fig. 10.2. Also, assume that the signal is within the passband of the amplifier and it is not distorted.

The signal power at the output of the amplifier is $P_{s,o} = GP_{s,i}$, while the noise power is as given in (10.6). Therefore, the output signal-to-noise ratio (denoted by SNR_o) becomes

$$\begin{aligned} \text{SNR}_o &= \frac{P_{s,o}}{P_{n,o}} & (10.8)\\ &= \frac{GP_{s,i}}{GkTB_{neq}\left(1 + \frac{T_e}{T}\right)} & (10.9)\\ &= \frac{1}{1 + \frac{T_e}{T}}\text{SNR}_i, & (10.10)\end{aligned}$$

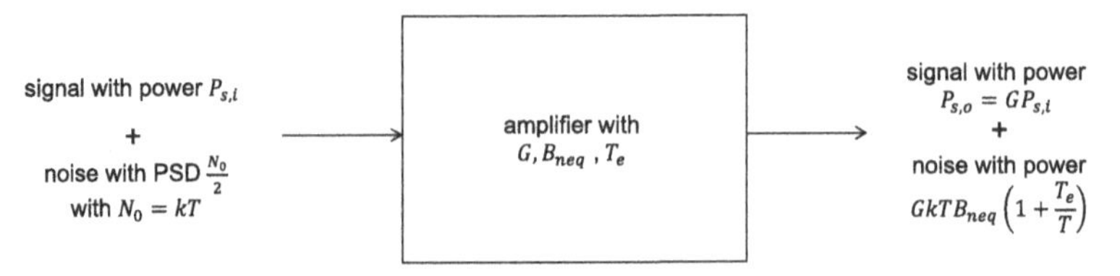

Figure 10.2 Signal and noise power levels at the input and output of an amplifier.

where SNR_i is the input signal-to-noise ratio, defined as

$$\text{SNR}_i = \frac{P_{s,i}}{N_0 B_{neq}}. \tag{10.11}$$

Note that SNR_i is also the signal-to-noise ratio at the output of an ideal amplifier (for which $T_e = 0$). We observe that while the output signal power is increased (by a factor of G), the output signal-to-noise ratio is degraded with respect to the input signal-to-noise ratio. In other words, by amplifying the signal to make sure that it can be processed by the communication circuitry, the signal quality gets worse, which may cause detrimental effects on the system performance.

Taking $T = T_o = 290$ Kelvin (i.e., room temperature), we also define the noise figure of an amplifier as

$$F = 1 + \frac{T_e}{T_o} \tag{10.12}$$

and write

$$\text{SNR}_o = \frac{1}{F}\text{SNR}_i. \tag{10.13}$$

We recognize that the noise figure is a direct measure of the quality of an amplifier. For a high-quality amplifier, the noise figure is closer to unity, and for an ideal amplifier, it is exactly unity (or 0 dB).

Cascade of K Amplifiers. We now consider a cascade of amplifiers with different gains and noise figures, as depicted in Fig. 10.3. We are interested in the overall gain and the noise figure of this system.

Let the gain and the noise figure of the lth amplifier in the cascade be G_l and F_l, respectively, $l \in \{1, 2, \ldots, K\}$. Then the overall gain is simply the product of the individual amplifier gains, that is,

$$G_{eq} = G_1 G_2 \ldots G_K, \tag{10.14}$$

and the equivalent noise figure is given by

$$F = F_1 + \frac{F_2 - 1}{G_1} + \frac{F_3 - 1}{G_1 G_2} + \cdots + \frac{F_K - 1}{G_1 G_2 \ldots G_{K-1}}. \tag{10.15}$$

To see how this result follows, we first consider the case of $K = 2$ amplifiers. Denoting the signal power at the input by P_s, the signal power at the output of the two-amplifier cascade becomes $P_s G_1 G_2$. Assuming white noise at the input with $N_0 = kT_o$, the average noise power at the output of the first amplifier is given by

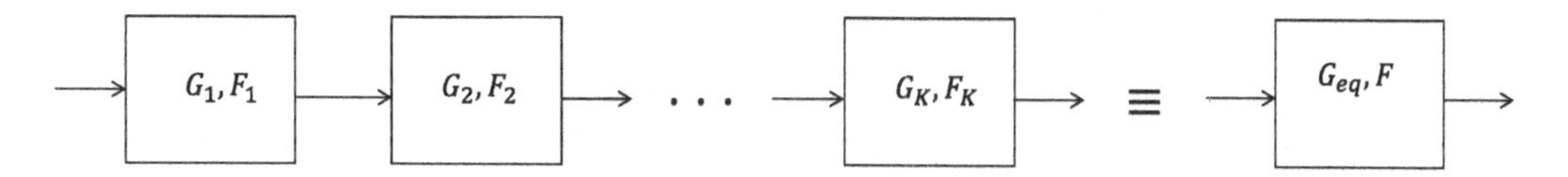

Figure 10.3 Cascade of amplifiers.

$$G_1 k T_o B_{neq} \left(1 + \frac{T_{e,1}}{T_o}\right), \tag{10.16}$$

and, hence, the noise power at the output of the second amplifier becomes

$$G_1 G_2 k T_o B_{neq} \left(1 + \frac{T_{e,1}}{T_o}\right) + G_2 k B_{neq} T_{e,2}. \tag{10.17}$$

Therefore, the output signal-to-noise ratio is given by

$$\text{SNR}_o = \frac{G_1 G_2 P_s}{G_1 G_2 k T_o B_{neq} \left(1 + \frac{T_{e,1}}{T_o} + \frac{T_{e,2}}{G_1 T_o}\right)}, \tag{10.18}$$

which can also be written as

$$\text{SNR}_o = \frac{\text{SNR}_i}{F_1 + \frac{F_2 - 1}{G_1}}, \tag{10.19}$$

establishing the validity of the expression in (10.15) for $K = 2$.

The result for an arbitrary number of amplifiers in the cascade follows by a repeated application of this result.

Example 10.1

We would like to use three amplifiers in cascade to improve the signal strength at a receiver. Amplifier A has a gain of 20 dB and a noise figure of 10 dB; amplifier B has a gain of 20 dB and a noise figure of 6 dB; amplifier C has a gain of 10 dB and a noise figure of 3 dB.

Determine the order in which these three amplifiers should be connected to ensure the highest signal quality at the output of the cascade. Also, determine the output signal-to-noise ratio assuming that the input signal-to-noise ratio is 40 dB.

Solution

Using the result in (10.15), for a cascade of three amplifiers, the equivalent noise figure is given by

$$F_1 + \frac{F_2 - 1}{G_1} + \frac{F_3 - 1}{G_1 G_2}. \tag{10.20}$$

Therefore, a simple calculation shows that the optimal cascade is $C \to B \to A$, resulting in a noise figure of

$$F_{eq} = 2 + \frac{4 - 1}{10} + \frac{10 - 1}{10 \times 100} \approx 2.3, \tag{10.21}$$

or 3.62 dB. Notice that using the best amplifier (with the lowest noise figure) as the initial stage is advantageous.

The output SNR corresponding to an input SNR of 40 dB is simply $40 - 3.62 = 36.38$ dB.

When we employ a cascade of amplifiers, the order in which they are connected matters in terms of the output signal-to-noise ratio. Considering that the typical amplifiers have large gains, a good rule of thumb is to place the amplifier with the lowest noise figure (i.e., the highest quality one) as the first in the cascade. The rest of the ordering is typically not critical. As an example, this is precisely what is done in the reception of satellite signals, where a low-noise amplifier is used as the first stage, and additional amplifiers are employed in subsequent stages.

10.3 Combatting Transmission Losses via Repeaters

Transmission losses over long distances may become so high that it would be impossible to communicate over a single hop. In such cases, we would need to employ one or more repeaters between the transmitter and receiver. The repeaters can be simple *analog repeaters* whose functions are simply to amplify and transmit the signal, or *regenerative repeaters* which are employed with digital transmission whose functions are to demodulate the transmitted digital data and then generate and transmit a clean version. We go over these two types of repeaters separately.

10.3.1 Analog Repeaters

Assume that transmission from a source to a destination is performed via analog repeaters over K segments as depicted in Fig. 10.4. Assume that the lth segment has a loss of L_l, and the lth amplifier (repeater) has a gain of G_l and noise figure F_l. Let us determine the signal-to-noise ratio at the destination (denoted by SNR_o).

Let us focus on one of the segments as depicted in Fig. 10.5. Let us denote the transmit power by P_T, the channel loss by L, and the amplifier gain and noise figure by G and F, respectively. The signal power at the input to the amplifier is P_T/L; hence the output SNR is

$$\frac{1}{F}\frac{P_T/L}{N_0 B_{neq}} = \frac{1}{FL}\frac{P_T}{N_0 B_{neq}}. \tag{10.22}$$

In other words, the equivalent gain of the segment is G/L and its equivalent noise figure becomes LF.

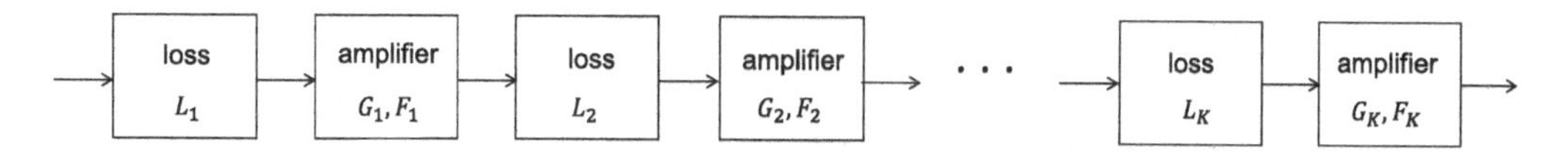

Figure 10.4 Cascade of K analog repeaters to combat transmission losses.

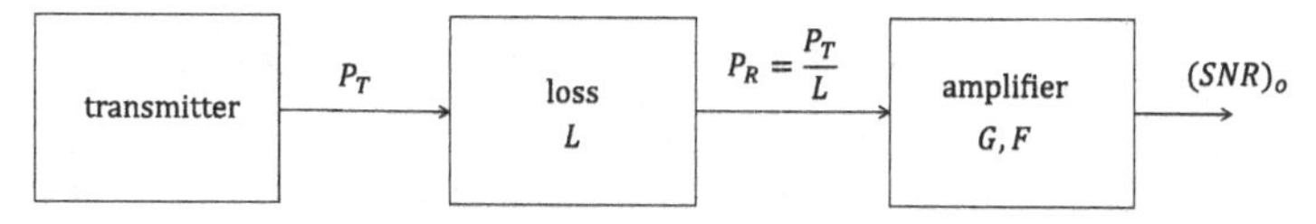

Figure 10.5 A segment of transmission loss and the amplification stage.

Going back to the use of K repeaters over K segments, we conclude that the received power is

$$\frac{G_1 G_2 \dots G_K}{L_1 L_2 \dots L_K} P_T, \tag{10.23}$$

and the overall noise figure becomes

$$F_{overall} = L_1 F_1 + \frac{L_2 F_2 - 1}{G_1/L_1} + \dots + \frac{L_K F_K - 1}{(G_1/L_1)(G_2/L_2)\dots(G_{K-1}/L_{K-1})}. \tag{10.24}$$

Hence, the output signal-to-noise ratio is given by

$$\text{SNR}_o = \frac{1}{F_{overall}} \frac{P_T}{N_0 B_{neq}}. \tag{10.25}$$

We may select the amplifier gains equal to the loss of the corresponding segments, that is, $G_l = L_l$, for $l = 1, 2, \dots, K$. In this case, the signal power at the output of each amplifier becomes P_T, and the overall noise figure can be written as

$$\left(\sum_{l=1}^{K} L_l F_l \right) - K + 1. \tag{10.26}$$

If, further, the segment losses (denoted by L) and the amplifiers are identical (each with gain $G = L$), the equivalent noise figure becomes $KLF - K + 1$, F denoting the noise figure of each amplifier. Since the losses are typically high, $KLF \gg K - 1$, and the output signal-to-noise ratio is well approximated by

$$\text{SNR}_o \approx \frac{1}{KLF} \frac{P_T}{N_0 B_{neq}}. \tag{10.27}$$

Example 10.2

We would like to transmit a signal over a distance of 100 km using a wireline channel. Assume that the signal attenuation is 2 dB/km. Assume that five analog repeater stages are to be employed to combat the transmission losses where the gain of each amplifier is 40 dB (which is equal to the loss in each of the five segments). Assume also that the noise figure of each amplifier is 6 dB. Determine the transmit power necessary to achieve an output signal-to-noise ratio of 30 dB assuming that the noise-equivalent bandwidth $B_{neq} = 10$ kHz and $N_0 = 2 \times 10^{-21}$ W/Hz.

Solution

For the scenario where the amplifiers are identical, and each segment's loss is the same as the amplifier gain, the output signal-to-noise ratio is given by (10.27). Written in dB units, we have

$$\text{SNR}_o(\text{dB}) \approx 10 \log_{10} \left(\frac{P_T}{N_0 B_{neq}} \right) - 10 \log_{10} K - L(\text{dB}) - F(\text{dB}). \tag{10.28}$$

Noting that the loss in each segment is 40 dB and there are five segments, we obtain the required transmit power as

$$P_T \approx 30 + 10\log_{10}(5) + 40 + 6 + 10\log_{10}(10 \times 10^3 \times 2 \times 10^{-21})\ \text{dBW}, \tag{10.29}$$

namely, $P_T \approx -84$ dBW, or only 4 pW (picowatts).

Note also that if repeaters were not employed, the required transmit power would have been 69 dBW, or roughly 8 MW, which is extremely high and, hence, highly impractical.

10.3.2 Regenerative Repeaters

A superior but more costly alternative to simple amplify and forward-type analog repeaters is the use of regenerative repeaters. Used with digital transmission, regenerative repeaters demodulate the transmitted digital symbols and then form a clean signal corresponding to the demodulated data and transmit it to the next repeater.

As an example, consider the use of binary phase-shift keying and K identical repeater segments. Denoting the signal-to-noise ratio for each segment by $\gamma_b = E_b/N_0$, we can write the probability of bit error for each segment as

$$Q\left(\sqrt{2\gamma_b}\right). \tag{10.30}$$

For transmission over the K regenerative repeater stages, the probability of bit error can be approximated as

$$P_b \approx 1 - \left(1 - Q\left(\sqrt{2\gamma_b}\right)\right)^K \approx KQ\left(\sqrt{2\gamma_b}\right) \tag{10.31}$$

for high signal-to-noise ratios ($\gamma_b \gg 1$).

This is a vast improvement over the use of analog repeaters, since in this case the error probability is linearly increased by the number of repeater stages while for the case of analog repeaters, the signal-to-noise ratio is reduced by a factor of $1/K$, which would correspond to a bit error probability of $Q\left(\sqrt{2\gamma_b/K}\right)$. For instance, with $K = 20$ repeaters, the use of analog repeaters will be around $10\log_{10}(20) \approx 13$ dB inferior to the case of regenerative repeaters at high signal-to-noise ratios.

10.4 Multiple-Access Techniques

In Chapters 5–8, we have focused on communications between one sender and one receiver over a dedicated channel. In this section, we will discuss how different transmitters can send their data to a common receiver using the same communication medium. Transmission from multiple geographically separated users to a common destination over the same medium is referred to as *multiple access*. While there are different multiple-access techniques, we will describe only time-division multiple access (TDMA), frequency-division multiple access (FDMA), and code-division multiple access (CDMA).

10.4.1 Time-Division Multiple Access and Frequency-Division Multiple Access

In TDMA, different users access the common communication medium at different times. Time is divided into frames of length T_f seconds, and each user accesses the channel for its designated portion of each time frame. For instance, if the channel is to be equally shared among N users, each user gets to transmit for $T_u = T_f/N$ seconds with the first user transmitting during the interval $[0, T_u)$ within the given frame, the second one transmitting in the interval $[T_u, 2T_u)$, and so on. This process is illustrated in Fig. 10.6.

FDMA uses a similar idea, except that the channel is divided among different users through frequency division. If the overall frequency band allocated for communication is $[f_1, f_2]$ with a bandwidth of $W = f_2 - f_1$, assuming that the channel resources are to be shared equally, each of the N users is allocated a frequency band of $W_u = W/N$. The first user transmits in the frequency band $[f_1, f_1 + W_u)$, the second one transmits in the frequency band $[f_1 + W_u, f_1 + 2W_u)$, and so on. FDMA is depicted in Fig. 10.7.

10.4.2 Code-Division Multiple Access

In code-division multiple access, the idea is vastly different from TDMA and FDMA. There is no clear separation of time or frequency resources; instead, the users effectively transmit simultaneously using the same frequency band. However, their signals are spread in frequency in a way that reduces or even eliminates the interference among them. To understand how CDMA works, we first need to talk about spread-spectrum communications.

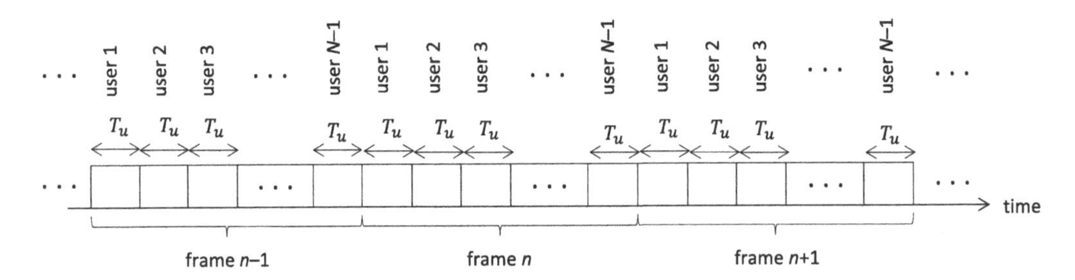

Figure 10.6 Illustration of time-division multiple access.

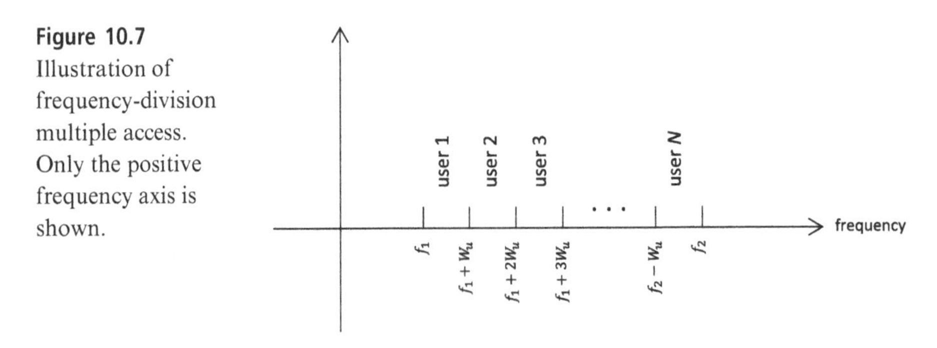

Figure 10.7 Illustration of frequency-division multiple access. Only the positive frequency axis is shown.

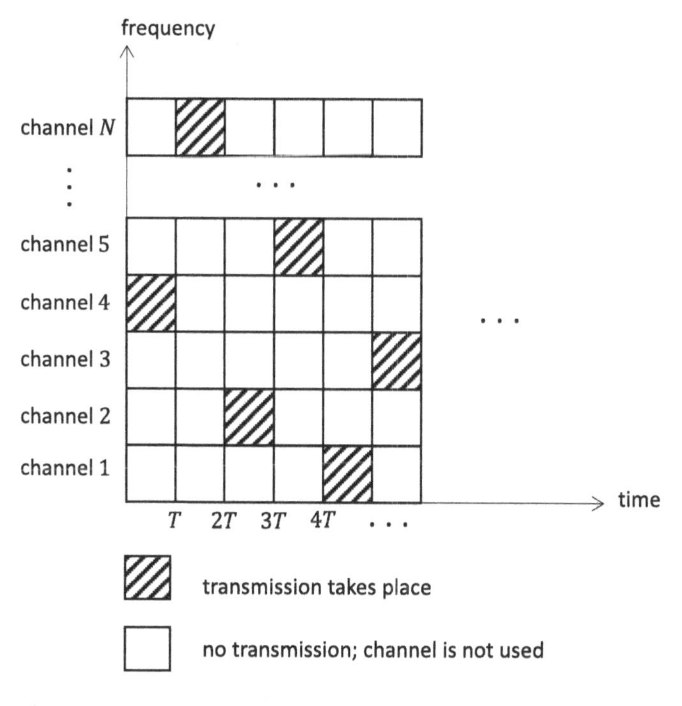

Figure 10.8 Illustration of frequency-hopping spread-spectrum.

Spread-Spectrum Communications. The idea in spread-spectrum communications is to spread a narrowband communications signal over a wide frequency band and then transmit it. There are two common ways of accomplishing this: frequency-hopping spread spectrum (FHSS) and direct sequence spread spectrum (DSSS), as will be described below. The benefits of spread-spectrum communications include low probability of intercept and low probability of detection (as the power density across different frequencies is reduced), high tolerance to narrowband interference, and high tolerance to channel fading (as the channel characteristics over some frequencies may be favorable even if it is poor for other parts of the transmission band).

In FHSS, the overall frequency band is divided into many different subbands, and each subband is used for transmission for only a certain amount of time before switching to a new subband. This is illustrated in Fig. 10.8, assuming that the entire bandwidth is W, there are N subbands, each with a bandwidth of W/N, and the transmitter switches to a new subband every T seconds. The order in which the subbands are selected for transmission is determined by a pseudo-random sequence available both at the transmitter and the receiver. Therefore, the receiver can synchronize with the transmitted signal and demodulate the data. There is no restriction on the modulation scheme to be used for transmission over each subband.

For DSSS, the narrowband signal is multiplied by a chip sequence, and the resulting sequence is transmitted. This is illustrated for BPSK signaling in Fig. 10.9, where the lowpass-equivalent BPSK signal with bit period T_b is multiplied by

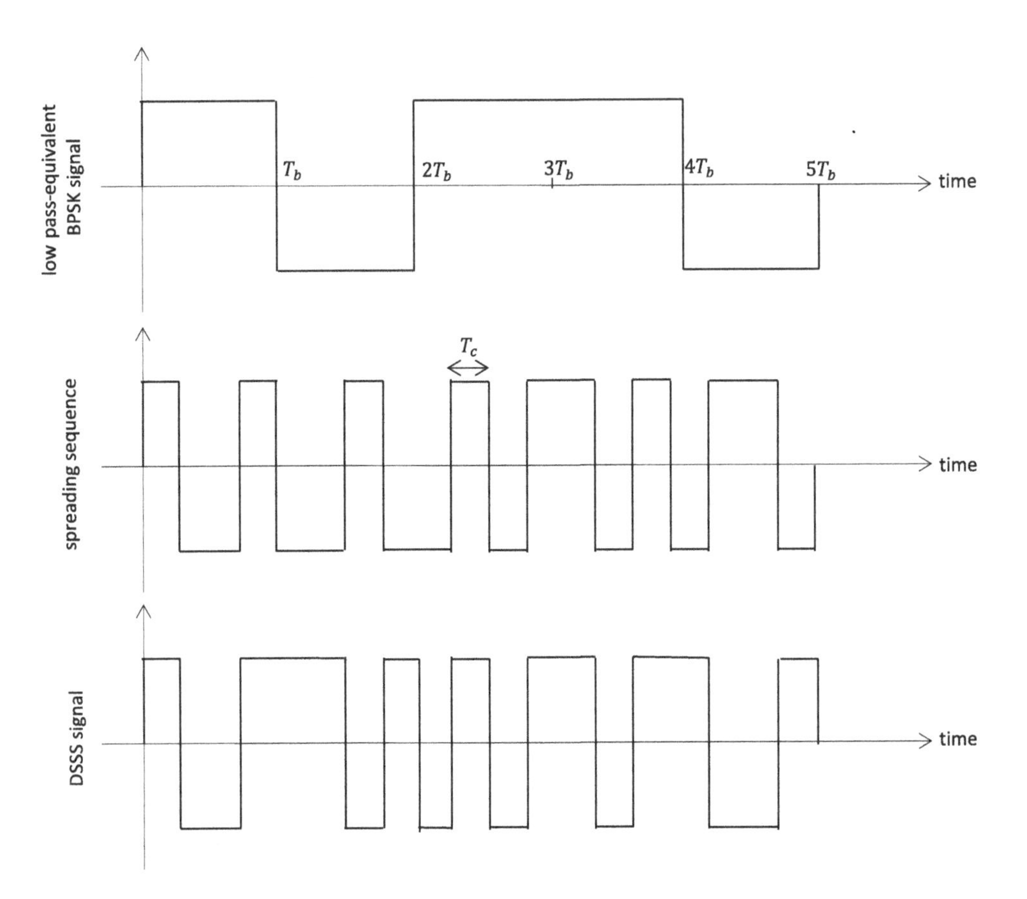

Figure 10.9 Illustration of direct-sequence spread spectrum.

a pseudo-random chip sequence (of $+1$ and -1) with chip period T_c. Selecting $N = T_b/T_c \gg 1$, the narrowband waveform is spread over a significantly wider frequency band, resulting in the spread-spectrum signal. The pseudo-random chip sequence is known at the receiver. Therefore, the same chip sequence can be constructed, and the received signal can be multiplied with it to obtain a *despread* signal, which is used for the demodulation of each bit. Without the knowledge of the pseudo-random chip sequence, all the receiver sees is a noise-like signal; hence, it is not possible to recover the transmitted bits.

Let us denote the lowpass-equivalent information-bearing waveform by $x(t)$, and the spreading waveform as $s_c(t)$. The transmitted signal is then given by $x(t)s_c(t)$. The receiver has access to the chip sequence, hence $s_c(t)$; therefore, it simply multiplies the received signal by $s_c(t)$ and obtains

$$(x(t)s_c(t) + \text{noise})s_c(t) = x(t) + \text{noise}, \tag{10.32}$$

and applies the standard demodulation method for the underlying modulation scheme used at the transmitter.

Code-Division Multiple Access. Spread-spectrum communications can be used to accommodate multiple users over the same communication medium, leading to *code-division multiple access*. The idea in CDMA is to assign a different pseudo-random sequence to each user to construct the spread-spectrum signal in such a way that it is possible to obtain the individual users' data separately from the superposition of the transmitted signals. Either FHSS or DSSS can be used with similar results.

Let us assume that DSSS is used and denote the individual users' lowpass-equivalent information-bearing signals by $x_i(t)$ and the spreading waveforms assigned to them by $s_{c,i}(t)$, for $i = 1, 2, \ldots, N$, where N is the number of users sharing the medium. The transmitted signals are of the form $x_i(t)s_{c,i}(t)$, hence the receiver observes their superposition

$$x_1(t)s_{c,1}(t) + x_2(t)s_{c,2}(t) + \cdots + x_N(t)s_{c,N}(t) + \text{noise}. \tag{10.33}$$

To demodulate the jth user's signal, the receiver simply multiplies the received signal by $s_{c,j}(t)$, and obtains

$$x_j(t) + \text{interference} + \text{noise}. \tag{10.34}$$

The interference is the sum of signals of the form

$$x_i(t)s_{c,i}(t)s_{c,j}(t) \tag{10.35}$$

over all the users except the jth one. If the spreading sequences are selected to be (near) orthogonal, then the interference level is low, and the jth user's data can be demodulated.

Spreading Sequences. Both with frequency-hopping spread spectrum and direct-sequence spread spectrum, we need to design spreading sequences with desirable properties. For instance, with DSSS we need spreading sequences with good autocorrelation properties and good cross-correlation properties. Good autocorrelation properties refer to small off-peak values, and they enable easier synchronization and help minimize self-interference due to the reception of the transmitted signal through multiple paths with different delays. Good cross-correlation properties of the spreading sequences for all possible delays ensure that after despreading by using a particular user's spreading sequence, the other users' signals do not cause excessive interference, and the desired user's data can be demodulated.

There are different ways of designing spreading sequences for spread-spectrum communications, including m-sequences (generated by maximal linear-feedback shift registers), Gold sequences, and Walsh–Hadamard sequences, among others. Coverage of these constructions is beyond our scope.

10.5 Chapter Summary

In this short chapter, we discussed the effects of transmission losses and noise on a communication system and ways of remedying them. We introduced common terminology including receiver sensitivity, link margin, and fade margin. We then studied the use of non-ideal amplifiers and determined the signal quality at the amplifier output in terms of the signal-to-noise ratio at the input and the noise figure of the amplifier. As a way of combatting transmission losses and communicating over long distances in a power-efficient manner, we reviewed the use of analog and regenerative repeaters. We also studied multiple-access techniques that enable sharing of a common communication medium among multiple users.

PROBLEMS

10.1 For a wireless communication channel modeled as Rayleigh fading, the received signal power can be modeled as an exponential random variable with probability density function

$$f(x) = \frac{1}{\alpha} e^{-x/\alpha},$$

where α is the average received power level.

Compute the probability that the received power level is 10 dB less than its average value. How about the probability that the received power level is less than 1% of the average power?

10.2 Consider Rayleigh fading as in Problem 10.1. What should be the difference (in dB) between the average received power and the receiver sensitivity (i.e., the fade margin) so that the link is available for at least 90% of the time? How about the availability with 99% probability?

10.3 Consider BPSK transmission over an AWGN channel. The received signal is input to an amplifier with a noise figure of 5 dB whose output is fed to a demodulator. Assuming that the input signal-to-noise ratio (E_b/N_0) is 15 dB, determine the bit error probability at the demodulator output.

10.4 We use three amplifiers in cascade to improve the signal strength at a receiver. Amplifier A has a gain of 30 dB and a noise figure of 15 dB; amplifier B has a gain of 20 dB and a noise figure of 10 dB; amplifier C has a gain of 10 dB and a noise figure of 5 dB.

Determine the order in which these amplifiers should be connected to ensure that the resulting signal quality is the highest at the output.

Assuming that the input signal-to-noise ratio is 50 dB, what is the resulting SNR at the output?

10.5 We want to transmit a signal over a distance of 200 km using a wireline channel, for which the signal attenuation is 2 dB/km. Assume that 20 analog repeater stages are employed to combat the transmission losses. The amplifiers are identical, and each has a gain of 20 dB. Assume also that the noise figure of each amplifier is 4 dB.

Determine the transmit power necessary to achieve an output signal-to-noise ratio of 50 dB, assuming that the noise-equivalent bandwidths of all the amplifiers are $B_{neq} = 50$ kHz and $N_0 = 5 \times 10^{-21}$ W/Hz.

10.6 Consider Problem 10.5. In this case, regenerative repeaters are used instead of analog repeaters, and PAM is selected as the modulation scheme.

How would you compare the resulting performance with that of the analog repeaters for high signal-to-noise ratios?

10.7 There are 30 users sharing the same frequency band and transmitting simultaneously with CDMA (with a processing gain of 500). At a particular receiver (which tries to detect only a specific user's signal, say user 1), the received power levels for all the transmitted signals are identical. Assume also that the system is interference-limited, that is, the additive noise is negligible.

(a) Determine the signal to interference plus noise ratio at the receiver output (after despreading and demodulating the desired user's signal).

(b) We would like to double the number of users. What should be the new processing gain if we want to make sure that the same error rate as in part (a) is maintained?

COMPUTER PROBLEMS

10.8 The model for a Rayleigh fading channel is given in Problem 10.1. The average power level is $\alpha = 0.0001$ mW, that is, -40 dBm. Assume that a received power level of -50 dBm is sufficient for proper demodulation.

(a) Generate a large number of realizations of the received power level (expressed in dBm) and plot its histogram.

(b) Using the realizations in part (a), estimate the probability that the link is available.

10.9 Consider the Rayleigh fading channel model in Problem 10.1. Assume that the receiver sensitivity is -60 dBm. Using MATLAB, generate a large number of received power levels to estimate the outage probability (i.e., the probability that the link is not available) for the following fade margins:

(i) 5 dB;
(ii) 10 dB;
(iii) 20 dB.

We consider the fade margin as the difference (in dB) between the average received power level and the receiver sensitivity.

10.10 Write a MATLAB code to simulate the effects of a non-ideal amplifier on an input signal for a given amplifier gain and a noise figure and use this code to generate samples of signals in AWGN at the input and output of the amplifier. Use a sinusoidal wave of magnitude 0.001 depicting five periods of the signal as the clean waveform, and assume that the input signal-to-noise ratio is 50 dB.

Plot the signal at the amplifier output for the following amplifier gain and noise figure values:

(i) $G = 20$ dB, $F = 0$ dB;
(ii) $G = 30$ dB, $F = 0$ dB;
(iii) $G = 30$ dB, $F = 10$ dB;
(iv) $G = 30$ dB, $F = 20$ dB.

Further Reading

Abramson, N. (1963), *Information Theory and Coding*, McGraw-Hill, New York.

Amoroso, F. (1980), 'The bandwidth of digital data signals', *IEEE Communication Magazine* **18**, 13–24.

Anderson, J. B., Aulin, T. and Sundberg, C.-E. (1986), *Digital Phase Modulation*, Plenum, New York.

Anderson, J. B. and Mohan, S. (1991), *Source and Channel Coding: An Algorithmic Approach*, Kluwer, Boston, MA.

Arens, R. (1956), 'Synchronous communications', *Proceedings of the IRE* **44**, 1713–1718.

Arens, R. (1957), 'Complex processes for envelopes of normal noise', *IRE Transactions on Information Theory* **3**, 204–207.

Arikan, E. (2009), 'Channel polarization: A method for constructing capacity-achieving codes for symmetric binary-input memoryless channels', *IEEE Transactions on Information Theory* **55**(7), 3051–3073.

Arthurs, E. and Dym, H. (1962), 'On the optimum detection of digital signals in the presence of white Gaussian noise – a geometric interpretation and a study of three basic data transmission systems', *IRE Transactions on Communications Systems* **10**, 336–372.

Ash, R. B. (1965), *Information Theory*, Interscience, New York.

Bahai, A. R., Saltzberg, B. R. and Ergen, M. (2004), *Multicarrier Digital Communications: Theory and Applications of OFDM*, Springer, New York.

Barry, J. R., Lee, E. A. and Messerschmitt, D. G. (2004), *Digital Communication*, 3rd ed., Kluwer, Boston, MA.

Benedetto, S. and Biglieri, E. (1999), *Principles of Digital Transmission: With Wireless Applications*, Springer, New York.

Berger, T. (1971), *Rate Distortion Theory. A Mathematical Basis for Data Compression*, Prentice Hall, Englewood Cliffs, NJ.

Berger, T. and Tufts, D. W. (1967), 'Optimum pulse amplitude modulation Part I: Transmitter–receiver design and bounds from information theory', *IEEE Transactions on Information Theory* **13**, 196–208.

Berlekamp, E. (2015), *Algebraic Coding Theory*, 3rd ed., World Scientific, Singapore.

Berrou, C. and Glavieux, A. (1996), 'Near optimum error correcting coding and decoding: Turbo-codes', *IEEE Transactions on Communications* **44**(10), 1261–1271.

Bertsekas, D. P. and Tsitsiklis, J. N. (2002), *Introduction to Probability*, 2nd ed., Athena Scientific, Belmont, MA.

Bingham, J. A. (1990), 'Multicarrier modulation for data transmission: An idea whose time has come', *IEEE Communications Magazine* **28**(5), 5–14.

Black, H. S. (1953), *Modulation Theory*, Van Nostrand, Princeton, NJ.

Blahut, R. E. (1983), *Theory and Practice of Error Control Coding*, Addison-Wesley, Reading, MA.

Blahut, R. E. (1990), *Digital Transmission of Information*, Addison-Wesley, Reading, MA.

Bliss, D. W. (2022), *Modern Communications: A Systematic Introduction*, Cambridge University Press, Cambridge.

Carlson, A. B. and Crilly, P. B. (2010), *Communication Systems*, 5th ed., McGraw Hill, New York.

Carson, J. R. (1922), 'Notes on the theory of modulation', *Proceedings of the IRE* **10**, 57–64.

Cherubini, G., Eleftheriou, E. and Olcer, S. (2002), 'Filtered multitone modulation for very high-speed digital subscriber lines', *IEEE Journal on Selected Areas in Communications* **20**(5), 1016–1028.

Chow, P. S., Cioffi, J. M. and Bingham, J. A. (1995), 'A practical discrete multitone transceiver loading algorithm for data transmission over spectrally shaped channels', *IEEE Transactions on Communications* **43**(2/3/4), 773–775.

Clark, G. C. and Cain, J. B. (1981), *Error-Correction Coding for Digital Communications*, Plenum, New York.

Couch, L. W. I. (2013), *Digital and Analog Communication Systems*, 8th ed., Prentice Hall, Upper Saddle River, NJ.

Cover, T. M. and Thomas, J. A. (2006), *Elements of Information Theory*, 2nd ed., Wiley-Interscience, Hoboken, NJ.

Davenport, J. B. J. and Root, W. L. (1958), *An Introduction to the Theory of Random Signals and Noise*, McGraw-Hill, New York.

Duman, T. M. and Ghrayeb, A. (2007), *Coding for MIMO Communication Systems*, Wiley, Chichester.

Elias, P. (1954), 'Error free coding', *IRE Transactions on Information Theory* **4**, 29–39.

Elias, P. (1955), 'Coding for noisy channels', *IRE Convention Record, Part 4*, pp. 42–47.

Fano, R. M. (1957), *The Transmission of Information*, MIT Press and Wiley, New York.

Feller, W. (1957), *An Introduction to Probability Theory and Its Applications*, 2nd ed., Wiley, New York.

Forney, J. D. (1973), 'The Viterbi algorithm', *Proceedings of the IEEE* **61**, 268–278.

Franks, L. (1980), 'Carrier and bit synchronization in data communication – a tutorial review', *IEEE Transactions on Communications* **28**(8), 1107–1121.

Gagliardi, R. M. (1988), *Introduction to Communication Engineering*, 2nd ed., Wiley, New York.

Gallager, R. G. (1962), 'Low-density parity-check codes', *IRE Transactions on Information Theory* **8**(1), 21–28.

Gallager, R. G. (1963), *Low-Density Parity-Check Codes*, MIT Press, Cambridge, MA.

Gallager, R. G. (1968), *Information Theory and Reliable Communication*, Wiley, New York.

Gallager, R. G. (2008), *Principles of Digital Communication*, Cambridge University Press, Cambridge.

Gardner, F. M. (2005), *Phaselock Techniques*, 3rd ed., Wiley-Interscience, Hoboken, NJ.

Gersho, A. and Gray, R. M. (1992), *Vector Quantization and Signal Compression*, Kluwer, Boston, MA.

Gibson, J. D. (1993), *Principles of Digital and Analog Communications*, 2nd ed., Macmillan, Cambridge, MA.

Golay, M. J. (1962), 'Notes on digital coding', *Proceedings of the IRE* **37**, 657.

Gold, R. (1967), 'Optimal binary sequences for spread spectrum multiplexing (corresp.)', *IEEE Transactions on Information Theory* **13**(4), 619–621.

Gold, R. (1968), 'Maximal recursive sequences with 3-valued recursive cross-correlation functions', *IEEE Transactions on Information Theory* **14**(1), 154–156.

Goldsmith, A. (2005), *Wireless Communications*, Cambridge University Press, Cambridge.

Golomb, S. W. (1982), *Shift Register Sequences*, revised ed., Aegan Park Press, Walnut Creek, CA.

Gronemeyer, S. and McBride, A. (1976), 'MSK and offset QPSK modulation', *IEEE Transactions on Communications* **24**(8), 809–820.

Gupta, S. C. (1975), 'Phase-locked loops', *Proceedings of the IEEE* **63**(2), 291–306.

Hamming, R. W. (1950), 'Error detecting and error correcting codes', *The Bell System Technical Journal* **29**(2), 147–160.

Harman, W. W. (1963), *Principles of the Statistical Theory of Communication*, McGraw-Hill, New York.

Hartley, R. V. L. (1928), 'Transmission of information', *The Bell System Technical Journal* **7**, 535–563.

Haykin, S. and Moher, M. (2005), *Modern Wireless Communications*, Prentice Hall, Upper Saddle River, NJ.

Haykin, S. and Moher, M. (2009), *Communication Systems*, 5th ed., Wiley, New York.

Helstrom, C. W. (1968), *Statistical Theory of Signal Detection*, 2nd ed., Pergamon, New York.

Helstrom, C. W. (1991), *Probability and Stochastic Processes for Engineers*, Macmillan, New York.

Holmes, J. K. (1982), *Coherent Spread Spectrum Systems*, Wiley, New York.

Huffman, D. A. (1952), 'A method for the construction of minimum-redundancy codes', *Proceedings of the IRE* **40**(9), 1098–1101.

Hwang, T., Yang, C., Wu, G., Li, S. and Ye Li, G. (2009), 'OFDM and its wireless applications: A survey', *IEEE Transactions on Vehicular Technology* **58**(4), 1673–1694.

Ingle, V. K. and Proakis, J. G. (2012), *Digital Signal Processing Using MATLAB*, 3rd ed., Cengage, Stamford, CT.

Johannesson, R. and Zigangirov, K. S. (2015), *Fundamentals of Convolutional Coding*, 2nd ed., Wiley, Hoboken, NJ.

Johnson, C. R. and Sethares, W. A. (2003), *Telecommunications Breakdown: Concepts of Communication Transmitted Via Software-Defined Radio*, Prentice Hall, Upper Saddle River, NJ.

Johnson, J. (1927), 'Thermal agitation of electricity in conductors', *Nature* **119**, 50–51.

Kasami, T. (1966), *Weight Distribution Formula for Some Class of Cyclic Codes*, University of Illinois Urbana-Champaign, Champaign, IL.

Kay, S. M. (1993), *Fundamentals of Statistical Signal Processing: Volume I, Estimation Theory*, Prentice Hall, Upper Saddle River, NJ.

Kay, S. M. (1998), *Fundamentals of Statistical Signal Processing: Volume II, Detection Theory*, Prentice Hall, Upper Saddle River, NJ.

Kay, S. M. (2006), *Intuitive Probability and Random Processes Using MATLAB*, Springer, New York.

Kieffer, J. (1983), 'Uniqueness of a locally optimal quantizer for log-concave density and convex error weighting function', *IEEE Transactions on Information Theory* **29**(1), 42–47.

Kotel'nikov, V. A. (1947), *The Theory of Optimum Noise Immunity*, PhD thesis, Molotov Energy Institute, translated by R. A. Silverman, Dover Publications, New York.

Lapidoth, A. (2017), *A Foundation in Digital Communication*, 2nd ed., Cambridge University Press, Cambridge.

Lathi, B. P. and Ding, Z. (2019), *Modern Digital and Analog Communication Systems*, 5th ed., Oxford University Press, New York.

Leon-Garcia, A. (2008), *Probability, Statistics, and Random Processes for Electrical Engineering*, 3rd ed., Prentice Hall, Upper Saddle River, NJ.

Lin, S. and Costello, D. J. (2004), *Error Control Coding*, 2nd ed., Prentice Hall, Upper Saddle River, NJ.

Linde, Y., Buzo, A. and Gray, R. (1980), 'An algorithm for vector quantizer design', *IEEE Transactions on Communications* **28**(1), 84–95.

Lindsey, W. C. (1972), *Synchronization Systems in Communication and Control*, Prentice Hall, Englewood Cliffs, NJ.

Lindsey, W. C. and Chie, C. M. (1981), 'A survey of digital phase-locked loops', *Proceedings of the IEEE* **69**(4), 410–431.

Lindsey, W. C. and Simon, M. K. (1973), *Telecommunication Systems Engineering*, Prentice Hall, Englewood Cliffs, NJ.

Lloyd, S. (1982), 'Least squares quantization in pcm', *IEEE Transactions on Information Theory* **28**(2), 129–137.

Lucky, R. W., Salz, J. and Weldon, E. J. (1968), *Principles of Data Communication*, McGraw-Hill, New York.

MacKay, D. J. (1999), 'Good error-correcting codes based on very sparse matrices', *IEEE Transactions on Information Theory* **45**(2), 399–431.

Mackay, D. J. C. (2003), *Information Theory, Inference, and Learning Algorithms*, Cambridge University Press, Cambridge.

MacWilliams, F. J. and Sloane, N. J. A. (1977), *The Theory of Error-Correcting Codes*, North Holland, New York.

Madhow, U. (2008), *Fundamentals of Digital Communication*, Cambridge University Press, Cambridge.

Max, J. (1960), 'Quantizing for minimum distortion', *IRE Transactions on Information Theory* **6**(1), 7–12.

Mengali, U. and D'Andrea, A. N. (1997), *Synchronization Techniques for Digital Receivers*, Plenum Press, New York.

Meyr, H., Moeneclaey, M. and Fechtel, S. A. (1998), *Digital Communication Receivers: Synchronization, Channel Estimation, and Signal Processing*, Wiley, New York.

Middleton, D. (1960), *An Introduction to Statistical Communication Theory*, McGraw-Hill, New York.

Molisch, A. F. (2011), *Wireless Communications*, 2nd ed., Wiley, Chichester.

Mueller, K. and Muller, M. (1976), 'Timing recovery in digital synchronous data receivers', *IEEE Transactions on Communications* **24**(5), 516–531.

Nguyen, H. H. and Shwedyk, E. (2009), *A First Course in Digital Communications*, Cambridge University Press, Cambridge.

North, D. O. (1963), 'An analysis of the factors which determine signal/noise discrimination in pulsed-carrier systems', *Proceedings of the IEEE* **51**(7), 1016–1027.

Nyquist, H. (1924), 'Certain factors affecting telegraph speed', *The Bell System Technical Journal* **3**, 324–346.

Nyquist, H. (1928a), 'Certain topics in telegraph transmission theory', *Transactions of the American Institute of Electrical Engineers* **47**(2), 617–644.

Nyquist, H. (1928b), 'Thermal agitation of electric charge in conductors', *Physical Review* **32**, 110–113.

Odenwalder, J. P. (1970), *Optimal Decoding of Convolutional Codes*, PhD thesis, University of California, Los Angeles.

Oliver, B. M., Pierce, J. R. and Shannon, C. E. (1948), 'The philosophy of PCM', *Proceedings of the IRE* **36**, 1324–1331.

Oppenheim, A. V. and Willsky, A. S. (1997), *Signals and Systems*, 2nd ed., Prentice Hall, Upper Saddle River, NJ.

Papoulis, A. and Pillai, S. U. (2002), *Probability, Random Variables and Stochastic Processes*, 4th ed., McGraw-Hill, New York.

Pasupathy, S. (1979), 'Minimum shift keying: A spectrally efficient modulation', *IEEE Communications Magazine* **17**(4), 14–22.

Peterson, R. L., Ziemer, R. E. and Borth, D. E. (1995), *Introduction to Spread-Spectrum Communications*, Prentice Hall, Upper Saddle River, NJ.

Pickholtz, R., Schilling, D. and Milstein, L. (1982), 'Theory of spread-spectrum communications – a tutorial', *IEEE Transactions on Communications* **30**(5), 855–884.

Poor, H. V. (1994), *An Introduction to Signal Detection and Estimation*, 2nd ed., Springer, New York.

Proakis, J. G. and Manolakis, D. G. (2007), *Digital Signal Processing: Principles, Algorithms, and Applications*, 4th ed., Prentice Hall, Upper Saddle River, NJ.

Proakis, J. G. and Salehi, M. (2002), *Communication Systems Engineering*, 2nd ed., Prentice Hall, Upper Saddle River, NJ.

Proakis, J. G. and Salehi, M. (2008), *Digital Communications*, 5th ed., McGraw-Hill, New York.

Proakis, J. G. and Salehi, M. (2014), *Fundamentals of Communication Systems*, 2nd ed., Prentice Hall, Upper Saddle River, NJ.

Proakis, J. G., Salehi, M. and Bauch, G. (2013), *Contemporary Communication Systems Using Matlab and Simulink*, 3rd ed., Thomson, Stamford, CT.

Rappaport, T. S. (2002), *Wireless Communications: Principles and Practice*, 2nd ed., Prentice Hall, Upper Saddle River, NJ.

Rice, S. O. (1944a), 'Mathematical analysis of random noise, Parts I and II', *The Bell System Technical Journal* **23**, 282–332.

Rice, S. O. (1944b), 'Mathematical analysis of random noise, Part III', *The Bell System Technical Journal* **24**, 46–156.

Rice, S. O. (1982), 'Envelopes of narrow-band signals', *Proceedings of the IEEE* **70**, 692–699.

Richardson, T. and Urbanke, R. (2008), *Modern Coding Theory*, Cambridge University Press, Cambridge.

Ross, S. (2014), *A First Course in Probability*, 9th ed., Pearson, Upper Saddle River, NJ.

Ryan, W. E. and Lin, S. (2009), *Channel Codes, Classical and Modern*, Cambridge University Press, Cambridge.

Sakrison, D. J. (1968), *Communication Theory: Transmission of Waveforms and Digital Information*, Wiley, New York.

Saltzberg, B. (1967), 'Performance of an efficient parallel data transmission system', *IEEE Transactions on Communication Technology* **15**(6), 805–811.

Sarwate, D. V. and Pursley, M. B. (1980), 'Crosscorrelation properties of pseudorandom and related sequences', *Proceedings of the IEEE* **68**(5), 593–619.

Scholtz, R. (1982), 'The origins of spread-spectrum communications', *IEEE Transactions on Communications* **30**(5), 822–854.

Schwartz, M. (1980), *Information Transmission, Modulation, and Noise: A Unified Approachto Communication Systems*, 3rd ed., McGraw-Hill, New York.

Shanmugam, K. S. (1979), *Digital and Analog Communication Systems*, Wiley, New York.

Shannon, C. E. (1948a), 'A mathematical theory of communication', *The Bell System Technical Journal* **27**(3), 379–423.

Shannon, C. E. (1948b), 'A mathematical theory of communication', *The Bell System Technical Journal* **27**(4), 623–656.

Simon, M. K. and Alouini, M.-S. (2005), *Digital Communication over Fading Channels*, 2nd ed., Wiley-Interscience, Hoboken, NJ.

Sklar, B. and Harris, F. (2020), *Digital Communications: Fundamentals and Applications*, 3rd ed., Prentice Hall, Upper Saddle River, NJ.

Slepian, D. (1962), 'The threshold effect in modulation systems that expand bandwidth', *IRE Transactions on Information Theory* **8**, 122–127.

Slepian, D. (1976), 'On bandwidth', *Proceedings of the IEEE* **64**, 292–300.

Smith, D. R. (1993), *Digital Transmission Systems*, 2nd ed., Springer, New York.

Stark, H. and Woods, J. W. (2012), *Probability, Statistics, and Random Processes for Engineers*, 4th ed., Prentice Hall, Upper Saddle River, NJ.

Stremler, F. G. (1990), *Introduction to Communication Systems*, 3rd ed., Addison-Wesley, Reading, MA.

Sundberg, C.-E. (1986), 'Continuous phase modulation', *IEEE Communications Magazine* **24**(4), 25–38.

Taub, H. and Schilling, D. L. (1986), *Principles of Communication Systems*, 2nd ed., McGraw-Hill, New York.

Tranter, W. H., Shanmugam, K. S., Rappaport, T. S. and Kosbar, K. L. (2003), *Principles of Communication Systems Simulation with Wireless Applications*, Prentice Hall, Upper Saddle River, NJ.

Tse, D. and Viswanath, P. (2005), *Fundamentals of Wireless Communication*, Cambridge University Press, Cambridge.

Tufts, D. W. (1965), 'Nyquist's problem–the joint optimization of transmitter and receiver in pulse amplitude modulation', *Proceedings of the IEEE* **53**(3), 248–259.

Van Trees, H. L. (2001), *Detection, Estimation, and Modulation Theory, Part I*, Wiley-Interscience, New York.

Viterbi, A. (1967), 'Error bounds for convolutional codes and an asymptotically optimum decoding algorithm', *IEEE Transactions on Information Theory* **13**(2), 260–269.

Viterbi, A. J. (1995), *CDMA, Principles of Spread Spectrum Communications*, Addison-Wesley, Reading, MA.

Viterbi, A. J. and Omura, J. K. (1979), *Principles of Digital Communication and Coding*, McGraw-Hill, New York.

Vucetic, B. and Yuan, J. (2000), *Turbo Codes: Principles and Applications*, Springer, New York.

Weinstein, S. and Ebert, P. (1971), 'Data transmission by frequency-division multiplexing using the discrete Fourier transform', *IEEE Transactions on Communication Technology* **19**(5), 628–634.

Welch, T. A. (1984), 'A technique for high-performance data compression', *Computer* **17**(06), 8–19.

Wicker, S. B. (1995), *Error Control Systems for Digital Communication and Storage*, Prentice Hall, Upper Saddle River, NJ.

Wilson, S. G. (1996), *Digital Modulation and Coding*, Prentice Hall, Upper Saddle River, NJ.

Wozencraft, J. M. and Jacobs, I. M. (1965), *Principles of Communication Engineering*, Wiley, New York.

Yates, R. D. and Goodman, D. J. (2014), *A First Course in Probability*, 3rd ed., Wiley, New York.

Ziemer, R. E. and Peterson, R. W. (2000), *Introduction to Digital Communication*, 2nd ed., Prentice Hall, Upper Saddle River, NJ.

Ziemer, R. E. and Tranter, W. H. (2015), *Principles of Communications: Systems, Modulation, and Noise*, 7th ed., Wiley, New York.

Ziv, J. and Lempel, A. (1977), 'A universal algorithm for sequential data compression', *IEEE Transactions on Information Theory* **23**(3), 337–343.

Ziv, J. and Lempel, A. (1978), 'Compression of individual sequences via variable-rate coding', *IEEE Transactions on Information Theory* **24**(5), 530–536.

Index

' in the United States
· & Taylor Publisher Services